Britannica Yearbook of Science and the Future

Encyclopædia Britannica, Inc.

Helen Hemingway Benton, Publisher

Chicago Toronto London
Geneva Sydney Tokyo Manila
Johannesburg Seoul

1974
Britannica Yearbook of Science and the Future

EDITOR
David Calhoun

ASSOCIATE EDITOR
Mary Alice Molloy

EDITORIAL CONSULTANT
Howard Lewis, Director, Office of Information,
National Academy of Sciences—National Academy
of Engineering—National Research Council

STAFF EDITORS
Judy Booth, Daphne Daume

ART DIRECTOR
Will Gallagher

ASSOCIATE ART DIRECTOR
Cynthia Peterson

DESIGN SUPERVISOR
Ron Villani

SENIOR PICTURE EDITOR
Holly Harrington

PICTURE EDITORS
Susan Cremin, Charles Hannum

LAYOUT ARTIST
Richard Batchelor

ART PRODUCTION
Richard Heinke

EDITORIAL PRODUCTION MANAGER
J. Thomas Beatty

PRODUCTION COORDINATOR
Lorene Lawson

PRODUCTION STAFF
Necia Brown, Charles Cegielski, Barbara Wescott Hurd,
Lawrence Kowalski, Ruth Passin, Susan Recknagel,
Cheryl M. Trobiani

COPY CONTROL
Felicité Buhl, Supervisor; Mary K. Finley,
Shirley Richardson

INDEX
Frances E. Latham, Supervisor; Virginia Palmer,
Assistant Supervisor; Gladys Berman, Grace R. Lord

SECRETARY
Fleury Nolta

DIRECTOR/YEARBOOKS
Margaret Sutton

ADMINISTRATIVE EDITOR/YEARBOOKS
Nellie L. Gifford

Library of Congress Catalogue Number: 69-12349 International Standard Book Number: 0-85229-287-2

THE UNIVERSITY OF CHICAGO
The Britannica Yearbook of Science and the Future
is published with the editorial advice of the faculties of
the University of Chicago

Encyclopædia Britannica, Inc.

CHAIRMAN OF THE BOARD
Robert P. Gwinn

PRESIDENT
Charles E. Swanson

VICE-PRESIDENT, EDITORIAL
Charles Van Doren

Editorial Advisory Board

Science and the Quality of Life

Research and achievement on the frontiers of science—work that will have a significant effect on human life in both the near and far future—have been the special concern of the *Britannica Yearbook of Science and the Future* since its inception. The effort has been to describe in layman's terms and with the aid of illustrations—for it is the layman who will bear the brunt of the changes—those developments that in the opinion of the editors are most important for the world of tomorrow, and the day after tomorrow.

In discharging this task, the editors have concentrated on research that is likely to have a positive effect on the quality of life in the future. But they have never shied away from problems as they arose, and from work that reflected the fact that mankind in the last third of the 20th century faces severe difficulties. Indeed, one measure of the importance of ongoing work is that it attacks an important problem.

This volume follows in that tradition. The 1974 edition of the *Britannica Yearbook of Science and the Future* presents a series of feature articles covering a wide range of subject matter, all written by authorities in their fields. Of paramount interest in this time of ever-increasing automobile traffic, rapidly rising gasoline prices, and gasoline rationing is "Engines of the Future," in which Donald N. Frey, chairman of the board of Bell & Howell Company, discusses the Wankel rotary engine and other devices—electricity, gas turbines, steam, fuel cells—and compares them to our present gasoline piston engine. He evaluates all these alternatives in the light of two major goals: less pollution and greater economy.

Pollution of another and most insidious kind is the subject of "The Rising Decibel Level," in which Toba Cohen, a medical researcher at the American Medical Association, tells of our growing problem with excessive noise and discusses some of the steps, both legal and technical, being taken to deal with it. And while we are on the subject of noise, we should remember that not all noise is unpleasant. In fact, one of the most intriguing fields of current research concerns stringed instruments. Many of us will find it difficult to believe that the superb violins of such virtuoso craftsmen as Antonio Stradivari, who made his most famous instruments during the early years of the 18th century, can possibly be bettered. But a number of physicists, including author Michael Kasha, are at least keeping an open mind on the subject. Using the methods and tools of modern science, they are working to develop instruments that may turn out to have tonal qualities unequaled in man's history.

Another of man's works—the giant multipurpose dam—is viewed with a critical eye by crusading journalist Claire Sterling in "The Trouble with Superdams." Far from being a panacea for all ills, a giant dam, Ms. Sterling argues, can sometimes be an economic and environmental disaster. She cites persuasive evidence to this effect from Egypt's Aswan High Dam and warns policymakers and engineers to proceed with caution before undertaking one of these huge projects.

Engineering on a different front—that of making structures, from homes to public buildings, as earthquake-proof as possible—is the subject of a feature article by George W. Housner and Paul C. Jennings. Both are Californians, and so are fully aware of the importance and urgency of this work. We are not yet sure how to predict earthquakes; but everything we know points to the likelihood of a major earthquake disaster in California within the present decade. It may already have happened by the time you read this! We can only hope that it holds off long enough for the advances described by Housner and Jennings to be effective.

For those readers interested in man himself, a number of feature articles treat various aspects of man's physical nature and behavior. In "The New Science of Human Evolution," anthropologists S. L. Washburn and E. R. McCown describe how researchers in molecular biology and immunology are shedding new light on the beginnings of man. Gay Gaer Luce, a previous contributor to the *Yearbook*, tells of the biological clocks within us and within other animals that account for such phenomena as "jet lag"—the uneasy feeling that we experience after a long trip by air—and for our differing responses to stress, depending on the time of day. (This article confirms my feeling that there really are "night-people" and "day-people"—that is, people who prefer nighttime and are wider awake at night, and people who prefer daytime and just can't keep their eyes open after sunset. I always thought this was true; and now science supports me.) Psychiatrist Jan Fawcett discusses schizophrenia, the most common of the severe mental disorders, and concludes that a combination of treatments—drugs, individual therapy, and family therapy—rather than any single treatment, offers the most hope of combating it. The mysterious and fascinating claims of acupuncture are probed by prize-winning physician Emanuel M. Papper, who places the ancient Oriental practice within the larger framework of anesthesiology in general. And in "Death and Dying," Diana Crane shows how our greater control over the timing and circumstances of dying has raised new problems both for physicians and for patients and their families. How typical of technological progress that is!

The age of space flight has given man some new tools for probing the frontiers of the universe. In "UHURU: The First Orbiting X-Ray Laboratory," scientist-engineer Riccardo Giacconi describes how a small and unheralded space satellite has discovered hitherto unknown X-ray sources in our own galaxy and in other galaxies and in so doing has caused a revolutionary change in our view of the cosmos. Closer to the earth, the mysteries of the aurora polaris—the Northern and Southern lights—are being probed by spacecraft and by ground instruments, as explained by geophysicist T. Neil Davis. Providing an overview of these and other discoveries, astronomer Jesse L. Greenstein in "Astronomy for the '70s" calls the past decade "one of many Galileos" and urges that the United States and other nations continue their financial support of such efforts.

All of the feature articles are abundantly illustrated, most in full color. And "Special Effects in the Movies" and "Man's New Underwater Frontier" are pictorial essays in which the numerous illustrations tell the stories.

In addition to the feature articles, the *Britannica Yearbook of Science and the Future* contains factual year-in-review treatments of individual disciplines within the area of science and technology. Such distinguished authorities as Nobel Prize winner Joshua Lederberg are among the authors.

All in all, it adds up to one of the most interesting editions of this *Yearbook*. I wish my late husband were here to read it, and write this Publisher's Message. He liked this book so much; and I hope you do, too.

Helen H. Benton

PUBLISHER

Contents

Contributors to the Science Year in Review

Joseph Ashbrook *Astronomy.* Editor, *Sky and Telescope,* Cambridge, Mass.
William J. Bailey *Chemistry: Structural Chemistry.* Research Professor of Chemistry, University of Maryland, College Park.
Jeremy E. Baptist *Molecular Biology: Biophysics.* Assistant Professor of Radiation Biophysics, University of Kansas, Lawrence.
Hyman Bass *Mathematics.* Professor of Mathematics, Columbia University, New York, N.Y.
Louis J. Battan *Atmospheric Sciences.* Professor of Atmospheric Sciences and Associate Director of the Institute of Atmospheric Physics, University of Arizona, Tucson.
Harold Borko *Information Science and Technology.* Professor in the School of Library Service, University of California, Los Angeles.
George M. Briggs *Foods and Nutrition.* Professor of Nutrition, University of California, Berkeley.
D. Allan Bromley *Physics: Nuclear Physics.* Henry Ford II Professor and Chairman, Department of Physics, and Director, A. W. Wright Nuclear Structure Laboratory, Yale University, New Haven, Conn.
John M. Dennison *Earth Sciences: Geology and Geochemistry.* Professor and Chairman, Department of Geology, University of North Carolina, Chapel Hill.
F. C. Durant III *Astronautics and Space Exploration: Earth Satellites.* Assistant Director (Astronautics), National Air and Space Museum, Smithsonian Institution, Washington, D.C.
Robert G. Eagon *Microbiology,* Professor of Microbiology, University of Georgia, Athens.
H. L. Edlin *Environmental Sciences.* Publications Officer, Forestry Commission of Great Britain and author of *Trees, Woods and Man.*
Gerald Feinberg *Physics: High-Energy Physics.* Professor of Physics, Columbia University, New York, N.Y.

Joseph Gies *Architecture and Building Engineering.* Consulting engineer, Chicago, Ill.
James G. Greeno *Behavioral Sciences: Psychology.* Professor of Psychology, University of Michigan, Ann Arbor.
Clayton H. Heathcock *Chemistry: Chemical Synthesis.* Associate Professor and Vice-Chairman, Department of Chemistry, University of California, Berkeley.
L. A. Heindl *Earth Sciences: Hydrology.* Executive Secretary, U.S. National Committee for the International Hydrological Decade, National Academy of Sciences—National Research Council, Washington, D.C.
Robert L. Hill *Molecular Biology: Biochemistry.* Professor of Biochemistry, Duke University, Durham, N.C.
Barton L. Hodes *Medicine: Ophthalmology.* Practicing ophthalmologist and Instructor of Ophthalmology, School of Medicine, Northwestern University, Chicago, Ill.
King Holmes *Medicine: Venereal Diseases.* Chief, Division of Infectious Diseases, U.S. Public Health Service Hospital, Seattle, Wash.
Richard S. Johnston *Astronautics and Space Exploration: Manned Space Exploration.* Director of Life Sciences, NASA Manned Spacecraft Center, Houston, Tex.
Lou Joseph *Medicine: Dentistry.* Assistant Director, Bureau of Public Information, American Dental Association, Chicago, Ill.
William L. Kissick *Medicine: Community Medicine.* Professor of Community Health, School of Medicine, University of Pennsylvania, Philadelphia.
Walter S. Koski *Chemistry: Chemical Dynamics.* Professor of Chemistry, Johns Hopkins University, Baltimore, Md.

Ernest R. Kretzmer *Communications.* Director, Data Communications Technology & Applications Laboratory, Bell Telephone Laboratories, Holmdel, N.J.

Joshua Lederberg *Molecular Biology: Genetics.* Professor of Genetics, Stanford University, Stanford, Calif.

John G. Lepp *Zoology.* President, Marion Technical College, Marion, Ohio.

Howard J. Lewis *Science, General; Astronomy: The Mayall Telescope.* Director, Office of Information, National Academy of Sciences—National Academy of Engineering—National Research Council, Washington, D.C.

Samuel P. Martin III *Medicine: Community Medicine.* Professor of Community Medicine, School of Medicine, University of Pennsylvania, Philadelphia.

N. F. Maude *Photography.* Editor, *Photographic Processor;* Consultant Editor, *British Journal of Photography;* Author of *Take Better Photos.*

Marcella M. Memolo *Agriculture.* Public Information Officer, Information Division, Agricultural Research Service, U.S. Department of Agriculture, Washington, D.C.

Norman Metzger *Chemistry: Applied Chemistry.* American Association for the Advancement of Science, Washington, D.C.

Raymond Lee Owens *Behavioral Sciences: Anthropology.* Assistant Professor of Anthropology, University of Texas, Austin.

Robert A. Page *Earth Sciences: Geophysics.* Geophysicist, National Center for Earthquake Research, U.S. Geological Survey, Menlo Park, Calif.

Alan J. Perlis *Computers.* Eugene Higgins Professor of Computer Science, Yale University, New Haven, Conn.

Willard J. Pierson, Jr. *Marine Sciences.* Professor of Oceanography, University Institute of Oceanography, The City College of New York, New York, N.Y.

Froelich Rainey *Archaeology.* Professor of Anthropology and Director of the University Museum, University of Pennsylvania, Philadelphia.

Richard Saltonstall, Jr. *Environmental Sciences.* Author of *Your Environment and What You Can Do About It;* co-author of *Brown Out and Slow Down.*

Flip Schulke *Photography.* President, Flip Schulke Underwater Films.

Mitchell R. Sharpe *Astronautics and Space Exploration: Space Probes; Medicine: Room W-204.* Science writer, author of *Satellites and Probes* and *Living in Space,* Huntsville, Ala.

M. Michael Sigel *Medicine: Immunology.* Professor, School of Medicine, University of Miami, Miami, Fla.

Albert Smith *Botany.* Assistant Professor of Biology, Wheaton College, Wheaton, Ill.

David L. Smith *Environmental Sciences.* Staff member, Centre for Environmental Studies, London, Eng.

Frank A. Smith *Transportation.* Senior Vice-President-Research, Transportation Association of America, Washington, D.C.

J. F. Smithcors *Veterinary Medicine.* Editor, American Veterinary Publications, Santa Barbara, Calif.

William E. Spicer *Physics: Solid-State Physics.* Professor of Electrical Engineering and Materials Science, Stanford University, Stanford, Calif.

Peter Stoler *Medicine: General Medicine.* Medicine Editor, *Time* Magazine.

Charles Süsskind *Electronics.* Professor of Engineering Science, University of California, Berkeley.

James A. West *Fuel and Power.* Staff Assistant, Mineral Resources Office of the Secretary, Department of the Interior, Washington, D.C.

A Pictorial Essay

Special Effects in the Movies

by Wayne M. Smith

Spacecraft that tumble headlong through fantastic fields of color and shape—ocean liners that turn upside down—orbiting geodesic domes that move away from one another—these are a few of the scenes that have startled and amazed viewers at recent movies. How are these "special effects" created?

Courtesy, Universal Pictures

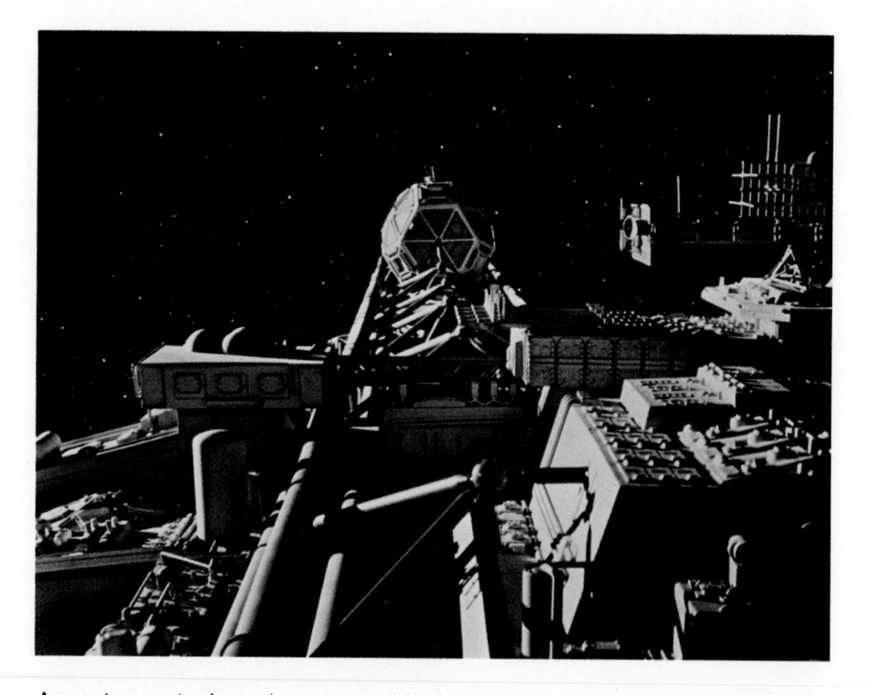

An astronaut aboard a spaceship begins to orbit Jupiter. Suddenly, he sees in the distance a giant slablike monolith, also in orbit. As he moves toward it, the whole shape of space appears to change, and the pilot and his craft are drawn headlong through a fantastic array of colors and shapes that form, dissolve, and re-form around them. During this "light trip" known physical laws are broken, as the astronaut finds himself and his ship abruptly in a Louis XVI drawing room and later views another man who turns out to be himself. Finally, this first man to explore the outer regions of the solar system returns as a newborn infant to contemplate from space an earth on which no traces of man are visible.

The viewer of this trip has just witnessed in the film *2001: A Space Odyssey* one of the outstanding examples of a rapidly developing motion-picture art, special-effects cinematography. More recently, several other films have joined *2001* in successfully employing a variety of special effects. They include *The Poseidon Adventure, The Andromeda Strain,* and *Silent Running.*

Special effects are divided into two general categories: photographic and mechanical, each with inherently different problems and each requiring totally different skills. Photographic effects are those requiring special camera operations in order to achieve the desired imagery on the film. They might be as simple as a dissolve from one scene into another or as complicated as the animated streak photography used in the light trip of *2001*. Mechanical effects are used when a director wants some unusual physical action such as fire, rain, snow, or explosions on the set. These are often dangerous to set up, and the utmost in skill and knowledge is required when they are used. The physical-effects man sometimes has the life of others in his hands and cannot make mistakes.

WAYNE M. SMITH, *a cinematographer, worked on special effects for* The Andromeda Strain *and* Silent Running.

Courtesy, Wayne M. Smith

Exterior parts of the space freighter "Valley Forge" (above and opposite page) are photographed for the movie Silent Running. *By means of photographic special effects the 25-ft-long model shown here appeared in the film to be a giant spaceship, large enough to house a city.*

Courtesy, Irwin Allen Productions, 20th Century Fox

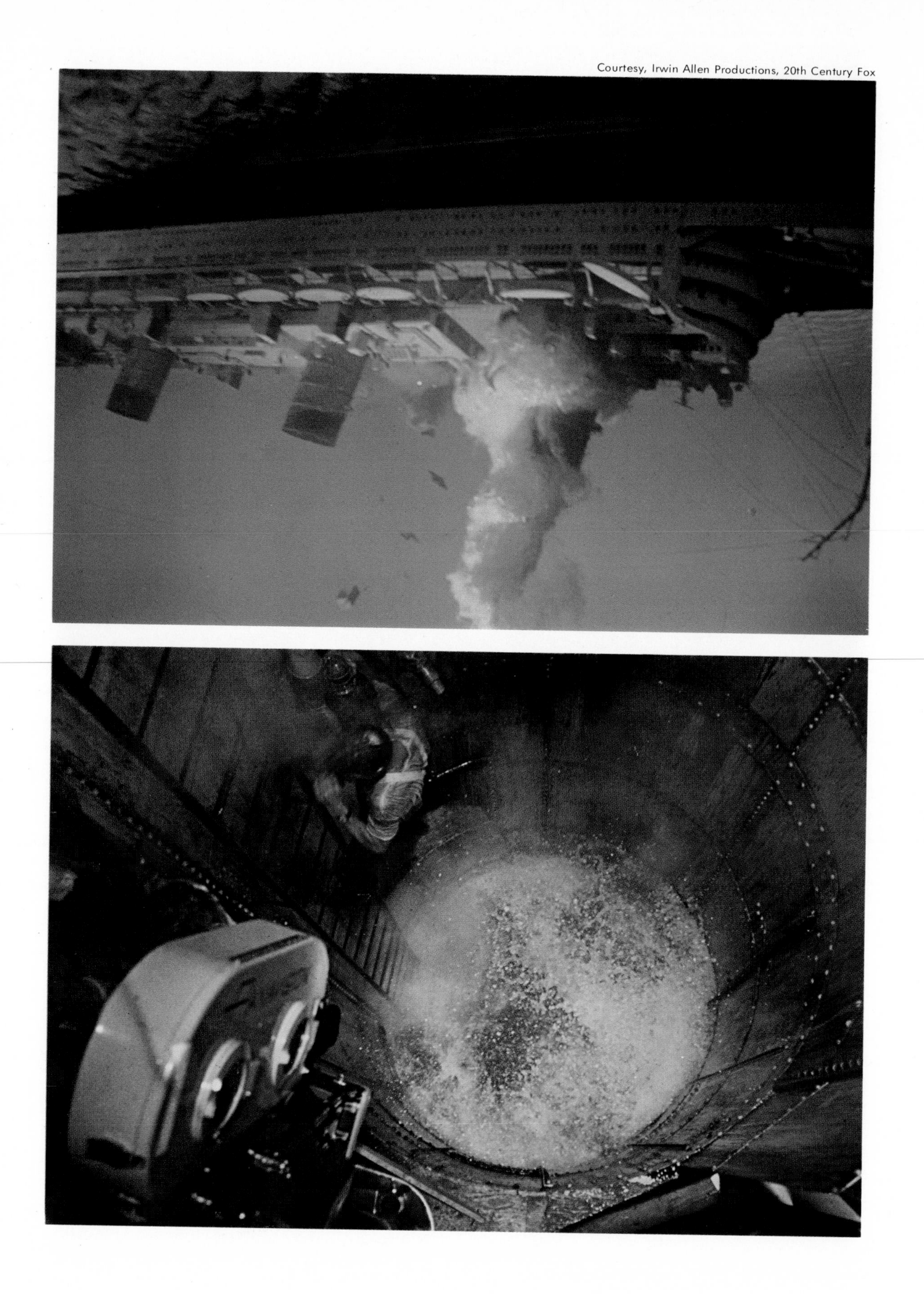

The optical printer

One of the most commonly used tools in special-effects photography is the optical printer. Nearly every 35-mm film undergoes some form of optical printing, whether it is simply for the insertion of titles or the more complicated combination of photographic elements. The optical printer is basically a camera mounted so as to rephotograph a transparency from behind. Although simple in theory, it becomes complex when one considers all the variables that must be working properly in order for it to function correctly. Not only must the projecting camera be in synchronization with the taking camera, but all movements relative to the picture frame and perforations on that frame must be absolutely registered to one another (in agreement with respect to position) so that the picture remains steady as new elements are added on successive multiple exposures. For this reason optical printers are built to close tolerances.

As an example, let us see how a particular effect was achieved through the use of an optical printer in the science fiction film *Silent Running*. The film required a shot depicting a geodesic dome moving away from another dome that housed the observer. In the background were stars moving in another direction. The shot was lined up with a model geodesic dome at one end of a camera track. In front of the camera a section of a geodesic dome was placed in such a manner as to make a foreground that would track back with the camera as the sequence was photographed. The shot was first photographed exactly as directed but without the stars, the background being completely black. In order to put in the star background it was necessary to make a matt to mask out everything but the background. This was done by wrapping the dome at the end of the track in white paper and tape, and relighting the front dome section and plants so as to produce no shadows. The shot was then rephotographed exactly as before to match the first move. A precise tracking dolly was necessary so that all moves would later match when optically combined in the optical printer.

In the preceding example the use of a black background was employed in order to combine the two elements in a traveling matt shot. However, this is not suitable for live-action photography because of the problems of superimposition of shadows and backgrounds. For live action a common method is the "blue screen" system, which takes advantage of the film's ability to record different colors in foreground and background during original photography. The actors are lighted conventionally with white light in front of a deep-blue backing. The color negatives that result are then contact printed onto different black-and-white stocks. This results in separate foreground and background elements.

By using various printing procedures with the optical printer, a traveling (moving images) matt is produced, having a clear area where the actors and foreground objects appear and an opaque one where the background is. The matt is then printed in conjunction with the

In The Poseidon Adventure *the ocean liner "U.S.S. Poseiden" is capsized by a giant wave and floats bottom side up (opposite, top). The model of the ship had to be large enough so that the waves would be in scale but small enough so that it would not dwarf the largest wave that special-effects men could produce. Up-ended air shaft in the ship (bottom), through which passengers made their way to safety, measures 30 ft in height and 14 ft in diameter.*

To achieve the effect of one geodesic dome moving away from another in space while stars travel in another direction, a model dome was lined up at one end of a camera track. In front of the camera a section of the other dome made a foreground. By wrapping the background dome in white paper and specially lighting the foreground, photographers made a color positive print (top) This was then printed onto a black-and-white high-contrast negative matt contact print (center), from which was made a high-contrast positive matt contact print (bottom). A color negative (opposite, top) was made by photographing everything but the stars, the background being completely black. This was then combined with the star background (opposite, center) and the high-contrast positive matt contact print to achieve the final, shadowless effect (opposite, bottom).

Courtesy, Universal Pictures

desired background for the final composite. This is an oversimplification of the method and does not account for such factors as the coordination of foreground and background elements, the elimination of matt line fringes, or all the laboratory procedures required to produce complex matts.

Computer synchronization

Plexiglas hexagons attached to rigs allowing them to rotate 360° on both horizontal and vertical axes were used to form the "Andromeda strain" bacteria. A computer controlled the apparatus and also caused a strobe light to flash whenever a rotating hexagon reached its desired position, thus building up three-dimensional shapes. The process began with relatively few hexagons (opposite, left), but more were added in each successive frame, giving the bacteria the appearance of growing. These shapes were then processed through a special high-resolution video system (diagram below) that added color and visually altered their static quality.

The movie *The Andromeda Strain* employed computer-synchronized equipment to achieve some of the most complex cinematographic special effects ever seen in a film. The "Andromeda strain" of the title was an unknown bacterium that had a crystalline structure. A truncated tetrahedron was selected as the basic shape for the structure because it offered many exciting possibilities for design and photography. A Plexiglas hexagon with a diameter of 3 in. and ⅛ in. thick was attached to a specially constructed rig by means of an arm allowing it to rotate 360° in both an X and a Y axis. This made it possible to position the hexagon exactly within an infinite volume of space and also to duplicate a position more than once.

The entire rig was controlled by digital pulse motors driven by a computer. In addition, a strobe light also controlled by the computer was set up to flash whenever the rotating hexagon reached its proper position in space. With the computer controlling the hexagon's positioning and strobe light, it was possible to build up three-dimensional hexagonal shapes within a relatively short period of time. The process began with relatively few hexagons, and then as each successive frame appeared the preceding arrangement would be repeated and a few

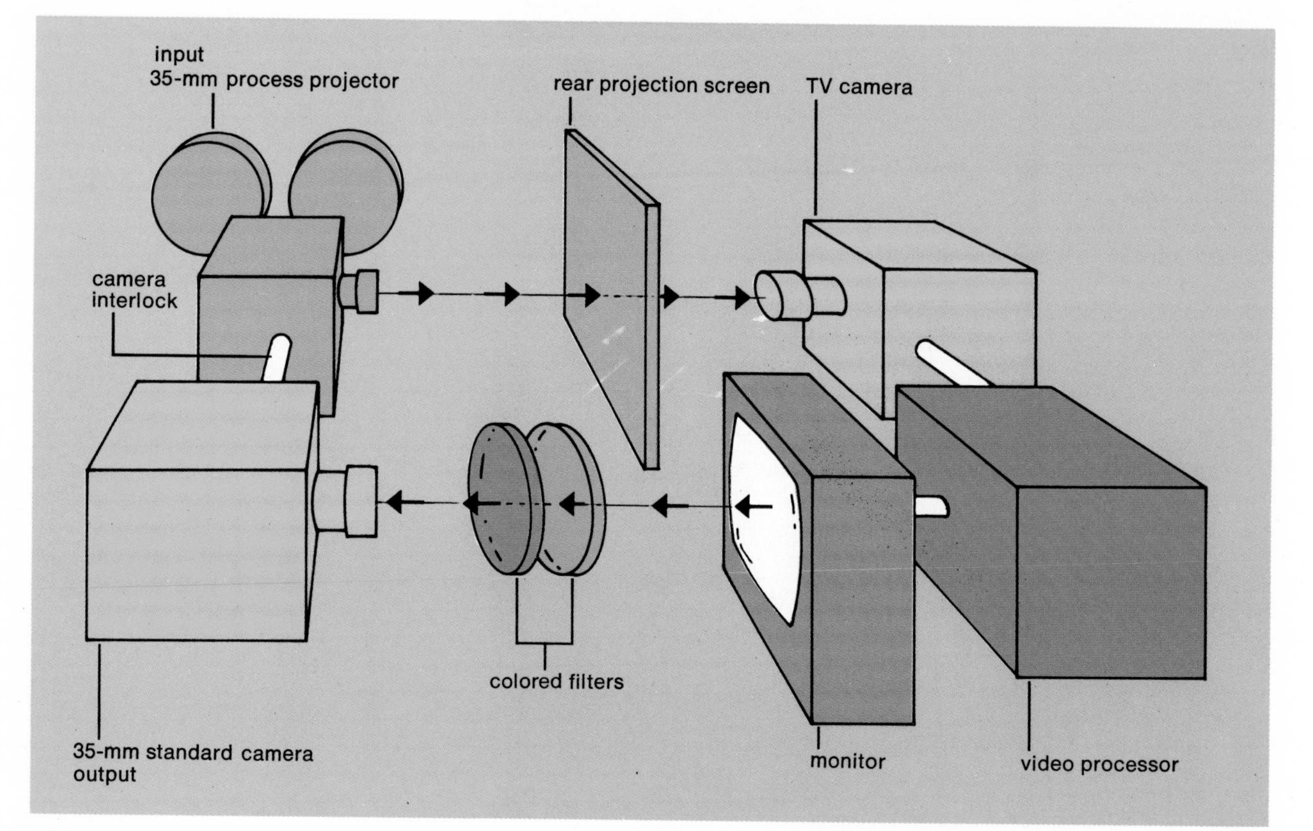

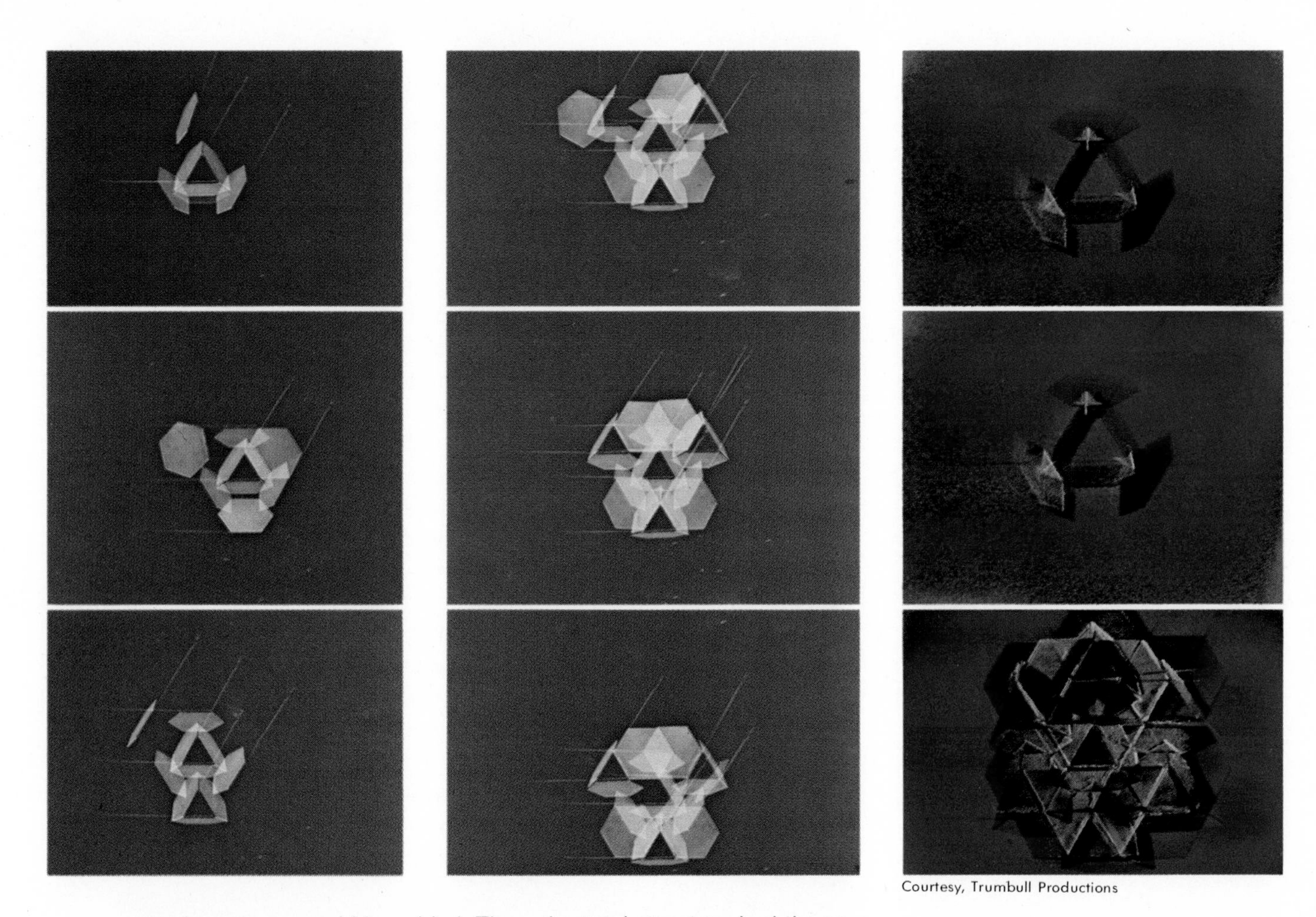

Courtesy, Trumbull Productions

more hexagons would be added. Thus, the total structure had the appearance of growing. The result was a series of black-and-white images of hexagonal shapes that were then taken and processed through a special high-resolution video process system adding color and edge generation and visually altering its static quality.

The high-resolution video process system had a resolution capability of 2,000 lines, a 30-megaHertz bandwidth, a facility for switching from positive to negative, and the option of being able to select narrow gray scales. It enabled the operator to put a 35-mm film through the system being picked up by a vidicon camera, electronically altering the image which was then shown on a high-resolution monitor from which it could be filmed. Color was added to the black-and-white system through the use of colored filters inserted in front of the 35-mm camera filming off the vidicon screen. In function it was very similar to an optical printer.

Electronic manipulation of imagery is a relatively new process. One firm uses a machine called "Scanimate" to animate two-dimensional graphic art in real time. What would formerly have taken days to produce can now be done in hours. "Scanimate" requires high-contrast negatives of an artist's work; these are placed in front of a specially designed television camera from which animation is created by adjusting appropriate knobs that control movements of the video image in three dimensions. The results are filmed or videotaped off a television screen. A more complex machine, "Caesar," allows cartoon characters to be animated in real time without the thousands of drawings required for earlier conventional animation.

The "Valley Forge"

One of the most detailed models ever constructed for a film was the space freighter, "Valley Forge," for *Silent Running.* Over 25 ft in length, it took more than six months to design and construct. The designers used hundreds of small parts glued on the surface of the model in series in order to achieve believable detail. Many of these parts were obtained from 850 Japanese model kits of German tanks purchased for this purpose. In addition, hundreds of small parts were cast from molds in epoxy resins. Results were always checked with 35-mm slides and projected to see if the detail remained believable.

The photography of the model was unique in that the front-projection matting system was used for all of it. This allowed the model to be combined with its background in the camera without requiring a traveling matt. Normally one cannot photograph a background image projected from the front because the projected image would fall upon the actors and set, who would then cast visible shadows on the screen behind them.

With the use of a special reflex reflection screen, it is possible to circumvent these problems. By locating the projector and camera

Front-projection matting system allows photographing of a background image projected from the front without the image falling on actors and sets and thereby causing them to cast visible shadows on the screen. A semitransparent mirror causes the projector and camera lens to be located in the same axis, completely hiding any shadows cast on the screen. By use of a special reflex reflection screen projected light is reflected back toward its source.

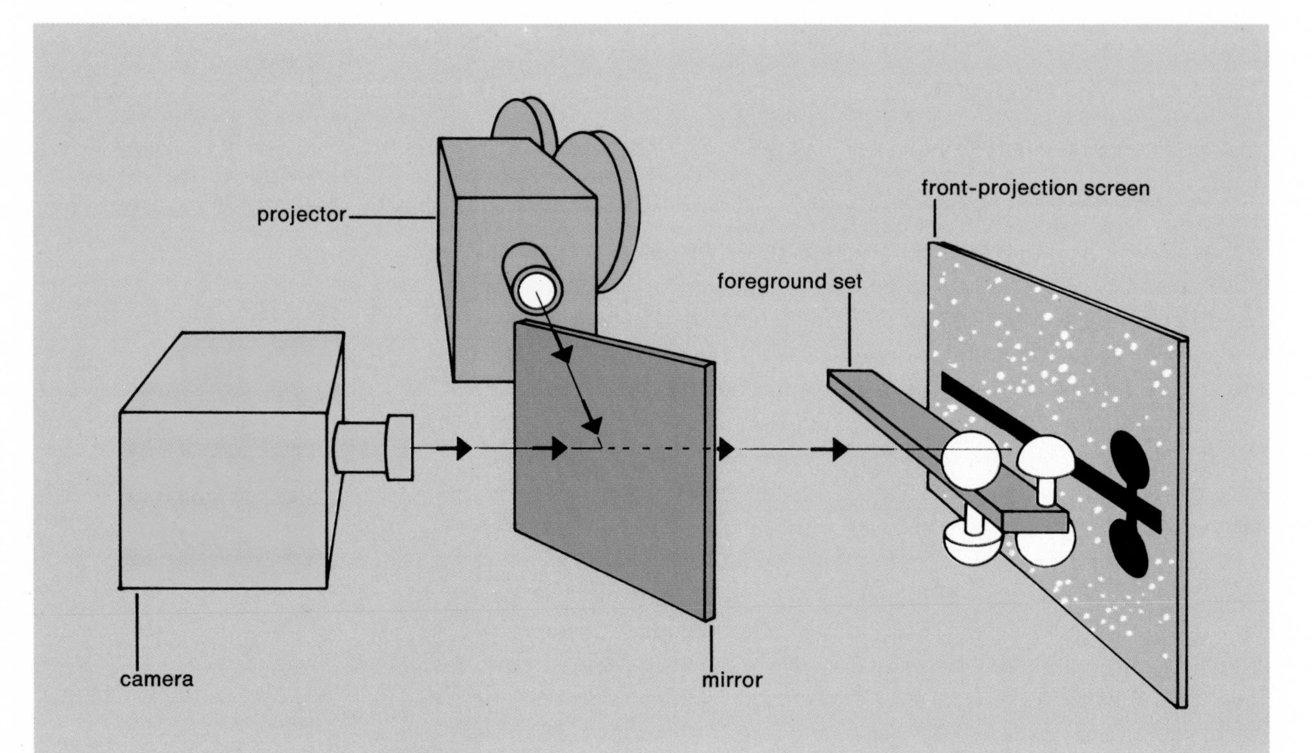

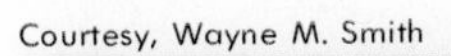
Courtesy, Wayne M. Smith

Courtesy, John Dykstra

Spaceship "Valley Forge" in Silent Running *(above) is photographed by front-projection method described in the diagram on the opposite page. The rig used to shoot the above picture is mounted on a special boom (left).*

lens in the same axis through the use of a semitransparent mirror, the shadows cast by actors on the screen are completely hidden. The reflective properties of the screen are such that the light projected is reflected back toward its source, being substantially brighter than that which falls upon the actors or other foreground objects.

Maintaining depth of field is imperative when shooting miniatures. This was maintained when shooting the spaceship model "Valley Forge" by shooting at *f*/22 on 35-mm film (*f*/22 indicates that the camera's lens has an effective aperture 1/22 that of its focal length). Exposures at *f*/22 required a camera speed between 10 and 30 seconds per frame. Background plates projected in the front projection machine were 4 in. × 5 in. color transparencies of stars, other ships, and Saturn.

Moving the model was impractical, and so the camera/front projector unit was mounted on an articulating boom arm which, in turn, was mounted on a 12-ft-long dolly track. Track speed had to relate to the camera speed, requiring the use of synchronous motors to dolly past the model as slowly as an inch a minute. Miniature vehicles are

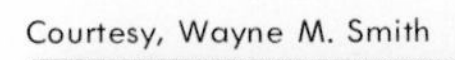

Courtesy, Wayne M. Smith

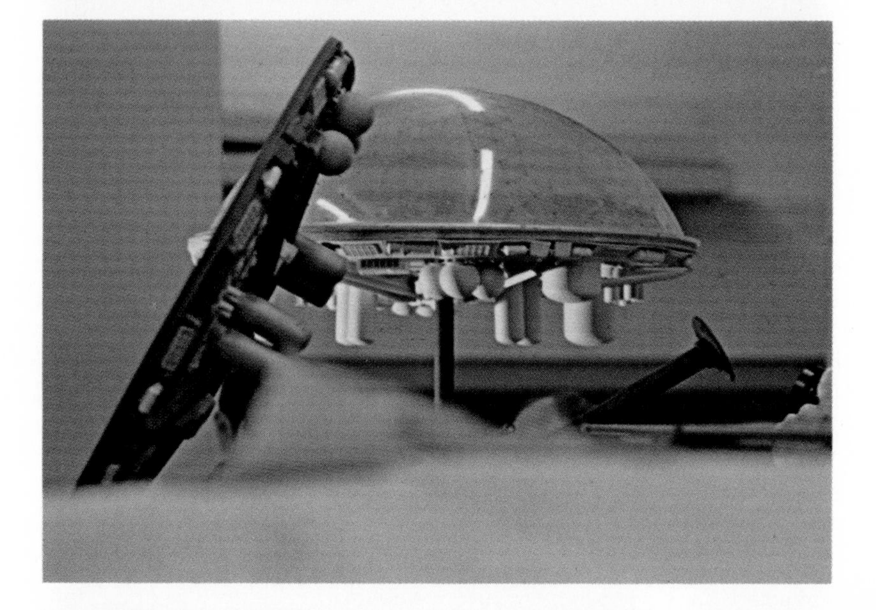

To make the spaceship "Valley Forge" (opposite page) designers used plastic parts from 850 Japanese model kits of German tanks (below, left). Many of the parts were cast in epoxy resins using silicon molds (below). Various stages in the construction of "Valley Forge" are shown at left.

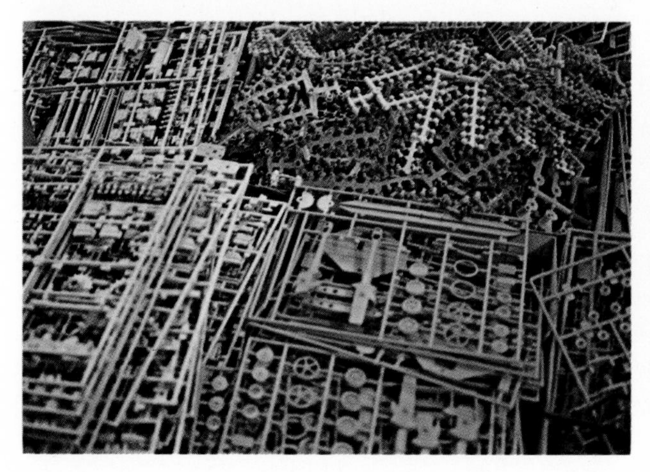

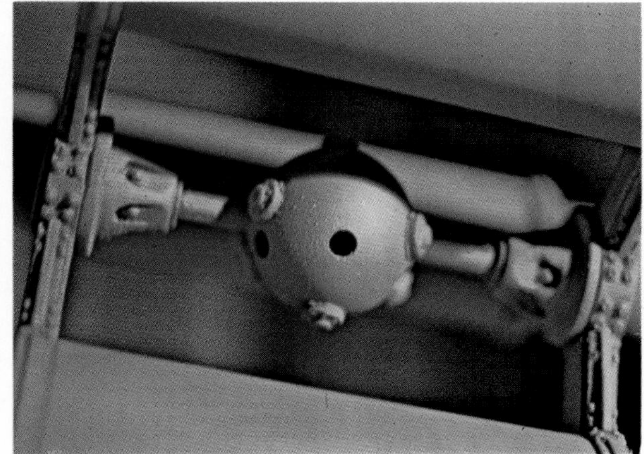

Front-projection method is used as cameramen prepare to film a shot of Saturn through a section of geodesic dome in Silent Running.

usually shot at high speed (normal speed being 24 frames per second) in order to smooth out the jerky action associated with miniatures.

Mechanical effects

Physical-effects men are sometimes called upon to help solve some unique problems. Such was the case in *Silent Running*. The director wanted three very unusual robot maintenance drones. Six months went into the research, design, and construction of the three drones. Requirements dictated that they be small and machinelike in appearance, with good maneuverability, but still have humanlike characteristics. After considerable experimentation, it was found that this result could be accomplished if "drone" suits were fitted around actual people.

And so began a project in which bilateral amputees, all in their teens, were fitted with custom-tailored drone suits that walked on their hands. The suits were made from ABS plastic, vacuum-formed in pieces and assembled. ABS is a heat-forming plastic capable of being mistreated without breaking. The weight of the suit was 20 lb, including a manipulator arm that was attached to the front.

The manipulator arm gave the drones the dexterity to perform the necessary tasks called for in the script. Five months' work went into the design and construction of the two small, light, and maneuverable pneumatically controlled aluminum devices. Arm control for positioning was provided remotely by radio through servo valves in the pneu-

Robot maintenance drones in Silent Running *were devised by fitting custom-tailored plastic suits around young amputees. Drones are seen performing their routine checkouts of the "Valley Forge."*

Courtesy, Universal Pictures

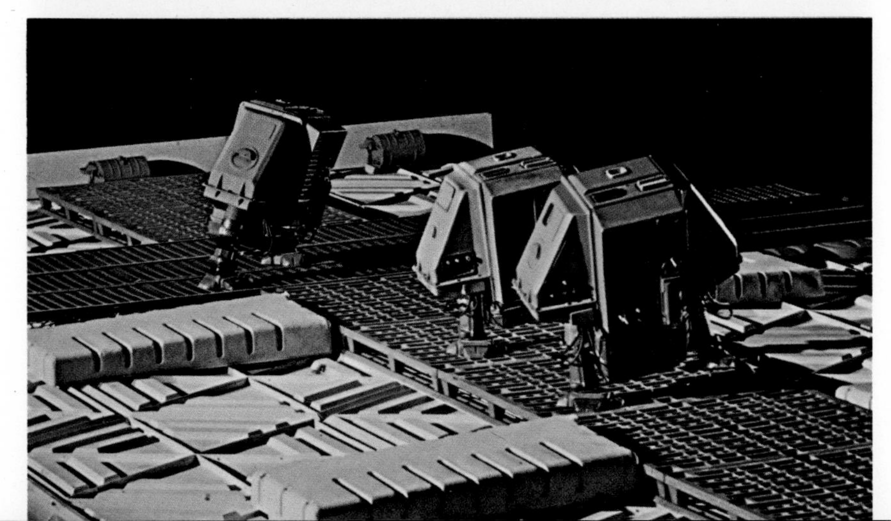

A trip through the solar system in a spaceship is the subject of a film at the San Diego Planetarium. Made possible by special-effects equipment and techniques, the movie was filmed in 70 mm with a 160° spherical lens to match the projected-upon concave ceiling surface of the planetarium. All elements were shot in the camera on different passes by masking the lens off with tape where one did not want another image to fall.

Courtesy, Graphic Films

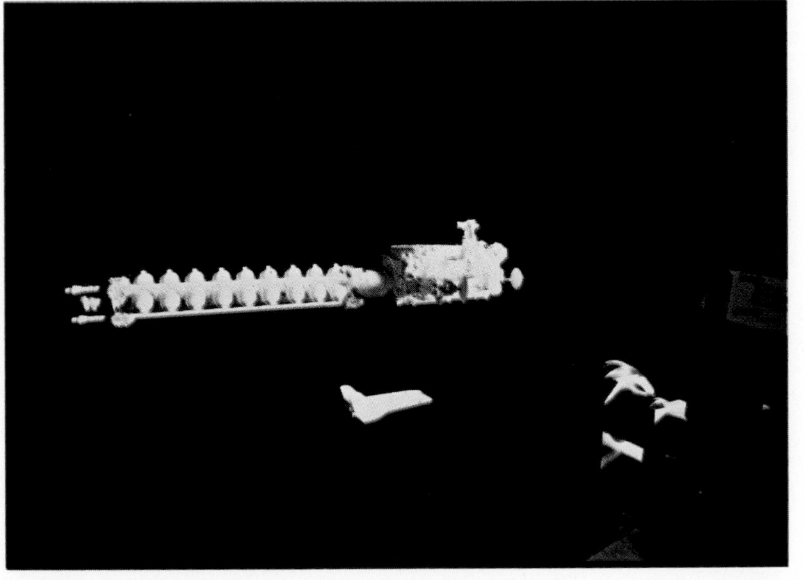

matic cylinders. Developments such as this are necessary when comparable devices cannot be purchased or made by modifying equipment already on the market.

The spectacular light-trip sequence of 2001: A Space Odyssey *(opposite) was achieved by using a new process called slit-scan. In the diagram below, a thin vertical slit on an otherwise blacked-out glass pane generates a flat plane in perspective onto a single frame of film. Artwork on the rear pane of glass is then moved slightly to animate the image, after which the whole process is repeated.*

Photos, Courtesy, Douglas Trumbull

The *2001* light trip

To achieve the light-trip sequence in *2001* a new process called "slit-scan" was introduced. An outgrowth of time-lapse photography, it is essentially controlled streak photography. In simple terms, when one exposes a piece of film to a moving, two-dimensional, flat image, or injects movement in the camera itself, a streaked image results. By controlling the streaking process, a third dimension can result with infinite visual possibilities. Therefore, the "slit-scan" process is basically a mechanical and optical method of producing a controlled image streaking and image movement onto motion picture film one frame at a time. The movement must be exactly repeated and incremented in a precise manner.

The equipment to do this consists of a camera and lens, which has, attached to the lens, a cam controlling focus in sync with the relative position of the dolly to which the camera is attached and the 16-ft track on which the dolly is riding. At one end of the track the two-dimensional artwork is placed in a vertical position on the rear piece of one of two 3 ft × 5 ft glass panes held in sliding tracks perpendicular to the dolly track. The front glass pane is completely blacked out, except for a single vertical slit 3/16 in. wide. The artwork consists of high-contrast negatives backed with color gels and light from behind. A synchronous motor ties the camera functions, dolly movement, and glass pane movements electronically to a control box that sequences all the equipment so that exact repetitive moves can be made frame by frame.

The image is generated by placing the camera lens close to the artwork, opening the shutter of the camera, and tracking back, while at the same time moving the front pane of glass with the slit on it over the artwork. This generates a flat plane in perspective onto a single frame of film. When the camera reaches the end of the track, the shut-

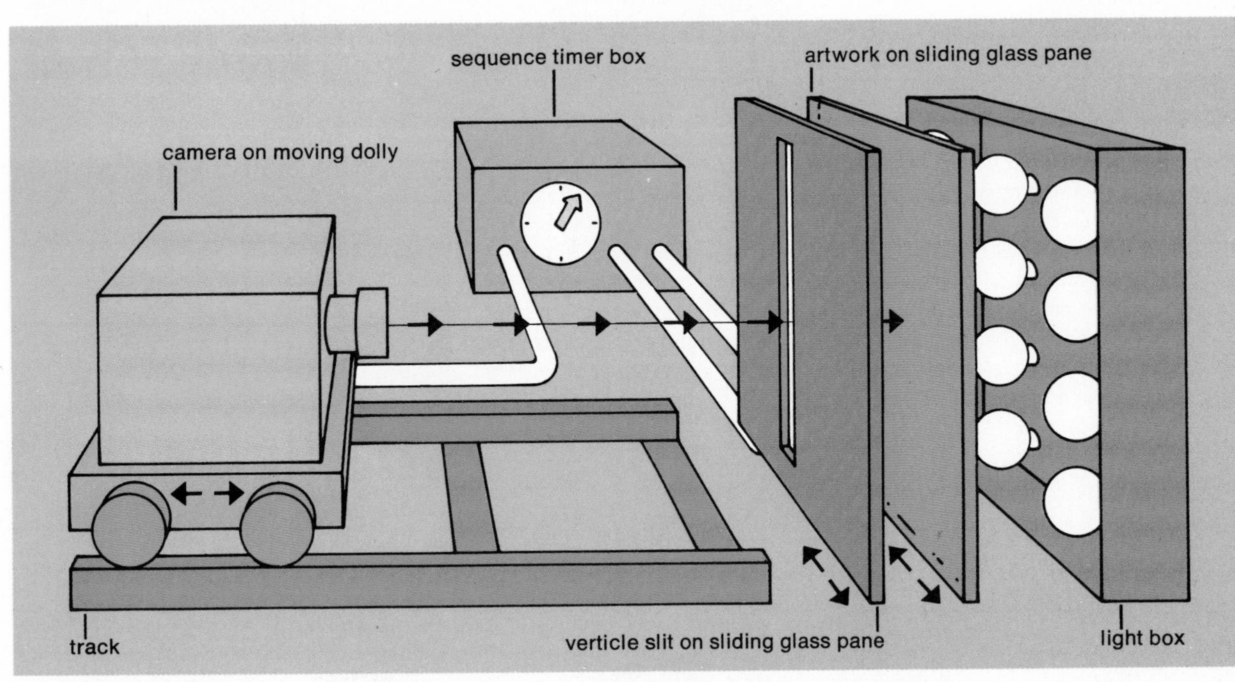

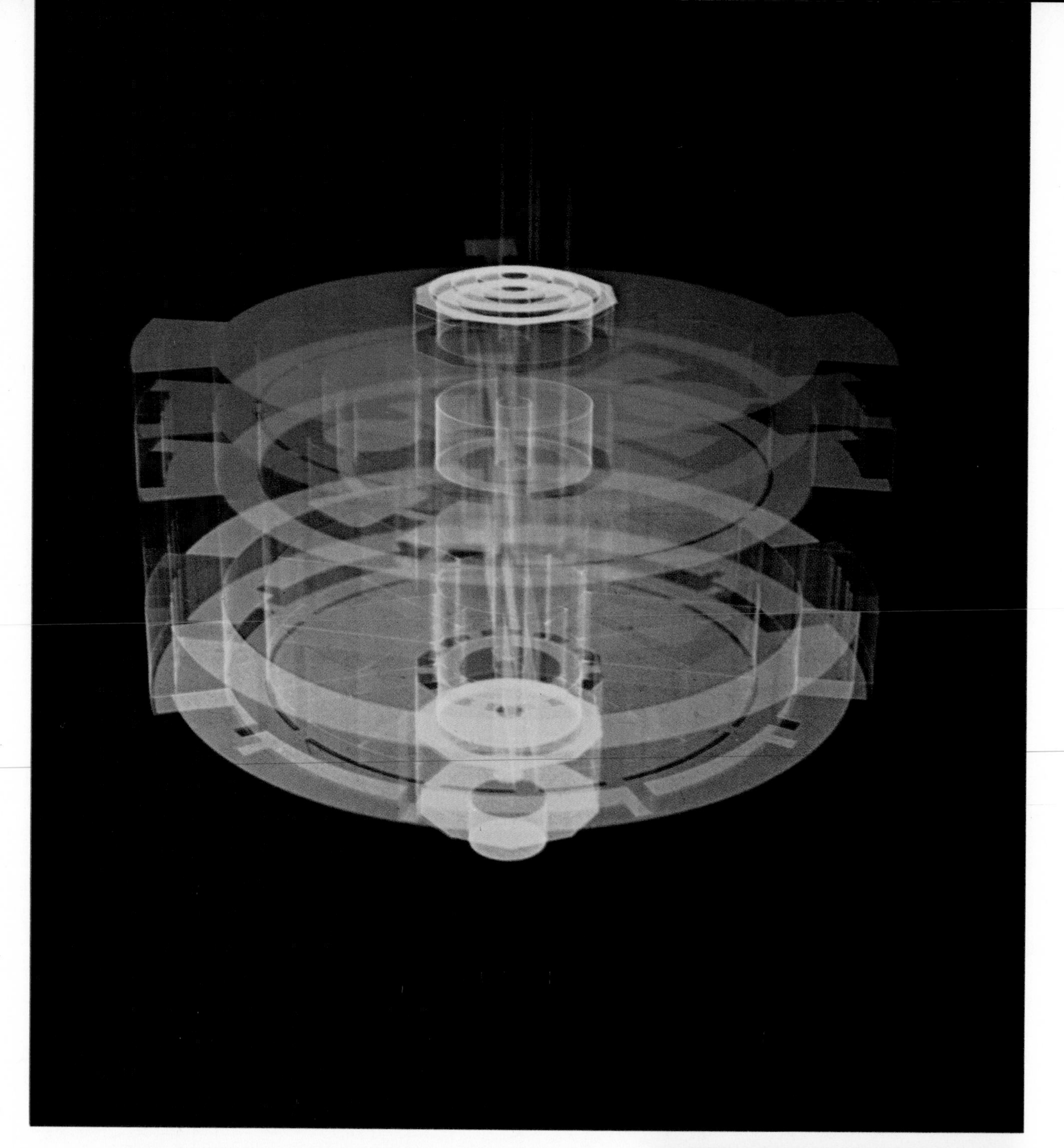

Animated light streak photography produced the effects above, at the right, and at the top and center of the opposite page. For the three-dimensional rotating "Wildfire" model in The Andromeda Strain *on this page, 28 pieces of two-dimensional artwork were projected on a screen off of which they were filmed. The camera remained in a fixed position, while the screen and projector moved the proper distance for the required streak. The two-dimensional artwork on the opposite page consisted of high-contrast negatives on a moving light box. The logo (opposite, bottom) was created by the slit-scan process.*

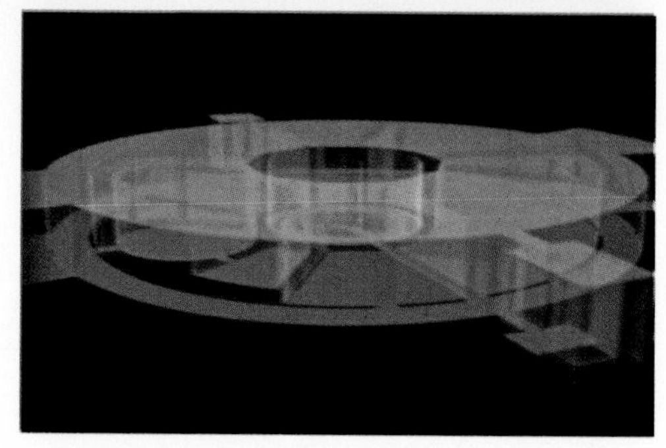

ter closes. Next, the rear pane of glass with the artwork on it is moved slightly so as to animate the image, and the whole process is repeated. This type of photography is very slow, with each frame taking as much as 30 seconds to expose.

Courtesy, Trumbull Productions

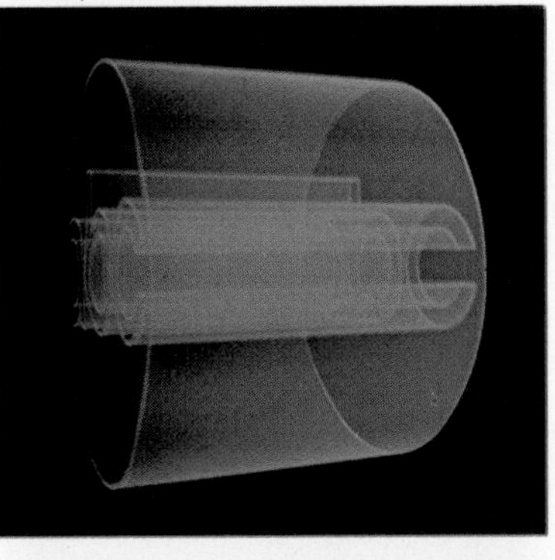

Future prospects

As of this writing, new television processes are being developed, which, when taken together, will give special-effects capabilities never before possible with film. The outgrowth of this development may shortly make film obsolete, and all or most "motion pictures" may become electronic.

One newly developed technique is improved electronic television tape-to-film transfers. Until recently the resolution quality of pictures transferred from videotape to film was not as good as from film to film, and, therefore, was not adequate for general theatrical release. However, recently developed equipment is soon expected to provide even better resolution than film-to-film transfers, opening up a new range of possibilities for special-effects cinematography. Also, electronic matting, editing, special effects, and other image manipulations can be done electronically in real time, with results instantly available and with much less effort than in conventional film production.

Another new technique is enabling the industry to matt electronically (combine more than one image element) moving and static images. This produces totally believable results, even including the casting of shadows from one image onto another.

With the improvement of many of these new processes, another new method of film production has emerged that solves a matting problem when shooting miniature sets. Because of the difficulty of matching camera moves, it had been impossible to dolly through a miniature set and matt in live action as if it were a full-size set. A recent development solved this by using a television matting system employing two synchronous cameras, one for live action and the other for the miniature set. Live action is shot on a blue stage with boundaries marked on the floor corresponding in scale to the miniature set. A snorkel camera in the miniature set is connected electronically to the live-action camera and dolly in such a way that the moves made by the live-action camera and dolly are exactly duplicated in scale in the miniature set. Thus, large elaborate sets can now be miniatures, resulting in considerable cost savings for the producer.

The New Science of Human Evolution

by S. L. Washburn and E. R. McCown

With only a few fossils of early man to use as guides, efforts to trace man's origins went in many directions. Now, modern techniques are putting the search back on course.

Recently there have been major developments in the understanding of human evolution. These have come from biochemistry, geology, and paleontology, and from studies of animal behavior. Changes in radioactivity have allowed the age of many of the fossils to be determined in years, something that would have seemed pure magic to the founders of evolutionary theory. Yet, in spite of the lack of modern techniques, the founders of theories that explain evolution came to conclusions that are supported by all the recently discovered techniques and evidences. When Charles Darwin offered the theory of natural selection to explain the process of biological evolution in 1859, he was careful not to discuss the problem of the origin of man. But the question was immediately raised, and Thomas Henry Huxley, the great exponent and public defender of Darwin's views, published a book on *Man's Place in Nature* only four years later in 1863. Darwin himself published a much longer discussion on human evolution, *The Descent of Man*, in 1871. Darwin believed that Huxley had "conclusively shown" that the ape that most nearly approached man in its biological organization was either the chimpanzee or the gorilla.

This point of view was widely held, but it came increasingly under attack. During the first half of this century almost every major group of primates was proclaimed as the one most closely related to man. A brief look at the primates, the order of mammals to which man belongs, will show how this could be the case.

Man's place among the primates

Man differs from the contemporary apes (the chimpanzee, gorilla, orangutan, and gibbon) in his way of locomotion. The bones and muscles that make bipedal walking possible are quite different from those of any other animal, but, in a general way, the anatomy of the arms and trunk of man is very like that of the apes. Other characteristics of the teeth and skull suggest the same conclusion, and it was this general similarity that led to the classic view of Darwin and Huxley. But, as more thorough studies were made, it became apparent that there were many differences between man and ape and that in some ways man seemed less specialized than the apes. It was suggested, therefore, that the line leading to man must have separated before that leading to the contemporary apes—a separation of 20 million to 25 million years.

But even this degree of separation did not seem enough to many scientists, and it was postulated that our nearest relatives are monkeys, quadrupedal forms rather than creatures that climb in the manner of apes (reaching, hanging, sometimes bipedal, often upright). The disagreements came from two sources. First, there were very few hominid fossil types and most of those that existed consisted only of parts of jaws and teeth. Second, there was no agreement as to how the anatomy of the contemporary primates was to be interpreted. Two competent scientists could look at the same anatomical characteristics and one could conclude that man and chimpanzee were very similar and the other that they were radically different and nothing properly called an ape could ever have been in human ancestry.

It was agreed that the primates separated from other kinds of mammals at the very beginning of the Age of Mammals, or even at the end of the Age of Reptiles, some 70 million years ago. The primates were arboreal and adapted to life in the trees by the evolution of hands and feet adapted to grasping. Primitive mammals had short fingers and sharp claws, probably not very different from those of a contemporary tree shrew or squirrel. In primates, the digits became long and the thick claws evolved into thin nails. This way of climbing is clearly reflected in the bones and was established at the beginning of primate evolution. The primates were also distinguished by features of the skull and teeth. The order was successful and evolved into many distinct groups. Some became extinct, and some of these early primates (prosimians) still survive, avoiding competition with the more advanced forms by being nocturnal or being removed from competition on the island of Madagascar. The scientists who believe that our nearest living relatives are to be found among the prosimians (possibly the tarsiers) think that certain features of the skull and teeth are so important that the whole course of evolution may be judged from them. All the apparent similarities of man and ape are then attributed to parallel evolution. That is, for example, the similarities in the arms would be regarded as being acquired independently long after the two groups were separated. Such an early separation would require a vast amount of parallel evolution, and this extreme point of view has never been held by a large number of scientists.

S. L. WASHBURN *and* **E. R. McCOWN**
are members of the anthropology faculty of the University of California at Berkeley. Washburn is also an editor of Perspectives on Human Evolution.
The authors are indebted to Alice Davis for editorial assistance.

The fossil prosimians had very small brains, eyes directed to the sides, and a sense of smell that was very important, as in primitive mammals generally. From the contemporary forms, it is known that tactile hairs were important. Although the limbs had adapted to arboreal life, the special senses remained those of primitive mammals. With the monkeys, the special senses adapted to arboreal life. Tactile hairs and sense of smell were reduced, and binocular, stereoscopic color vision evolved. The brain increased in size, and the number of teeth was reduced. It is this combination of features that gives the similarity to the heads of monkeys, apes, and man. This complex had probably evolved by 30 million years ago, and it is seen in the monkeys of South America and the monkeys of Africa and Asia. With the exception of some South American forms, the monkeys remained quadrupedal. The apes (gibbon, orangutan, chimpanzee, and gorilla) evolved a highly specialized way of climbing. They could reach in almost any direction, enabling them to feed in small branches. It is this locomotor adaptation that distinguishes the apes from the monkeys.

In summary, it can be seen that, if man evolved from a prosimian, all the anatomy of the special senses and nervous system that we share with the monkeys and apes must have evolved independently. If man evolved from a monkey-like ancestor, then all the anatomy we share with the apes must be due to parallel evolution. If man evolved from an ape-like ancestor, then only bipedal locomotion, some features of the teeth, and the large brain are evolutionarily new and unique to man.

Sweeping away the controversy

In spite of an enormous amount of comparative anatomy, many fragmentary fossils, and the efforts of many scientists, there was no agreement on the course of human evolution and the various theories suggested that man had had a separate ancestry for anywhere from 5 million to 50 million years. To overcome these traditional difficulties, methods are needed that are objective and that give the same results regardless of who performs the tests. Over the last few years such methods have become available, and it is now possible to supplement the fossil record, reassess comparative anatomy, and gain a new perspective on human evolution.

A thorough reorganization of our understanding of human evolution is now in progress. This must be seen as part of modern science, employing methods and understandings very different from those of only a few years ago. For example, it is a common belief that the New World monkeys were descended from now extinct North American prosimians, that they must have had a separate ancestry for at least 50 million years, and that their similarities to the Old World monkeys were a remarkable case of parallel evolution. But recent evidence, summarized by a scientist in Europe, Robert Hofstetter, makes it probable that the ancestors of the New World monkeys were in Africa, and that they rafted across the Atlantic to South America before the two lands had drifted far apart. Acceptance of the concept of continental drift, which was laughed at only a few years ago, changes the whole geography of evolution. Relationships among the primates that were thought to be impossible turn out to be most probable.

Just as the geography of evolution has changed dramatically, so has our ability to measure time more precisely given a new framework to evolutionary studies. Numerous methods of determining the age of rocks have now been devised, the most useful of which for human

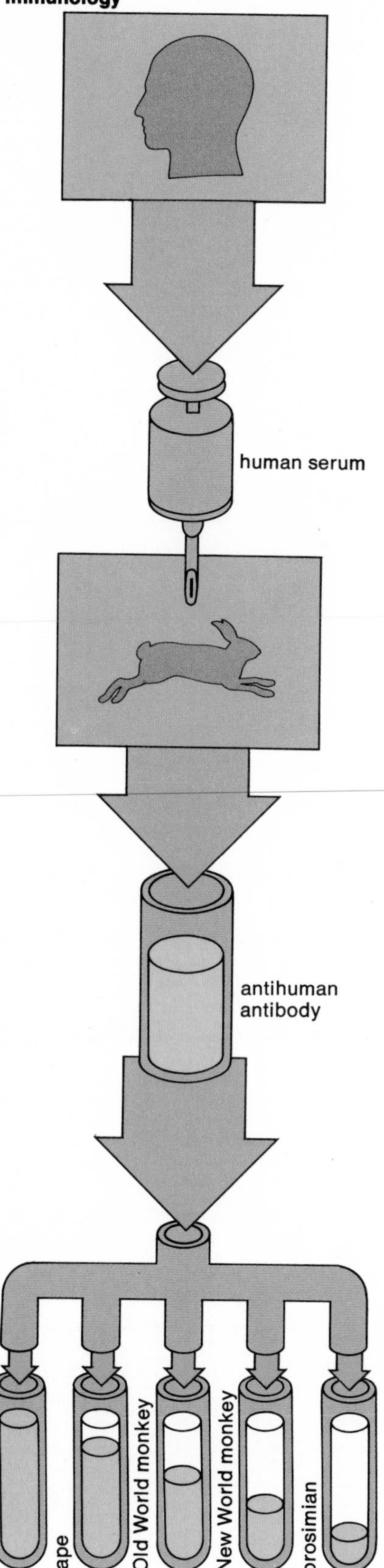

evolution has been the one measuring the decay of radioactive potassium ions into argon ions in rock formations. Formerly, the geologic epoch called the Pliocene was considered to have started about 12 million years ago, and man was considered to have been present at that time. Many dates now show that the great change at the beginning of the Pliocene was the filling of the Mediterranean from the Atlantic Ocean, changing it from a desert 10,000 ft below sea level to its present form, and that this occurred 5.5 million years ago. The point is not that any one particular date can be determined, but that the establishment of many dates by radiometric means is beginning to yield a chronological framework that is independent of human judgments. For example, potassium-argon dating established the date of 1.8 million years for a hominid skeleton from Olduvai Gorge, Tanzania, a date four times the age that had been previously estimated.

New techniques allow the measurement of the rates of the drifting continents, the temperatures of ancient seas, or the age of volcanic eruptions millions of years ago. Similarly, it is possible to measure the genetic differences between the various forms of life. Our view of the world should start with geographical scenes that are very different from those of today and the study of evolution should be approached with the new insights that come from molecular biology. Starting in this way will sweep away a whole century of controversies and provide a quantitative foundation from which we may view the problems of human evolution.

New insights from molecular biology

Traditionally, evolution was studied on the basis of the direct evidence of the fossils and the indirect evidence of comparative anatomy. In neither case could the results be quantified, counted, or managed in ways that would force different scientists to agree. Molecular biology provides a very different kind of evidence that overcomes the traditional difficulties and settles evolutionary problems in ways that are independent of any particular scientist or laboratory. The chemical basis of life, DNA, is composed of two strands, or chemical chains, with nucleotide bases arranged in a complementary manner. The genetic information is carried in the sequence of these bases, and the quantity of genetic information is made possible by the number of the pairs, over two billion in mammals. The important point about the double-stranded nature of the DNA for the study of human evolution is that the two strands may be experimentally separated. The separated DNA strands of one animal form may then be mixed with single strands from a different kind of animal and made to reassociate with them to form a hybrid DNA. The reaction is dependent on temperature, and the hybrid form is less stable than the pure form. The less related two animals are, the less stable is their hybrid DNA. The relative thermostabilities of some primates have been found to be: human, 0; chimpanzee, 0.7; gorilla, 1.4; gibbon, 2.7; and Old World monkey, 5.7. For the New World monkey and the various prosimians the figures would be even higher.

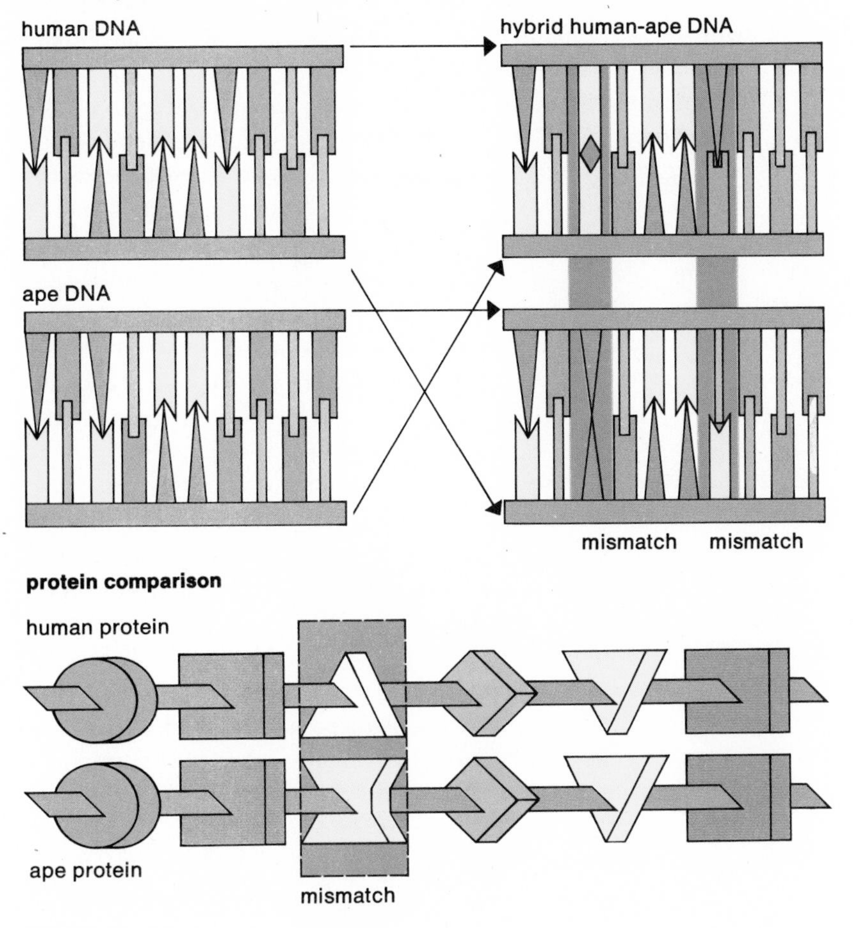

This means that the view held by Darwin and Huxley and the phylogenetic tree of the primates based on these DNA studies, which was published by David Kohne of the Carnegie Institution of Washington, D.C., in 1970, show the same relationships among the primates.

The direct comparison of the genetic substance shows that man is particularly closely related to the African apes—the chimpanzee and the gorilla. This conclusion may be checked by the study of other molecules, particularly proteins. For example, hemoglobin, the red blood protein, is composed of chains of amino acids whose sequences have been worked out in detail for many forms of life. There are no differences in structure between human and chimpanzee hemoglobin, two between human and gorilla hemoglobin, and 12 between the human and rhesus monkey molecules. Fibrinopeptides, short amino acid chains discarded in the process of blood clotting, are particularly rapidly evolving molecules. For them, the sequence differences have been found to be: from man to chimpanzee, 0; to gorilla, 0; to orang, 2; to gibbon, 3–5; to monkey, 5–8; to New World monkey, 9–10; and to prosimian, 18. These results show the same order of relationships as the DNA studies and emphasize that man is particularly close to the African apes.

Noticeable differences between living creatures are the result of mutations in the chemical structure of their molecules. The fewer the molecular variations, the shorter the time two animals have been separated. Three techniques measure these evolutionary distances. Single strands of DNA from two animals will match up and recombine, except at the points where mutations have occurred (above, top). Differences in amino acid sequences of proteins from the same animals can be counted (above, bottom). Or, the varying degrees of reaction to an antiserum produced in a test animal can be measured (opposite page). All three methods established an especially close relationship between man and the African apes.

The determinations of the sequences in protein molecules was going on in many laboratories in 1973, with the result that this kind of data would eventually be available in large quantities. But such sequences are difficult to determine and the work is very time-consuming; it is highly desirable to have a faster method that gives comparable results. Immunology provides such a method. It has been known for many years that the injection of one animal with a serum from another sufficiently different from itself (as in the injection of a rabbit with human serum albumin) causes the animal to respond by making antibodies (an antiserum) to the foreign antigen. The antiserum is specific against the animal that supplied the antigen; it will produce a slightly different reaction in a different animal. The degree of difference in the reaction can indicate the degree of relationship between the animals. Thus, when the rabbit's antiserum against human albumin is tested with other primate sera, the order of decreasing reactions (and, hence, of relationship) is chimpanzee, gorilla, orang, gibbon, Old World monkey, New World monkey, and various prosimians. It is an interesting comment on the progress of science that although this method has been known since 1900 and has been used by several scientists, it was never influential in evolutionary thinking.

Recently, however, the immunological methods have been improved so that the specific amino acid sites involved can be determined and so that the tests can be made in ways that enable the results to be quantified and expressed in terms of immunological distance units. If the amino acid sites in the primate serum prove to be identical to the corresponding sites in the antiserum, the two proteins will be immunologically identical and the immunological distance between them will be zero. The less closely the antigen-antibody sites correspond the greater will be the number of immunological distance units between them. Vincent Sarich of the University of California at Berkeley, in his extensive studies of primate albumins, has reported that the units of immunological distance between some of the primates are: man to chimpanzee, 7; to gorilla, 9; to orangutan, 12; to gibbon, 15; to the macaque, an Old World monkey, 32; and to New World monkey, 58.

A phylogenetic tree of the living primates has been reconstructed from the findings of immunological studies of Vincent Sarich at the University of California at Berkeley. Numbers throughout the chart are the times in millions of years that the species or families indicated are estimated to have begun divergent existences. Modern primates fall into five major groups that are thought to exhibit throughout their range all the forms primate evolution has taken. Typical members of each group are illustrated on pages 40–41.

What the molecules show

On the basis of these immunological distances, a phylogenetic tree may be reconstructed for the primates. In general, the tree will be the same whether it is based on immunological distances or on the DNA or amino acid sequence data. Other immunochemical studies in progress in 1973 indicated that they would support a comparable ordering of the data. It should be stressed that the general arrangement of the primates indicated by the molecular findings is in accord with traditional taxonomy, the attempts to classify living organisms in some meaningful sequence. It is only the very close relation of man to the African apes that is, perhaps, unexpected—and it is shown in every method of comparison. The molecular studies are settling many questions about the relationships of all the major groups of primates, but particularly they

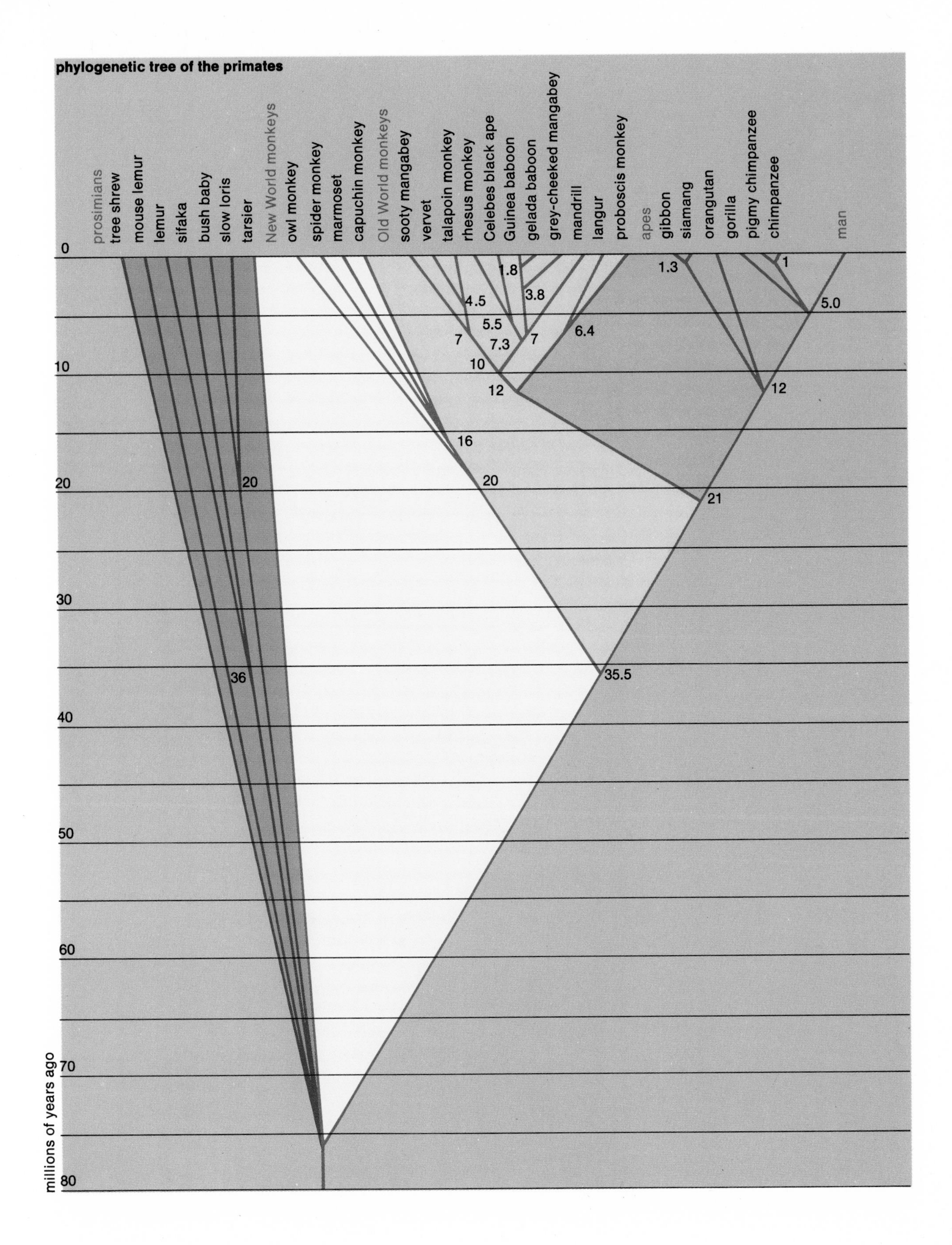

phylogenetic tree of the primates
prosimians
tree shrew
mouse lemur
lemur
sifaka
bush baby
slow loris
tarsier
New World monkeys
owl monkey
spider monkey
marmoset
capuchin monkey
Old World monkeys
sooty mangabey
vervet
talapoin monkey
rhesus monkey
Celebes black ape
Guinea baboon
gelada baboon
grey-cheeked mangabey
mandrill
langur
proboscis monkey
apes
gibbon
siamang
orangutan
gorilla
pigmy chimpanzee
chimpanzee
man
0
10
20
30
40
50
60
70
80
millions of years ago
1.8
3.8
4.5
5.5
7
7.3
7
6.4
10
12
16
20
20
21
35.5
36
1.3
1
5.0
12

PROSIMIANS
tree shrew
slow loris
tarsier
lemur
NEW WORLD MONKEYS
pigmy marmoset
wooly monkey
squirrel monkey

OLD WORLD MONKEYS
olive baboon
rhesus monkey
proboscis monkey
langur
APES
MAN
gorilla
gibbon
chimpanzee

are demonstrating man's closeness to the African apes and indicating that man shared a long period of common ancestry with the apes after both had separated from the evolutionary lineages leading to the monkeys.

The structure of the pelvis and leg bones of the African apes, left, is adapted for their characteristic hunched-over, knuckle-walking form of locomotion. The human bones, right, have evolved so that man can assume the upright, bipedal walk. Molecular data prove man's common ancestry with the chimpanzee; it is hoped that studies comparing anatomy and behavior will determine how a knuckle-walking ape became a biped.

The advances in molecular biology and immunology not only establish phylogenies but also raise the possibility of using such information to determine the time of the separation of the various forms of life. For example, Richard E. Dickerson of the California Institute of Technology used the known amino acid sequences of four molecules from throughout the range of life forms in which they are found and established the rate of change that has occurred in each and expressed it in terms of unit evolutionary periods, which are the lengths of time in which a 1% change in amino acid sequence has occurred after two evolutionary lines have separated. The fibrinopeptides had changed the most rapidly, then hemoglobins, then cytochrome *c*, a protein in mitochondria, the principal energy-producing units in cells; histone IV, one of the proteins that bind DNA in chromosomes, has hardly changed at all.

Changes in immunological distance units occur still more rapidly than amino sequence changes and can be used to determine an evolutionary clock, as has been done using the albumin studies of Sarich and Allan Wilson. To use immunological distance units to estimate time, the groups under study must be compared to animals of a different group. When studying primates, a good choice for comparison is the carnivores, and from them the distances have been found to be: to man, 169; to gibbon, 169; to two New World monkeys, *Cebus*, 169, and *Aotes*, 150; to macaque, 169; and to two prosimians, slow loris (*Nycticebus*), 143 and *Lemur*, 150. Sarich has made many more comparisons but these figures show the main results. Man, apes, and Old World monkeys have evolved the same amount (169 units). Among the New World monkeys, *Cebus*, the capuchin monkey, has evolved the most and *Aotes*, the owl monkey, the least. It should be noted that when there have been differences in the rate of evolution the clock method detects them: *Cebus* has evolved almost 20 units more than *Aotes*. Prosimians have evolved the least. These small forms, which have short generation times and some of which have multiple young, have changed less than the large forms with long generations. Man has changed more than any of the prosimians and as much as the small monkeys.

If the evolution of albumin had been slower in the line leading to man, then the distance unit figure would be much less than 169; if it had been very rapid, the figure would be larger. But, in fact, the rate of evolution seems to have been remarkably similar in monkeys, apes, and man. The more branches existing in phylogeny, the more the evolutionary time units can be checked. For example, it could be expected that the distances from the Old World monkey to man, chimpanzee, and gibbon would all be the same—and this is the case. If man had evolved much more slowly than the other primates, his total distance from the nonprimates would be less than it is for the apes. If the rate were fast at first and then slow, his distance from the monkey should be less.

Because the primate phylogenetic tree can now be described in terms of immunological distance units, and because these units change at rates that can be shown to be quite regular, it should be possible to determine the time of origin of the various groups. The problem is to calibrate the rates of change. Some scientists believe it is already possible to do this; others believe that it is impossible. The authors believe that it is now possible, at least within rather wide limits, and that these limits will be narrowed rapidly by research.

It is probable that in the near future the times of separation of the principal groups of mammals will be determined more accurately from the molecular and immunological information than from the fossil record. As far as the origin of man is concerned, all the lines of biochemical evidence (whether based on DNA, amino-acid-sequence, or immunological data) suggest that a short interval of time separates the human species from the species of African apes. The fossil record makes a separation of less than five million years improbable, while the other lines of evidence make a separation of more than ten million years very unlikely. The main point is not that precise conclusions have been made, but that there are, at last, objective and quantifiable methods that have solved some of the questions of human evolution and that give promise of solving many more. The conclusions that have been reached provide a framework within which the other lines of evidence may be used efficiently.

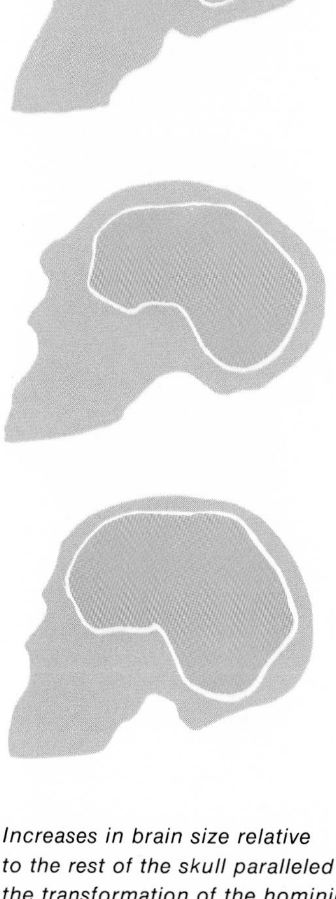

Increases in brain size relative to the rest of the skull paralleled the transformation of the hominids from an ape-like existence to a fully human way of life. Fossils indicate that four stages along this route were, from top to bottom, those of Australopithecus, Homo erectus, *Neanderthal man, and Cro-Magnon man, who was a true* Homo sapiens sapiens.

Field studies of behavior

The problem of the origin of man has been considered to be one of how some arboreal primate came to the ground. Field studies show, however, that many kinds of primates have come to the ground (as have the langurs, patas monkeys, baboons, and macaques), but without becoming bipedal or in other ways paralleling human evolution. Our nearest living relatives, the African apes, are primarily ground-living and walk in a unique way—that is, on the backs of their knuckles. Since both the chimpanzee and the gorilla walk in this manner, it is probable that their common ancestor did also. This probability is increased by the fact that orangutans occasionally knuckle-walk. Since the molecular data show that man shared a common ancestor with the chimpanzee after the gorilla had branched off, our ancestors were, very likely, knuckle-walkers, and the central problem of human origin becomes not how an arboreal ape came to the ground but how a ground-living, knuckle-walking ape became a biped. It should be noted that in this reconstruction the molecular information and the behavioral information supplement each other. Knowledge of the closeness of the human-ape relationship and the order of separation comes from the molecular information, but the behavioral stages involved in the change are inferred from knowledge of the behavior (and the anatomy) of the contemporary forms.

The field studies of the chimpanzee made by Jane van Lawick-Goodall show that these apes behave in many ways that were thought

Geza Teleki, University of Georgia

to be unique to man. Their behavioral repertoire includes such activities as hunting (sometimes cooperatively), the throwing of stones, the use of sticks for many purposes, and "fishing" for termites and ants. Chimpanzees mature much more slowly than monkeys, and the bond between mother and young is much closer and of longer duration.

The behavior of chimpanzees, rather than that of gorillas, is stressed only because much more information has been published on that species. The results of what field studies have been done, however, fit the other lines of investigation. The behavioral differences between man and the African apes are far less than between man and the more arboreal Asiatic apes, or the other primates.

New uses for anatomy studies

Some of the behavioral similarities in man and ape are the result of their having certain anatomical features in common. Short, wide trunks and long arms with similar joints and muscles make equivalent actions possible. It is obvious which aspects of anatomy fit in with the molecular and immunological data and which will lead to very different conclusions. If anatomists had remained interested in whole animals and in their behavior, as the early anatomists were, then the similarities of man and ape would have remained stressed. When, on the other hand, the details of a single part of the body were stressed and studied intensively, everything could be interpreted differently. It was the detailed study of separate parts that led to the belief that man could not be descended from the apes and must have had a very long period of separate evolution.

If the evolutionary relations of the major groups of primates have now been largely explained by the biochemical information, the study of anatomy can be useful in understanding what happened during evolution; it may no longer be used to determine phylogeny. In the matter of walking, man walks bipedally and this peculiar method of locomotion is made possible by extensive anatomical changes in the pelvis and associated musculature. Anatomy studies help us to see the nature of the evolutionary changes that separated man from ape, but alone they are insufficient to "prove" that man could not have descended from apes or that this transition must have taken place many millions of years ago. With the phylogenetic questions settled, anatomical and behavioral information may be used to understand the adaptations that determined the course of evolution.

Reinterpreting the fossils

Even if the relations of all the living forms and the times of separation of the major groups were determined from molecular and immunological evidence, without fossil evidence nothing would be known of all the kinds of life that had become extinct or of the nature of the events that have occurred in the history of life. Fossils show what ancestral forms were like, how they adapted, and the ecology of times past. In the case of man, the fossil record shows that the adaptation to bipedal locomotion occurred long before the large brain developed. The small-brained biped (*Australopithecus*, using the term in the broadest possible sense) made stone tools, hunted, and lived in the savanna. This stage of human evolution appears to have lasted from five million years ago to one million years ago, when one species of the genus evolved into *Homo erectus*. Because one form evolved into *Homo*, it can be inferred that there were transitional forms, and if these are classified in the genus *Homo*, then *Australopithecus* and *Homo* existed side by side for at least a million years.

Although hundreds of fossil men have been unearthed from the period between three million and one million years ago, there is not a single moderately complete skeleton. The remains are so fragmentary that almost every aspect about the genus they represent is controversial—whether it is the number of species, diet, locomotion, or distribution. The very number of these unanswered questions shows why fossils are necessary for understanding the nature of our ancestors and why interpretations based on incomplete fossil information leads to so much debate. The important reality for evolutionary study is the populations living in times past, populations now reduced to bones. To understand the roles of these fragments in the evolutionary process requires a reconstruction of the functioning populations from their preserved remains. Even with extensive knowledge of the contemporary primates, any such reconstruction is highly subjective, with the result that every potentially important conclusion is controversial. A brief outline of what *may* have happened can be presented, but it must be emphasized that the evidence is changing all the time. Each year new human fossils are found, methods of dating are improved, and new fossil-bearing locations are discovered. The rate of discovery is slow only because so few scientists are involved in the search, and because the fieldwork receives so little financial support.

The coming of man: a reconstruction

After the separation of the monkeys (Old World, or Cercopithecidae) and the apes (Pongidae), man's ancestors shared a long period of common lineage with the apes, a period in which the many observable anatomical, behavioral, and molecular similarities evolved. The part of human anatomy that has changed least since that time is the complex of arms, trunk, and viscera. Based on the very fragmentary fossil evidence of arm bones, the features man shares with the contemporary apes had not evolved in *Proconsul* (a Miocene ape). In late Miocene times (ten million years ago) some apes evolved the peculiar locomotor habit of knuckle-walking, and pre-hominids continued to share this adaptation with other apes. About ten million to five million years ago, the lineage leading to man separated from that leading to the African apes. The complex of adaptive conditions that led to the separation may have been the tool-using, bipedal, savanna-living complex. This development opened the way to a new kind of life for apes and was reflected in changed natural selection pressures on the dentition, resulting in the

Field studies show that apes display traits once thought to be uniquely human. An adult male chimpanzee (opposite page, top) has white eyes, uncommon in chimps but seen by some as an evolutionary trend toward a human characteristic. The strong bond between a chimp mother and her child is typically human. In a family portrait (middle), a male grooms a mother with a baby on her lap. An adult male (bottom) demonstrates the precision with which a grass stem can be inserted into a termite mound and the food treat extracted.

D. Durrance from Photo Researchers

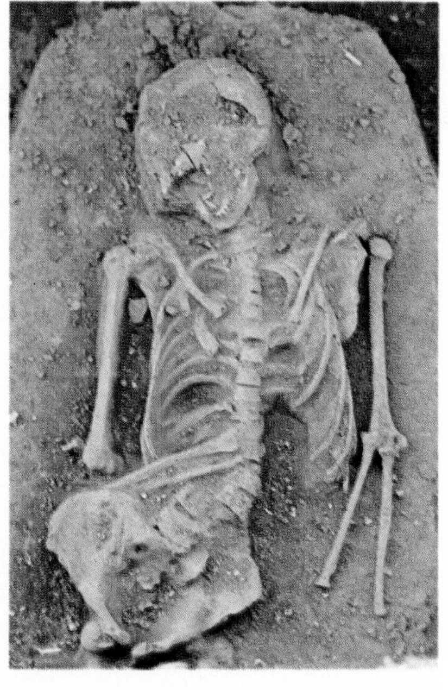

reduction of the anterior dentition, particularly of the canine teeth. The behavior of the contemporary chimpanzee gives one set of clues as to what the basis for such a life may have been. Major adaptive change is shown in the fossils of the species *Australopithecus*. Fossils in the ten million to five million-year range (unfortunately absent) might provide enough information to permit the complex of adaptive changes to be followed in some detail.

Progress in making stone tools (the only tools that are preserved and, therefore, known but presumably representative of only a small part of the tools actually used) appears to have been exceedingly slow. Simple forms of these tools persisted for some millions of years, certainly for two million, and probably for at least twice that long. Then much more complex tools (Acheulian) appear in the fossil record. These tools are exceedingly difficult to make, and it takes archaeologists today many weeks of practice to learn to chip stone as efficiently. Along with the complex stone tools are found evidences of habitual hunting of large animals (including elephants), of the use of fire, and of living sites throughout the tropical and temperate regions of Europe, Africa, and Asia. This was the way of life of the genus *Homo*—in the traditional sense (not including the forms transitional to *Australopithecus*). Judging from the fossil record, this was a fully human way of life, but change was still exceedingly slow.

Little change in stone tools used by small-brained man *(Australopithecus)* occurred during a period covering millions of years. In a later period, the larger-brained fossil men (*Homo erectus* and later primitive forms) used the same kinds of tools for hundreds of thousands of years. But once anatomically modern man *(Homo sapiens sapiens)* was on the scene, rapid changes appeared everywhere. In this period, from 40,000 to 30,000 years ago, the whole rate of cultural evolution accelerated. Tools changed rapidly, with much regional specialization. New techniques for hunting and food collecting were developed, as evidenced by the appearance of the bow and arrow, the spear, the fish trap, and the sickle. Settled communities, with such features as houses, boats, art, and sculpture, developed. Ritual burial practices came into being. Clothes and ornamentation were prevalent. It was during this period that population movements became extensive. Migrations began into previously uninhabited areas, including the Arctic, the New World, and Australia. All of these events did not take place at one time but they did occur over a period of a few thousand years. If some millions of years of human history is being considered, all of these events—including the origin of agriculture—occurred in the last 1% of the history.

The role of language

The principal cause of this great acceleration was probably the origin of languages. Recent experiments show that the sounds made by the nonhuman primates are controlled by phylogenetically very old parts of the brain (limbic system). These old parts also control emotions and, therefore, the sounds made by monkeys and apes primarily convey the emotional states of the animal. For example, fear and threat may be indicated by sounds, accompanied by special postures, gestures, and actions. The noise or sounds are part of the body's whole communication complex, and the system carries very little information. In contrast, human language is controlled by the cortex on the dominant side of the brain. From an evolutionary point of view, it is a new adaptive system, and human speech is different, not just an increase in a behavior that is present in nonhuman primates. Speech is made possible by a sound code, the elements of which (phonemes) have no meaning but can be combined in an almost infinite number of ways that do have meaning. It is the presence of this code (made possible by evolutionary changes in the brain and articulatory mechanisms) that makes possible human language.

Modern man has the ability to learn one or more languages, and the communication resulting from linguistic codes is essential for the complex social systems and technologies he encounters. It is likely that the biological base for language evolved during the period of *Homo erectus,* with the entire adaptation being completed some 40,000 years ago. The complex bio-social nature of language as we know it today must have been preceded by simpler forms of communication. Unfortunately, there is no direct clue to language in the fossil bones; the origins of language must remain speculative. However, the major evolutionary change is clear. Man evolved a new system of communication based on a sound code and requiring a much larger brain for its control. There is no comparable system in any nonhuman primate.

Erich Lessing from Magnum Photos, Inc., courtesy, Henri Lohte expedition

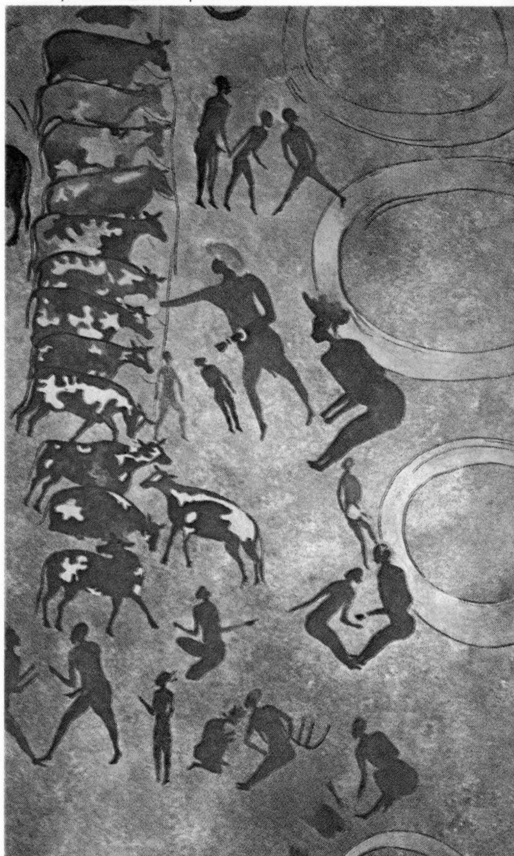

Only slow cultural progress occurred in the period between five million and one million years ago when man's large brain was evolving. But once anatomically modern man was present, rapid cultural changes occurred over a few thousand years. Some features of this life were the settlement into communities, domestication of cattle, and making and wearing of ornaments depicted in cave drawings (above), and the practice of ritual burial, as found at an archaeological site in Romania (opposite page).

A new outlook

Some of the major controversies of the past century have been resolved by the recent major advances in our understanding of human evolution. The closest living relatives of man are indeed the chimpanzee and the gorilla. The relationship is proved by recent studies using DNA, sequence data on amino acids in proteins, and immunological evidence. With the phylogenetic issues settled, behavioral and anatomical information may be used to clarify the events of human evolution. The rapid discovery of fossils has confirmed a record of human existence beginning more than three million years ago. It may be that it was the adaptation of object-using that led to the separation of the human lineage; if this is the case, stone tools may be as old as man. After separation from the apes, human evolution may be divided into three stages; a first, long phase of slow evolution in which several species of early man existed (family Hominidae, genus *Australopithecus*), beginning between ten million and five million years ago and ending about one million years ago; a second phase of faster cultural evolution associated with a variety of ancient men (*Homo erectus* and related forms); and a final short stage associated with *Homo sapiens sapiens*, one in which change occurred at the unprecedented rate we consider normal.

How certain can we be of these "facts"? Clearly, the major outline of human evolution is on far surer ground than ever before. None of the

Courtesy, Playboy Productions

Reconstructions based on current findings show how early forms of man looked. Australopithecus *(left) had heavy jaws and big grinding teeth but a cranial capacity greater than any earlier form.* Homo erectus *(right), the first member of genus* Homo, *retained primitive, ape-like features of the hand and the brain. Neanderthal man (opposite page, left) was one of the forms of* Homo *that appeared in the Mediterranean area about 100,000 years ago. He was replaced in Europe by Cro-Magnon man (far right), who was physically similar to modern man but did not display all of his cultural traits.*

recent information suggests a radically new interpretation. In fact, as already noted, the latest methods of evolutionary study, those based on the analysis of DNA, lead scientists to the same conclusions as those of Huxley and Darwin. Although such agreement can hardly be called new, it must be emphasized that what is new is the removal of many competing theories, theories that denied that the contemporary apes are closely related to man. The molecular and immunological facts provide a setting in which anatomical and behavioral studies may be used efficiently; they represent a major clarification of the issues.

The new methods have been criticized, mainly to point out that it is too soon to judge their effectiveness. It is true that comparisons based on DNA are new, but the immunological methods are 70 years old and gave the same conclusions when first used by Nuttall in 1902, conclusions that were correct but ignored. There is no reason to consign the more recent advances to a similar fate. It is a new scientific world. Knowing that continents drift provides evolution with a new geographical base. The ages of fossils can be determined with degrees of accuracy unsuspected a few years ago. Quantifiable, objective comparisons can be made of contemporary animals. To see human evolution in perspective, one must start with what we know now. The mistakes and controversies of the past have largely been settled.

See also *Encyclopædia Britannica* (1973) MAN, EVOLUTION OF; PRIMATES.

FOR ADDITIONAL READING:

Chiarelli, B. (ed.), *Comparative Genetics in Monkeys, Apes and Man* (Academic Press, 1971).

Dickerson, R. E., "The Structure and History of an Ancient Protein," *Scientific American* (April 1972, pp. 58-72).

Dolhinow, Phyllis (ed.), *Primate Patterns* (Holt, Rinehart and Winston, 1972).

Dolhinow, Phyllis, and Sarich, Vincent M. (eds.), *Background for Man* (Little, Brown, 1971).

Hsü, K. J., "When the Mediterranean Dried Up," *Scientific American* (December 1972, pp. 26-36).

Kerwin, C. (ed.), *The Emergence of Man* series (Time-Life Books, 1972).

Napier, J. R., and Napier, P. H. (eds.), *Old World Monkeys* (Academic Press, 1970, pp. 175-226).

Pilbeam, D., *The Ascent of Man* (Macmillan, 1972).

van Lawick-Goodall, J., *In the Shadow of Man* (Houghton Mifflin, 1971).

Washburn, S. L., and Dolhinow, P., *Perspectives on Human Evolution,* vol. 2 (Holt, Rinehart and Winston, 1972).

Acupuncture: Medicine or Magic?

by Emanuel M. Papper

There is no "scientific" explanation for acupuncture, but we do not know just how our commonly used anesthetics work either. Perhaps the first order of business is to find out whether acupuncture really does what its proponents claim.

Acupuncture dolls of the type shown on pages 50–51 are used as guides by practitioners of this age-old form of medicine. The lines on this Chinese-made model indicate the 12 channels or meridians, which serve as pathways for the body's vital energies. The dots are the acupuncture points where needles are inserted to correct an imbalance in these energies. In traditional Chinese medical theory, such imbalances are the cause of pain and disease.

Pain has been fought in many ways over the centuries, beginning with the performance of primitive rites and the ingestion of special herbs. Modern science has developed more rational procedures for dealing with this unrelenting foe of man but, despite the undisputed successes of anesthesiology, many kinds of pain still cannot be reached in a predictable, secure fashion. It is no wonder, then, that the claims of Eastern practitioners—widely publicized as a result of the U.S.-Chinese political détente—have brought an upsurge of Western interest in an ancient but allegedly effective way of treating pain: the Oriental practice of acupuncture.

An age-old practice

Acupuncture is an ancient Chinese therapy used for the treatment of illness and the relief of pain. In the traditional practice of acupuncture, needles are inserted into the body at one or several of 365 specified points (nearly 1,000 in modern practice), which may not—to Western eyes, at least—be related anatomically to the site of the disease or pain being treated. The depth of penetration varies. Sometimes the needles are rotated, occasionally they are simply left in place and, in modern practice, weak electrical currents are often passed through them. The earliest needles were made of flint. Later gold and silver were used, and today the most common material is stainless steel.

The theoretical basis of acupuncture lies in the Chinese concept that man is a small image of the universe and is subject to identical—and immutable—laws. These universal laws were predicated on the coexistence of two forces known as yin and yang. Yin, the female element, was said to possess negative properties, while yang, the male element, had positive qualities. The two forces were admixed in both sexes—a concept not altogether unfamiliar to the modern psychologist or endocrinologist. Yin and yang were conveyed through the body along 12 hypothetical channels or meridians. Health resulted from a balance of yin and yang, and all diseases were thought to be due to an imbalance of these forces. Thus the Chinese saw all illnesses as resulting from a single cause. The variety of diseases and different organs and parts of the body that were affected were only specific manifestations of a general malady. The needles, inserted at points along the 12 channels, served to correct the imbalance between yin and yang, perhaps by permitting the dissipation of whichever force was excessive. All organs could be reached via these channels, so the needle did not necessarily have to be placed near the target organ.

The practice of acupuncture spread to France via Indochina and from there to Germany. Interest in the subject has surfaced in Britain and also in the Soviet Union, where attempts are being made to determine, under scientific conditions and in accordance with the behavioral theories of Pavlov, whether acupuncture is deserving of Western attention. In China itself acupuncture fell into some disrepute during the late imperial period because the emperor would not allow royal personages to have their skin "injured." Subsequently, there

***EMANUEL M. PAPPER** is vice-president for medical affairs and dean of the School of Medicine at the University of Miami.*

The Bettmann Archive

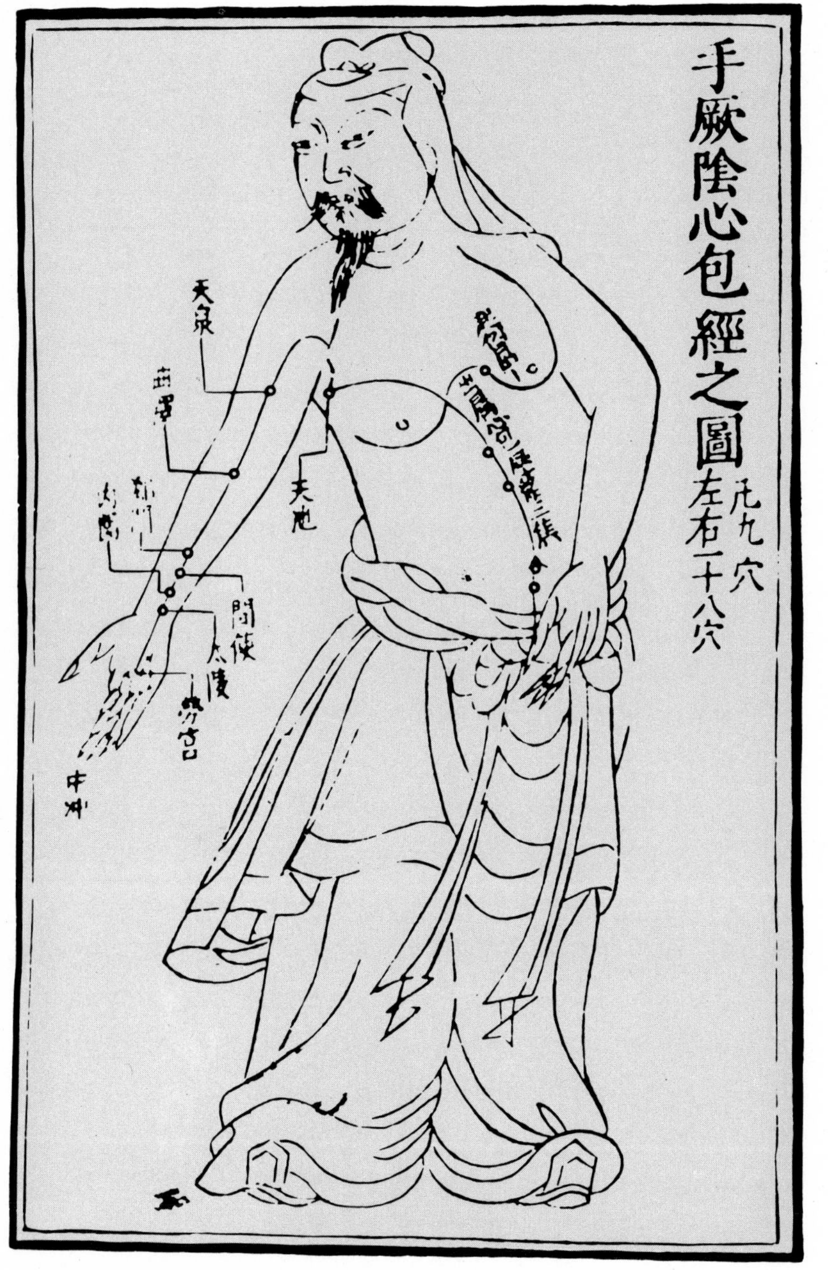

was a gradual resurgence of the practice until the Nationalist government, with its partiality toward things Western, forbade most traditional Chinese medicine. Many Chinese physicians went to the United States, the United Kingdom, and Canada to learn Western medicine, while the Japanese sent young physicians to Germany and Austria for similar purposes. There was, therefore, a strong Western medical tradition in the Orient when the Communists came to power in China in 1949.

According to legend, acupuncture originated when an ancient emperor discovered that soldiers wounded by arrows were cured of ailments in other parts of their bodies. Thus, the acupuncture points, shown here on a Chinese anatomy drawing, may seem to bear no relation to the site of the illness being treated.

For centuries, acupuncture has been used in the Orient as a form of curative medicine. The woman at right, in Saigon, South Vietnam, is receiving treatment for vision problems. The needles are left in place for about 20 minutes. In modern practice, a weak electric current is sometimes passed through the needles, as with the patient, below, being treated in Saigon for arthritic pains. The use of acupuncture as an anesthetic during surgery is relatively new and seems to have originated when China's Chairman Mao Tse-tung ordered traditional and Western-trained medical personnel to work together. The patient, below, right, at Peking Medical College, is undergoing surgery for removal of a tumor of the esophagus. Three needles have been inserted in his left ear and left forearm.

Harold Ellithorpe from Black Star

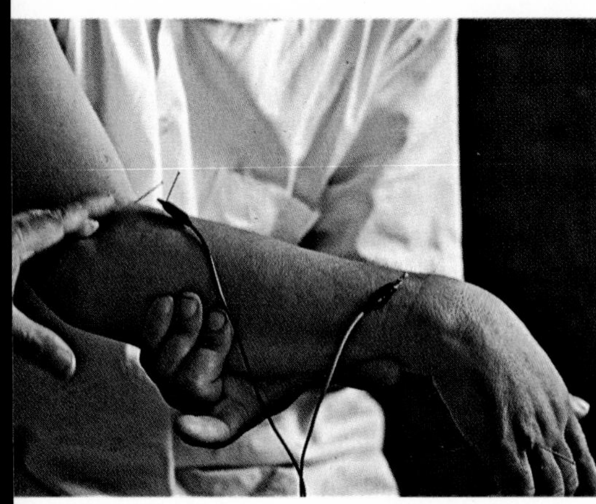

Harold Ellithorpe from Black Star

Andy Berhaut Archive from Photo Researchers

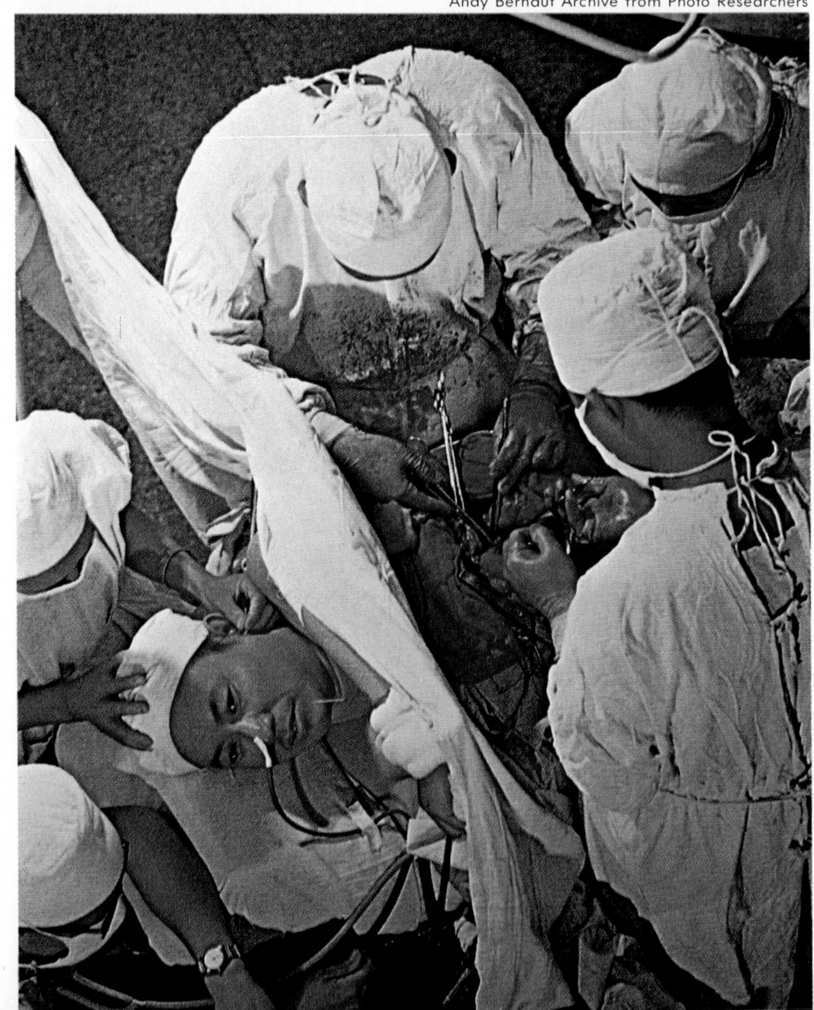

There were, however, only about 70,000 Western-trained physicians in China in 1949, obviously not enough to cope with the health needs of the country. Accordingly, Mao Tse-tung pressed into service the 500,000 or so practitioners of traditional Chinese medicine who had, to some extent, gone underground. In order to reconcile these two potentially divisive forces, Western-educated physicians were required to undergo training in traditional medicine, including acupuncture. Many of the westernized physicians became convinced of the efficacy of this ancient practice, not only as a method of treating many ailments but also as an anesthetic used in conjunction with modern surgical procedures.

G. Damian Loescher from Pictorial Parade

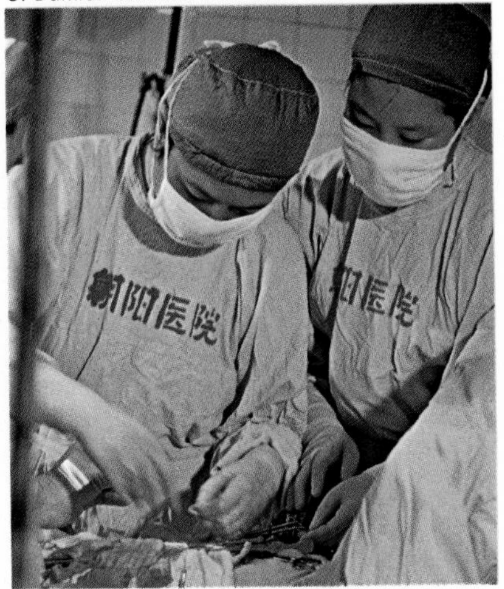

Practitioners at Eastern Peking Hospital anesthetize a surgical patient by acupuncture. Official Chinese sources claim that more than half a million surgical procedures have been carried out using acupuncture anesthesia. Western witnesses report that the patients remain conscious during surgery, are able to breathe normally, converse with the doctors, and even drink tea.

Science and the role of culture

Although the traditional Western methods of alleviating pain are not totally satisfactory, there are proven scientific approaches to the therapy of pain, and they may be summarized as follows:

1. Eliminating the disease or stimulus that causes the pain.
2. Interrupting the pain pathway between the receptor site and the central nervous system.
3. Altering the receptivity of the central nervous system. This is the classic approach embracing drugs, medication, and pain-relieving operations. Under this heading come many of the familiar Western medications (*i.e.*, anesthetic agents such as ether and halothane) that relieve the pain sensation so completely as to permit surgical operations.
4. Interrupting the pathways that go through the spinal cord to the brain, either surgically or chemically.
5. Influencing areas in the brain itself by drugs, surgery, or psychiatry.

Since acupuncture apparently fits none of these patterns, the question arises: what does it really do? Most Westerners in the medical and biological sciences and most practicing physicians feel almost intuitively that the practice of acupuncture cannot succeed because there is no "scientific" basis for its action. On the other hand, long experience in the Orient, especially in China, indicates that acupuncture has a clinical value, and in recent months reputable Western observers have testified that it relieves even the intense pain that accompanies surgical procedures.

Does acupuncture "work" and, if so, what relationship does it have to Western medicine? Before attempting to explore this fascinating therapeutic method further, it would be worthwhile to examine some important interpretive problems concerning health and disease.

All men everywhere have the same basic physiology and the same nerve endings through which pain is transmitted. Yet much of our attitude toward illness and pain is culturally determined, and there are many obstacles that prevent an easy transposition of attitudes and practices regarding pain from one culture to another. A conflict between faith and "science" exists even within our own culture. Those who believe in faith as a modality for the treatment of disease point to the many healing acts performed by Jesus, including the raising of

Lazarus from the dead. They argue that these are true, and that, if there were another Jesus, or if the Second Coming occurred, or if we were simply worthy of it, there would be further healing miracles. Although there is no scientific verification for any of these miracles, the possibility of miraculous healing cannot be firmly denied.

Crossing into different cultures, one encounters fantastic problems of both interpretation and acceptance. For example, among blacks raised in the West Indies, *e.g.*, the Bahamas and Haiti, there is an illness known as the falling-down disease which appears to be related to emotional stress. This illness results in loss of consciousness and in states that resemble major psychoses. The head injuries incurred during the "spells" may be a cause or an effect of the illness; which is not known.

In New York City and Miami, medical personnel have found young Cuban and Puerto Rican mothers to be remarkably resistant to the prescriptions of pediatricians with respect to such matters as diet and control of fever. Only now is it becoming evident that this resistance stems from an unconscious but inevitable conflict between Western scientific medicine and the poorly understood taboos of a powerful religio-medical inheritance.

There is much that we do not understand. Accordingly it should be possible for a Western-oriented society to view acupuncture with some objectivity, even though it cannot be scientifically explained by Western methods. It is possible to agree that acupuncture may achieve some of the clinical and therapeutic benefits claimed for it even though we do not know why.

The difficulties of evaluation

It should be emphasized that experiments bearing on the question of whether acupuncture does, in fact, work can be devised and conducted. However, any studies of pain are extraordinarily difficult, and this kind of Western scientific verification of a transcultural method will take time. The question of whether acupuncture is useful in providing surgical anesthesia should be the most amenable to experimental verification, but even this must be tested under carefully controlled conditions if a clear answer is to be obtained. And even here there are certain problems that must be kept in mind. Without question, people differ in their ability to withstand pain and in their ability to respond to psychological or hypnotic suggestion. There is also a strong possibility that some people are more susceptible to certain kinds of drugs and medication than others.

It has been shown that persons who have a strong orientation toward authority respond more readily to hypnotic suggestion than those who do not. Chinese culture, with its emphasis on respect for elders and for authority, would seem to predispose people to accept suggestion. (Some Eastern religious practices—Yoga is an example—appear to contain a strong element of autohypnosis, although this is not necessarily pertinent to China.) Even in the U.S. it has proved possible to provide total surgical anesthesia by hypnosis in approximately 15–20% of subjects. Studies of this kind have not been done elsewhere in the Western world, so we cannot make comparisons among the various Western peoples. Clearly, however, any method that purports to provide surgical anesthesia without the use of chemical anesthetics would have to succeed in more than 25% of the subjects in order to convince Western-oriented scientists that it was anything more than a particularly effective way of inducing hypnosis.

A similar situation pertains with respect to the relief of pain. In approximately 30–35% of patients, relief from any kind of chronic or severe pain can be obtained by administration of a placebo, a non-analgesic medication or treatment that the patient believes will be effective. This includes even the pain caused by cancer. Thus, in order to prove that acupuncture—or any other pain-relieving method, for that matter—has more than a placebo effect, it is necessary to prove that it is effective in more than 35% of cases.

Finally, there is the problem of biological differences among patients. Although studies in this area are inadequate, it does appear that Chinese subjects are more sensitive than Westerners to the depressant effects of analgesic drugs such as meperidine (Demerol), the barbiturates, alcohol, and the anesthetics. Further and more definitive studies are needed but they would be difficult to conduct. One would have to study, in some quantitative way, the effects of anesthetic drugs on native-born Chinese living in China, on first- or second-generation Orientals living in Hawaii or California, on transplanted third-generation subjects in whom the ethnogenetic lines have remained unmixed, and finally on hybrids in order to determine the possible effect of intermarriage between races.

In short, one cannot discount the possibility that cultural attitudes, placebo or hypnotic effects, biological differences among peoples, or some combination of all these factors may account for the recorded successes of acupuncture in dealing with the pain problem. Even when all these reservations are taken into account, however, the fact remains that there is substantial first-hand and anecdotal evidence that acupuncture is, in fact, capable of producing surgical anesthesia in Chinese patients, treated by Chinese acupuncturists and physicians and surgeons. This is corroborated by reports in the Chinese medical literature.

One is forced to conclude, therefore, that acupuncture as a method for producing surgical anesthesia is a real possibility, although as yet there is no convincing Western "proof" of its efficacy. Obviously, further studies are needed, including the exchange of physicians between China and the West in order to acquire more information than is possible from the observation of isolated cases. If acupuncture does prove to be an effective method of surgical anesthesia, it will be a boon of vast proportions. If it turns out not to be successful, the investigations themselves will have been worthwhile insofar as they provide additional data on the problem of pain.

Courtesy, National Library of Medicine

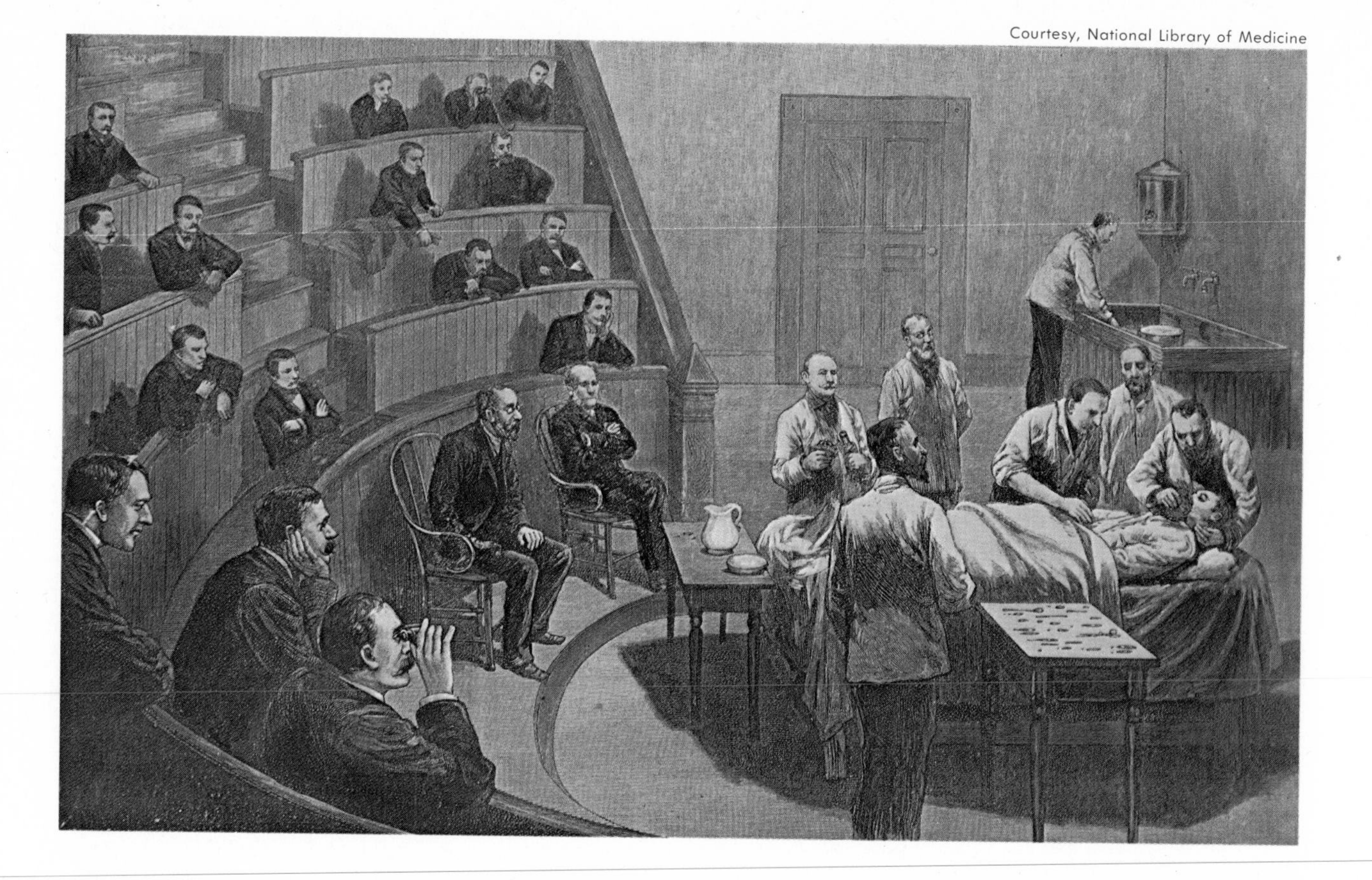

Modern anesthesiology dates from the discovery of the anesthetic properties of ether in the 1840s. This wood engraving, executed about 1889, shows ether being administered during a surgery lesson at Massachusetts General Hospital, Boston.

The mystery of anesthesia

Although much more information exists concerning the mechanism of action of chemical analgesics and anesthetics than on acupuncture, scientists still do not have clear-cut answers. They are still arguing about how aspirin works. The use of inhalation anesthetic agents such as ether is approximately 130 years old, and for at least a century, scientists and physicians have been attempting to find out how they actually exert their effect. Some of the scientists who have studied this problem believe that if one understood the mechanism of action of anesthetics, one could get closer to the fundamental biological process we call life.

Many theories about the action of general anesthetics have been advanced. Some are based on the physical properties of the anesthetics, others on the chemical properties. Some theories have taken into account the biochemical activity of these drugs and others have dealt with the physiological changes that occur during the anesthetic process. At present no one theory has gained absolute acceptance. Many of the phenomena associated with general anesthetics can be explained by certain aspects of their action; others cannot.

In 1875 a distinguished physiologist, Claude Bernard, theorized that certain elements in the cells, called colloids, form an aggregation that either causes or is associated with anesthesia. Lower forms of life in the presence of anesthetics behave as though this theory were correct, but it has not been demonstrated as having any validity for higher species.

Around the turn of the century, an attractive theory was proposed independently by H. H. Meyer and C. E. Overton. They indicated that there was a positive relationship between the affinity of a general anesthetic for lipid, or fat, substances and its ability to produce the depressant action known as anesthesia. There is certainly good experimental evidence for such a relationship, but there are also exceptions in that some fat-soluble substances do not depress the central nervous system and do not cause anesthesia. However, many researchers believe that the fact that the membranes or outer linings of nerve cells do contain lipids may at least explain how an anesthetic gets inside a nerve cell in the brain or the central nervous system to produce anesthesia.

Other investigators propose a theory that relates the power of anesthetic agents to their ability to lower surface tension, much as soap lowers the surface tension of water. The distinguished scientist Otto Warburg suggested that if an anesthetic accumulated at the surface of the cell, it could change the metabolism of that cell and hence the transmission of nervous impulses, thereby causing anesthesia. Modern support for this theory was provided in the work of J. A. Clements and K. M. Wilson in 1962, in which they showed that lowering surface tension changed the intimate physical chemistry of cells.

Since 1950, several scientists have suggested that anesthetics change the permeability of cells in the nervous system, thus interfering with the movement of ions in and out of the cells. Local anesthetic agents appear to have this effect, and there are general anesthetics that also change the permeability of cells to sodium. However, a true cause-and-effect relationship has never been established to the satisfaction of all concerned.

There are biochemical theories of the mechanism of action of anesthetic agents. For instance, it is known that the consumption of oxygen in the body is reduced during general anesthesia; similarly, oxygen consumption in cell cultures or slices of brain is lowered when they are exposed to general anesthetic agents. Certain anesthetic agents also interfere with the power-generating parts of cells, and conceivably this could be the cause of the anesthetic state. Other theories deal with the physiology of the nervous system and its alteration by anesthetic agents. According to these views, which have not gained widespread acceptance, such agents act by disorganizing the transfer of impulses along nervous tissues. Some physical theories relate anesthetic action to the reorganization of cells in the central nervous system and particularly of the pericytial materials around the cell. Linus Pauling and others have suggested that anesthetics might act by forming microscopic water crystals in the central nervous system that interfere with the conduction of impulses, but there is no evidence that such crystals actually exist. The neurophysiologist Patrick Wall suggested in 1967 that anesthetics have a multiple action, rather than a single mechanism, but again the evidence is unsatisfactory.

Courtesy, National Library of Medicine

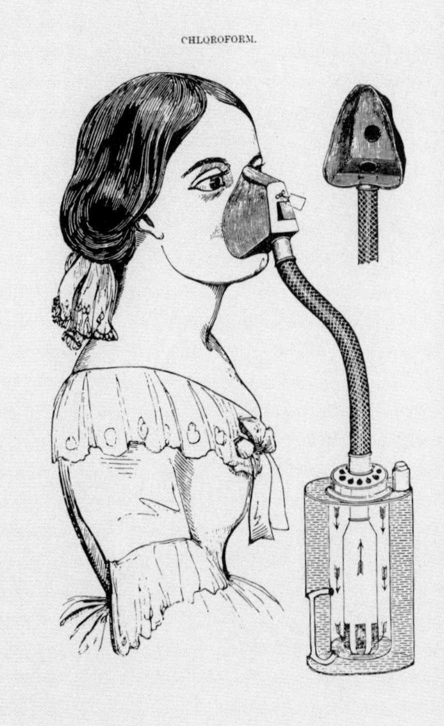

This illustration of a chloroform inhaler appeared in On Chloroform and Other Anaesthetics *(1858) by John Snow, the first full-time anesthetist. Snow helped to make anesthetics respectable when he administered chloroform to Queen Victoria during the birth of her eighth child in 1853.*

The most widely accepted view at present is that the mechanism of action of anesthetics is probably based on physical-chemical alterations and very likely is associated with lipid solubility, either in terms of the actual mechanism of action or by allowing anesthetic agents access to the cells through a lipid-containing membrane or envelope. Theories concerning spinal and local anesthesia are generally extensions of one or another of those proposed for the general anesthetic state.

A pragmatic approach

Thus, after approximately a century of scientific observation concerning the mechanism of action of anesthetics, and after the collection of innumerable data, there is still no universally accepted theory of how anesthetics work. It is not surprising, then, that there is no accepted "scientific" theory of the mode of action of acupuncture, which has been seriously studied in the West for a comparatively short time.

Some efforts have been made to find a theoretical basis in Western terms for the presumed effects of acupuncture. Such efforts presuppose that acupuncture does have the effects claimed for it, and that these were discovered, accidentally or by trial and error, by ancient Chinese practitioners who then constructed a theoretical basis for them in accordance with their own world view. If this is the case, acupuncture should be explainable in Western terms, just as the effects of folk medicines, attributed by their original users to "spirits" or similar phenomena, can be explained by Western pharmacological concepts.

One theory of pain that has been widely cited as a possible explanation of acupuncture is the so-called gate theory, developed by Wall and Ronald Melzack. This theory rejects the long-held idea that sensations of pain are transmitted directly from the receptors to the brain via nerve structures. Instead, it postulates the existence within the nervous system of "gates" that can modulate or close off pain signals, either as the result of neurophysiologic changes or through brain activity. It has been used to explain such phenomena as the failure of some traumatically injured persons to feel severe pain and the persistence of "pain" in amputated limbs. Applying this theory to acupuncture, some observers believe that the procedure somehow closes the gates through which pain signals reach the brain, though there is a difference of opinion as to whether this is a physiologic effect of the insertion of the needles or whether it results from brain activity, based on the expectations of the patient. Wall himself believes that the effects of acupuncture result from hypnosis, adding, however, that this does not necessarily diminish its potential therapeutic importance.

But before a satisfactory theory of acupuncture can be formulated, it will be necessary to determine whether or not it actually does produce the effects claimed for it. For all the disagreement about the whys

Charles Harbutt from Magnum

Today's anesthesiologist has at his fingertips a wealth of complex equipment and a constantly growing armamentarium of chemical agents. Yet scientists still do not know exactly how anesthetics work.

of Western forms of anesthesia, one thing is clear: anesthetic drugs produce the anesthetic state. In the case of acupuncture, the question of whether it does or does not produce surgical anesthesia has not yet been resolved to the satisfaction of the Western scientific world.

If it does, and this is demonstrated in a convincing manner, then the fact that its mechanism of action is somewhat fuzzy should provoke fewer problems among Western scientists and clinicians than it does now. As we have seen in the case of anesthetics, there are many empirically successful practices in Western medicine whose mechanisms of action cannot as yet be completely explained.

Historically, Western science has worked on the assumption that, once a fact is established beyond doubt, a study of its mechanism of action is indicated and eventually will yield precise information. It seems, therefore, that the first order of business is to establish whether or not acupuncture produces surgical anesthesia. If it does, anesthesiology will have gained a new and valuable tool, no matter how stubbornly the procedure yields its secrets to scientific scrutiny.

FOR ADDITIONAL READING:

Adriani, John, *The Pharmacology of Anesthetic Drugs*, 4th ed. (Charles C. Thomas, 1960).

Adriani, John, *The Chemistry and Physics of Anesthesia*, 2nd ed. (Charles C. Thomas, 1962).

Beecher, Henry, K., "Pain," *The Surgical Clinics of North America* (June 1963, p. 613).

Schizophrenia

by Jan Fawcett

Of all the diseases that afflict mankind, few appear in so many guises, have been traced to so many causes, or provoke such heated argument among those who would treat it.

A quiet, seemingly well-adjusted college freshman, with a good scholastic and athletic record in high school, is referred to the psychiatric ward of a general hospital by college officials. When his parents visit him, they find him confused, distracted, and disheveled. He hardly reacts to seeing them, and when he does talk his conversation is a stream of unrelated ideas—religion, outer space, Communism, and sexual themes—colored with bizarre, obscene terms.

A prim young girl living with her parents while beginning her career as a teacher spends increasing amounts of time alone in her room. When her bewildered parents ask her why she no longer goes to work, she replies that she has discovered an electronic mind-control ray emanating from the educational TV sets at school. She begins to accuse her mother of being part of a plot against her.

Both these young people are suffering from schizophrenia, a mysterious condition that can distort virtually every aspect of the personality, resulting in tormenting anxiety, total apathy, and inability to participate in life. Its chief victims are men and women between the ages

of 25 and 40, with a peak incidence between 25 and 35. Schizophrenia accounts for occupancy of approximately 10% of all hospital beds in the U.S. and almost half of all psychiatric beds. It has been estimated that more than 500,000 individuals in the U.S. have symptoms of this illness, and that one person in a hundred will develop the symptoms during his lifetime (2,080,000 based on current estimated U.S. population). Available evidence suggests that the rate of occurrence of schizophrenia has not changed over the years. The illness has been found in all cultures, although the symptoms vary according to cultural background.

Schizophrenia is an important cause of an array of destructive processes, including alcoholism, suicide, and homicide. Probably more schizophrenics commit suicide than all other types of psychiatric patients combined. The percentage of diagnosed schizophrenics who commit murder is not high; nevertheless, a significant number do commit homicides, often without warning or obvious reason. Closer examination leads to the conclusion that many of these patients felt rejected, depressed, or threatened.

A "cancer of the mind"

What is schizophrenia? What are its symptoms? What causes the condition and what can be done to combat it? Is the prognosis hopeless or can the victim be helped? There is probably no other major, common illness, save cancer, that appears in such a diversity of forms or that has inspired so many contradictory theories as to its etiology and treatment. Indeed, the analogy readily suggests itself: schizophrenia, a cancer of the mind and personality.

The term schizophrenia was coined in 1911 by the Swiss psychiatrist Eugen Bleuler. The word means, literally, "split-mindedness." It has often been interpreted to mean "split personality," erroneously implying that its victim develops multiple personalities—a rare, albeit dramatic, symptomatology occasionally seen in hysterical individuals. What Bleuler meant to imply was a split between the various mental functions such as logical thought, volition, and feelings or affect. He also introduced the concept of "the schizophrenias," indicating a group of conditions with a variety of causes but with generally similar symptoms.

What was henceforth called schizophrenia was by no means a "new" disease. Descriptions of it have been found in Hindu writings dating back to 1400 B.C. A Greek physician in the 1st century A.D. described a group of patients exhibiting regression and deterioration of the personality. Late in the 19th century, Emil Kraepelin, a German psychiatrist, called the illness dementia praecox ("early insanity"), conveying his impression that the disease began early in youth and led inexorably to a state of regression and mental deterioration then termed idiocy or dementia. In so doing, Kraepelin overlooked the fact that 13% of his own group of patients improved, but he set the tone that dominated the public's view of emotional illness until very recently. To the

JAN FAWCETT *is professor and chairman of the department of psychiatry at Rush Presbyterian-St. Luke's Medical Center, Chicago, Ill.*

fear and disgust that could be traced back to the Middle Ages when "insanity" was attributed to possession by the devil, Kraepelin added hopelessness.

In the 20th century, psychiatry began to take a more constructive and helpful approach to the entire range of emotional disorders. Rather than defining schizophrenia in terms of its purported outcome, as Kraepelin had done, Bleuler attempted to distinguish its characteristics as a first step toward a scientific understanding of the condition and its possible alleviation. He began by carefully describing the psychological processes that appeared to be basic to "the schizophrenias."

The visible symptoms

In Bleuler's view, the most obvious symptoms of schizophrenia were secondary to more basic psychological processes, but it is these secondary symptoms that strike the layman most forcefully. Two of the most common are delusions and hallucinations. A delusion is a persistently held belief, contrary to reality as it is perceived by most people, that is not dispelled by logical arguments. Delusions in schizophrenia often have a grandiose or a persecutory theme: the individual may express a belief that he is Jesus Christ or that he is the object of a worldwide search by a supersecret organization. A hallucination is a perception of sight, sound, smell, or touch that is not "real." In schizophrenia, hallucinations most frequently take the form of voices. A patient may be tormented by voices ordering him to jump out of a window or accusing him of heinous crimes. Sometimes the hallucination consists only of noises or isolated words, or the patient may seem to "hear his thoughts." Other hallucinations include frightening visions, strange smells, and odd bodily sensations. Delusions and hallucinations are not limited to schizophrenia; they may occur in a wide range of organic conditions (*e.g.*, infections of the brain substance or a decreased flow of blood to the brain caused by arteriosclerosis) as well as in other emotional disorders. Similar phenomena can be induced by hallucinogenic drugs such as LSD (lysergic acid diethylamide).

Most of the diverse and puzzling symptoms of schizophrenia can be accounted for in terms of Bleuler's "primary symptoms," known to medical students as the "four As": (1) loosening of logical thought *associations;* (2) disturbance of feelings or *affect;* (3) *autism;* and (4) *ambivalence* or contradictory feelings.

In a conversation with a typical schizophrenic whose symptomatology is fully developed, a loosening of associations becomes obvious. A polite comment about the weather may evoke a jumbled discussion of philosophical concepts, progressing to a tirade on the role of vitamins in promoting sexual morality and the possibility of world destruction. The disturbance in affect may also be apparent, with the unexpected, bizarre response accompanied by inappropriate laughter, unpredictably shifting to furious rage. Another manifestation of disturbed affect seen in schizophrenia is a blunting or flattening. An individual with this characteristic seems to be devoid of emotions, even when some emotional response would be appropriate. This is not a simple holding back of emotions, as in the case of a bereaved wife "bearing up" in public; instead, the schizophrenic individual appears to have no feelings at all.

Autism contributes to the impression that the schizophrenic individual is uninvolved with the outside world. Autism implies that an inner world of thought has been substituted for the experience of reality. In varying degrees, the schizophrenic establishes a universe peculiar to himself, and his speech may contain a strange vocabulary that he has constructed to describe the world in his own terms. This apparent withdrawal, however, should not lead to the assumption that the patient is unaware of his surroundings. Actually, he may be extremely sensitive to them, while his autistic interpretation of events may lead to odd and unpredictable responses.

Ambivalence, the fourth of Bleuler's primary symptoms, often takes the form of rapidly alternating contradictory feelings toward others. A schizophrenic may veer from childish admiration to violent rage toward the same person, without any obvious reason for the change. This ambivalence often renders a schizophrenic individual incapable of choosing between alternatives.

The clinician rarely sees patients with such clear-cut symptoms, but Bleuler's classification is still useful in helping us to understand the schizophrenic thought process. Various investigators have described other aspects of that process. One is a concretization of thought, an inability to deal with abstract concepts. Another is an egocentric overinclusiveness, in which the individual assumes that everything occurring in the environment has a direct connection with him; thus he may believe that someone coughing in his presence is doing it to make fun of him. The schizophrenic may be unable to ignore unimportant sense perceptions, so that he feels overwhelmed with visual and auditory stimuli.

A frequently observed abnormality is a difficulty in conceptualizing the boundaries of the self as separate from others. Often the schizophrenic feels depersonalized and has no sense of being himself. On occasion he may feel that he is an inanimate object. At other times he may seem to be watching himself and telling himself what to do. Usually the schizophrenic, to some degree, feels he is being controlled by an outside force. This may be little more than a vague feeling, or it may take the more concrete form of a delusion that his mental processes are being directed by some sort of "thought-control machine" or ray.

Some of the major symptoms of schizophrenia are depicted opposite and on the following pages. They include, in order, the loosening of logical thought associations, autism, grandiose delusions, and ambivalence.

This feeling of outside influence is often helpful in differentiating a schizophrenic process from other psychiatric illnesses with similar symptoms. Many clinicians believe that an inability to trust others and to experience closeness while maintaining the ability to define one's self is a basic and pervasive experience of the schizophrenic individual.

Attempts at classification

Because schizophrenic illness takes so many forms, efforts have been made to divide it into subtypes, based on the presence or absence of certain characteristics. The traditional division includes four main types. Hebephrenic schizophrenia is characterized by childish regression and silly affect with a progressive deterioration of behavior patterns. The catatonic form of the illness is characterized by a stuporous state in which the patient becomes mute and immobile. During these periods he appears totally insensible to outside reality, but experience has shown that he is actually keenly aware of everything that is taking place. In the excited phase a catatonic patient may become highly agitated and aggressive. Simple schizophrenia most closely resembles what Kraepelin described under the name dementia praecox. Typically, the illness comes on insidiously, without such symptoms as delusions or hallucinations. The patient gradually withdraws from his activities and interests, turning in on himself and avoiding communication with others. As the disease progresses, he loses concern with bodily functions and personal hygiene, and eventually reaches a state resembling idiocy. This type traditionally has a poor prognosis. A paranoid form of the illness is defined by a later onset, generally in the late 20s or early 30s, and by paranoid delusions, usually centering about themes of grandiosity and/or persecution. The individual does not suffer the deterioration of personality and intellectual functions seen in other types of schizophrenia. On the other hand, because he tends to see his well-being as threatened by others, he may strike out violently against his "persecutors." A fifth subtype, schizoaffective schizophrenia, has severe depression or mania as a major symptom and is sometimes difficult to distinguish from a pure depressive illness.

The classic forms of schizophrenia are rarely diagnosed today, reflecting a tendency on the part of contemporary psychiatrists to avoid this sort of descriptive classification. In addition, modern treatment methods tend to prevent the disease from progressing to the point where the typical symptom complexes become apparent. The most frequent diagnosis at present is of chronic undifferentiated schizophrenia, in which symptoms associated with the classic subtypes may be present, but the picture is mixed and no one form of the illness predominates.

Another, more general, method of classification divides schizophrenic illness into two major types: process and reactive. The process form of the illness is characterized by a slow, insidious onset in the absence of an identifiable life stress and tends to have a poor prognosis. The reactive form occurs in the presence of an identifiable life stress, coming on in acute form in an individual with an apparently intact personality; the outlook for recovery is much better. European psychiatrists use the term schizophreniform psychosis to describe an acute psychosis occurring in the presence of stress and followed by rapid and complete recovery. It is a matter of definition whether a

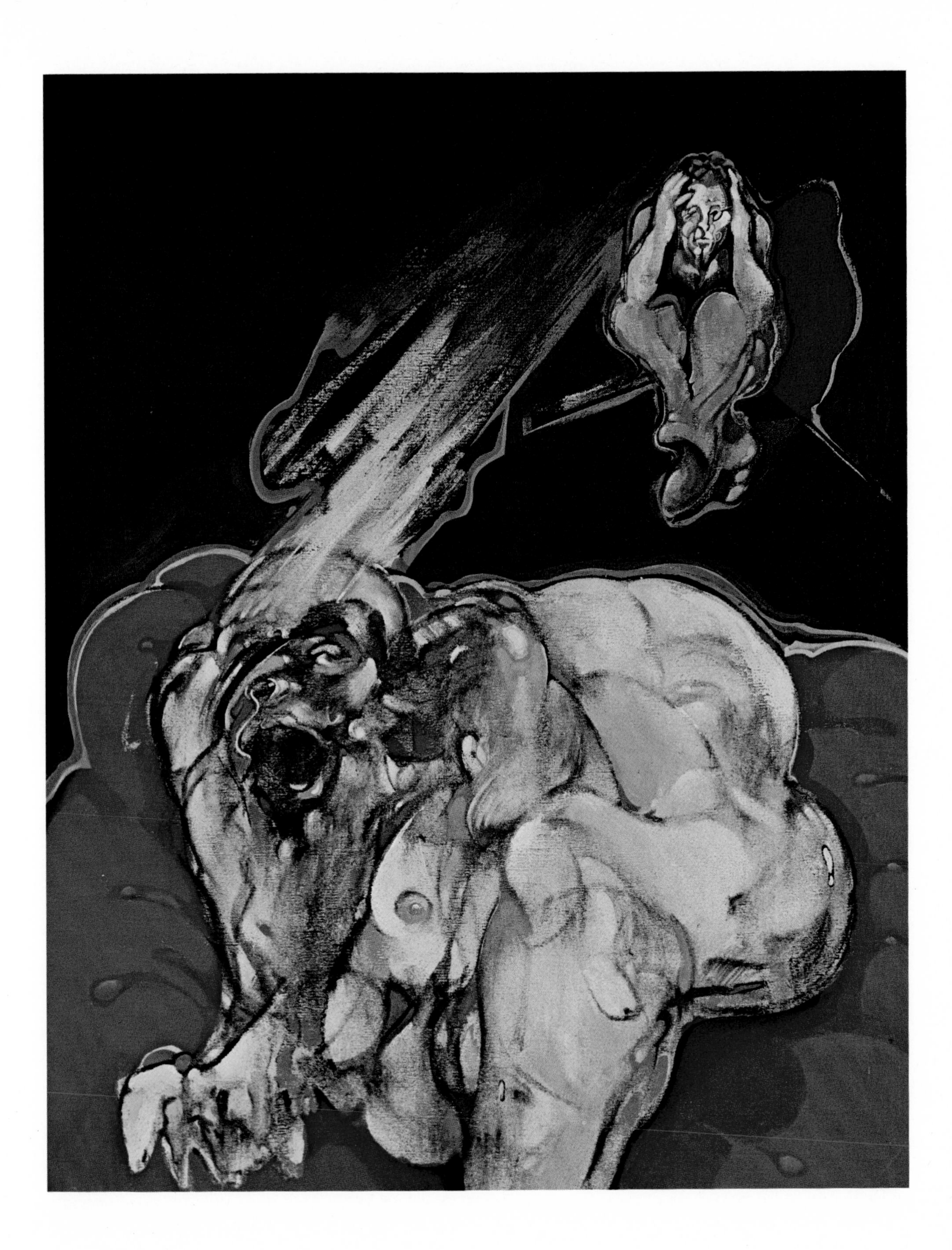

brief illness with acute schizophrenia-like symptoms should be considered as a mild form of the disease or simply a temporary schizophrenia-like state.

Child schizophrenia is diagnosed when the symptoms occur before puberty. The differences between this rare form of the illness and the adult form relate to the developmental processes of the infant. There may be abnormalities in the development of muscle tone and motility, seen as early as the first to third year, and failure to develop normal communicative behavior, especially speech. The affected child tends to ignore the presence of others, even its mother, and to be preoccupied with repetitive, stereotyped behavior patterns, such as spinning a top or tapping on nearby objects. Over half of these autistic children are believed to have normal or better intelligence. The outlook for them has been considered poor, although recent experience with long-term, intensive residential care has provided some grounds for optimism.

Searching for causes

Despite the vast amounts of clinical description and theorizing, complete understanding of the schizophrenic process still eludes psychiatrists and behavioral scientists. Experienced clinicians often disagree even in the matter of diagnosis. For instance, when videotaped interviews of the same patient were shown to British and American psychiatrists, the Americans proved much more likely to make a diagnosis of schizophrenia than their British counterparts. Different patients show different symptoms, and one patient may show widely varying symptoms during the course of his illness.

It is apparent that we must go beyond the mere description of psychological symptoms if we are to approach an understanding of the schizophrenias. What do we know, at this time, about their causes? Few illnesses of comparable importance have been traced to such a variety of etiological factors, ranging from a distorted family environment to abnormal genes to a chemical imbalance in the brain.

Over $10 million annually is being spent to study schizophrenia. The research falls into three general categories. Biological studies include the examination of evidence for genetic factors in this illness, as well as studies of abnormal biochemical processes associated with the symptoms of schizophrenia and research in psychophysiology, including perceptual studies. Psychological-interpersonal studies focus on the influence of early experiences and the dynamics of family relationships. Social-epidemiologic studies examine the effects of social conditions.

In the field of biological research, recent studies have provided increasingly strong evidence for the presence of a genetic factor in schizophrenic illness. Earlier studies had indicated that if one of a pair of fraternal twins develops schizophrenic illness, the probability that the other twin will also develop it is greater than would be true of the population at large. Among identical twins (developed from the same ovum and therefore having identical genetic material) the probability is even higher. A high correspondence was also found between schizophrenic parents and the illness in their offspring. These studies were criticized, however, because they did not rule out the environmental influence of a schizophrenic person in the household.

More recently, a study of identical twins raised apart by nonschizophrenic foster parents showed that, even under these circumstances, there is a high rate of correspondence between the appearance of schizophrenia in one twin and its appearance in both. Similarly, a relatively high proportion of the children of schizophrenic mothers raised by nonschizophrenic foster parents develop the disease. Even so, none of these studies (except those involving identical twins) indicated a rate of correspondence higher than 20%. Still unexplained is why, if a tendency toward schizophrenia is inherited, the illness appears in some "vulnerable" individuals and not in others.

Biochemical studies involving schizophrenia have been carried out for some 20 years and a vast array of findings has been produced, much of it pointing to differences in body chemistry between groups of schizophrenic patients and groups of "normals." The significance of this is far from conclusive. Many of the earlier studies suffered from faulty methodology in terms of the clinical definition of schizophrenia and control of the subjects' nutritional state. Later studies have been more sophisticated in design, but the question remains as to whether a biochemical "difference" is a cause of the condition or a by-product of it. The studies are limited to the detection of chemicals in the blood, urine, or other body fluids of live patients, and it is technically very difficult to determine whether the chemical events under study are occurring in the brain or in the entire body. Moreover, the studies are largely confined to the breakdown products of vital biochemicals, which is somewhat analogous to attempting to understand the operation of an internal-combustion engine through the study of its exhaust.

Nonetheless, determined efforts to pinpoint biochemical factors in schizophrenia have continued, and several show promise. It has been suggested that schizophrenic individuals, perhaps in times of stress, may produce abnormal biochemicals that lead to disordered brain function, including delusions and hallucinations. Another possibility that continues to find support is that proteins in the blood of schizophrenic individuals provoke an autoimmune or allergic response, resulting in the production of abnormal biochemicals that may cause localized damage to strategic areas of the brain. Other studies have reported deficiencies in hormonal stress response, abnormalities of energy metabolism, and the presence of abnormally high levels of proteins ordinarily associated with rare wasting diseases of the muscles, but the significance of these findings remains undetermined.

Recent psychophysiological studies have provided evidence that schizophrenic individuals suffer from a disturbance in the neurological processing of incoming information. Again, however, the relevance of these results has not been established.

Psychological-interactional studies of schizophrenia have a long history, dating back to the pioneer efforts to treat the illness through psychotherapy. In the 1930s and 1940s Harry Stack Sullivan and others proposed that severe difficulties in the mother-child relationship might be a major factor in the genesis of schizophrenia. Unfortunately, this was frequently interpreted to mean that the mother was "to blame" for "making" her child ill, an oversimplification that has given rise to tremendous guilt and defensiveness on the part of parents of schizophrenic patients. Such significant difficulties may play at least a partial role in the development of the illness, but it must be understood that these difficulties are not consciously created by the parent. Only harm can result from communicating the notion that the parent, without help, could have prevented what was taking place. In assessing such parent-child relationships, it must be remembered that the child's response affects parental behavior, as well as the other way around, and that parents have problems of their own, resulting, perhaps, from their relationships with their own parents.

Later studies have focused on the role of the entire family system in the development of schizophrenic illness. Based on the observation of schizophrenic patients and their families, several characteristic patterns that seem to foster the development of schizophrenic thinking have been described. One such situation portrays a mother who is unable to distinguish between her own needs and those of her developing son in a family where the father is either passive or absent. The child fails to separate from the mother psychologically, and the result is a symbiotic dependency on the mother, oversensitivity and subsequent fear of peer relationships, failure to develop a sense of identity, and, finally, the feeling of being controlled in thought and behavior by others. In another pattern, the parents are locked in conflict and undercut one another, while their daughter is caught between their conflicting demands. It is believed that this situation may lead to a fragmentation of logical thinking and failure to develop a healthy identification with either parent.

Social and epidemiologic studies add another important dimension to our understanding of schizophrenia. A number of studies have reached the same conclusion: the incidence of schizophrenia varies with socioeconomic status and tends to be highest in the lower socioeconomic groups, especially where there is social strife or disorganization. There has been considerable disagreement among researchers as to whether simply living under such conditions promotes the development of schizophrenia or whether individuals with schizophrenic tendencies "drift downward" socially as a result of their illness. It seems clear, however, that while "downward drift" may occur, the pressures of living under conditions of social disorganization do provoke the development of schizophrenic symptoms.

This sample of the research into the etiology of schizophrenia only begins to convey the complexity of the problem and the challenge of integrating the various types of information into a unified theory. While

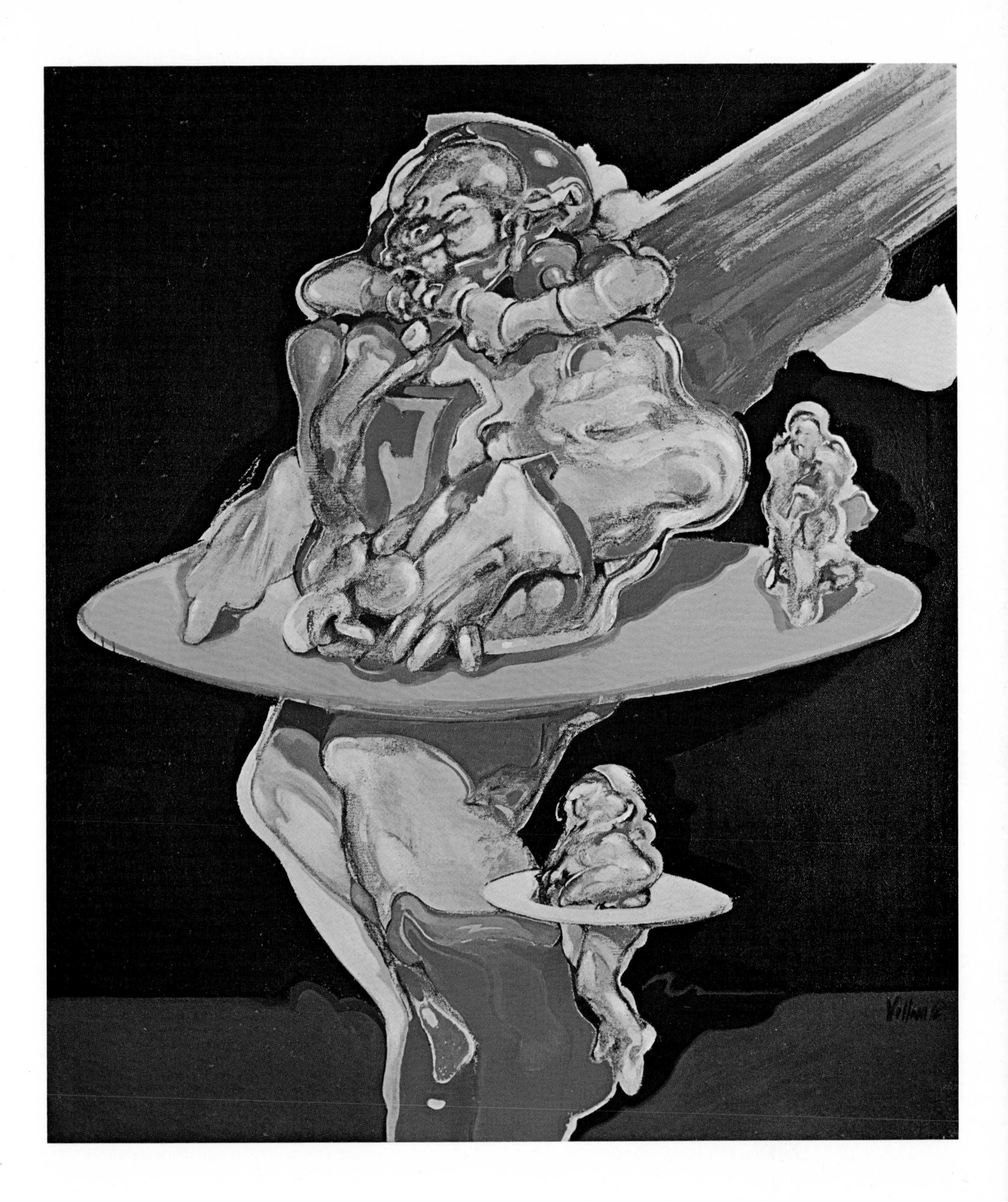

many authorities take an either/or, mind versus body approach, it seems evident that neither the psychological nor the biological "explanation" alone can provide the basis for an adequate theory of the illness. In our present state of knowledge, we may postulate the existence of a biological vulnerability to schizophrenia, involving a tendency to produce abnormal brain chemicals, which is not, however, sufficient to bring on the disease by itself. In the presence of stresses provided by distorted family relationships or social pressures, biologically vulnerable individuals will tend to develop the overt symptoms characteristic of schizophrenia. Nonvulnerable individuals, faced by the same pathogenic family environment or social milieu, will not become schizophrenic, although they may develop other forms of illness, such as sociopathic or neurotic behavior patterns. Much more must be known before a theory of schizophrenia can be developed to the point of permitting more effective intervention or, possibly, prevention. These goals may be within reach, however, provided that society is willing to support the necessary research.

Current treatment methods

As in the case of etiology, authorities disagree widely on the subject of treatment. One extreme view, put forward most forcefully by the British psychiatrist R. D. Laing, contends that the psychosis of the schizophrenic individual is a developmental process which, though a risk to the patient, should be permitted to proceed without interference. Laing characterizes the experience of the schizophrenic as one of being "torn apart" or "destroyed" by the contradictory and hypocritical demands of society and interpersonal relationships. Describing the psychosis as a voyage from which the individual may not return, Laing at the same time considers it a vehicle through which he may reach a more advanced level of human development. Even with its pain and danger, it is viewed as a positive experience that can be understood only by those who have been through it. Hospitalization and treatment with standard methods are likened by Laing to interrogation and imprisonment of the patient as a punishment for his different perceptions of reality. Instead, the sensitive, undefended schizophrenic should be supported through the psychotic experience by a "guide" who understands his pain and can relate to what he is undergoing. Reading Laing, one gets the feeling that society is viewed as psychotic, while the schizophrenic is seen as trying to deal with the demands this sick society places on him. Open residence houses designed as havens for schizophrenics going through this process have been established in Britain and, more recently, in the United States.

Intensive psychotherapy is a more traditional way of dealing with schizophrenia, and for many years it was almost the only active treatment method. One of its chief goals is to find a basis for relating to an individual whose experience and distorted development have left him terrified of close relationships and unable to trust others. Many advocates of intensive psychotherapy categorically reject the use of medication. While this method may be applicable to individuals with mild illness or to those with the financial resources to support long-term hospitalization and frequent psychotherapeutic sessions (often three to five times a week), it is not really feasible for the person with average resources. Moreover, the few careful studies that have been completed have not demonstrated that this approach is effective with severely ill patients.

Family therapy, a technique wherein the patient is seen, together with his family, by one or more psychotherapists, was developed to deal with the disturbed communications commonly present in the family of the schizophrenic individual, especially those involving the parents. Studies of its therapeutic value are still incomplete, but it appears to have the advantages of maintaining the organization of the family and clarifying the family's response to the patient's illness. As a therapeutic mode it seems promising, particularly when used in combination with medication and social rehabilitation techniques.

The introduction of psychotropic medication in the early 1950s revolutionized the treatment of schizophrenia. The first such drug, Reserpine, was soon replaced by chlorpromazine (Thorazine) and subsequently by a flood of more advanced medications. Almost overnight, mental hospitals ceased to be overcrowded human storage bins requiring the constant efforts of staff to manage violent and regressed patients.

With the florid symptoms of schizophrenia brought under control, efforts could be made toward the creation of a therapeutic milieu. These medications do not cure the condition by any means, but they do promote a remission of primary and secondary symptoms and pave the way for resocialization, group therapy, and rehabilitation that can help restore the individual to a life in society.

The mechanism whereby these medications act is not thoroughly understood, but their efficacy is clear. With present-day treatment, the prognosis of the schizophrenic patient has improved tremendously. In 1913 Kraepelin reported that only 13% of his patients showed improvement, and most of these relapsed. Today, 40 to 50% of patients who have had acute schizophrenic episodes can be discharged symptom-free within six months, and 75 to 80% of these patients can return home with, at most, only residual symptoms. Given follow-up therapy and maintenance medication, only 10–15% will relapse, compared with the 35–40% who could be expected to relapse in a year without such aftercare. It is now realized that much of the deterioration noted in hospitalized patients resulted more from prolonged isolation and the lack of normal social and perceptual experience than from the illness itself. Today's medication permits more rapid discharge to the community and the use of social therapies and rehabilitation early in the disease process. This is not to say that medication is the whole answer. Many patients fail to respond to medication or achieve only partial improvement. There is still a great need for more effective treatment techniques.

A recent therapy of some interest is the utilization of massive doses of niacin, one of the B vitamins. This treatment, proposed by H. F. Osmond and A. Hoffer, is based on the research finding that niacin acts as an antagonist to certain abnormal biochemicals implicated in schizophrenia. The treatment has received considerable publicity, partly because it coincides with a movement toward the use of high doses of vitamins to treat various emotional disorders. Despite the current enthusiasm of its proponents, however, its beneficial effects remain unproven.

The future outlook

It appears highly probable that clearer findings of abnormal biological processes in schizophrenia will be made. This will be no simple matter, but the evidence continues to suggest that such aberrant processes exist. The development of more effective medications may precede more comprehensive understanding of the psychobiology of the illness. The possibility of early identification and prevention is a dream that may one day be realized.

Continuing social and rehabilitative efforts on behalf of schizophrenic individuals could lead to considerable improvement in the degree of their recovery. We have just begun to approach the task of finding a place for these patients in society where they can pursue their lives, protected, if need be, from stresses that they cannot tolerate. This includes provision of sheltered working and living facilities for schizophrenic individuals who cannot fully adapt to society. It may be that the development of more tolerant attitudes toward mildly divergent behavior patterns will be the most important method of ameliorating an illness that has haunted mankind for centuries.

More than 200 years of observation and research have brought vast changes in our outlook on mental illness. This is dramatically illustrated by the following article, which appeared in the first edition of Encyclopædia Britannica, *published in 1768.*

Of Melancholy and Madness

Melancholy and madness may be very properly considered as diseases nearly allied; for we find they have both the same origin; that is, an excessive congestion of blood in the brain: they only differ in degree, and with regard to the time of invasion. Melancholy may be looked upon as the primary disease, of which madness is only the augmentation.

When persons begin to be melancholy, they are sad, dejected, and dull, without any apparent cause; they tremble for fear, are destitute of courage, subject to watching, and fond of solitude; they are fretful, fickle, captious, and inquisitive; sometimes niggardly to an excess, and sometimes foolishly profuse and prodigal. They are generally costive; and when they discharge their excrements, they are often dry, round, and covered with a black, bilious humour. Their urine is little, acrid, and bilious; they are troubled with flatulencies, putrid and fetid eructations. Sometimes they vomit an acrid humour with bile. Their countenances become pale and wan; they are lazy and weak, and yet devour their victuals with greediness.

Those who are actually mad, are in an excessive rage when provoked to anger. Some wander about; some make a hideous noise; others shun the sight of mankind; others, if permitted, would tear themselves to pieces. Some, in the highest degree of the disorder, see red images before their eyes, and fancy themselves struck with lightning. They are so salacious, that they have no sense of shame in their venereal attempts. When the disease declines, they become stupid, sedate, and mournful, and sensibly affected with their unhappy situation.

The antecedent signs are, a redness and suffusion of the eyes with blood; a tremulous and inconstant vibration of the eye-lids; a change of disposition and behaviour; supercilious looks, a haughty carriage, disdainful expressions, a grinding of the teeth, unaccountable malice to particular persons; also little sleep, a violent head-ach, quickness of hearing, a singing of the ears; to these may be added incredible strength, insensibility of cold, and, in women, an accumulation of blood in the breasts, in the increase of this disorder.

These things being duly considered, together with the state of the brain in persons who died of this disease, we may conclude, that melancholy is a strong and lively working of the fancy, with a fixed attention of the mind to a particular object, which it continually dwells upon; together with a delirium, a long continual dejection, dread and sadness without any manifest cause, arising from a difficult circulation of blood through the vessels of the brain, where it is too copiously congested and becomes stagnant. Madness is a violent rage, attended with rashness and preternatural strength, caused by an impetuous motion of a thick melancholic blood through the vessels of the brain. It differs from a phrenzy, which is a delirium accompanied with a fever, and arises from an inflammatory stagnation of the blood in the brain: for we learn from experience, that all the shining faculties of the mind are changed or depraved, diminished or totally destroyed, when the blood and humours, receding from their natural temperament and due quantity, are not conveyed to the brain in a moderate and equable manner, but on the contrary with an impeded, flow, and languid motion, or with too brisk and violent an impetus.

Both these disorders suppose a weakness of the brain, which may proceed from violent disorders of the mind, especially long-continued grief, sadness, dread, uneasiness and terror; as also close study and intense application of mind, as well as long protracted lucubrations. It may also arise from violent love in either sex, especially if attended with despair; from profuse evacuations of the semen; from an hereditary disposition; from narcotic and stupefactic medicines: from previous diseases, especially acute fevers. Violent anger will change melancholy into madness; and excessive cold, especially of the lower parts, will force the blood to the lungs, heart, and brain; whence oppressive anxieties, sighs, and shortness of breathing, tremors and palpitations of the heart; thus vertigoes and a sensation of weight in the head, fierceness of the eyes, long watchings, various workings of the fancy intensely fixed upon a single object, are produced by these means. To these may be added a suppression of usual haemorrhages, and omitting customary bleeding; hence melancholy is a symptom very frequently attending hysteric and hypochondriac disorders.

The causes which contribute to the generation of a thick blood, are idleness and inactivity, which weaken the body, impair the functions, diminish the salutary excretions, and render the humours thick, viscid, and stagnant: All which are heightened by solitude, which is apt to give rise to various fantastic and gloomy ideas in the patient's mind.

Likewise acid humours in the stomach will increase the appetite, and tempt them to feed on coarse, gross, flatulent aliments, without drinking enough to dilute them sufficiently, whence a matter proper to nourish these diseases will proceed. It is evident from observation, that the blood of maniac patients is black, and hotter than in the natural state; besides, the serum separates more slowly and in less quantity than in healthy persons. The excrements are hard, of a dark-red or greyish colour, and the urine is light and thin.

Diseases of the mind have something in them so different from other disorders, that they sometimes remit for a long time, but return at certain periods especially about the solstices, the times at which they first appeared. It may likewise be observed, that the raving fits of mad people, which keep the lunar period, are generally accompanied with epileptic symptoms.

This disease, when it is primary or idiophatic, is worse than the symptomatic that accompanies the hysteric or hypochondriac passion, which is easily cured: as is that also which succeeds intermitting fevers, a suppression of the menses, the lochia, the haemorrhoids, or from narcotics. When the paroxysms are slight in the idiophatic kind, the cure is not very difficult: but if it is inveterate, and has but short remissions, it is almost incurable; which is often owing to this, that they reject physicians and their medicines as poison. It is a bad sign if, after a profound sleep, the patient still continues delirious, and is insensible of cold, or is unaffected with strong drastic medicines. If after want of sleep and long abstinence the patient is exceeding weak, or becomes epileptic, convulsive, or lethargic, death is not far off. Mad people are seldom subject to epidemic or other disorders, and some have lived seventy years and upwards in this unhappy state.

Sometimes this disease terminates by critical excretions of blood from the nose, uterus, or anus. Sometimes diarrhoeas and dysenteries will terminate these disorders. Pustules, the itch, and ulcers, have also done the same.

In the cure, bleeding is the most efficacious of all remedies; and where there is a redundance of thick, grumous blood, a vein is first to be opened in the foot, and a few days after in the arm; then in the jugular vein, or in the nostrils with a straw; and, last of all, the frontal vein with a blunt lancet, for fear of hurting the pericranium, a ligature having been first made round the neck to render the veins tumid.

Tepid baths are also convenient, to drive the blood from the head to the inferior parts; and before the patient enters the bath he should have cold water poured on his head, or it should be covered with a cloth dipt therein; for cold water pumped or poured on the head constringes and corroborates the vessels of the brain weakened with stagnant blood, and promotes the more easy discussion of the humours congested therein.

Purgatives are likewise useful; but the lenient are preferable to the drastic: Thus manna, cassia, rhubarb, cream of tartar, tartar-vitriolate, are most convenient when the disease arises from the hypochondriac passion, and a stagnation of the blood in the intestines, and the ramifications of the *vena portae*; especially when they are taken in decoctions and infusions, not all at once, but at repeated intervals, so as to operate in an alterative manner.

Some kinds of mineral waters are also highly efficacious in melancholy and madness; for since madness generally derives its origin from the melancholy, and melancholy from the hypochondriac passion, and the hypochondriac passion from impure and peccant fluids slowly circulating through the intestines and viscera of the abdomen the circulation of the blood ought to be rendered free and easy. It is no wonder therefore that mineral waters have been held in high esteem for the cure of these disorders for these being impregnated with a highly pure alkaline and neutral mineral salt, if they are drank in a due quantity, they not only change the peccant humours but incide such as are thick, render the glutinous fluid, and open the obstructions of the vessels; they also relax the tense fibres of the solids, and corroborate the weak and tender, as well as, by stimulating the emunctories, they restore all the salutary excretions. The waters of Selters mixed with asses or goats milk have not their equal in these cases. They should be drank in the spring and fall for five or six weeks. The proportion is one part milk to three of water.

But, after all, there is nothing better to remove the cause of these disorders than depurated nitre, but especially in that species of madness which inclines to melancholy; for it corrects the bilious acrimony of the humours, allays the tumultuous motions of the solids, by diminishing the preternatural heat. Sennertus and Riverius affirm, that nitre, given with a little camphor, is a specific in madness.

Particular medicines among vegetables are, balm, betony, vervain, brook lime, sage, wormwood, flowers of St John's wort, of the lime-tree and camphor: from animals, asses blood dried: among minerals, steel, cinnabar, sugar of lead, and the calx and tincture of silver.

Camphor is much praised by the moderns, particularly by Etmuller.

And Dr Friewald affirms, that with a few doses of camphor, of xvj grains each in pills, he has cured several mad patients even in inveterate cases.

Stahl recommends a powder of the following cephalic and nervine herbs: vervain, sage, betony, with plaintain and white maidenhair.

As to diet, the patient should carefully abstain from salt and smoak-dried flesh, whether beef or pork; from shell fish; from fish of a heavy and noxious quality; from aliments prepared with onions and garlick, all which generate a thick blood. In general, he should eat no more than is necessary to support nature. Small-beer or pure cold water are the best drink. Sweet and strong wines are highly prejudicial; as is also excessive smoaking tobacco; for it not only penetrates thick blood, but throws the fluids into preternatural commotions. Change of air and travelling may be beneficial.

Though in deliriums bleeding is highly useful, yet it agrees best with those that are plethoric, bilious, and in the vigour of youth: these likewise will bear frequent purges of corrected hellebore; but then the strength must be repaired by cordial, corroborating, and anodyne sedatives. When the patient is exhausted, bleeding is hurtful, and restoratives good.

As a high degree of the itch has terminated these diseases, it will be proper to make issues in the back, or to procure ulcers with a potential cautery near the spine of the back.

Sedative medicines are good; but not opiates and narcotics, for these induce stupidity and folly. Those that are good in an epilepsy, will be beneficial here; such as castor, shavings of hartshorn, the roots and feeds of piony, and anti-epileptic powders, the valerian root, flowers of the lily of the valley and of the lime tree.

And to the other sort of madness, which proceeds from being exhausted and weakened by autumnal, violent, and obstinate intermitting fevers, and from their being injudiciously treated with bleedings and purgings, it is only to be cured by restoratives, cordials, and corroboratives, long persisted in.

The Clocks Within Us

by Gay Gaer Luce

Man lives in a rhythmic universe and has a built-in biological clock tuned to the cycles of nature. The pace of modern life threatens to disrupt that clock, with injurious consequences to health and well-being.

A pilot on a long-run flight might take twice as long to react to a simulated emergency around 3 A.M. when he would ordinarily be asleep. Reports on industrial accidents of night-shift workers have shown clustering in the early hours of the morning, when bodily and mental functions are normally slowed for the sleep period. Research on rotating shifts has been inadequate to allow solid conclusions, but recent studies of task performance conducted in applied psychology laboratories in Great Britain indicated that workers do better at repeated arithmetical, attention-demanding, and decision-making tasks during the hours when body temperature is high.

Much of what man considers constant about himself and his world is actually pulsating in patterned cycles. Important cycles can be observed in plants, animals, and man's own body. Night follows day, the tides ebb and flow, the seasons change. All creatures wake and sleep, experience hunger cycles, menstrual and estrous rhythms, spring fever. Some rhythms, such as the period of the heartbeat, depend upon genetically inherited oscillators within the body. The universe is rhythmic and immersed in an endless sea of time information that influences rhythms of gravity or electromagnetic fields, barometric pressure, the electric charge of the atmosphere, sound or cosmic radiations, and the 11-year cycles of solar flares that are reflected in the growth rings of trees.

It is surprising how little attention has been paid to these rhythms considering that our lives depend upon them. There is a tendency to read the work of biologists with detachment, ignoring the messages relevant to human welfare. Man does not migrate like birds, hibernate, or spawn, yet these researches are beginning to emphasize that he, also, has a time structure tuned to these same natural cycles. Until recent times, man lived in consonance with the tempo of nature, working by day, resting by night, and traveling only as fast as animals or sails could carry him. Modern inventions—electricity, communications, jet flight—have changed that tempo: now the mechanization of industrial society sets up schedules that are more suited to the efficiency

Courtesy, National Park Service, Department of the Interior

Pioneering researchers Bruce Richardson (left) and Nathaniel Kleitman (right) shut themselves up for 32 days in 1938 in a chamber of Kentucky's Mammoth Cave to see if they could adjust to a 28-hour cycle of activity and sleep. Within a week Richardson had made the adjustment, but Kleitman, 20 years older than his colleaque, was apparently so tied to the 24-hour cycle that he could not.

(Opposite page) Cell division in animals and plants fluctuates in a daily cycle related to the rhythm of activity and rest. A whitefish cell within minutes undergoes the process of division in a series of stages, top to bottom, interphase, prophase, metaphase, anaphase, telophase, and, after division, interphase.

of machines than to the adaptability of the human body. Rotation of work shifts, rapid travel across time zones, and a general disregard for regularity have brought their own problems to 20th-century man.

Studies of biological rhythms are beginning to suggest that these self-inflicted time displacements may be injurious to health in the long run. Indeed, rhythms so permeate life that they should not be overlooked when planning a trip, diagnosing an illness, conducting an experiment, or even testing the safety of a drug.

The importance of intermeshed cycles is easy to understand by the simple analogy of the body as a factory. Imagine the production schedule of the body, in which numerous enzymes, hormones, and other substances must be available in tissues that need them exactly on time, without shortage or delay. The regularity of the heartbeat requires that a complex neurochemical be ready at the nerve ends precisely at the right moment to modulate the contraction. Countless subroutines like this are intermeshed to keep the body running smoothly. Some of the production lines are so microscopic that the body seems to be unaware of them. Animal and plant researches suggest that there are many levels of rhythmicity, but the most basic cycles follow from the process of cell division. Before a cell can replicate itself the template molecule in the nucleus, DNA, must unwind itself and be synthesized anew in a series of steps that, within minutes, ends in two cells where there had been one.

Cell division in any particular tissue does not occur steadily or evenly around the clock, but in each tissue rises and falls in a daily cycle that is related to the rhythm of activity and sleep. The skin, cornea of the eye, lining of the nose and mouth, eroded by contact with the environment, must always be renewed, with the renewal cells showing greatest division during sleep. Throughout the night man and other diurnal mammals undulate on waves of changing consciousness, for sleep is profoundly rhythmic. Human beings undergo cycles of vivid

GAY GAER LUCE, *a science writer, is the author of* Body Time.

dreaming and physiological changes every 90 to 110 minutes during the night. The same rhythms are evident in cats and rats, but in shorter cycles because of the faster metabolism in smaller animals.

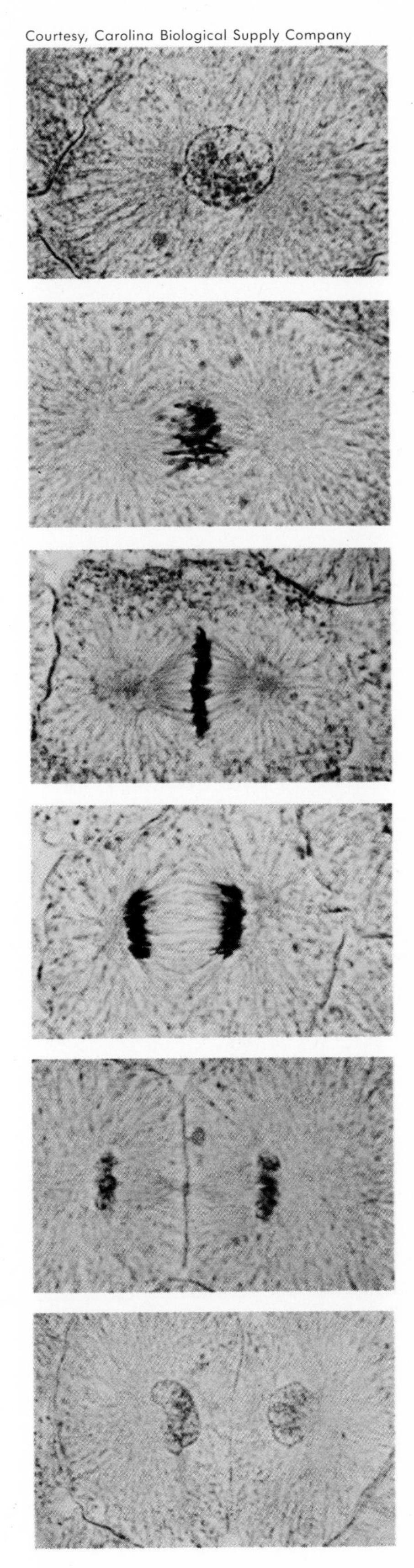
Courtesy, Carolina Biological Supply Company

Circadian rhythms

A great deal of the research on biological rhythms has focused on circadian rhythms. The term derives from the Latin "circa diem" meaning "about a day." When left in isolation from sunlight and darkness, and all social time cues, plants, animals, and humans show cycles of activity and rest that vary within a period of 23 to 27 hours. Since we live on a planet that is strongly influenced by its sun and moon, it is not surprising to find a combination of the solar and lunar days—somewhere between 24 and 24.8 hours—imbedded in earth creatures.

Biologists have found circadian rhythms of activity and rest in organisms as primitive as paramecium and as complicated as chimpanzees. Any living thing keeps its body and activity in the proper phase with the external world, usually through some rhythmic factor such as light, which stimulates diurnal creatures like man to be wakeful and active. However, plants that were kept in deep caves and animals that were blinded or were kept in constant environments maintained a circadian cycle, although they did show evidence of drifting out of phase with the day-night rhythm of the world.

Many attempts have been made to discover man's basic cycle of waking and sleeping in the absence of all time cues. Scientists have undergone the rigors of working under primitive and difficult conditions along with the men and women volunteers who consented to live for periods of from 2 weeks to 6 months in underground caverns or in laboratory capsules. Some volunteers controlled their own illumination, while others remained either in constant bright or dim illumination. Rarely did any of these volunteers show a "short" day of less than 24 hours. Most oscillated, often between 24.8 and 26 hours, and occasionally, for a two-week period, a 48-hour day was observed. Man's basic "free-running" rhythm (so-called when there are no external time givers such as light) seemed to be longer than 24 hours, and extremely persistent.

In rodents, even more persistently, the circadian rhythm of running and rest did not disappear even when the animals were blinded, stressed with drugs, physical injury, freezing, heart stoppage, and other extreme measures. A West German botanist, Erwin Bunning, observed that the rhythm of plants raised in a constant environment was never more than about 3 hours more or less than 24 hours. Moreover, each species demonstrated a typical rhythmic period in isolation. When Bunning crossed a plant showing a 24.2-hour rhythm with one having a 25.6-hour rhythm, the hybrids had an intermediate period. This seemed to indicate that the circadian rhythm could have been inherited.

Zoologists, experimenting with unicellular organisms, insects, and mammals, found considerable evidence for circadian rhythms. Colin

Pittendrigh observed, for instance, that a large batch of fruit fly larvae did not hatch randomly a few at a time, around the clock, but instead seemed to emerge from the pupae in groups at 24-hour intervals. If they were raised in continuous darkness there was no such rhythm, yet after a single flash of light into the darkness flies emerged together at 24-hour intervals. Pittendrigh concluded that the circadian rhythm must be the product of some basic molecular mechanism that has a periodicity of 23–25 hours and that, once started, would continue its oscillation in the absence of external synchronizers such as light or temperature change. This rhythm appears to be a basic adaptation to our rhythmic planet, in which the sun and moon create 24–25-hour periodic changes of light, temperature, and pressure. Indeed, rhythmicity must have been one of the first forces of natural selection some 200 million years ago when the first organisms tuned their metabolic changes to the recurrent changes of light, temperature, and humidity around them.

Many scientists believe that circadian rhythms are inherited, but a few argue that geophysical changes cause such a continuous inpouring of rhythmic information that organisms do not need inner clocks. Frank Brown, a biologist at Northwestern University and the most active proponent of a "cosmic receiving theory," pointed out that signals are beating upon us from the environment at all times.

Brown found that metabolism in crabs, and cycles of oxygen consumption in shellfish, rats, carrots, and potatoes underwent changes that correlated with unusual modifications in cosmic radiation. He showed that worms could orient themselves using the direction of a weak laboratory magnet. In 1962 Yves Rocard, a physics professor at the University of Paris, found that by holding his arm taut and balancing a stick, the nerves and muscles of the arm became sensitive to extremely small magnetic changes (gradients of 0.3–0.5 milligauss), which he measured with a magnetometer.

Isolation experiments conducted by Jürgen Aschoff and Rutger Wever in underground bunkers at the Max Planck Institute in West Germany indicated that human beings could be influenced by the earth's electromagnetic fields, although totally without conscious awareness. One experimental compartment was shielded with heavy metal plating against magnetic and electric field changes of earth, while the other compartment was not. Of the 75 volunteers who lived there none could have been aware of the difference between the compartments, yet those in the shielded unit lived an average day of 25.25 hours while those in the unshielded unit lived an average 24.84-hour day. The difference may seem negligible, yet the people experiencing the longer period had veered far enough from their usual 24-hour day to suffer notable desynchronization between the normally persistent, nearly 24-hour cycles of many body functions and their changed cycle of sleep and waking. The hours of sleep and waking tend to synchronize the phase of all the internal rhythms, and many of the volunteers noted in their diaries that they felt miserable and unhinged when their

Dan Morrill

Circadian rhythms in man and other living organisms seem to be adaptations to the rhythmic nature of the earth, in which the sun and moon create changes of light, temperature, and pressure every 24–25 hours.

bodies were out of synchrony. When Wever turned on alternating electric fields (ten Hertz), the shielded volunteers consciously felt nothing new, yet their behavior changed so that activity and internal rhythms were more nearly synchronous. If natural electromagnetic fields help to keep us synchronized, we may ask ourselves how good it is for us to work and live in even partially shielded buildings of glass and steel.

A brief summary of some of the circadian rhythms of man's physiology may indicate what the consonance between body state and activity can mean. It is known that body temperature rises and falls about 1 ½ degrees every 24 hours. Temperature rises in the morning to a daytime plateau that again may dip in midafternoon. In an experiment conducted by Michel Siffre of the French Institute of Speleology a volunteer who lived 4 ½ months in a cave involuntarily established a 48-hour day but nonetheless demonstrated a stubborn 24-hour rhythm of body temperature. When tested around the clock, it was found that man's time sense expanded during the temperature peak and that this was the best time for vigilance, quick reactions to signals, and decision-making. On the other hand, it was observed that the weakest performance came around 3 A.M., when body temperature was at its cyclic low.

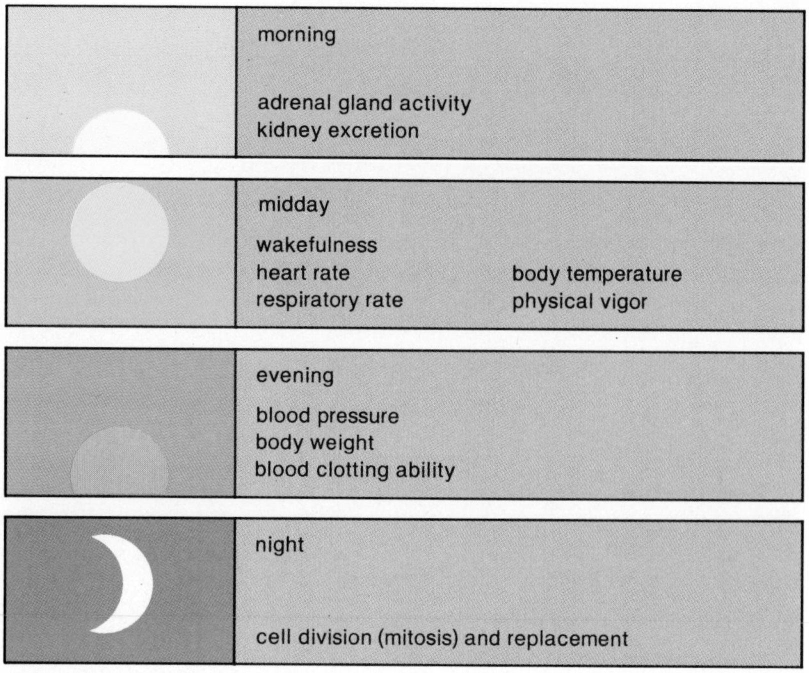

morning

adrenal gland activity
kidney excretion

midday

wakefulness
heart rate
respiratory rate
body temperature
physical vigor

evening

blood pressure
body weight
blood clotting ability

night

cell division (mitosis) and replacement

During the 24-hour day certain biological processes within man, the circadian rhythms, demonstrate a regularly recurring ebb and flow. Shown above is the time of peak activity for some of these processes.

Adrenal tides

A normal, healthy human being who sleeps at night will secrete more than 50% of his daily quota of adrenocortical hormones into his blood in the last three hours of sleep. Some acutely depressed persons observed in round-the-clock studies did not show this rhythm until health was restored; also the rhythm was less pronounced in persons with severe cataracts until the removal of the cataracts allowed them to experience again the influence of light and darkness.

These adrenal tides are responsible for differences in responses to stress, such as a quarrel or an emergency, occurring at 3 A.M. during low tide or at 7 A.M. at high tide. Sensations of vitality or fatigue, the speed with which food is metabolized, and the sensory perceptions are all influenced by the amounts of these important hormones available to the tissues and nervous system at a given time. An inverse relationship exists between sensory keenness and the concentration of steroid hormones such as hydrocortisone. In the morning when hormone levels are high, sensitivity to subtle gradients of taste, smell, or sound is not so acute as it is by evening. Indeed, low levels of adrenocortical hormones may be part of what usually is called fatigue, for low levels occur at the end of the day, causing the body to become oversensitive to sensory inputs.

One more possible corollary to this adrenal rhythm is worth mentioning particularly in the United States where so many people worry about overweight and where the heaviest eating and drinking occurs after 6 P.M. In round-the-clock studies in which heavy protein meals were eaten at different hours, blood levels of amino acids suggested that metabolism was more efficient in the morning. Therefore, morning or midday might be a better time to eat heavily.

A diagnostic tool

Circadian rhythms, of little interest to the healthy person in his daily activities, are a useful diagnostic tool to a physician. For instance, the constituents of a person's blood change significantly during the day and if an accident occurs in the mid-to-late day blood will clot faster than in the early hours of morning. Immunity to viruses and infections is not equal at all hours because the level of the blood's gamma globulin (that fraction of serum containing the most antibodies) is lower in the early hours of morning that are usually passed in sleep. Since virtually no part of the body is exempt from the tides of chemistry, it could be a matter of life or death in some cases, as when an anesthesiologist must adjust a dosage for a fragile patient without knowing how quickly the liver can detoxify the toxic substance. Animal studies have shown that the use of a muscle relaxant could be fatal at certain phases of the animals' cycles. Subsequent studies indicated that the fatalities occurred at the phase when the liver enzymes called upon for detoxification were least active.

Relevant as these rhythms may be to human medicine, they are practically uncharted for man, and most of what is known comes from studies of rodents. Such research on the phase of a particular target organ could determine whether X rays, drugs, surgery, or a trauma is helpful, harmful, or lethal. Lawrence Scheving, by giving identical doses to identical groups of rats at two-hour intervals around the clock, demonstrated that a heavy dose of amphetamines does not cause the same reaction at every hour. At 3 A.M., the peak of the nocturnal rat's activity cycle, 78% of the animals died—yet at 6 A.M. the same dose killed only 6%.

What does this imply about the testing and use of drugs? Prescriptions and instructions on over-the-counter drugs seldom mention hazards at particular times of day, and we assume that all available drugs are equally safe or effective at all hours. However, a better knowledge of circadian rhythms would allow the physician to understand possible variation. Some doctors became sensitive to the need for timing medications in the late 1920s knowing, for instance, that a large dose of insulin might be helpful to a diabetic in mid-morning but might send him into coma at midnight. In other illnesses, such as Addison's disease where the person's adrenals could not produce sufficient hormones, Alain Reinberg and his co-workers in Paris showed that replacement with synthetic hormones is most efficacious if the dosage mimics the body's normal rhythm. Thus Parisian patients began taking their hormones in the morning, instead of adhering to a former schedule of equal doses.

Treating or accurately diagnosing a patient's ailment without knowing his body time could be difficult—it might be 2 P.M. in the clinic but 9 A.M. in the night-owl patient who just awakened. In the future, doctors may require patients to precede their visits to a clinic by keeping a diary of bodily functions, energy, and moods. With the help of computers, information could be assayed to determine whether a

Dan Morrill

Disruption of biological rhythms can occur when many time zones are crossed rapidly on an east-west journey. Some studies indicate that aviation crews on east-west runs experience greater psychosomatic and mood disturbances than do those flying north-south routes.

person is showing cyclic changes, periodic symptoms, or mood cycles that he generally blames on some external cause. Such a diary, developed at the Institute of Living, a Hartford, Conn., psychiatric hospital, uncovered patterns of recovery and drug reactions which never before had been noticed in hospital patients. A detailed diary could acquaint an individual with his own rhythms and provide a way to watch changes that might predict an oncoming virus.

It seems inevitable that sophisticated modern medicine soon must incorporate concepts of rhythmicity in scheduling surgery and medication. Consider a heart transplant in which it has been discovered that the donor heart is beating with a circadian rhythm in a different phase than the tissue to which it is grafted. This raises a question as to whether organ transplants need a matching of recipient and donor rhythms. In the early 1960s a British biologist, Janet Harker, did experiments with cockroaches that suggested illness might be a consequence of conflicting internal rhythms. She hypothesized that the circadian activity rhythm of the cockroach was governed by a kind of "second brain," a neurosecretory ganglion, for when this was removed the creature no longer was active in a rhythmic manner relative to light and darkness. Moreover, when this second brain was transplanted into another cockroach, the recipient adopted the activity rhythm of the donor. When the two creatures were out of phase, the recipient would develop a tumor after the transplant; but this has not been replicated.

Cancer has been called a disease of mistiming by Franz Halberg of the University of Minnesota, and others, who observed that cancer cells not only multiply more rapidly than normal cells but also do not do so in the circadian rhythm of the surrounding tissue. A number of common ailments are related to failure of timing, the most common and recognizable being cardiac arrhythmia, and menstrual irregularity.

Periodic agitation among mental patients revealed underlying abnormalities in adrenal rhythms, and slight desynchronism of phase relations among body functions may be the subtle harbinger of depression.

Health care, thus, seems destined to concern itself with job schedules, travel, and the timing of meals. At present, the busy executive, or vacation traveler subjects himself to stress by flying across many time zones, and then forces himself to negotiate business at once or go sight-seeing in local time even when the body feels that it is the dead of night. Many airline pilots and seasoned travelers know that they can maintain their equilibrium if they hold to a stable home-time regimen on short trips, and on longer trips allow at least a couple of days for transition. Internal disruption is particularly unpleasant and lasting because body systems seem to take different amounts of time to return to their normal relationship with the activity and sleep cycle. Persons, such as students or shift workers, may disturb phase with no immediate consequences, but long-term effects are not known.

Biologists and physiologists have conjectured that long-term disruption or free running might cause later illness or shorten the life span. Halberg and his co-workers subjected a group of adult rats to a single inversion of their light-dark schedule each week (equivalent to a one-way flight to Japan) and found they had shortened the animals' life spans by an average of 6%. Pittendrigh and Dorothea Minis of Stanford University scheduled light and darkness so that groups of fruit flies lived 27-, 21-, and 24-hour days, while another group lived in continuous light. The flies on the 24-hour schedule lived the longest.

Larks and owls

It has been postulated that pilots and crews of aircraft flying east-west routes actually shorten their lifespans. There is no way of knowing whether this is true, although some aviation studies show greater psychosomatic and mood problems in such crews than in those flying north-south runs. A voluminous literature from aviation studies, isolation chambers, and Soviet and U.S. space studies suggests that certain individuals are much more resistant to the effects of rotating shifts or east-west travel than are others. Like Bunning's plants, humans do not show exactly the same free-running circadian period when they are in isolation—some living a 24.3-hour day, others a 25.6-hour day. Individuality of tempo is most noticeable in the way people distribute their energy across the day: some rise swiftly like larks in the early part of the day, full of vitality and cheer but then turn limp and sleepy at evening parties; others awaken slowly and reach a crescendo of activity at night, resembling the owl. Partly these patterns may be due to an inherited time structure, but they also could reflect early training or psychological factors. Yet rats in their cages show distinctive energy patterns similar to the human larks and owls. These patterns offer an individual a rough guide for timing events to his hours of strength, and avoiding strain during hours of weakness.

Dan Morrill

The accelerated pace of modern life sets up many situations of stress. These are often followed by serious illness, which some scientists believe are brought on by stress-induced disruptions of internal biological rhythms.

German studies of pilots doing simulated flight problems during transatlantic flights, demonstrated that those who made the most night errors also showed the greatest physiological and subjective day-to-night change. Phlegmatic, stable, individuals, heavy pilots who were impervious to changing mealtimes, suffered the least from irregular schedules. Ironically, the professional group that endures the most time disruption is the medical. Institutions concerned with health, the hospitals and medical schools, subject students and staff to inordinately difficult schedules, and not even the sick person is spared from a curious logistics of meals, surgery, and medication that has little regard for body time.

Stress and illness

A link between stress and illness always seemed plausible, but this theory went unverified until the mid-1960s when compelling statistics gathered by Thomas Holmes and Richard Rahe, of the University of Washington, Seattle, were published showing that life stresses of a certain magnitude (death, marriage, divorce, new job, buying a house) preceded illnesses of a commensurate severity. One link between stress and subsequent illness may be the disruption of internal cycles during stress, the uncoupling of rhythms from their former phase relationships in the body.

Curt P. Richter, an early worker in psychology at Johns Hopkins, suggested that the shock of a virus or trauma might cause certain cycles that were normally out of phase to become synchronized, setting up a periodic beat in the tissue. Illnesses such as a nine-day swelling of the joints, periodic fevers, edema, or the better known phenomenon of manic-depression, could be attributed to this disruption. Richter also observed that certain drugs (such as the female hormone estradiol, or sulfamerazine) did not affect the 4–5-day cycle of female rats until after the drug was discontinued. Then two-thirds of his animals began to show long cycles (25 days) of lethargy and weight gain, alternating with activity and weight loss. Richter and other researchers later wondered whether antibiotics, hormones, and other drugs might leave a similar cyclic aftermath in human beings, as, for example, the menstrual irregularity that sometimes follows the discontinuation of birth control pills.

The stress of psychological suffering may, itself, jostle internal rhythms so that their abnormal phase relationships place strain on organs or behavior. A series of experiments at the Institute of Living offered some insight into the way emotional and psychosomatic illnesses may develop. Charles F. Stroebel of the Institute of Living discovered that rats showed a fear response to a particular signal at a certain phase of their circadian adrenal cycle. After the fear training he noticed that a number of the animals had lost their regular 24-hour temperature rhythm and were free running, reaching their circadian temperature peak 15 to 20 minutes later each day. After about 20 days these nocturnal animals were in a state of readiness for sleep at the hour when darkness fell in the laboratory, putting them biologically at odds with the demands of their natural environment. Moreover, monkeys that had been exposed to stresses related to loud noise or shock later showed reverberations in their bodies when left alone in their cages. Even after 28 days they re-experienced the stresses, shallow breathing and blood acidosis, at roughly the time of the original experiment. Biologists conjectured that there must be some kind of time memory in tissue, and experimented with certain organs and glands to demonstrate this belief. An excised heart, kept in artificial nutriment, continued to accelerate and decelerate its beat, secretion, and neural responses as if it were continuing the circadian rhythm of the donor. Perhaps such memories of stress are also responsible for elusive medical symptoms, recurring at the hour when an unpleasant event had taken place.

Stroebel's experiments subsequent to the fear training in rats are worth citing in enough detail so that their possible interpretation for human illness is clear, since they demonstrate that a seemingly mild behavioral stress can cause a shift in rhythms resulting in illness. First the monkeys learned to obtain food pellets by pushing a lever to discriminate among images on a panel. Although subjected to loud noises or flashing lights, they quickly learned to turn them off by pressing another lever, clinging to the lever like a child clutching a security blanket. The annoyances then were stopped and the lever recessed in such a way that it could not be grasped. The removal of their "security lever" panicked the monkeys and 14 days later they became sick.

David Hughes from Bruce Coleman Inc.

Joe B. Blossom from the Natural History Photographic Agency

Evidence of biological rhythms can be found in the behavior of plants and animals. Ridley turtles (above) return to the sea after one of their periodic mass nestings on shore to lay their eggs in the warm sand. A chameleon (above, right) sheds its skin on an average of every one to three months in a process related to growth. Emerging from its larval case (opposite, top) is a periodic cicada (often mistakenly called a locust). For 17 years it lives underground in the larval stage before climbing out to shed its skin and become an adult. Leaves of the silk tree (opposite, bottom) open during the day and close at night.

About half of the animals resembled neurotic, psychosomatically ill humans, while the others seemed to resemble psychotics. The first sign of illness was that the reliable 24-hour brain temperature rhythm developed a longer period (24.5 hours) in the "neurotic" animals, who shifted slowly out of synchronization with their environment, doing poorly at their discrimination task. Sores on the skin and bloody stools appeared, and finally ulcers so severe that two of the monkeys later died. The other "psychotic" animals leaped from a 24-hour brain temperature rhythm to a cycle of 48 hours: they seemed out of contact with the environment, and spent their time catching imaginary insects, or masturbating, paying no attention to their tasks, to food, or to people. All of the animals suffered from insomnia, inevitable since their bodies were out of phase with light and darkness in the laboratory. When two animals from each group were given back the security lever, the neurotic animals began to recover and show their 24-hour temperature rhythm, but the psychotic animals remained out of touch, unchanged. Tranquilizers and antidepressant drugs, pulsed into the disturbed animals at 12-hour intervals, slowly began to bring them around to their 24-hour temperature rhythm. Once the temperature rhythm returned they began to act more normally.

Effects of light

Would it seem that the individual's harmony cannot be separated from the environment, particularly the alteration of night and day, the effects of light and darkness? Is it not apparent that man's own cycles are synchronized with those of earth, as are those of plants and birds through the potent effects of light? Plants count the seasons by registering the ratio of light and darkness as the days shorten into fall and lengthen into spring, a process known as photoperiodism. Some are triggered to bud by short days and long nights, some by long days and brief nights. Light inhibits or stimulates growth and budding, depending upon when it occurs in the plant's internal oscillation. Birds respond to the changing seasons by the ratio of light and darkness which penetrates the brain and affects a light-sensitive gland, the pineal. This mythical gland, sometimes known as a vestigial third eye that can be seen protruding on the skulls of lizards, responds to light by manufacturing certain neurochemicals, which in turn modify the reproductive system. Thus birds of temperate zones receive the message of increasing light, become sexually ready for mating, and migrate in a direction that is related to their hormonal state.

In the human being and other mammals, the pineal gland is buried deep within the brain, and may respond to indirect nerve messages of light. A decade of research on the pineal gland, and on the differential responses of birds to different wavelengths of light, suggests that human endocrine systems may be more influenced by seasonal light changes and by artificial light than might have been recognized. Sunlight, one of the most important cosmic forces, and artificial light, too, may guide man's physiology, for he is an open system, a being who resonates to the harmonies of earth and space whether or not his philosophy considers this possibility.

Technology is permitting man to venture into space, but he remains relatively ignorant about his own time structure and the extent to which he resonates to the clock of the earth and the solar system. What is called a physiological clock may be a collection of oscillators that tend to respond to the rhythms of earth. The time structure is only partly within the skin of the body, and the body cannot detach itself from the beat of the natural environment, even in cities of steel and glass where the time of day is a matter of choice. Internal timing may be the organization that knits together the human race, and binds it firmly to the earth and vast universe beyond.

John Kohout from Root Resources

Joy Spurr from Bruce Coleman, Inc.

FOR ADDITIONAL READING

Colquhoun, W. P. (ed.), *Biological Rhythms and Human Performance* (Academic Press, 1971).

Conroy, R. T. W. L., and Mills, J. N., *Human Circadian Rhythms* (J and A Churchill, 1970).

Fraser, J. T. (ed.), *The Voices of Time* (George Braziller, 1966).

Luce, Gay, *Body Time* (Pantheon Books, 1971).

Sweeney, Beatrice M., *Rhythmic Phenomena in Plants* (Academic Press, 1969).

Ward, Ritchie R., *The Living Clocks* (Knopf, 1971).

Death and Dying

by Diana Crane

Medical science is giving man ever greater control over the timing and circumstances of death. Such control raises new questions of responsibility—for physicians, patients, and relatives.

Because of advances in medical knowledge and technology, the nature of dying has changed in recent years. Chronic rather than acute diseases are now the most prevalent causes of death in industrial societies. Due to increasingly sophisticated technology, terminal illnesses last longer and the physician's control over the exact timing of death has increased. As a result, difficult problems are created for the physician, for the patient, and for his family. Should treatment be withdrawn so that the suffering patient or the severely brain-damaged patient can be permitted to die? If so, at what point should this decision be made? How much additional life justifies the use of a heroic procedure on a patient for whom long-term survival is out of the question? These problems are often intensified by the difficulty that most people experience in discussing the imminence of death frankly. Dying frequently disrupts the social relationships between spouses or between parent and child more severely than death itself.

There is growing concern about these issues among the public and among members of the medical and nursing professions. Articles in the popular press and in professional journals have been appearing at an increasing rate, and a spate of conferences has been organized. This popular interest is probably another expression of the desire to be autonomous in all areas of human experience. Increasingly, people are exerting their rights to control their own lives and their own deaths.

Current debate centers on three sorts of problems: (1) Under what conditions should lives be prolonged or terminated? (2) How does the dying patient respond to his situation and how is his response affected by attitudes that are prevalent in our society toward death and dying? (3) How do those involved in the care of a dying patient, either professionally or as members of his family, cope with the problems of communicating with the dying person and with each other when decisions have to be made?

To prolong or terminate?

Much recent speculation has involved the circumstances under which a physician would be likely to withdraw treatment from a dying patient. Some physicians argue that the patient's life should be maintained as long as the heart is still beating, regardless of the amount of physical or mental damage that the patient has sustained. Others argue that the patient's present or potential capacity to engage in social relationships is the important factor.

David Sudnow's study of the behavior of physicians toward critically ill patients who were brought to the emergency room of a large county hospital in California concluded that the patient's social worth or value had a noticeable effect upon the amount of effort which was made to save his life. Older persons were much less likely to be resuscitated than younger ones. Less vigorous efforts were also made to save the lives of persons who were deviant in some respect. The alcoholic was the prime example in this category, which included attempted suicides, dope addicts, prostitutes, vagrants, and criminals.

DIANA CRANE *is a visiting associate professor of sociology at the University of Pennsylvania and author of* Critical Decisions in Medical Practice.

Photos, Dan Morrill

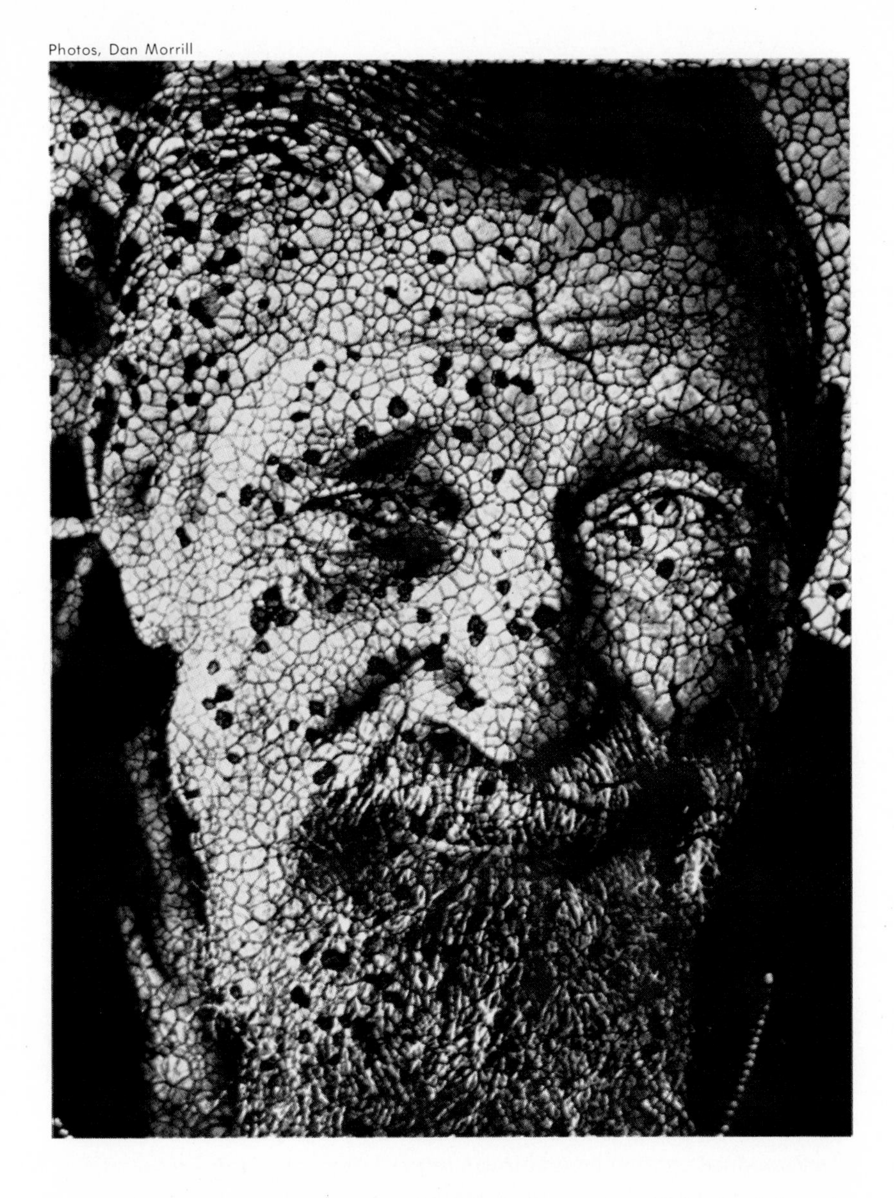

A questionnaire survey, made by Diana Crane, in which physicians in four medical specialties were presented with case histories of critically ill patients found that social aspects of the patients' conditions affected the doctors' expressed intent to treat the patient actively. For example, salvageable patients were treated more actively than those with a terminal illness, and patients who were physically damaged were treated more actively than those with brain damage.

Considerable controversy has arisen concerning the use of brain death rather than cessation of heartbeat as a criterion of death. A set of precise indications for determining the cessation of brain function was defined by an interdisciplinary committee at Harvard University. Laws have been passed in two states to legalize these or similar criteria as indications of death. In the survey mentioned above, respondents were presented with cases appropriate for their specialties which were described as having met the criteria of brain death, as defined by the

Harvard committee. From 70 to 76% of the respondents in the various specialties indicated that they would accept brain death as an indication of death. However, no more than 13% in any sample were willing to turn off the respirator without any consultation whatsoever. This suggested that these physicians have some reservations concerning the criteria.

The same survey also provided some indication of the extent to which physicians are willing to perform acts which fall under the definition of euthanasia (mercy killing). Specialists in internal medicine were asked to indicate whether or not they would increase the dosage of narcotics for a terminal cancer patient to the point where it might risk or would probably lead to respiratory arrest. Of these, 43% of the physicians and 29% of the residents were willing to incur the high risk of inducing respiratory arrest in such a patient by increasing his dosage of narcotics.

On the other hand, response to a question addressed to pediatricians concerning direct killing of an anencephalic infant (one born without a brain) was overwhelmingly negative. Among the respondents (both residents and physicians), only 1% said that they would be likely to give an "intravenous injection of a lethal dose of potassium chloride or a sedative drug" to an anencephalic infant; 3% said that they might do so.

The apparent contradiction between these two sets of findings can probably be explained by the fact that large doses of narcotics which both suppress pain and hasten death are not considered as true examples of euthanasia by many physicians since the physician's *intention* is to suppress pain and not to cause death. It appears that an older physician is more likely to make this distinction between intent and action than a younger one.

The dying patient

In a society where the average life expectancy is high, deaths of friends, relatives, and especially children are rare events. The healthy person apparently reacts to the idea of his own death either with indifference or with concern for his survivors. The sick person, faced with the realization that death is imminent, is believed to undergo a complex and difficult process of adjustment which begins with denial of the accuracy of the diagnosis. This is followed by a period of anger in which the patient asks himself why he should have to die and directs his aggressive feelings toward others in his environment. Some patients attempt to postpone their acceptance of the reality of their situation by bargaining for time to participate once more in some highly valued activity. This stage is followed by depression and eventually by a kind of acceptance in which the patient gradually loses interest in his surroundings.

Some evidence indicates that people would prefer to control their mode of dying. When asked where and how and under what conditions they would prefer to die, people respond that they would prefer to die quickly, painlessly, and with as little fuss or inconvenience as possible. This suggests that they would like to be able to maintain as much control as possible over their deaths. An additional finding in which people said that they would prefer to die in their own homes can also be interpreted in this manner.

A study of dying patients, patients who were not dying, psychiatric patients, and nonpatients found that the majority of each group favored euthanasia. It was most favored by the dying. However, the complexity of the problem is indicated by the fact that, in the same study, dying patients were less likely than healthy ones to want to be told whether they were going to die.

Attitudes toward suicide among the terminally ill reveal the conflicts in this area. While measures are usually taken to prevent such suicides, the patient's attitude toward his death sometimes influences the amount of effort the staff will make to save him. Although suicide by terminal cancer patients is often considered as justified and reasonable, when such suicides actually occur both family and hospital staff express guilt, surprise, and embarrassment. These contradictory types of behavior reflect a norm in the process of changing. The proportion of suicides among terminal cancer patients is believed to be small, suggesting that taboos against it are still strong. Since many such suicides are carefully concealed, however, the phenomenon is difficult to measure.

Suicides are easier to identify among patients undergoing dialysis for chronic kidney failure. A recent study showed that the incidence of suicide among 3,478 dialysis patients was 400 times the rate of the general population. This can probably be explained as the result of the highly stressful nature of this form of treatment.

Suicide as a response to the stress and discomfort of chronic illness may be more common than has been supposed. An examination of all suicides occurring in a town in Great Britain during a seven-year period found that only one-third had no organic disease; the remaining two-thirds had either severe hypertension or a wide variety of "painful, disabling, or fear-engendering diseases." A similar study in the Seattle area found a high incidence of peptic ulcer, cardiovascular disease, and malignancies in the people who had committed suicide.

Dying patients may have more control over the timing of their natural deaths than is generally realized. For example, mortality rates decline prior to important social events, such as presidential elections and one's own birthday. A. D. Weisman and T. P. Hackett described patients who correctly predicted their own deaths when medical personnel did not anticipate them.

Relationships with the dying

Communication between dying patients and both medical personnel and relatives is generally poor. The process of dying is affected by the fact that the majority of deaths now take place in hospitals. In the hospital setting, efforts at treatment tend to isolate the patient from his

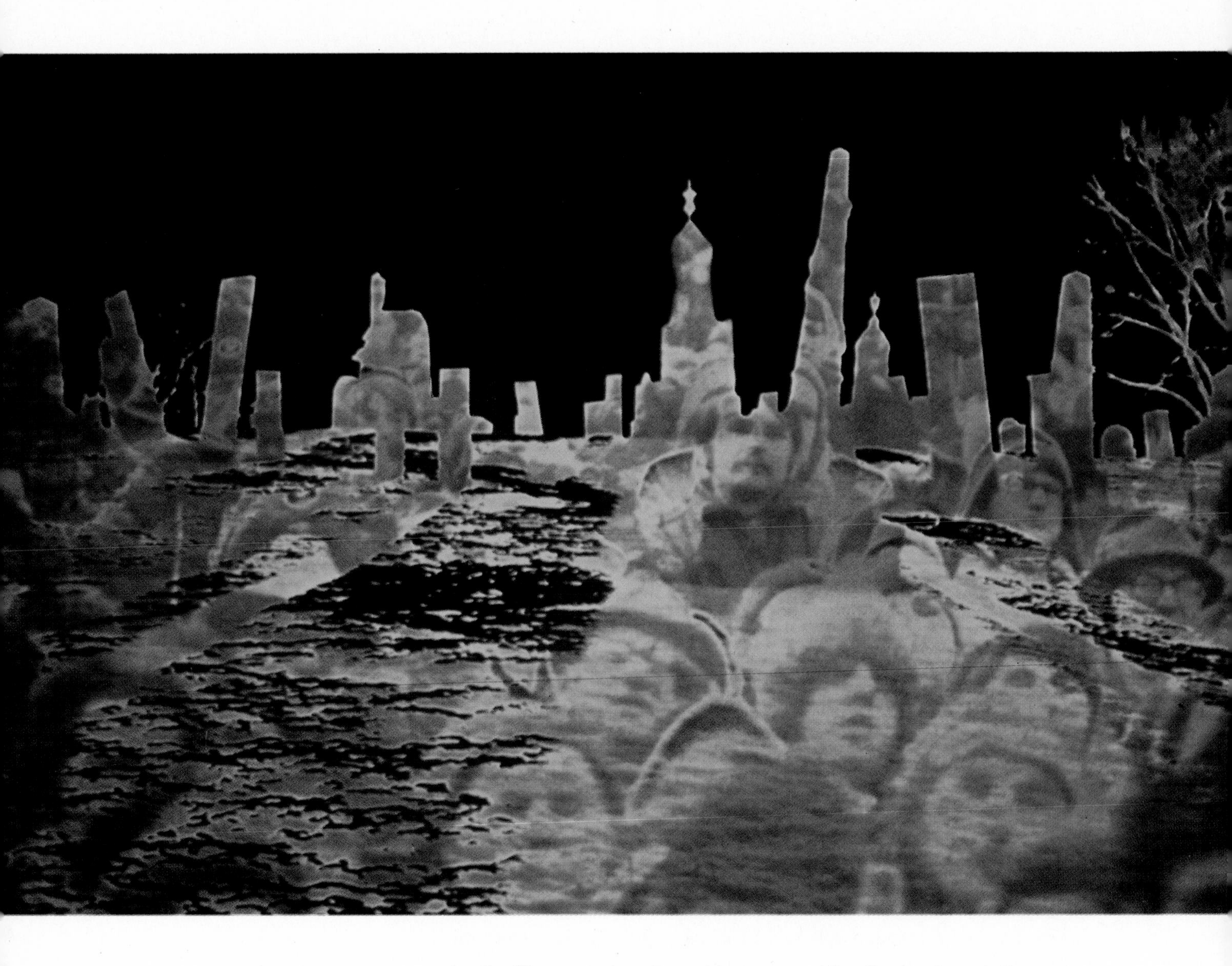

family. The more heroic and hence complex the treatment, the more difficult it is for the family to communicate effectively with the patient. This of course occurs at a time when communication is already extremely strained. B. G. Glaser and A. L. Strauss described different patterns of interaction that take place between relatives and dying patients. In some cases the family is aware of the patient's fatal prognosis, but the patient is not. In other cases both parties are aware, but neither is willing to admit it to the other. Even when both parties admit the truth, frank discussion is usually very difficult. Weisman and Hackett refer to the "bereavement of dying" and suggest that the dying patient may suffer more from abrupt emotional isolation and deprivation than from his illness.

Medical personnel prefer to treat dying persons as if they were expected to live. The motive, conscious or unconscious, may be that of maintaining their control over the doctor-patient relationship.

Donald Oken found that not only did 88% of the 219 physicians whom he studied prefer not to tell cancer patients their unfavorable prognoses but also that their commitment to this policy appeared to be largely based upon emotional reactions rather than upon a rational assessment of the situation. Their resistance to change in this area was such that a significant proportion indicated that their approach was not likely to be altered by the results of research.

Effective emotional support of dying persons by medical personnel apparently occurs relatively infrequently. Evidence shows strong emotional resistance to such interaction by medical staff. In one study, the length of time it took hospital nurses to respond to call lights for terminal cases was compared with the time for nonterminal cases. The nurses were startled to learn how much they delayed answering the ring of the dying.

Glaser and Strauss argued that medical personnel need to be systematically educated in the appropriate ways of handling dying patients. While such training would undoubtedly be useful, it seems unlikely that this would provide a complete solution to the problem. Effective interaction with dying patients is an exacting and delicate task, and, as Glaser and Strauss themselves pointed out, there are no rewards in the hospital for the quality of a staff member's interaction with dying patients. Such efforts are unlikely to be forthcoming unless there is a major change in the allocation of rewards by hospitals.

In recent years some notable attempts to deal with these problems were made. One example was the creation in London of a special hospital for terminal cancer patients with a prognosis of three months or less. Over a period of several years, Cicely Saunders developed an atmosphere in which the patient is given the maximum opportunity to express his needs, to relate to others, and to adjust to his pain without oversedation. In this setting, death is accepted and neither postponed by life-prolonging techniques nor denied with indifference.

Elisabeth Kübler-Ross spent many hours interviewing dying patients in an attempt both to understand the kinds of adjustments they make and to develop ways of communicating with them more effectively. Based on this research she suggested that different styles of communication are needed depending upon the length of time the patient has had to adjust to the fact that he is dying. Weisman also interviewed terminal patients in order to develop ways of helping them to adjust to their situation. He advocated granting autonomy to the patient to make decisions concerning his treatment, his contacts with others, and even his funeral.

Recent studies have also indicated that, particularly among older patients, mental changes occur prior to death that affect the individual's ability to interact with others. These changes, according to A. D. Weisman and Robert Kastenbaum, are indicative of "diminished competence, control, perception and performance." Morton Lieberman found that learning capacity diminished, mental organization

continued on page 108

Beyond dying

Dr. Pahnke: What else happened during the day?
Mrs. Proctor: I died.
Dr. Pahnke: What was that like?
Mrs. Proctor: Beautiful. It sounds vindictive, but it was beautiful. (Chuckles.)
Dr. Pahnke: How can that be?
Mrs. Proctor: I don't know. I felt like I was dying . . . I don't know how I came back. I don't remember that. I think I called for you—did I?
Dr. Pahnke: What does it feel like to die?
Mrs. Proctor: You're just like the thin air, that's it. You have no pain. No fear.
Dr. Pahnke: Did that scare you?
Mrs. Proctor: No. No fear at all. Very relaxed. If it's unusual maybe it's me. I don't know. Is it?
Dr. Pahnke: What?
Mrs. Proctor: Unusual to feel that way?
Dr. Pahnke: That it's relaxed to die? Other people have said the same.

For several years now, a little-publicized project has been going on at the Maryland Psychiatric Research Center in Catonsville, just outside Baltimore. The work embodies the assumption that it is at least as important to deal with a dying person's psychic anguish as with his physical pain. The research concerns the use of psychedelic drugs within the framework of psychiatric counseling to help patients achieve a transcendent level of awareness from which it may be possible for them to deal more easily with impending death.

Years before LSD became the illegal sacrament of the counterculture, psychiatrists at the Catonsville center, led by Dr. Albert Kurland, started administering the drug to mental patients and alcoholics to see if its profound mind-altering powers could possibly help in the therapy of those for whom altered states of consciousness were a way of life. Gradually, a remarkable phenomenon emerged: with striking regularity, and with no suggestion from the therapists, many of the LSD subjects would undergo an experience of "dying" and "being reborn" that often left them with a much more serene, well-integrated approach to life that persisted long afterward.

The possibility that LSD therapy might lighten both the physical and psychic anguish of dying became more immediate for the Catonsville group in 1965: a research assistant working there was diagnosed as having inoperable carcinoma of the breast. When it became clear that neither medical nor surgical therapy had more to offer, and the pain grew increasingly severe, she agreed to become the group's first terminal LSD subject. The encouraging results of her therapy led to a systematic study of the usefulness of LSD therapy for the dying.

Before drug-assisted therapy can start, the way must be cleared by conventional psychotherapy. In a culture that views death as something obscene, and that has endless ways of distracting our attention from it, the patient must first be brought to the point where he can admit the certainty of his own mortality. Often this is a major barrier, especially for the family. The late Dr. Walter Pahnke, who directed the early LSD-therapy research in Catonsville, came up hard against the problem in treating the woman whose words opened this article. "The family's greatest reluctance over having the treatment at all," he recalled, "was the fear that the patient would discover her diagnosis. They had decided in the past that she should never know, and that they would lie to her about her condition so as not to take away her hope and to keep her happy. However, in the first few months her psychological condition had worsened greatly and she was filled with anxiety and worry."

Often, denial of the truth is for the sake of the patient's family members, who cannot bear to confront the reality of a relative's—and hence their own—mortality. So, traditionally, the games continue around the dying patient, usually increasing his anxiety as the suspicion grows that no one is leveling with him. At a time when closeness is one of the only remaining sources of joy, it is undermined by an unacknowledged system of deceptions whose intricacy can reach Laingian dimensions.

Another patient, Mrs. Busen, a 50-year-old woman who knew her diagnosis of widely metastatic cancer of the ovaries, also had to overcome the denials of well-meaning relatives.

"It was very hard for some of my family to swallow the words *dead* and *dying*, and to say them. In fact, they can't say them yet. . . . My husband had gotten tickets for Acapulco next February, and he would say, 'Oh, you'll be well enough. Don't be silly—we got reservations,' and all this kind of baloney, and I knew darned well that I wouldn't be going to Acapulco next February. He's been playing games with me since April and because of the (pre-LSD) therapy the game stopped being played and we faced what we had—we learned to enjoy each day together. . . . And we've enjoyed each other so much more because I don't have the anxiety when I talk to him: the fear of hurting him, and he was always afraid of hurting me."

Once a groundwork of discussion and rapport has been laid, the therapy proceeds to the drug experience. Pahnke, who held a Ph.D. in religion from Harvard as well as an M.D. from its medical school, brought to the work a deep interest in the "peak experiences" described by humanistic psychologists like Abraham Maslow and by centuries of religious mystics. As described by Maslow, people who achieved spontaneous peak experiences frequently felt an ineffable joy that they attributed to a fuller understanding of the meaning of life.

Previous LSD work, too, had shown that subjects who underwent the death-rebirth-ecstasy experience were left with a "psychedelic afterglow" that seemed to infuse them with an exultant clarity that lasted long after the drug had left the body—and that seemed similar to the effects of other kinds of peak experience. Knowing the importance of "set and setting" (expectation and environment) to the character of the drug experience, the Catonsville group set about devising a therapeutic program that would maximize the chances of inducing a "peak." Now, in a modern new building, as part of their work with cancer patients less far advanced in the course of the disease, they use specially designed "psychedelic treatment rooms" that at first look like comfortable living rooms: carpeting, sofa, stereo. Only one element distinguishes the rooms from thousands like them in homes across the country—the small remote-controlled TV camera placed in a recess of the breakfront, for videotape recording of the drug sessions. Of the 60 patients who have volunteered for the cancer project, about one-third have reached what the Catonsville group considers a pure peak experience, with many more having less ecstatic but nonetheless therapeutic sessions.

What accounts for the ability of the psychedelic experience to ease the psychological pain of dying? Dr. Stanislav Grof, the Czech-born psychiatrist who succeeded Pahnke as head of the cancer project, offers one answer: "Death, instead of being seen as the ultimate end of everything and a step into nothingness, appears suddenly as a transition into a different type of existence for those who undergo the destruction-rebirth-cosmic-unity experience. The idea of possible continuity of consciousness beyond physical death becomes much more plausible than the opposite. The patients who have transcendental experiences develop a rather deep belief in the ultimate cosmic unity of all creation and experience themselves as part of it without regard to the situation they are facing."

While Drs. Pahnke and Grof both placed emphasis on helping subjects reach a peak unitive experience, not all of them have. There are, however, levels of transcendence, and the majority of patients do appear to experience some kind of heightened insight into the meaning of their own death, even if it falls short of cosmic consciousness. On the day of the session, patients are asked to bring photographs of family members or friends to whom they feel especially close. With these as a trigger, the drug may make possible vivid reliving of life experiences. Patients often are reimmersed in feelings of warmth and closeness normally too powerful to allow into consciousness.

Alternatively (and less commonly, because of the supportive setting of the sessions), the patient may recall long-repressed bitter or tragic moments. In this case, the reexperiencing is often accompanied by a sense of catharsis, of resolving painful intrapsychic conflicts that had been generating considerable tension (a phenomenon common in conventional psychiatry, elicited with its much weaker tools of spontaneous recall and free-association).

Frequently, one drug session is sufficient—followed by meetings with the therapist to integrate the meaning of the experience into the rest of the patient's life. Occasionally, one or two repeat LSD sessions may be arranged.

There is, additionally, the issue of bodily pain. Neurophysiologists believe that the perception of pain has two components: the physical sensation itself, and the psychological interpretation of that sensation. LSD may have an effect on the latter.

A different kind of benefit available from this therapy is explained by therapist William Richards. "There's a tremendous potential here for practicing preventive psychiatry. Mrs. Busen, for example, before therapy had not told her children of her illness, and hoped she would die while they were away at summer camp. If they had been called home unexpectedly to attend their mother's funeral, the trauma they experienced would have been much greater than coming home to be counseled by her about accepting her approaching death and continuing to live fully, the course she finally took. Much of the grief work can then be done in advance of the death instead of originating with the death or, worse, being repressed to cause internal conflicts in future years."

Or, in the words of the daughter of another patient, in a letter to Richards: "I had never watched anyone dying before. Thank you for giving me the chance to say and do so many things for my mother that others like me never have the chance to do for theirs. I'm still griefing for my mother. Maybe I'm griefing and crying for all the wasted times in our lives. I was the black sheep in my family, but in the end I had a place for my mother to come stay before going to the hospital. We became very good friends in the end. We both asked each other's forgiveness and we both gave it. . . . Out of all the sadness and pain, something good came of it."

—Jerry Avorn

A fourth-year student at Harvard University Medical School, Jerry Avorn based his article in part on a month he spent at the Maryland Psychiatric Research Center. He plans to devote an extra year of study to research on the medical implications of altered states of consciousness.

continued from page 103
became simpler, and efficiency declined in the same degree as the subject's proximity to death. There is also some evidence of a relationship between mental illness and certain types of physical illness. Anxiety and depression seem to be connected to heart disease, and schizoid and schizophrenic reactions to cancer.

These findings suggest that, if communication with dying patients is to be meaningful, skilled assessment of their psychiatric status will be necessary. It may be that in certain types of cases only a psychiatrist can interact with any degree of effectiveness. Unfortunately, psychiatrists apparently prefer not to work with dying patients, although a few are beginning to specialize in this type of practice. Hiring psychiatric social workers to perform as their primary task the role of helping dying patients cope with the reality of their situation also seems promising. Such interaction is resisted to such an extent that only as a kind of specialty is it likely to be satisfactorily and dependably handled. Hospital administrators are showing increased interest in improving the care that dying patients receive. Approximately 70 out of the 70,000 private hospitals in the United States in 1973 had active programs for dying patients.

Prospects for the future

As a result of publicity concerning the issues raised by dying patients, the autonomy that physicians have traditionally exercised in this area is being questioned. The problem that is beginning to be discussed is how patients can effectively exert their rights.

Legally the patient or his agent has the right to refuse life-sustaining medical procedures such as surgery or blood transfusions, but this right is not always respected. In fact, in recent court cases the right has been challenged. In one case a physician sought and obtained a court order permitting him to discontinue treating an elderly patient who was suffering great pain and who did not wish to be treated. In the opinion of the doctor, the patient's refusal was not sufficient grounds to withdraw treatment, and additional legal support was needed. In another case a wife's refusal to permit a new battery to be installed in the pacemaker that was maintaining the cardiac function of her irreversibly brain-damaged husband was overruled by a New York court at the request of the husband's physician.

The second case illustrates one of the difficulties involved in patient control. A patient who is comatose or irreversibly brain-damaged cannot express his wishes concerning treatment. It has been suggested that such patients could exercise some control over these decisions by means of "living wills" prepared in advance of illness. These documents typically state the conditions under which the future patient would like life-prolonging treatment to be omitted and death-hastening treatments (narcotics) to be used. Unfortunately, documents written in advance of illness must necessarily be vague and subject to different interpretations since the medical possibilities are so numerous.

DIED

Their usefulness also depends to a considerable extent upon the prior existence of a good doctor-patient relationship in which the physician is sympathetic to the wishes of the patient. Otherwise, such papers will be ignored. Given the conditions governing the delivery of medical care at the present time, the number of individuals who have family physicians is probably small.

In addition, if the patient is conscious but his physician is unwilling to abide by his wishes, stronger measures than his statement of preference are obviously needed. At least three legislative bills have recently been proposed: the Florida bill on "death with dignity" of 1970, the Idaho euthanasia bill of 1969, and the West Virginia bill of 1972. As of early 1973, none had been enacted into law. The three bills are quite different. The Florida legislation would give the individual the right to execute a document directing that he be permitted "death with dignity and that his life shall not be prolonged beyond the point of meaningful existence." The emphasis in the Idaho euthanasia bill of 1969 is upon active assistance of the dying process: "euthanasia shall be administered to a patient in accordance with the terms of his declaration." Finally, the West Virginia bill would strengthen the patient's right (or that of his relatives in cases of patient incompetency) to refuse "artificial, extraordinary, extreme or radical medical or surgical means or procedures."

All of these bills have defects. The Florida legislation is too general, due in part to the difficulties involved in defining terms such as "meaningful existence," and in the last analysis leaves the interpretation in an individual case to the physician. The Idaho bill is more precise but does not permit relatives a role in the decision if the patient is incompetent. It does, however, provide legal protection for the physician if he ceases treatment or actively hastens death, and permits him to withdraw from the case if he is opposed to euthanasia. The West Virginia bill is too narrow in its formulation of the treatment that the patient may refuse. Regardless of these defects, these bills represent an important beginning toward strengthening the autonomy of the patient.

Additional support for the patient was recently provided by the American Hospital Association, which published as a statement of its national policy a "bill of rights" for patients. Among these rights, the bill includes that of refusing treatment "to the extent permitted by law."

If these or any other proposals are to work effectively under the conditions of modern medical practice, it will be necessary for the public to be much better educated about these problems than it has been in the past. Not only must patients be aware of their rights but decisions that are both complex and emotion-laden can be made effectively only if the patient and his family have considered the issues in advance. Some hospitals have begun to hire ombudsmen to help patients obtain the kind of care they want. Physicians, too, are gradually beginning to realize that these issues cannot be neglected in the training of doctors. Courses in medical ethics are being introduced into medical school curricula. Actual cases are being discussed in terms of the ethical and social advisability of treatment and not just in terms of correct procedures or diagnoses. Thus, the introduction of new medical technology is making necessary a redefinition and restructuring of traditional values, attitudes, and behavior toward death and dying.

FOR ADDITIONAL READING:

Bowers, M. *et al.*, *Counseling the Dying* (Thomas Nelson, 1964).

Crane, D., *Critical Decisions in Medical Practice* (Russell Sage Foundation, 1973).

Fulton, R. (ed.), *Death and Identity* (Wiley, 1965).

Glaser, B. G., and Strauss, A. L., *Awareness of Dying* (Aldine, 1965).

Kübler-Ross, Elisabeth, *On Death and Dying* (Macmillan, 1969).

Oken, D., "What to Tell Cancer Patients: a Study of Medical Attitudes," *Journal of the American Medical Association* (April 1961, pp. 1120–28).

Saunders, C., "The Moment of Truth: Care of the Dying Person" in L. Pearson (ed.), *Death and Dying* (The Press of Case Western Reserve University, 1969).

Sudnow, D., *Passing On* (Prentice-Hall, 1967).

Weisman, A. D., *On Dying and Denying* (Behavioral Publications, 1972).

Weisman, A. D., and Hackett, T. P., "Predilection to Death," *Psychosomatic Medicine* (May-June 1961, pp. 232–256).

The Trouble with Superdams

by Claire Sterling

One after another less-developed countries have sought the benefits that come from giant dams—flood control, navigation, electric power generation, and recreation. But some of these huge projects have generated such unforeseen damage to the environment that according to the author they should never have been built.

On Jan. 15, 1971, the late Gamal Abd-an-Nasser's birthday, Egypt formally inaugurated the world's most famous superdam on the Nile River at Aswan. Built over a span of 11 years at a cost of nearly $1 billion, designed to generate 10,000,000,000 kw of power and store 163,000,000,000 cu m of water, the Aswan High Dam was modern technology's fulfillment of an age-old dream. President Nasser, who changed the course of 20th-century history to get it, had expected it to deliver his people from poverty, flood, and famine, pay for itself in two years and double national income in ten, industrialize his backward agricultural state in the same decade, and turn a vast, empty desert green. He was not the first, or last, leader of a less developed country to be sadly mistaken.

Someday the High Dam may ease the Egyptians' lot. But so far—it has been storing water since 1964 and generating power since 1967—it has cost at least double the original outlay in damage to the country's soil, riverbed, lakes, coastline, and people. This is not just Nasser's fault. The blame must be shared by the West German engineers who designed the High Dam, the World Bank which approved it, the U.S. State Department which agreed to finance it and backed out only for political reasons, the Soviets who finally let Nasser have it—in short, an entire generation distracted by politics and bemused by technology. They were all right about one thing: as a feat of engineering, the High Dam is stupendous. So are many other superdams, built by the dozen during the past decade or two from Pakistan, India, Thailand, Laos, and Iran to Ghana, Nigeria, Rhodesia, Colombia, and Brazil. Yet, as we are just beginning to learn, checking the timeless flow of a mighty river is bound to cause ecological upheaval. As the distinguished Egyptian ecologist Muhammad Kassas observes, the complexity of an ecosystem like that of the Nile "is often beyond the engineer's genius."

Larry Dale Gordon

Florita Botts from Nancy Palmer Photo Agency

Tor Eigeland from Black Star

Civilization was born in the valley of the Nile River (right), where, over thousands of years, silt deposited during the annual floods formed a narrow strip of the world's most fertile soil. Beyond the reach of the floods Egypt was barren desert, like that in Upper Egypt, above.

Hapi's revenge

When a big dam is built, the waters that back up behind it may flood thousands of square miles. Trees, flowers, crops, animals, insects are all drowned. Everything changes: the chemistry of the water, the habitat of river and coastal fish, the kinds and numbers of aquatic plants, the actuarial prospects of disease-carrying insects and the creatures that feed on them (not to mention those they feed on themselves), the weather and wind, the flights of birds, the pressures on the earth's crust and therefore the tendency to earthquakes and landslides, the levels and movements of underground streams and springs, the fertility and salinity of the soil downstream, the depth, speed, and course of the river, the formation of the coast where it empties into the sea, and, not least, the way of life of the people who once lived on the land where the lake was formed.

The changes may be small or momentous, mysterious or all too predictable, for better or worse or both, and months or years may pass before they become apparent. Sooner or later, however, any big river whose delicate natural balance has been disrupted will strike back. The Nile, with a watershed that drains half a continent and a four-thousand-mile course that is probably the longest on earth, is a particularly instructive example.

Unlike many superdams built for hydroelectric power alone, the one at Aswan had another primary purpose: to store enough of the Nile's yearly floodwaters to irrigate existing lands and reclaim still more from the desert that makes up about 96% of Egypt's territory. Water storage

***CLAIRE STERLING** is a special correspondent for the* Washington Post. *Her articles have been published in* The Atlantic *and* Reader's Digest.

was, and is, a vital question for the Egyptians, whose very existence depends on the Nile's bounty. Rain almost never falls on most of their country, though Herodotus tells us that a light shower did fall on Thebes once during the reign of Psammenitos: "an unparalleled event." The instinct to hoard every drop of Nile water has been strong since civilization began on the river's banks; and it is much stronger now that Egypt's 13,928 sq mi of habitable land must support 34 million people, doomed to become 60 million by the next generation. But the water they depend on is not just any kind. It is the kind that rises in the Ethiopian highlands and gathers incredibly rich soil nutrients as it flows down to the sea.

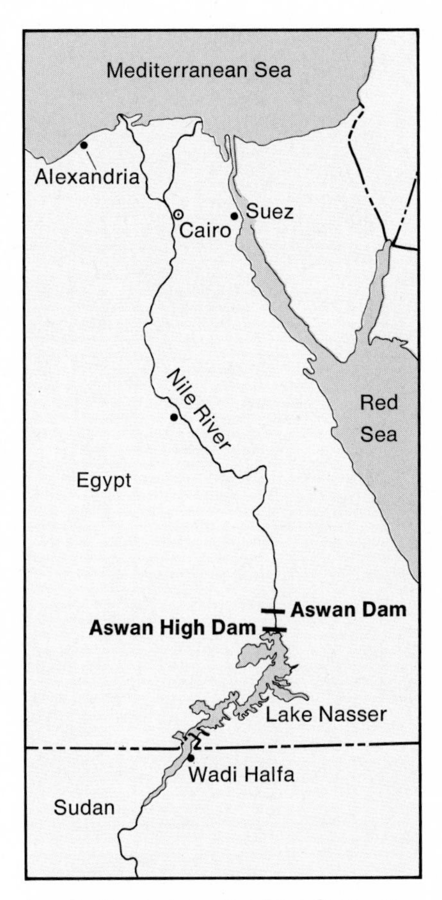

Located just above the First Cataract, the Aswan High Dam has altered the immemorial flow of the Nile throughout its course from Aswan to the Mediterranean Sea. Behind the dam the river backs up to form a vast lake, named for the late Egyptian Pres. Gamal Abd-an-Nasser who believed the dam would bring prosperity to his poor and overcrowded country.

Five thousand years ago, the people of the Nile Valley used to throw a beautiful virgin into the river every August to appease Hapi, god of the Nile, whose miraculous brown waters renewed their parched soil year after year. The annual flood was one of nature's exquisitely balanced wonders. Sometimes it brought too much water, engulfing villages, and sometimes it brought too little; even in normal years about 30,000,000,000 tons flowed unused to the sea. But the water was not wasted. Every sediment-laden drop that emptied into the Mediterranean strengthened the aquatic food chain, nourishing marine life, and maintained an exact balance of salinity. Every year the sediment added a little more land to the Nile Delta, sealed off underground seepage to sweet-water lakes, and shielded the sand dunes that served as dikes protecting the shore from the sea's powerful west-to-east currents. Before reaching the Mediterranean, moreover, the floodwaters flushed away accumulating soil salts that otherwise would have choked the life out of growing plants and dumped them at sea, and they left behind on the floodplain 130 million tons of that soil-building and fertilizing silt that has made the Nile Valley the agricultural marvel it is. Now, because the Aswan High Dam is made of rockfill without sluices, the Nile will never flood Egypt again. The swirling muddy water runs down as far as Lake Nasser, behind the dam, and there the priceless sediment sinks. Downstream from Aswan to the Mediterranean, about 600 mi away, the water flows limpidly clear and always will.

Since the floods stopped forever, in 1967, the aquatic food chain in the eastern Mediterranean has been broken along a stretch of continental shelf 12 mi wide and 600 mi long. The lack of Nile sediment has reduced plankton and organic carbons to a third of what they used to be, either killing off the sardines, scombroids, and crustaceans in the area or driving them away. The 18,000 tons of sardines that were once a fifth of Egypt's annual fish catch have disappeared. At the same time, salinity is rising as the Red Sea waters pour through the Suez Canal without the sweet Nile floodwaters to counteract them.

Erosion is crumbling the delta coastline, exposed now to the full force of marine currents. Some parts of the coast are receding several yards a year, while seawater is starting to intrude into five big sweet-water lakes indispensable to the cultivation of a million acres laboriously and expensively reclaimed over the last 20 years. The silt-free

Florita Botts from Nancy Palmer Photo Agency

The Aswan High Dam, officially inaugurated in 1971, was designed to supply power for Egypt's developing industry and water for its agriculture and to fulfill the age-old dream of controlling the Nile flood. With these benefits, as yet incompletely realized, have come unexpected side effects that threaten to upset the delicate balance of nature in Egypt and possibly in the entire eastern Mediterranean.

water flows downstream much faster, carrying off not just fringes of coast but a quantity of the riverbed. This scouring process is undermining 3 old barrages and 550 bridges built since 1952. To prevent their collapse, the Ministry of the High Dam plans to build ten new barrier dams between Aswan and the sea. The project, known as the Nile Cascade, will cost $250 million—a quarter of what the High Dam cost originally.

Without Nile sediment, many of Egypt's six million cultivated acres need chemical fertilizer already, and all the farmland will need it in the next few years. A team of Egyptian soil experts has found that potassium, phosphorus, nitrogen, calcium, and magnesium will have to be added to the delta's once inexhaustible soil within 4 or 5 years, and copper, zinc, molybdenum, boron, and manganese in 15 or 20. Even apart from these micro- and macro-nutrients, Egyptian farmers are now using 2,350,000 tons of artificial fertilizer annually, an estimated two-thirds of which go to make up for lost fertility and mineral content once supplied by the silt. The cost comes to more than $100 million a year, cutting average income per acre by about a fifth—quite a drop for a farmer earning perhaps $75 a year. And this will be a recurrent annual expense, for eternity.

Moreover, without the annual floods to wash salts away, soil salinity is reaching ominous levels, not just in the heavily waterlogged delta but throughout Middle and Upper Egypt. In the delta itself, the world's most ambitious and costly project of underground tile drainage is already under way; agronomists say that unless this is extended to cover all cultivated land, at a cost of over $1 billion, millions of acres will be reduced to barren rubble in barely a decade.

Reclamation of new land with the dam's waters has been a disappointment. President Nasser had hoped to reclaim 1.3 million ac, but soil studies showed that only about 750,000 ac were suitable for reclamation at anything remotely resembling reasonable cost. Then Prime Minister Mahmoud Fawzi announced a cutback in the official goal in January 1972. The new plan called for reclamation of about 519,000 ac. Reclamation projects thus far add up to well under half a million acres. The one undisputed merit of the dam's stored water has been the conversion of 700,000 ac from flood to canal irrigation. This permits double and even triple cropping that has added 50% to the yearly yield on these lands.

But there is more to the story. Since there are no more floods, there are also no more dry periods to help limit the population of water snails. And water snails are hosts to the blood flukes that carry schistosomiasis, the scourge of Egypt. Schistosomiasis (or bilharziasis) is an intestinal and urinary disease so debilitating that a sufferer can rarely work more than three hours a day. The snail host thrives in placid irrigation canals, where an infected human need only urinate or bathe to spread the microscopic larvae, and a healthy human need only set foot to pick up the infection. One in every two Egyptians suffers from it; one in every ten deaths in the country is caused by it; and where new canals have been constructed since the High Dam was built, the infection rate has shot up from zero to 80%.

Making the desert bloom?

These are terrible afflictions for a nation perennially threatened by famine as its people multiply, bursting the bonds of their narrow green

Don North from Nancy Palmer Photo Agency

Georg Gerster from Rapho-Guillumette

Water provided by the High Dam has permitted double cropping and improved yields on some land (left), but the increases in productivity have been far below those visualized when the dam was being planned. Placid irrigation canals such as that above are ideal breeding grounds for the snails that carry the disease schistosomiasis.

K. Scholz from Shostal Associates

Lake Nasser was expected to be full by 1970, but two years later it held less than half the planned amount of water, and some estimates placed the completion date as much as 200 years in the future. Planners apparently failed to take into account the high evaporation rate and loss by seepage through the porous rock.

valley only to be driven back by an implacable desert. Yet they might bear it all if the High Dam could give them the one thing they prize most: limitless water. It has not. For all the billions of dollars spent and yet to be spent, there is less water flowing downstream from Aswan today than there was before the dam was built.

Nilologists had estimated that a minimum storage requirement of 163,000,000,000 cu m would be needed to provide for irrigation and reclamation, river transport, and hydroelectric power and to guarantee against drought. Lake Nasser, forming behind the High Dam, was supposed to hold that much, and it should have filled by 1970. But it is not yet half full, and according to Abdel Fattah Gohar, Egypt's foremost limnologist, it may not fill up for 200 years, if then.

One reason, presumably, is an elementary error in arithmetic. Calculating probable evaporation losses in a large body of water is supposed to be child's play, even if the body is as large as Lake Nasser, some two thousand square miles, and is located in the hottest and driest corner of the globe. Yet while the High Dam's planners allowed for a colossal evaporation loss of 10,000,000,000 cu m per year, the figure turned out to be 15,000,000,000. What the planners apparently overlooked was the effect of a high wind velocity over such a vast expanse of water.

Nor is that all. Three quarters of a century ago, an eminent hydrologist named John Ball demonstrated that, for some 200 mi between Wadi Halfa and Aswan, the Nile cut across an immense underground bed of water, part of an aquifer of porous Nubian sandstone 386,000 miles square, underlying the Libyan Desert from Egypt and the Sudan

R. Eith from Shostal Associates

Waterweeds form a solid mat in the Zambezi River at Rhodesia's Kariba Dam. Tough, fast-growing weeds, like this "Kariba weed" or water lettuce, threaten to block waterways and clog the turbines of hydroelectric projects throughout the tropics.

to Chad and Libya. Incalculable quantities of water were feeding into the Nile, he said, from the permeable bed beneath it and the sandstone lining its western banks. Then, in 1902, the first Aswan Dam was built. In a remarkably short time, the age-old movement of this groundwater was reversed. Under counterpressure from the first Aswan reservoir, the water began to move elsewhere through countless fissures in the sandstone, and a tremendous quantity also began to seep out through the porous rock from the reservoir itself. From 1902 to 1964, when the new High Dam was sealed, the Aswan reservoir stored some 5,000,000,000 cu m of water a year for seasonal use. At the same time, the Egyptians lost 12,000,000,000 cu m a year from the Nile through seepage and reversed underground flows.

So far, no one knows exactly how much more than that is escaping from Lake Nasser. But with a far bigger dam, an incomparably bigger lake, that water-bearing bed, those sandstone banks, and that colossal evaporation rate, the lake appears to be losing at least 30,000,000,000 cu m a year, more than a third of the water flowing into it. Downstream, the Egyptians are getting almost 10,000,000,000 cu m less than they used to: 53,700,000,000 instead of 63,000,000,000. Things may improve in time, as the fine clay suspended in Lake Nasser's waters starts to seal the porous banks. But if the fissures in the banks are large and the escaping water finds an outlet to the sea, to another lake, or to a land depression, it could flow downhill forever.

It need hardly be added that these waters are not bringing the surrounding desert into bloom. At the Lake Nasser Development Center in Aswan, devoted Egyptian and UN scientists continue to study and

Hamilton Wright from Photo Researchers

Nile waters backing up behind the Aswan High Dam inundate a Nubian village. Some 90,000 Egyptian and Sudanese peasants had to be relocated from land that was flooded when the High Dam was sealed.

map the dry, treeless land bordering the new lakeshore, against the time when the lake may fill. But that time is remote. "Some day," an Egyptian geologist remarked, "somebody will make use of our soil maps. Not now. The turbines have to run; we can't spare another drop of water."

Yet all but 2 of the High Dam's 12 magnificent turbines are idle. The grids have been built, strung across 550 mi of desert to the delta, and Cairo is ablaze with light. But the country as a whole cannot yet absorb even a third of the dam's 10,000,000,000-kw capacity.

Unwanted side effects

The Aswan story is worth telling at length because it embodies so much of what has happened elsewhere. But the trials of a superdam owner do not stop there.

Once, for instance, the prospect of vast new lakes evoked extravagant dreams of fish. The dream may even come true for a while, especially in the tropics. At first, rotting vegetation makes the water black and so short of oxygen that most of the fish die off. But then the water clears and other species come along to feed on the decaying plants, decomposing animals, and each other. In Thailand, 33 species disappeared from the lake behind Ubolratana Dam within a year of its formation, while the population of predatory murrals doubled. In Ghana's Lake Volta behind the Volta Dam at Akosombo, six types of mormyrids became extinct within a year while year-old cichlids and *Lates*—species that had formerly lived in the river's still pools and backwaters—showed up in fishermen's nets for the first time. The fish catch soon rose to 60,000 tons a year. Yet this largesse does not always last, and sometimes it does not come at all.

Lake Nasser was formed in the same month and year as Lake Volta, and it is larger and much richer because of the marvelous Nile silt that drops into it, but it has yet to yield more than a tenth of Lake Volta's fish catch. Fish landings in Lake Volta itself fell by one-fifth in 1971, and the lake behind Rhodesia's monumental Kariba Dam appears to be getting nearly fishless. Formed five years before Lakes Volta and Nasser, Kariba seemed so promising that experts predicted a 20,000-ton catch in short order. Yet the catch declined to 4,000 tons in 1963 and to half that by 1964. By 1967 it was so small that the corps of 2,000 fishermen had dwindled to 500.

Experts from the Food and Agriculture Organization (FAO), summoned to the rescue, can only guess at some of the reasons: overfishing, undercropping, poor gear, too many drowned trees, too many predatory fish. The one thing they are sure of is the waterweed.

Waterweeds can be very pretty, especially the lush pink water hyacinth, but not to anyone who tries to get rid of them. Wherever the weather is warm enough, they reproduce with incredible rapidity, forming mats thick enough to support the weight of a man. Today waterweeds are clogging rivers, lakes, canals, and even ponds in several southern U.S. states, most of central and South America, nearly all of Africa from Zaire to Rhodesia to Tanzania and the Malagasy Republic, and eastward to India, Pakistan, Indonesia, Thailand, Vietnam, and Australia. The whole Congo River has been clotted with them. Half the flow of the White Nile—14,000,000,000 cu m—is lost every year in the Sudan's great Sudd swamps through their transpiration. Thailand's three newest man-made lakes are failing to fill for the same reason. Waterweeds block irrigation ditches and stall riverboats. Fish trapped under them die for lack of food and oxygen. And because the weeds can easily become caught in a dam's turbines—the water lettuce very nearly put the Kariba Dam out of business—almost every hydroelectric project built in the tropics since World War II is menaced by them.

Waterweeds such as the hyacinth have survived every effort to wipe them out: boats equipped with saws, leaf-eating beetles, weed-eating fish, herbicides poured, painted, and sprayed from boats, the shore, and the sky. If only for that reason (there are others), development planners have learned not to count on all that fish—or on all that water. Lake Nasser provides just one example of how water can get away. Through transpiration, water hyacinths can use up ten times as much water, meter for meter, as Lake Nasser's evaporation rate. Silting cut the storage capacity of Algeria's Habra reservoir by 58% in 22 years and of Colombia's Anchicayá reservoir by 21.4% in 21 months; it now appears that Pakistan's $600 million Mangla Dam in the Himalayan foothills, planned to last 100 years, will silt up in 50. And any reservoir with big cracks or open passages in the earth around or under it can leak. The United States has had to abandon several

K. Scholz from Shostal Associates

Temple at Abu Simbel is shown here as it stood for more than 3,000 years, from the time it was carved out of the Nile cliffs in the reign of the pharaoh Ramses II. Threatened with submersion by the Aswan High Dam, the Abu Simbel temples were cut from the solid rock and reconstructed on higher ground.

for this reason: Cedar Lake reservoir in Washington state, McMillan and Hondon in New Mexico, Tumalo in Oregon, and Jerome in Idaho.

On the other hand, if the backed-up waters do not find some way to escape, they create enormous fluid pressures that might well lead to earthquakes. Scientists are still not sure of the precise relationship between an immense new body of water and seismic disturbances, but they do point out the likelihood that one as large, say, as Lake Volta, covering 3,088 sq mi, might bring about sinister changes in the relative positions of land and sea. Papers presented at a recent international symposium on man-made lakes in Accra, Ghana, suggested that substantial changes in the loading of underlying rocks there could mean "a tendency for the lightened areas to rise and more heavily laden areas to sink," possibly resulting in "faulting and concomitant earthquakes." Such suspicions have been deepened by the alarming increase in earthquakes around Denver, Colo., in the 1960s. Geologists determined that injections of liquid wastes from a U.S. Army poison-gas plant into a deep well caused an old locked fault to slip. Since the injections began in March 1962, 1,700 earthquakes have been recorded in the Denver area, most of them with epicenters within a five-mile radius of the well. Lest this still appear to remain in the realm of theory, it should be noted that an earthquake did occur at India's

Hamilton Wright from Photo Researchers

Pieces of the Abu Simbel temples are raised to their new site (left). A worldwide fund-raising campaign was waged to finance salvage of these and other Nubian archaeological treasures, such as the temples on the island of Philae (below).

Koyna Dam in 1967, with a magnitude of 6 on the Richter scale, killing 200 people.

A cost in more than money

For a number of reasons, then, it is not often that a superdam can turn a desert green. More often than not, indeed, such dams drown more and better land than they can reclaim.

It is impossible to estimate in money the value of the land, plants, timber, ancient monuments and artifacts, villages, and towns that now lie at the bottom of man-made lakes. The ancient site of the Abu Simbel temples in Egypt and the large Sudanese town of Wadi Halfa are among them. So are some of the richest cocoa farms in Ghana, fertile alluvial lands along the Zambezi's banks in Zambia, giant stands of teak in the Bandama Valley of Ivory Coast. Millions of people have been forced to move as a result of dam construction: 100,000 in Egypt and Sudan, 80,000 in Ghana, more than 100,000 in Ivory Coast, to list only a few. Governments impatient for progress regard them as "an expensive nuisance," in Thayer Scudder's words. Even when their governments spend a thousand dollars a head to resettle them—usually moving them from good bottomland to stony, arid ground higher up—they are miserable.

Florita Botts from Nancy Palmer Photo Agency

I have visited displaced Thais, Baoules in the Bandama Valley, Ghanaians in their "gone elsewhere" villages, and Nubians of Upper Egypt in their dreary cement huts at Kom Ombo. Some were better off than others, but nearly all showed signs of the classic resettlement syndrome: they were unself-reliant, passively aggrieved, and dispirited.

The Baoules were still in their thatched village of Angouassi waiting for the rising waters to overtake them, not quite believing that it would happen. They were full of complaints and premonitions. Their elders had visited the new settlement prepared for them. It was much grander than the old one, with $1,600 houses and corrugated tin roofs, but it would be hot, they said, since bulldozers had knocked down all the trees. They thought there would be no food until the new crops came in (though in fact they were to be fed for two years by the FAO's World Food Program). They had been urged to learn how to fish, but they were afraid of the water. They had been advised to raise cattle on the poorer soil around their new homes, but they did not care for cattle. What they did care about was land, and they were to receive only an acre apiece, half as much as they had now.

The Thais near Khon Kaen, in northeast Thailand, had already been through it all. Their rice fields were at the bottom of Nam Pung reservoir and their resettlement money had been spent. For two back-breaking years they had cleared the harsh upland, with no help from the government and almost no water. About one in five had drifted away, and many had joined the Communist guerrillas. The rest were trying to grow soybeans and kenaf, a kind of jute. Some were beginning to make out, but others never would.

Everywhere the wrench of departure had left lasting scars. People forced to leave the land where their families had lived for centuries may sicken and die of unbearable psychological shock. Mortality rates rose sharply among Ghanaians evacuated when Lake Volta formed, especially among the very young and the old. Or sickness may occur for other reasons.

Whenever a superdam has gone up in the tropics, it has been followed by an explosion of water-borne disease. Even a small dam can have this effect, as Ivory Coast has learned. Within a year of the completion of a small dam at Bia in 1964, there were epidemics of river blindness (800 cases), sleeping sickness (200 cases), and schistosomiasis (four out of every five inhabitants). Fortunately, the area was sparsely populated. When Lake Kossou began to form, however, World Health Organization officials braced for "a very serious threat" of sleeping sickness, "endemic" schistosomiasis and malaria, a "special danger" of yellow fever, a "violent outbreak" of guinea worm, and chronic "excreta illnesses" such as dysentery, cholera, and typhoid, brought on partly because people accustomed to relieving themselves in the bush do so in the water instead.

Much of this sickness results from changes in the environment that favor the breeding of disease carriers. The black fly *Simulium,* which causes river blindness, requires fast-moving water, such as that flowing through a dam's sluices. Nearly all other carriers of tropical water-borne diseases, however, find ideal breeding conditions in placid lakes with shallow, twisting shorelines. These include not only the snail that carries schistosomiasis, but the mosquitoes that bear malaria, yellow fever, dengue fever, and elephantiasis, the water flea host of the guinea worm that can grow a yard long inside the human body, and the Chinese liver fluke that hides inside raw fish (a favorite northern Thai dish is raw fish dipped in fresh cow bile).

The worst of a bad bargain

Supposedly, the land enriched by some dams, at least, should make up for such incidental human suffering. Not every dam withholds silt like Aswan, and some do impound enough water to leave farmers knee-deep. But irrigation is no simple matter. In Thailand, for instance, irrigation made possible by the Nam Pung Dam should theoretically permit rice farmers to grow two or three crops a year, with double the yield. But the poor lateritic soil in the northeast is soon exhausted under intensive cultivation; the monsoon-farmers are not accustomed to the work irrigation farming requires; and the farm extension service is hopelessly inadequate. All in all, Thailand's Royal Irrigation Department has spent $180 million to develop irrigation in the last three years and has only about 400 successful acres to show for it.

In this and several other ways, the Nam Pung Dam is an illuminating case. Built in 1966 at a cost of $60 million, it now earns $1 million a year by generating electricity and another $500,000 from a one-ton fish catch (originally projected at eight tons). Irrigation was to have brought in profits of $3.5 million a year, but these have yet to materialize. When and if they do, 10,000 families living on the irrigated land stand to benefit; 4,000 families were forced to move to the barren uplands when the lake was formed.

Even in countries beautifully equipped for large-scale irrigation, terrible accidents can happen. Pakistan, in its Indus River basin, inherited the most extensive irrigation system on earth from the British. But it faced calamity a few years ago when waterlogging and salinity were found to be seriously damaging 5 million of its 23 million canal-irrigated acres, with 50,000 to 100,000 ac a year passing out of crop production altogether. The cost of repairing the damage by canal remodeling, drainage works, and tube wells has come to just over $1 billion, roughly the cost of the Tarbela Dam, which itself represented about half the cost of the Indus Basin development project.

The real payoff, of course, is supposed to come from hydroelectric power, and the installed capacity of a superdam can be enormous. But installed capacity does not necessarily equal power consumed, as the Egyptians, among others, have discovered. Hydroelectric power is cheaper than thermal or nuclear power, but only if a dam is used to capacity over a span of many years (and even this does not take into account the social and ecological costs). To use the power a dam can supply, a less developed country must also provide everything an ad-

Larry Dale Gordon

Aerial view of Guri Dam on the Caroni River in Venezuela gives some hint of the immensity of modern superdams and their potential, through sheer size alone, for changing the environment. Guri was designed solely to generate electricity for the development of Venezuela's mineral-rich Guiana region. Its reservoir includes the flow from the world's highest known waterfall, Angel Falls, in the remote Guiana Highlands.

vanced industrial state already has: grids, factories, transport, communications, trained workers, seasoned managers, markets. This can multiply the original investment in the dam by a factor of perhaps 10 or even 20.

Second thoughts

Curiously enough, such calculations simply were not made until recently. Engineers honestly did not know they were committing ecological sins. The rich countries that put up the money sincerely thought of themselves as benefactors. The poor countries were suitably grateful. The displaced people, plainly, were expendable.

But times have changed. Everywhere, I have found officials eaten by doubt. In Yamoussoukro, Ivory Coast, operating center for the Kossou Dam, harassed social workers coping with refugees dreamed of putting up stickers saying, "Is This Dam Really Necessary?" The Volta Dam is an economic success because Valco Aluminum provides a built-in market for its power, but the curse of the refugees still lies heavy. "I sometimes wonder if the whole thing was worth it," says Letitia Obeng, co-manager of the UN Development Program-Ghana Volta Lake Research Project. In Egypt, where the spell of the High Dam is still strong, only the elite know the extent of the damage. Privately, they admit that probably they would not build it if they had it to do over again.

Disenchantment is strongest in Bangkok, where for nearly two decades 500 UN experts have been working with the four states of the Mekong Committee (Laos, Thailand, Cambodia, and South Vietnam) to harness one of the world's last great undammed rivers. Not long ago the U.S. Bureau of Reclamation retired from the field in confusion. Since 1964 it had worked under extremely trying conditions to determine whether or not the first of four or five projected mainstream dams on the Mekong, at Pa Mong, would be technically feasible. The answer was yes, but by the time it was reached, the question had changed. "Who needs Pa Mong, really?" demanded Kasame Chatikavanij, director of Thailand's Electricity Generating Authority, in an interview. "Of course we could use the power eventually; we have a very nice growth rate. But we are not rich. I would rather put our money into smaller projects, a little at a time, as the need arises, and not gamble it all on one big throw—like Aswan." Some of his countrymen may still disagree. But such unclouded faith in Pa Mong as still exists is to be found more among would-be givers than receivers.

Not all poor countries are turning down superdam offers, but they are learning to ask more questions. After all, thermal power stations can be built for a fraction of what a dam costs, and small ones can be built one at a time, as consumption expands. Nuclear power plants, cleaner though costlier, can also be built in fairly small units. But once a giant dam is built it is *there,* evaporating, transpiring, leaking, silting up, weeding over, scouring the riverbed, driving away or killing off the fish, drenching or salting the soil, breaking natural laws, uprooting people.

A thousand years ago, the Khmer emperors of what later became Cambodia built a splendid system of hydraulic works in the Mekong River basin, enabling their farmers to grow four crops a year and their kingdom to grow very rich. Knowing nothing of ecology and electricity, they were not even aware of the trouble they were avoiding. All things considered, the contemporary Khmers—not to mention Pakistanis, Indians, Egyptians, other Africans, and South Americans—might do worse than to fall back on such ancestral wisdom.

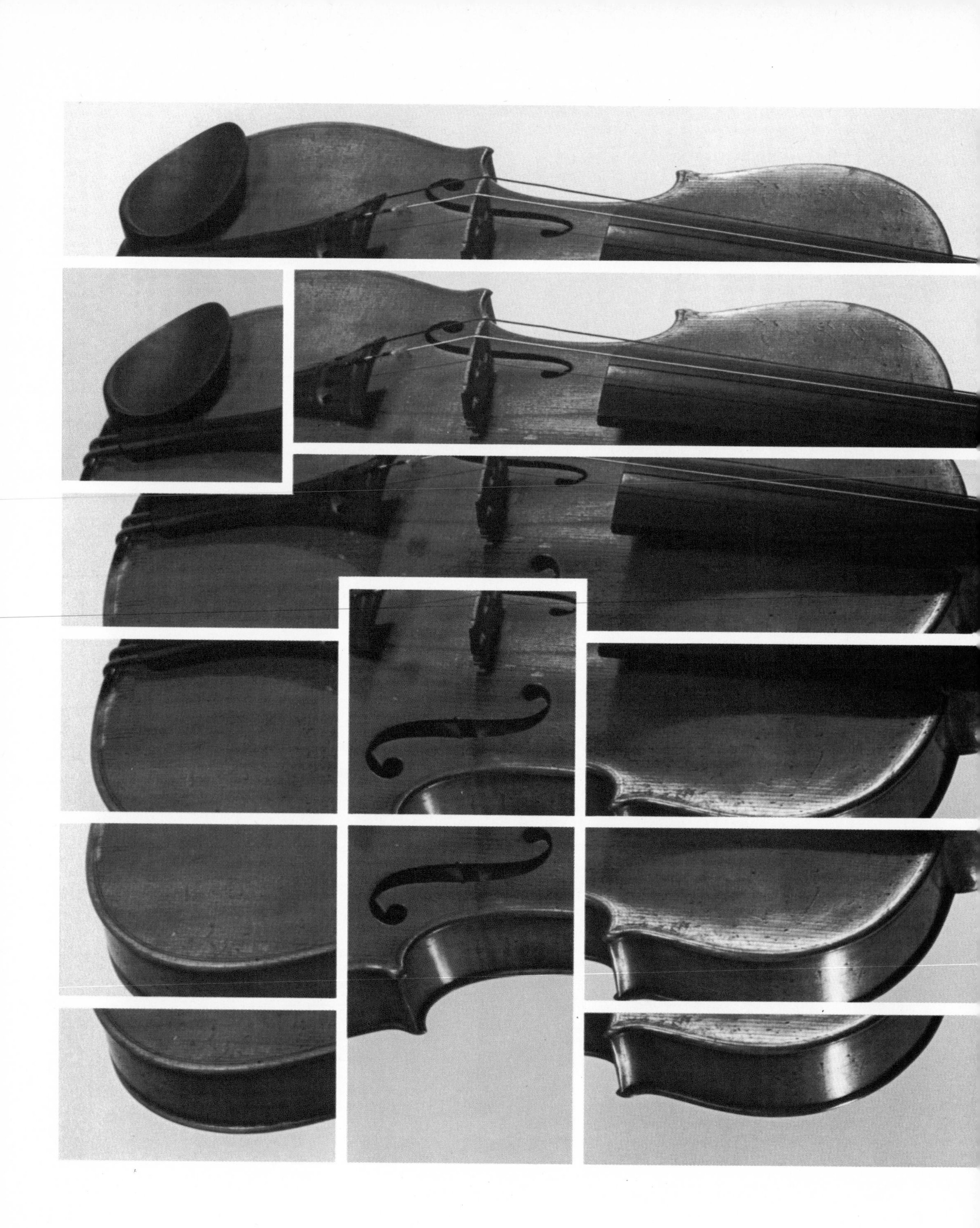

Physics and the Perfect Sound

by Michael Kasha

The violins and other stringed instruments of such old master craftsmen as Antonio Stradivari have long seemed the ultimate in perfection. Some scientists now believe, however, that modern physics offers a way to improve on them.

Musical instruments through the ages have contributed as much to the enjoyment of life in all human cultures as have any other man-made devices. Stringed instruments have risen to an especially prominent role. The violin—supported by the viola, violoncello, and double bass as bow instruments—is the foundation of the symphonic ensemble. The guitar has undergone a resurgence of popularity both as a classical concert instrument and as the ubiquitous folk instrument of the day.

These instruments have evolved over millennia by a combination of accident and the intuition of such superb virtuoso craftsmen as Antonio Stradivari. Have they now reached the ultimate form of acoustical perfection? Can scientists using the methods of modern physics improve on them in any way?

The development of classical instruments

Through museum collections of surviving instruments of the Sumerian, the ancient Egyptian and Chinese, and other cultures, we learn of the high attainment of craftsmen, and the important role of musical instruments in ancient societies. In archaeological remains—stone carvings, sculpture, cave and tomb frescoes—we see further evidence of the diversity and antiquity of folk and professional instruments. Although only the gross structure of ancient instruments can be deduced from this evidence, nevertheless we can build up a picture of the slow evolution of their forms. Two typical and unusually interesting old sculptures serve as examples of the kind of analysis possible.

One of the earliest authentic representations of a guitar appears in a 1st century A.D. Hellenistic sculpture frieze found in a Buddhist monas-

Courtesy, State Hermitage Museum, Leningrad

One of the earliest representations of a guitar appears in a 1st-century A.D. sculpture frieze near Termez in Soviet Central Asia. The arrangement of crescent-shaped sound holes on both sides of the strings persisted for more than 1,500 years afterward.

tery in Ayrytam, near Termez, in what is now the southern tip of Uzbekistan in Soviet Central Asia. The towns and way-stations of central Asia were built up from the rich profits of the old silk routes from the Orient to Persia and the Mediterranean countries, and their artistic life developed a rich hybrid of Oriental, Indian, Mongol, Persian, and Hellenistic cultures.

It is surprising to find such an early representation of a true guitar. In 13th-century Spain the guitar appears as a four-stringed instrument similar to present-day models. In fact, the Old Spanish word *quitarra* is thought to be derived from the Sanskrit *catur-tar* (literally, "four strings"). The Termez guitar clearly shows four strings attached to a terminal bridge. The instrument shows characteristic incurved sides, and must have had a comparatively flat body. Also characteristic is the set of crescent-shaped sound holes arrayed on the face of the instrument on both sides of the strings. This type and disposition of sound hole persisted on Mediterranean and other European instruments for more than 1,500 years afterward. It is evident from the location of the bridge near the end of the instrument that the Termez guitar would have exhibited a good treble response, with a highly quenched bass response.

Another especially interesting stringed instrument representation is the three-string Viol of St.-Denis, from the great tympanum of the abbey church of St.-Denis north of Paris. Acoustically, the Viol of St.-Denis is particularly interesting in relation to the evolution of European bow instruments. Such three-string viols are often seen in representations in church sculpture and frescoes. Evidently they were played either by plucking or bowing. The normal (vertical) disposition of tuning pegs persisted for a long period, as did the D-shaped sound-hole arrays. The strings overpass the bridge in typical bow-instrument fashion and end in a typical tailpiece. The acoustical features of greatest interest are the relative locations of the bridge, the sound holes, and the incurved sides. In this instrument the bridge is placed at the stiffest, most restricted part of the top, the incurved side cusp. For this reason the Viol of St.-Denis would be a very high-pitched, even shrill, instrument with negligible middle-range and bass fundamental response.

As we follow the evolution of bow instruments through sculptures, frescoes, miniatures, paintings, and manuscripts for 500 years from the medieval to the late Renaissance periods, we see the slow optimization of the bow-instrument mechano-acoustical structure. The D-holes become C-holes and then *f*-holes. The incurved sides lose their cusp-character and become gracefully curved and more deeply cut for bowing facility. And steadily but surely over 500 years the sound holes are moved to parenthetically occlude the bridge, finally giving relaxation of bridge vibration to counteract the growing stiffness of the top under the incurving sides. At this point, the instrument remained to be acoustically perfected internally by accident and by craftsman's intuition, until it became mechanically asymmetric and optimized bass to treble response over the whole instrument range.

***MICHAEL KASHA** is professor of chemistry and director of the Institute of Molecular Biophysics at Florida State University. His publications include* Complete Guitar Acoustics.

Model physics

When confronted with the problem of understanding the mechanical action of a stringed instrument, a physicist is first tempted to consider subjecting the complex and beautiful object to minute analysis in its final perfected state. But trying to follow the complex action of the evolved stringed instrument generally prevents an analysis in terms of elementary principles. Instead, researchers have used a "model physics" approach, attempting to identify the essence of each component and its action by first constructing a primitive model that corresponds to the essential mechanical-acoustical action. All aesthetic refinement is omitted. Next they analyze the expected performance of the model theoretically and compare it and its expected performance with the real instrument in all its complexity and aesthetic refinement. When that has been done, one is in a position to understand if the real instrument is optimized with respect to the model. If not, structural changes can be made in the real instrument and then tested, both by laboratory measurement and by virtuoso performance and audition, to determine whether a desirable change has occurred.

Using model physics to analyze a guitar (figure 1), we visualize as Oscillator I a string whose length is defined by the peghead nut and bridge saddle, coupled via a terminal bridge (having mechanical impedance Z) to a vibrating diaphragm, Oscillator II. The diaphragm is considered to be approximated by a vibrating circular plate clamped at the edge. Of course, the diaphragm represents the face of the lower bout of the guitar, and only partially represents the vibration of the guitar body, although a principal vibrating part. Finally, Oscillator III is the resonant cavity or air chamber inside the instrument, and is driven by body vibration modes which are in turn driven principally by the diaphragm.

In the guitar model we recognize that the vibrating string will tend to impart a rocking motion to the bridge, which will tend to reinforce particular normal modes of vibration of a circular plate or diaphragm. However, since we are dealing with a wood diaphragm with grain lines (nonisotropic) and peculiar elastic properties, we expect nonlinear effects to be important. In other words, we expect the normal, independent vibration modes of an ideal circular plate to become coupled to each other in the case of the real top plate of the guitar.

In a violin or bow-instrument model (figure 2), we envisage as Oscillator I the string, whose length is defined by the peghead nut and the bridge, coupled to Oscillator II, the top plate of the instrument, via the overpassing bridge. As an approximation to the vibrating top plate we choose a vibrating elliptical plate clamped at the edge. We consider in the model that the stiffening of the top plate by the incurved sides is compensated by the relaxation offered by the *f*-hole cuts. Obviously, we abandon all complexity in the model, including the carved contour and thickness grading of the top—part of the highest art of the violin maker. Finally, as Oscillator III we have the resonant air cavity enclosed within the body of the instrument.

Three-string Viol of St.-Denis appears on the great tympanum of the abbey church of St.-Denis, north of Paris, built in the 12th century. Played either by plucking or bowing, the viol had a high-pitched, shrill sound because the bridge was placed at the incurved side cusp, the stiffest and most restricted part of the top.

Figure 1

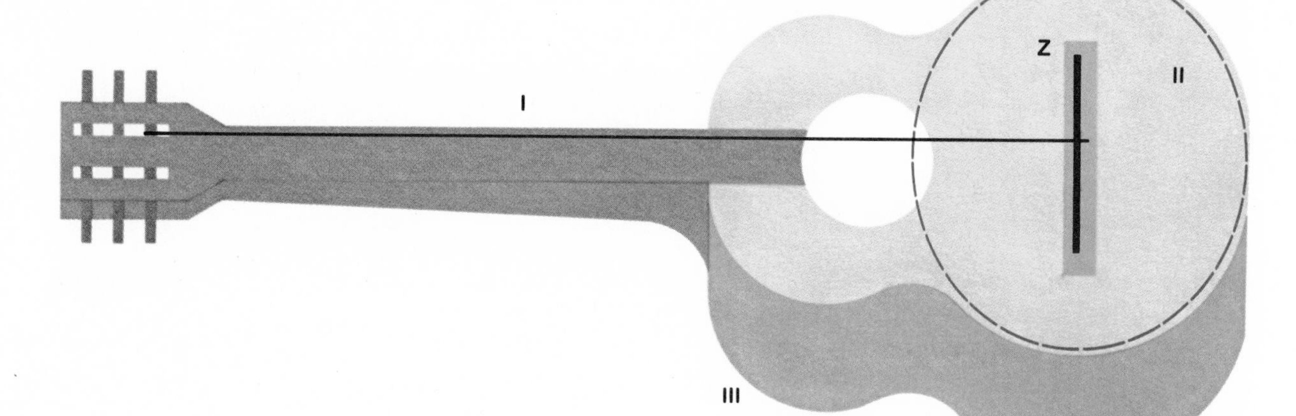

Figure 1. Guitar is diagrammed as analyzed by using the model physics method. Oscillator I is a string connecting the peghead nut with the bridge saddle and is coupled to Oscillator II, a vibrating diaphragm, by a terminal bridge having mechanical impedance Z. Oscillator III, the air chamber inside the instrument, is driven by body vibration modes generated mainly by the diaphragm.

In the violin model we consider the normal modes of vibration of an elliptical isotropic plate. The bridge action is principally a rising-and-falling motion, pivoting on the treble foot because of the internal stiffening provided by the sound post. The feet of the bridge are normally made equivalent, neglecting mechanical impedance considerations. The rising-and-falling action of the bridge would tend to reinforce favored normal modes of the elliptical plate in the model. Since the top plate is not an ideal isotropic plate in the real instrument, the modes of vibration will not be independent and will all be coupled together. Thus, unfavorable modes of vibration would still be driven, although less efficiently than those conforming directly to the primary vertical motion of the bridge.

Physics of coupled oscillators

Once we have reduced the complex musical stringed instrument to an essential model, we can diagram the nature of the coupling model involved. A systems analysis model for a stringed instrument allows us to examine the elements of theory that will govern ultimate mechanical and acoustical properties of the instrument. Traditionally, discussions of the physics of stringed instruments offered mainly information about the physics of string vibration. More recently, some attention has been given to detailed physical or engineering measurement of output curves of real instruments, or their components. But an analysis of the simple physical theory of coupled oscillators is essential to the understanding of the mechanical action of stringed instruments. Only by understanding the fundamental action of the elemental components of the coupled oscillator system can we expect to understand the variations in the output spectrum and expect to control them.

The analysis begins with Oscillator I, the vibrating string. The essential physics of the vibrating string is known to every music student. Generally, a vibrating string is a harmonic oscillator; the frequencies of higher harmonics (overtones) are simple integral multiples of the

Figure 2

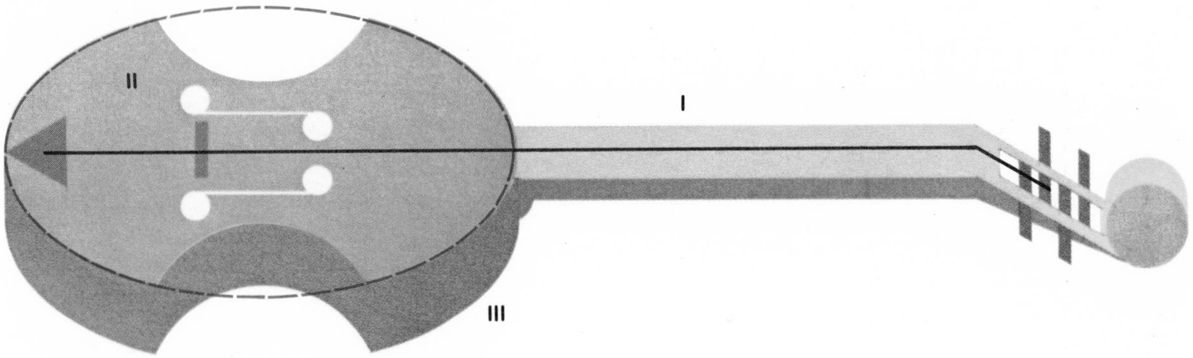

Figure 2. Diagram of a violin according to model physics shows Oscillator I as the string connecting the peghead nut with the bridge. It is coupled to Oscillator II, the vibrating top plate of the instrument, by the bridge. Oscillator III is the resonant air cavity enclosed within the violin.

frequency of the first harmonic (fundamental). When heard together, such integrally related harmonics make a consonant sound. Strings made of proteinaceous materials, such as animal gut, and nylon tend to be good harmonic oscillators. Metallic strings, vibrating under very high tension, tend to make nonharmonic oscillators; the higher harmonics of these strings depart positively from integral multiples of the first harmonic, producing a twangy sound. String vibration alone is largely inaudible, mostly because of air circulation around the vibrating string.

The bridge mechanically couples the string motion to instrument soundboard motion. The efficiency of this coupling is governed by mechanical impedance, the complex ratio of the applied alternating force to the resulting velocity experienced by the system at the point at which the force was applied. Mechanical compliance is the reciprocal of the mechanical impedance. In practical terms, if we wish the bridge to transmit a significant vibrational amplitude to the soundboard, the mechanical compliance must be high. But the mechanical impedance increases with frequency. Therefore, the bridge should be asymmetric from bass to treble end to provide for large-amplitude, low-frequency motion on the bass side and, at the treble side, very low amplitude with high frequency. Since guitar bridges and bow-instrument bridges classically are symmetrical, mechanical compliance is not optimized.

Turning to Oscillator II, the vibrating top plate, a study of the normal modes and frequencies of clamped isotropic plates quickly reveals that they are nonharmonic oscillators of a special type: they have nonintegral, irregular frequency ratios. In practical terms, a vibrating diaphragm or plate, circular or elliptical, of any material—metal, elastic quartz, wood—makes a drumlike, dissonant sound when several nonintegrally related frequencies are heard together. Moreover, the characteristic frequencies of a vibrating diaphragm are fairly widely and irregularly spaced. How then can the soundboard render audible the

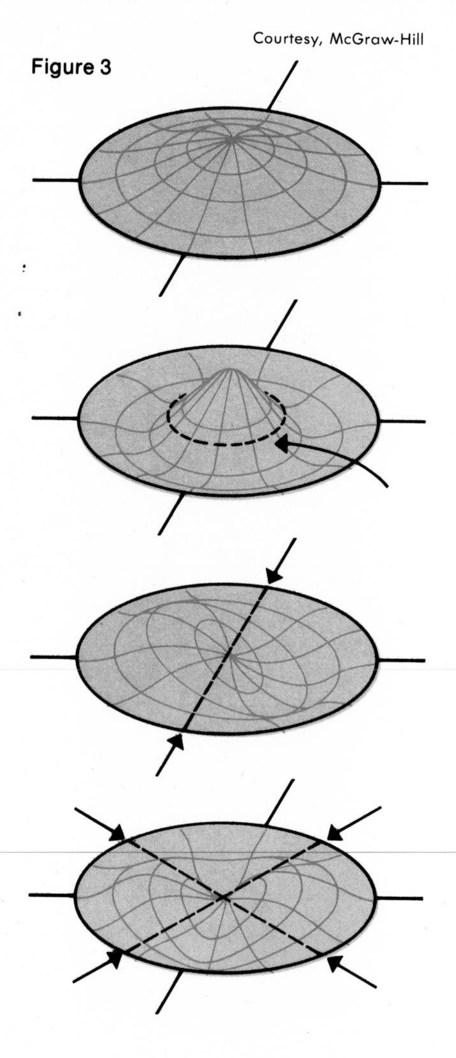

Figure 3. Schematic diagrams indicate the nature of the vibrational distortion for some of the normal modes of motion of a clamped circular plate. The dotted straight lines indicate linear nodes (lines of immobility), while the dotted circles delineate circular nodes. The solid lines show the direction and extent of the distortion.

musical consonances of string vibrations? Furthermore, since many string frequencies fall between the existing diaphragm frequencies, those notes are weakly reproduced. There is a natural tendency to generate an output spectrum of very uneven intensity distribution when a string vibration is coupled to a "singly-structured" or simple soundboard-diaphragm. However, the clamped-plate frequency spectrum is diameter-dependent; that is, frequency varies as the inverse square of diameter for the case of a clamped isotropic circular plate. There has been an intuitive historical tendency to use this dependence in evolving an asymmetrically structured diaphragm as a soundboard—at least in the bow instruments, if not in guitars. We shall return to this topic in the following sections on the physics of vibrating plates and real instruments.

In regard to Oscillator III, the resonant cavity, the physics of this component is as well understood as the physics of vibrating strings, and applies in unidimensional form to the organ pipe. A cylindrical resonant air cavity has a simple harmonic spectrum, with a powerful first harmonic, and rapidly descending intensities for the higher harmonics.

The resonant cavity of a stringed instrument is three-dimensional, and if the instrument box were a rectangular parallelepiped we could resolve the cavity resonances into independent modes in the Cartesian coordinates x, y, and z. (A Cartesian coordinate is the distance of a point from either of two intersecting straight-line axes measured parallel to the other axis.) Each of these would then have independent harmonic progressions whose first harmonic would occur at a frequency inversely proportional to the Cartesian dimensions. However, the complexity of the instrument box shape precludes resolution by Cartesian coordinates, and nonlinear coupling of the air cavity modes complicates the output spectrum of the complete instrument.

The soundboard as a vibrating plate

A soundboard of a musical instrument can be analyzed theoretically by analogy to an idealized vibrating plate of suitable geometry. After studying the normal modes of motion of a simple isotropic clamped plate of a particular shape, we then can venture to understand the actions of a plate of varying dimension and, finally, structured (ribbed) plates with introduced asymmetry.

In a vibrating string there are simple point nodes (points of immobility) for the normal modes of vibration, but in a vibrating plate there can be both linear and circular nodes. The actual vibrations of an ideal clamped circular plate can be resolved into independent (normal) component vibrational modes, just as a general vibrational motion of a string can be resolved into independent string harmonic modes. The nature of the vibrational distortion for some of the normal modes of a clamped circular plate is indicated by the schematic diagrams of figure 3.

An experimental study of circular plate normal modes by the British

Courtesy, Mary D. Waller

Figure 4

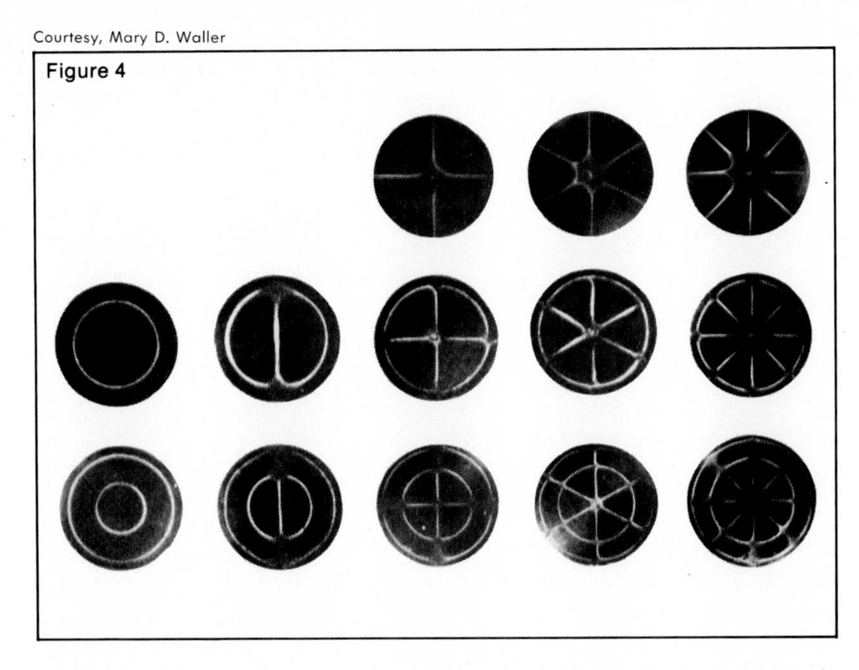

Figure 4. Normal modes of vibration of a delicately suspended circular plate are photographically summarized. The plate is caused to vibrate by tickling an antinode, a region of maximum vibration, with dry ice. Fine sand on the face of the plate is then driven to the nodes to form varying patterns, depending on which antinode is stimulated.

physicist Mary Waller is photographically summarized in figure 4. In the experiment a fine sand or powder is driven to the nodes, as places of immobility, while the delicately suspended plate is set into vibration by tickling an antinode or region of maximum vibration with a sliver of dry ice. The result is a beautiful and simple demonstration of the reality of circular plate normal modes.

The great musicologist Curt Sachs reported that Hindu secular drums were sanctified for religious ceremonies by decorating the drum membrane or diaphragm with simple geometrical circles, lines, and crosses, using heavy pigment pastes. Such drums then exhibited pure tones according to the pattern used: the heavy pigment paste defined a nodal position in exactly analogous fashion to Mary Waller's defining of nodal positions by placement of rubber feet under the vibrating plates. This is one of the many instances in the development of musical culture where intuitive discoveries conform exactly to a theoretical analogue.

Physics of real instruments

Musical acoustics has been the subject of many books by physicists attempting to bridge the gap between musical aesthetics and elementary physics. The following analysis of bow instruments and their parts, offered by the Dutch physicist J. W. Giltay, seems to be the most operationally accurate.

Giltay used loading experiments to demonstrate the action of the bridge as primarily rising and falling under string vibration, with the bridge pivoting on the stiffer part of the top plate conditioned by the placement of the sound post near the treble foot of the bridge. The sound post is a dowel of wood vertically wedged between the top plate and back of the instrument (figure 5). Bow-instrument craftsmen and players know well that the sound post location is critical for attaining

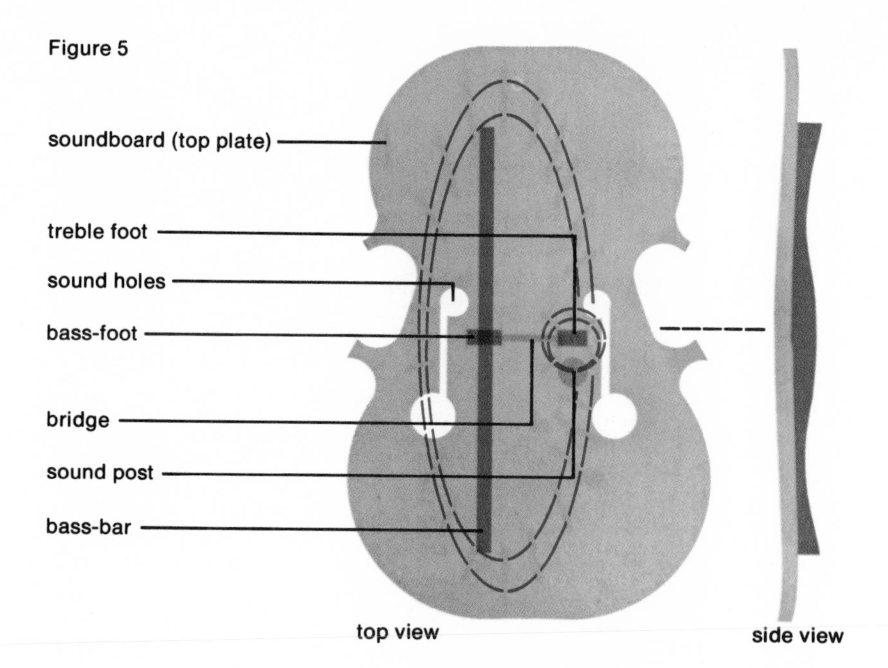

Figure 5. Top view of a bow instrument reveals its major acoustical and mechanical components. The dimensions and arrangement of these parts provide the instrument with its particular tonal quality.

the best treble response. The ideal position of a sound post is found by trial-and-error adjustment, and in the violin it is usually fixed about 1 cm away from the treble foot of the bridge in the direction of the tail-piece. It is often stated that the purpose of the sound post is "to transmit vibrations from the front to the back" of the instrument. Giltay recognized that the sound post has the opposite function: it defines a nodal point for top-plate vibration. The treble foot of the bridge consequently has defined for it a top-plate region of very small diameter (figure 5) for extreme high-frequency response; presumably, such a region spreads out somewhat, with the treble foot of the bridge at the antinode (region of maximum displacement).

The bass-bar is the other structural component of bow instruments (figure 5) that confers mechanical asymmetry on the top plate and, hence, on the body of the instrument. The bass-bar is a bar of wood glued to the underside of the top plate of the instrument and is almost parallel to the longitudinal grain lines of the top. The bass-bar passes directly under the bass-foot of the bridge, and extends almost the entire length of the instrument.

The bass-bar is alleged to have been discovered accidentally about 1600 when Gasparo da Salò introduced a longitudinal repair splint to overcome a carving error in the underside of a top plate; the splint caused a deepening of the bass response. The carved ridge was later replaced by a glued-in "bass resonance bar," which, by the time of virtuoso craftsman Antonio Stradivari, was elegantly streamlined. As string tensions increased in the two centuries since Stradivari, the bass-bar thickness (height) has steadily increased.

The mechanical action of the bass-bar is simply explained in terms of normal motions of an elliptical top plate in the elementary bow-instrument model (figure 2). The rising-and-falling motion of the bridge would tend to drive the soundboard only locally if no bass-bar were present. With a bass-bar affixed to the top plate, the bridge motion is extended over the longest dimension, assuring some low-frequency components in the vibration of the top plate when played in bass register. This helps satisfy the criterion of brilliance of tone in stringed instruments, which requires the development of a powerful first harmonic, or fundamental tone, for each note.

In terms of the model of figure 2 and the simplicity of the two elementary internal mechanical accouterments, the sound post and bass-bar, the violin and other bow instruments constitute a marvel of cultural evolution. The craftsman's selection of woods, the shaping of components, choices of dimensions and geometry, and intuitive grading of thicknesses and contours constitute the artistry that distinguishes an outstanding instrument from a mass-produced mediocre one.

Guitars are much older than the bow instruments. Comparing the representations of guitars in archaeological remains with those of medieval Europe and today, we see that the principal structural features were developed very early. But in spite of its earlier origin the guitar did not reach as high a stage of development as have the bow instruments. The latter in the 17th and 18th centuries became popular in performances for large audiences, and so the demand for their perfection was accelerated. In contrast, the guitar in the same period remained largely a personal adornment of royalty. Stradivari made a dozen or so guitars. The rare examples that survive are curious museum pieces of little musical importance, a striking contrast to the continuing role of the Stradivari violin.

Although the limited demands on the guitar as a concert instrument helped prevent its early mechanical and acoustical perfection, a primary reason for this lack is intrinsic. A bow-instrument player feeds power continuously to the strings through the bow motion. By contrast, a guitar player relies on a single finger stroke to deliver power to a string. Thus, to convert the guitar into a concert instrument, the maximization of every power-generating component action and the minimization of all power-dissipating component actions must be attempted. The guitar poses one of the most severe tests of correlation of physical theory with musical performance demand.

Transverse bracing perpendicular to body axis and wood grain is the common type of bar structure seen in ancient guitars, including the slender-bodied guitars of the Renaissance (figure 6, center). From the symmetrical placing and more or less random location of such bars, it is apparent that the purpose of such bracing was structural reinforcement rather than vibrational function. From the mechano-acoustic viewpoint, such bars would act primarily as nodal bars imparting good treble response but generally inhibiting that of the bass. In the hands of some notable craftsmen, a slight tilting of one of the transverse bars with respect to the normal alignment to the instrument axis introduced a slight asymmetry in structure (figure 7, center) extending the range of good response of the diaphragm.

Figure 6

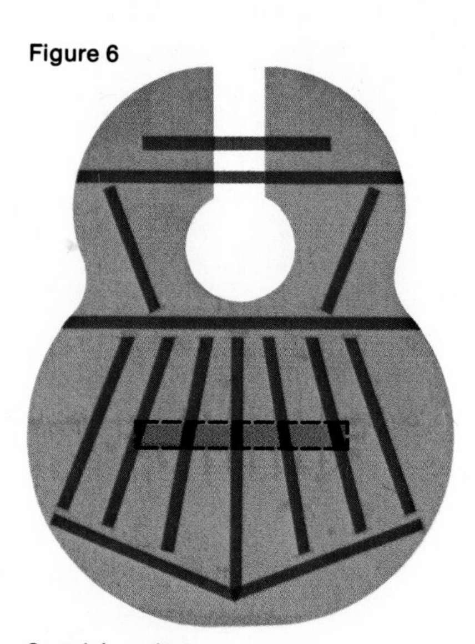

Spanish guitar
Torres fan bracing

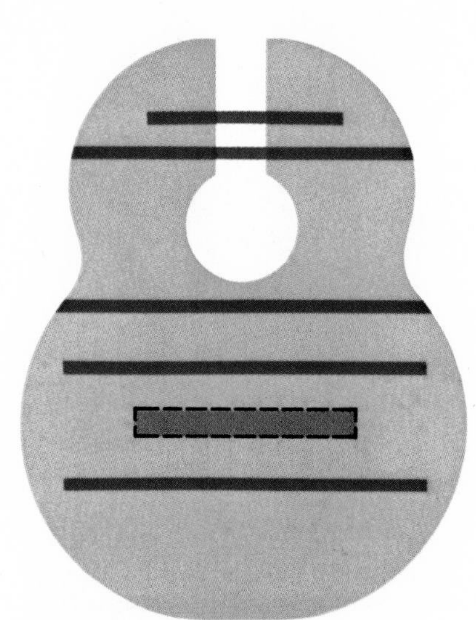

Viennese-Italian guitar
transverse bracing

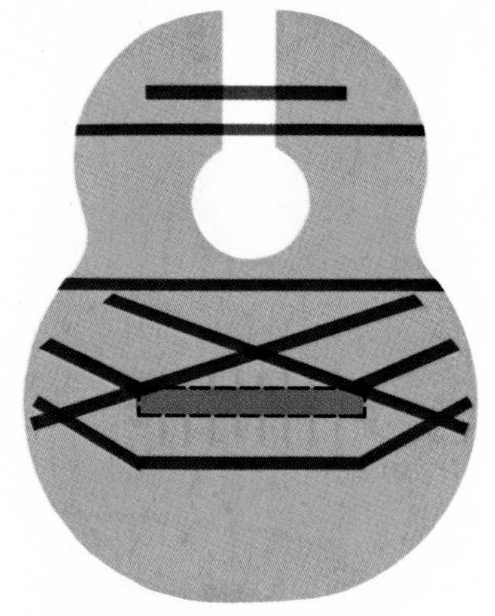

Austrian guitar
x-bracing

Figure 6. Bar bracing patterns for the guitar evolved over centuries to provide structural reinforcement and confer particular tonal qualities. Ancient and Renaissance guitars used transverse bracing (center), which primarily supplied reinforcement and imparted a good treble but a weak bass response. The bass tones were strengthened by the Torres fan arrangement (left), introduced in the mid-1800s. Crossed bracing, an old pattern, was revived to strengthen the guitar top against the high tensions imposed on it by steel strings. It yields a good treble response, but is relatively weak in the bass register.

Longitudinal bracing in fanlike arrays constituted one of the main innovations introduced in guitar construction by Antonio de Torres in the mid-1800s (figure 6, left). Torres greatly lengthened and widened the guitar body over that of the slender Renaissance instrument, and he recognized the need of much more powerful support for the bridge. Also Torres recognized that longitudinal bars act to improve bass resonance, transmitting local vibrational motion of the guitar bridge over the entire diaphragm zone of the guitar top plate. It seems quite possible that Torres emulated the practice of violinmaking in the use of the bass-bar. However, he introduced seven such resonance bars in a perfectly symmetrical fanlike array. Obviously, such a symmetrical structure could not be optimum over the entire four-octave range of the guitar. Moreover, such a "singly-structured" diaphragm would be expected to have a frequency spectrum essentially analogous to that of a simple clamped circular plate. The ribbed plate could be expected to emphasize some vibrational modes and to quench others whose nodal properties cannot be freely excited owing to the stiffening effect of the entire ribs. The Torres bracing pattern would tend, therefore, to afford a soundboard rather uneven response: a good bass efficiency with sufficiently delicate construction and a quite possibly restricted treble response. The abrupt evolution of the Torres bracing is in the sharpest contrast to the centuries of slow evolution which the bow-instrument structure had undergone. Recently, there has been a trend to make the longitudinal fan bracing somewhat asymmetric (figure 7, left).

The crossed-bracing pattern illustrated in figure 6, right, represents another older pattern revived to cope with the high tensions imposed on the guitar top by the recent development of the steel stringed guitar. Such a structure imparts a good treble response but yields a

Figure 7

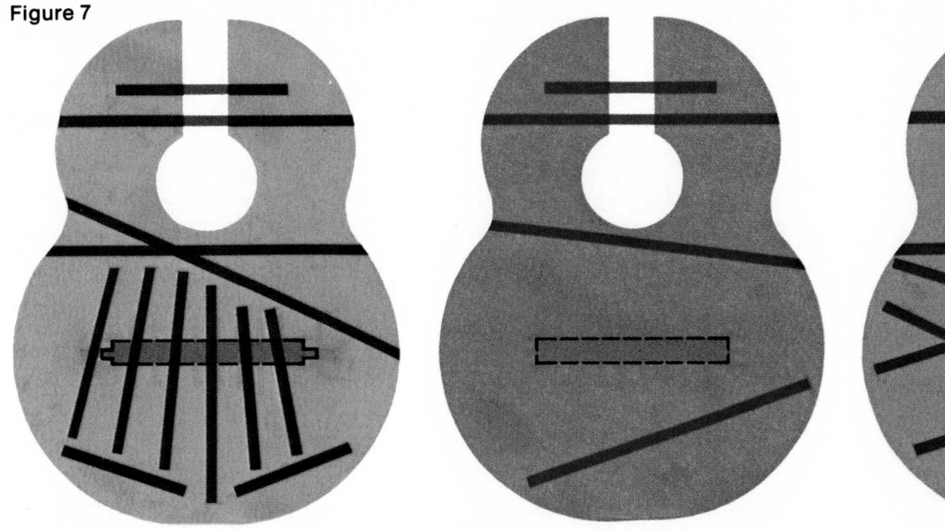

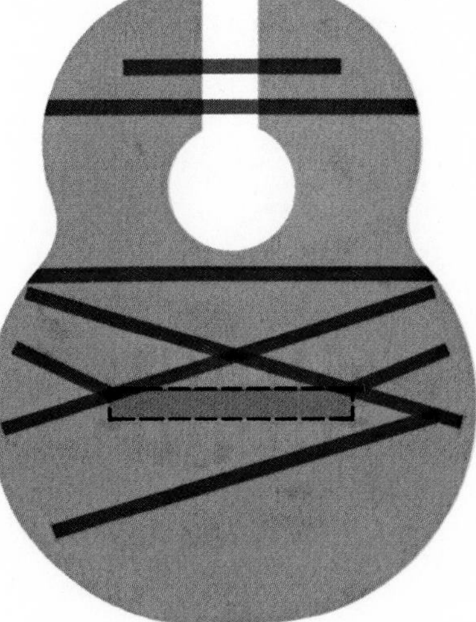

J. Ramirez (<1965) Stauffer (~1830) C. F. Martin (~1875)

Figure 7. By rearranging one or more bars in the bracing patterns shown in figure 6, designers have tried to achieve improved tonal qualities. The slight tilting of the lowest bar in the transverse pattern (center) produced a better bass response. An asymmetrical arrangement of the Torres fan (left) was designed to improve the treble sound of the original, while asymmetry in the cross-braced pattern (right) was an attempt to produce a better response in the bass. All three instruments appear as seen from the top, with the bass side to the left and the treble to the right.

relatively deficient bass response. There is also a tendency in contemporary guitars to introduce considerable asymmetry in steel guitar construction (figure 7, right). But the fundamental effect of the x-bracing pattern as a bass-register inhibitor is inexorably present.

Thus, the fundamental structure of the bow instrument and the guitar evolved in a relatively accidental manner. We may marvel at the essential mechanical simplicity and economy of functioning components of the bow instrument. At the same time, in terms of physical theory, we may be chagrined at the relative retardation of guitar mechano-acoustical design.

Modern physical measurements on string instruments

Modern technology provides several measurements that offer an objective picture of an instrument's basic characteristics. They include the power spectrum, characterizing acoustical power output in relation to human ear sensitivity; the harmonic spectrum, characterizing the harmonic intensities of each note; and the onset and decay transient curve, characterizing intensity versus time under a particular sound-generating technique.

The power spectrum, adapted to human perception, offers evenness of response to the performer so that playing dynamics can be more readily controlled. The harmonic spectrum indicates the relative brilliance of a particular string note, as well as qualitatively characterizing the voice of the instrument. The onset and decay transient curve indicates the tone characteristic, over the range "tympanic" to "round" as an expression of white-noise component on one hand to tonal definition on the other.

Two scientific measurement techniques, laser holographic interferograms and power spectrum measurements, are currently reaching

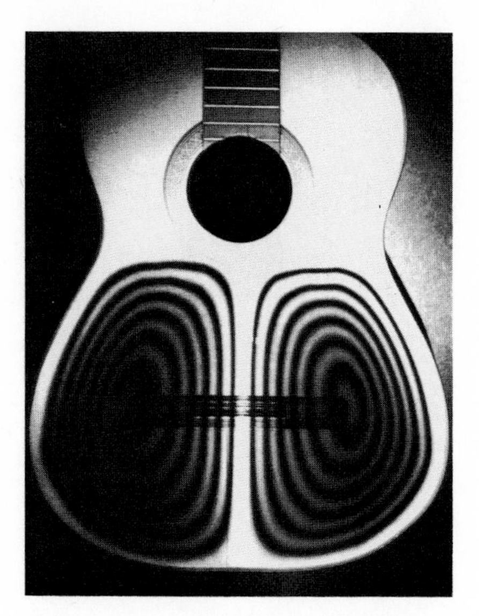

Figure 8 (pages 140–141). Interferograms of the top plate of a guitar permit the visualization and measurement of the distortional motion associated with normal modes of vibration. The interferograms represent the lowest two characteristic normal frequencies for this top plate, which is on a guitar with Torres fan bracing. The depression and corresponding elevation on each side of the figure above indicate a planar mode.

practical application in musical acoustical studies. Laser holographic interferograms of real musical instruments driven mechanically were explored as a new technique recently by Karl A. Stetson and co-workers. At the Institute of Science and Technology of the University of Michigan, R. L. Powell and K. A. Stetson developed a time-average holographic interferometric technique for the study of vibrational stationary-state patterns. Holographic interferograms permit one to visualize and measure the distortional motion associated with normal modes of simple and complex vibrating objects. In this technique, the vibrating object may have a diffuse reflecting surface and need not be perfectly flat, making the method exceedingly powerful for the study of vibrational modes of real objects of complex shape and surface characteristics.

Guitar top-plate interferograms were studied by N.-E. Molin and K. A. Stetson at the Institutet für Optisk Forskning, directed by Erik Ingelstam, in Stockholm, Swed. Two of these interferograms are shown in figure 8. The guitar used was one made by Georg Bolin of Stockholm, with the symmetrical Torres bracing pattern. For the experiment the instrument was glued to a rigid frame, the guitar having no back and no strings, and was driven mechanically at selected points from the underside.

The interferograms of figure 8 represent the lowest two characteristic normal frequencies for this guitar top plate. The interference bands correspond to contour lines of equal displacement like the contour lines on a geodetic survey map; the illustration on the right in the figure obviously contains a planar mode, so that on one side there is a depression and on the other a corresponding elevation in the vibrational mode. However, higher frequency modes depart significantly in pattern from those observed for simple circular plate modes. Moreover, a complete guitar under string tension may be expected to behave somewhat differently from the guitar top plate studied in these first experiments.

Nevertheless, the Molin and Stetson guitar interferograms offer reassurance that the vibrations of a complex real musical instrument can be understood with a simple theoretical analogue oscillator, such as a circular plate, as a starting point. As more complex patterns are analyzed, suggestive structural ideas for guitar construction should be deducible. As the time-average holographic interferogram technique is applied to modern asymmetric guitars, the expected growth in complexity and asymmetry in the interferograms, together with the accompanying richness of the characteristic frequency spectrum, should become observable.

Viola da gamba interferograms were studied by Carl-Hugo Ågren and K. A. Stetson in the Stockholm laboratory. The interferograms for this instument show a high degree of asymmetry and complexity compared with those obtained for the symmetrically-structured classical guitar. The characteristic frequency spectrum deduced from this study permitted Ågren to design a more powerful new instrument, the magnum gamba, whose body frequencies more suitably bracketed the string frequencies.

Power spectrum measurements on the guitar were studied by E. E. Watson and Michael Kasha at Florida State University. A large sound tunnel or "impedance tube" 16 ft long and 2 ft in diameter was devised to permit the measurement of the sound pressure level produced by the passage of a sound wave along the tube from the plucked instrument. A guitar was mounted in rubber mounts inside the tube at one end, and was plucked by a "standard nylon finger" or rounded plectrum, driven mechanically. A sound pressure probe was mounted in the center of the tube, and at the far end bags of fiberglass wool acted as total sound absorbers. The acoustical properties of the tube permitted measurements from 80 to 700 Hz directly (1 Hz = 1 cycle per second), and up to about 3,000 Hz by using a concentric insert tube to block transverse standing waves at high frequencies. Since the guitar has a fundamental frequency range of 82.41 to 987.8 Hz, the acoustical characteristics of the tube permitted accurate and reproducible measurements over the instrument's entire range.

Courtesy, K. A. Stetson

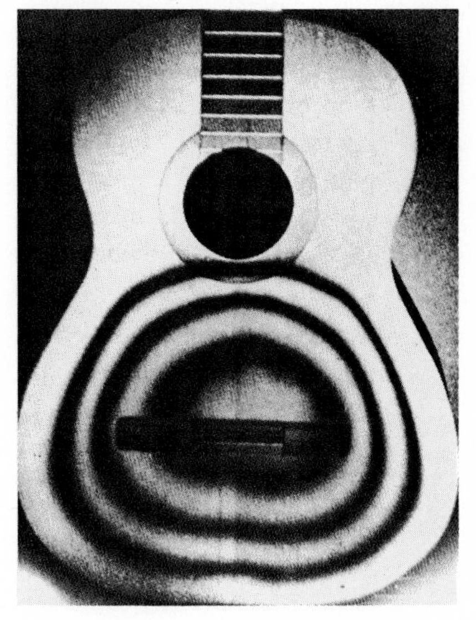

The purpose of power spectrum measurements is to compare the output of an instrument over its working frequency range with that required by the hearing characteristics of the human ear. In general, traditional instruments are found to have a rather erratic, uneven, power output spectrum. However, measurements on some early prototype guitars of novel asymmetric design indicated a strong parallel to the human hearing curve requirement. The relative insensitivity of the human ear to low-frequency and very-high-frequency power requires a compensating behavior in the power output spectrum of the instrument.

Modern instruments: the challenge of aesthetic criteria

One of the greatest challenges to the acoustician-designer of musical instruments is to cope with the subtle and variable vocabulary of musical aesthetics. The virtuoso performer speaks of the "brilliance," the "roundness," the "silkiness" of a tone as qualities he seeks, and of "cloudiness" and "dryness" as those he may wish to avoid. How can these terms be translated into physically tractable forms?

Time-dependent variations of such instantaneous measurements as power spectrum and harmonic spectrum offer room for exploration of subtleties. Beyond this, characteristics of the elastic properties of component materials such as the woods used in the construction of materials allow at least empirical bases for discrimination: spruce tops yield bright-toned guitars; cedars, mellow-toned guitars; redwood, deep-toned instruments.

There are many current efforts in the physical investigation of musical stringed instruments. The successful interferogram viola da gamba studies by Carl-Hugo Ågren mark one type of effort: the evolution of instruments new in size and tonal range. An especially effective and broad program of this type is the well-known major effort by Carleen

Figure 9

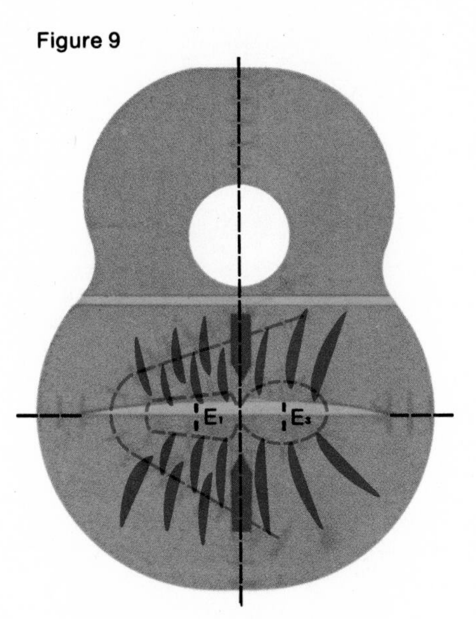

Figure 9. Experimental radial bracing pattern for a guitar attempts to resolve a dilemma long faced by designers: how to transmit bridge motion over the entire diaphragm region without simultaneously inhibiting the amplitude of that motion. The underside of the guitar is shown, with the bass to the right and the treble to the left.

M. Hutchins. Mrs. Hutchins, following the early violin studies and tutelage of Harvard University physicist Frederick A. Saunders, worked to develop a whole new "family of fiddles" whose body dimensions are scaled to the desired tonal range of the instrument. Many new instruments with remarkably beautiful sound have resulted.

The older problem, however, is not to design entirely newly scaled instruments but to characterize and optimize the classical instruments having classical dimensions. Such a goal is what most performers hope to see fulfilled.

In one current line of research model physics for the bow instrument revealed a mechanical paradox in bass-bar function. Considering that the bass-bar should act to distribute bass-foot vibrations of the bridge over the entire top plate of the instrument (guaranteeing a strong fundamental for low frequency string-tones), it is a paradox that the bass-bar is thickest (highest) at the very point at which the bridge vibration is applied. Thus, the bass-bar would tend to inhibit the very bridge motion that it should transmit to, or distribute over, the top plate. This problem grows more severe in the order violin-viola-cello-bass, owing to instrumental dimensions and bass-bar stiffness. The simplest of three envisaged solutions was to make an arch cut in the underside of the bass-bar in alignment with the bridge bass-foot. More than a dozen pairs of violas and violins were subjected to this simple change, with carefully controlled pairs of instruments in each experiment and arch cuts of different shape and size. The results, judged by experienced musicians and virtuoso performers, were almost uniformly approved as a remarkably improved "brilliance" in the low register. The degree of change in the bass register sound was found to be reproducibly dependent on the depth of the arch cut.

The bass-bar arch cuts resulted in a deeper, richer tone for notes played on the C and G strings (viola), which was accompanied by an increase of depth of tone of notes played on the D and A strings. However, there was some loss of incisiveness, which led some listeners, perhaps 25%, to prefer the unmodified instrument.

The entire-bar structure used traditionally in guitar construction led to a characteristic dilemma. If the bars are made heavy enough to transmit bridge motion over the entire diaphragm region of the guitar top plate, they simultaneously inhibit the amplitude of the very bridge motion which is to be coupled. If the bars are made much thinner, affording a fair amplitude of vibrations, and hence instrument power, then they fail to couple to the lowest frequency (nodeless) normal modes of the top plate regions driven. The guitar craftsman faces this dilemma at the hazards of poor power (and good strength) or poor strength of top (with good power).

The acoustician-designer faced with the guitar structure dilemma must use his intuition to bridge the gap between simple theory and the real instrument if the model physics approach is to be used. The author worked with a number of guitar craftsmen, trying out a number of different intuitive solutions to the problem of greater amplitude of motion with greater coupling of the bridge to the fundamental normal modes of a frequency-dependent diaphragm. Figure 9 exemplifies the essential bracing system used. Over the years, approximately 25 models have been built, tested, and subjected to concert performance by virtuoso players. Although one or two paths of evolution were abandoned as unrewarding, most of this long series of experiments using the model physics approach appeared to be leading to a highly desirable musical instrument for concert performance: a powerfully projecting guitar, with a solid brilliant tone, and unusually fine balance throughout its entire four-octave range.

The concert artist constantly searches for the perfect sound, the sound he dreams of and strives for. The model physics approach may suggest bold new steps to advance toward this goal, offering possibilities for further development in the guitar, the viola, and the violin. These may lead to a new cello and double bass freed from violin acoustical scaling. The message of the future for musical stringed instruments must surely be that the old instruments will not survive forever, and as the correlation of modern science with virtuoso performance expands the modern instrument can be expected to reach a new level of perfection.

FOR ADDITIONAL READING:

Backus, John, *The Acoustical Foundations of Music* (W. W. Norton, 1969).

Boyden, David, *The History of Violin Playing* (Oxford University Press, 1965).

Hutchins, C. M., "The Physics of Violins," *Scientific American* (November 1962, pp. 78–84), "Founding a Family of Fiddles," *Physics Today* (February 1967, pp. 23–28).

Kasha, M., *Complete Guitar Acoustics* (Cypress Cove Press, 1973).

Perlmeter, Alan, "Redesigning the Guitar," *Science News* (Aug. 22 & 29, 1970, pp. 180–181).

Sachs, Curt, *The History of Musical Instruments* (W. W. Norton, 1940).

Taylor, C. A., *The Physics of Musical Sounds* (American Elsevier Publishing Co., 1965).

Waller, Mary, *Chladni Figures; A Study in Symmetry* (G. Bell and Sons, Ltd., 1961).

Probing the Mysteries of the Aurora

by T. Neil Davis

As man learns more about the aurora, he becomes better able to understand the regions of space near the earth. Brilliant displays of the northern and southern lights (aurora borealis and aurora australis) signal rumblings in the magnetosphere and earlier outbursts on the sun.

Always an enigma to man, the aurora polaris recently began giving up some of its secrets and is proving to be a major tool in the study of the atmosphere and other regions near the earth. The shimmering auroras that play across the dark polar skies are visible indicators of the locations where energetic electrons and protons are entering the atmosphere and causing it to glow. Like returning spaceships, the energetic particles plunge through the high atmosphere and lose momentum as they travel through the increasingly dense medium. Unlike spaceships, these particles never reach the ground but instead come to rest at an altitude near 100 km (60 mi) after giving all their energy to the atmospheric gas. This energy heats the atmospheric gas and produces several effects, one of which is excitation of oxygen and nitrogen, which largely make up the atmosphere. Excitation is the process whereby an electron contained in the outer shell of an atom or molecule is raised to an energy state above its normal level. When excitation occurs, this additional energy is often reemitted from the atom or molecule as a packet of light called a photon. When enough photons are given off from a particular parcel of the atmosphere, that parcel becomes visible to the eye and is called an aurora.

Courtesy, A. Lee Snyder

The general nature of the excitation processes that cause an aurora has been understood for many years. But only since the late 1950s has there been much comprehension of where the primary auroral particles (the incoming electrons and protons) come from. That comprehension derives from the discovery by satellites that the earth is separated from interplanetary space by a protective shield called the magnetosphere. Now an identified entity, the magnetosphere is simply that region surrounding the earth wherein the earth's magnetic field dominates. If space were a complete vacuum, the earth's magnetic field would extend to infinity. However, an irregular flow of charged particles emerging from the sun (the solar wind) compresses the magnetic field and thereby gives the magnetosphere a definite boundary. Within this boundary the earth's magnetic field controls the motion of charged particles. By means that scientists do not yet understand, some of these particles are accelerated within the magnetosphere and then guided into the polar regions, where they interact with the atmosphere to cause auroras.

Auroral structures and their colors

One of the continuing mysteries of the aurora is its intricate structure. Why do the incoming energetic particles flow in confined streams and therefore produce auroral structures that are sharply defined and often multiple? Auroral forms typically exhibit remarkable intricacies of shape and internal structure. It is, of course, this intricacy that contributes much to the beauty of the aurora—this and the violent motions these structures at times undergo. The most commonly observed auroral forms are called arcs and bands. These usually are very thin (less than 100 m to several kilometers) compared with their height (in excess of 10 km or even 100 km) and their great length, which often is thousands of kilometers. To the eye, arcs and bands may appear homogeneous or they may exhibit striations called rays. Sometimes the arcs and bands are broken up into segments called draperies. At other times, the observed display may consist largely of diffuse patches scattered over the sky. Arcs and bands are in reality huge ribbons of light standing on the edge of the atmosphere; their great length usually extends in an approximate east-west direction. The ray structure often seen in these forms has a special significance, for it maps out the direction of the local magnetic field. A compass needle freely suspended in an aurora would orient itself exactly in the direction of an auroral ray at that location. Just as the rays are aligned along the magnetic field, the auroral forms also are so aligned, a consequence of the fact that the incoming protons and electrons causing them are constrained to move downward along the direction of the earth's magnetic field.

Although every aurora is a mixture of specific colors, the weaker auroras appear colorless because they are detected only by the rods in the human eye, which do not recognize color. As an aurora becomes brighter, the eye's cones come into use and then the aurora appears

T. NEIL DAVIS, *a professor of geophysics at the University of Alaska and deputy director of the university's Geophysical Institute, has written numerous articles on auroras for scientific journals.*

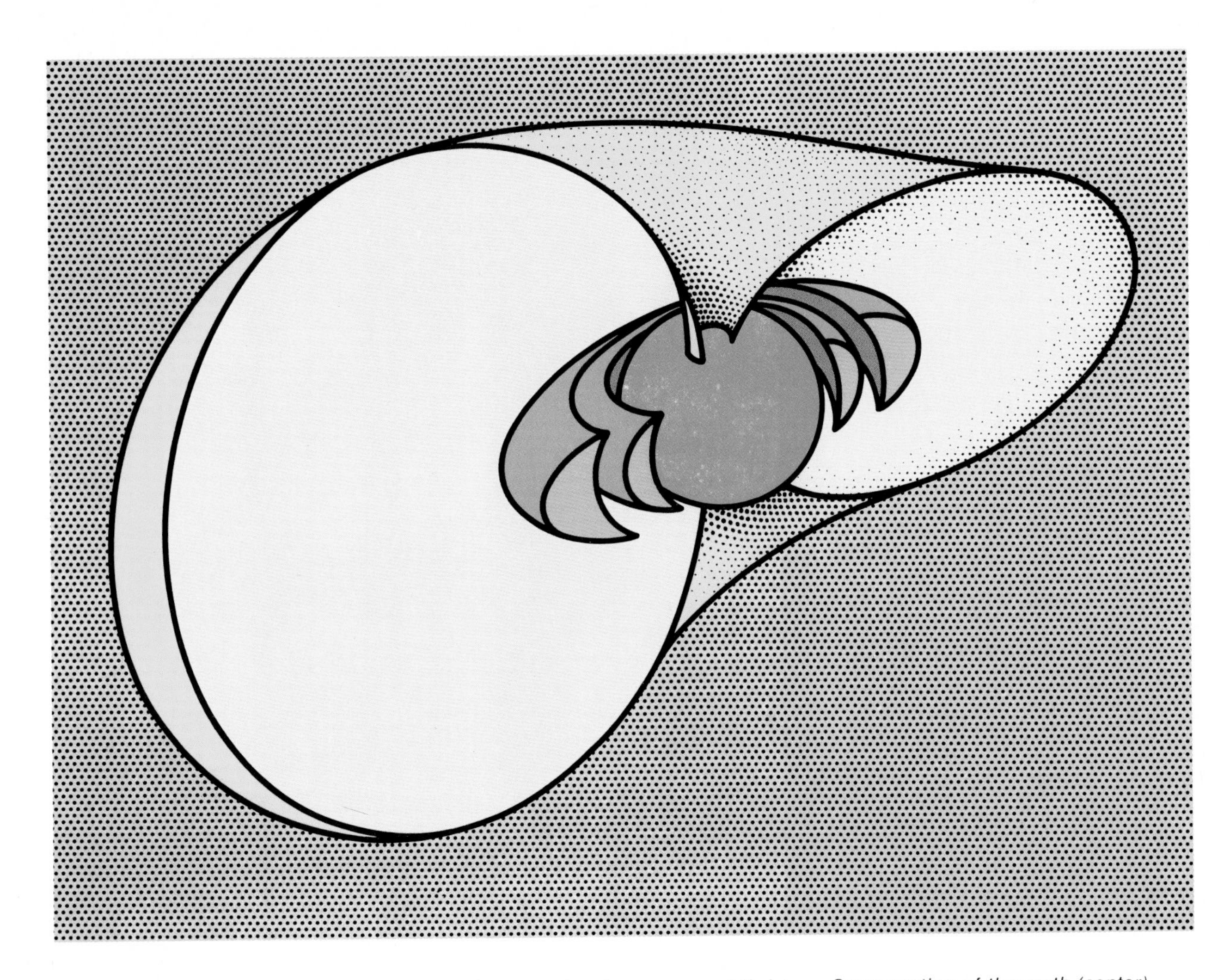

Cross section of the earth (center) is surrounded by the Van Allen radiation belts (inner shells) and the magnetosphere, that region within which the earth's magnetic field dominates. Charged particles from the sun (the solar wind) contain the magnetosphere and give it a definite boundary. Some of these particles are accelerated within the magnetosphere and guided to the polar regions, where they interact with atmospheric gases to cause auroras.

green. Unlike sunlight, which is a continuum of colors, auroral light is composed of specific lines and bands extending across the visible spectrum and to either side in the infrared and the ultraviolet regions. The green color in an aurora occurs because the eye is quite sensitive to green and there is a strong auroral green line at 5577 Å (angstroms; 1 Å = 10^{-8} cm) emitted when electrons excite oxygen atoms.

Rarely, at times when very large auroral disturbances occur, entirely blood-red auroras are seen. These are very-high-altitude auroras (300–400 km) caused when intense fluxes of low-energy particles impinge on the atmosphere and give rise to the so-called "red-line" emissions of oxygen at 6300 Å and 6364 Å.

The most beautiful of all auroras are those that are green and have red lower borders. These are caused when particularly energetic electrons penetrate closer to the earth's surface than usual to excite red-band emissions in the nitrogen gas of the atmosphere as well as exciting the oxygen green-line emission higher up. Invariably the red-lower-bordered auroras are bright and intricately rayed. The bright green rays with their red tips flash rapidly along the arcs and bands to provide a scene of splendor rivaling the grandest sunset.

Unfortunately, most color photographs do not show the true colors of auroras. This is because color film is balanced for sunlight and is unable to reproduce adequately for the eye the mixture of line and band emissions in the aurora. Many color photographs show bluish auroras because in the aurora there is blue light to which the film responds but the eye does not.

Where and when auroras occur

On any clear, dark night auroras can be seen in two regions of the earth called the auroral zones. These zones lie in a circle approximately 23° in size, centered on the north and south geomagnetic poles. The north geomagnetic pole is near Thule, Greenland, and the southern pole is in Antarctica at the Soviet scientific station Vostok. Auroras tend to be brightest and most numerous at the auroral zones, and they are most likely to be seen there in the hours near local midnight.

Major auroral displays occur during intervals known as magnetic storms because at these times large fluctuations are observed in the direction and strength of the earth's magnetic field. Storms usually last a day or two and tend to follow by one or two days the eruption of a flare on the sun. During the storms, auroras become very widespread in latitude; at such times, they may be observed overhead at middle latitudes. From 10 to 20 moderate or major storms occur in the minimum years of the 11-year solar cycle of activity; during years of maximum activity the number of storms may double.

A major advance in the understanding of the behavior of auroras and related phenomena began in the early 1960s as a result of the finding that there is, within a magnetic storm, a shorter sequence of activity called a substorm. Each substorm lasts one to three hours, and several of them usually occur during each storm. During the substorm, radical changes take place in the aurora. At the beginning of the substorm, quiet arcs and bands usually occur; as the substorm progresses, these tend to become more numerous and active until there develops an explosive change called the breakup. During it the auroras brighten and undergo violent motions. Then the aurora again becomes quiescent and often takes the form of weak, diffuse patches or segments of arcs and bands. The substorm is global in extent, in that the global pattern of auroras in both the northern and southern hemispheres is affected. The energy dissipated into the earth's atmosphere during a large substorm is approximately equal to that released in the explosion of an atomic bomb or in a damaging earthquake of magnitude 6 or 7 on the Richter scale.

Linked together by the earth's magnetic field lines passing high above the equator, auroras in the northern and southern hemispheres behave quite similarly. Simultaneous photographs taken at conjugate points (points on the same magnetic field line) show that identical auroras sometimes occur in the two hemispheres. Identical but displaced auroras are frequently observed, and these show that

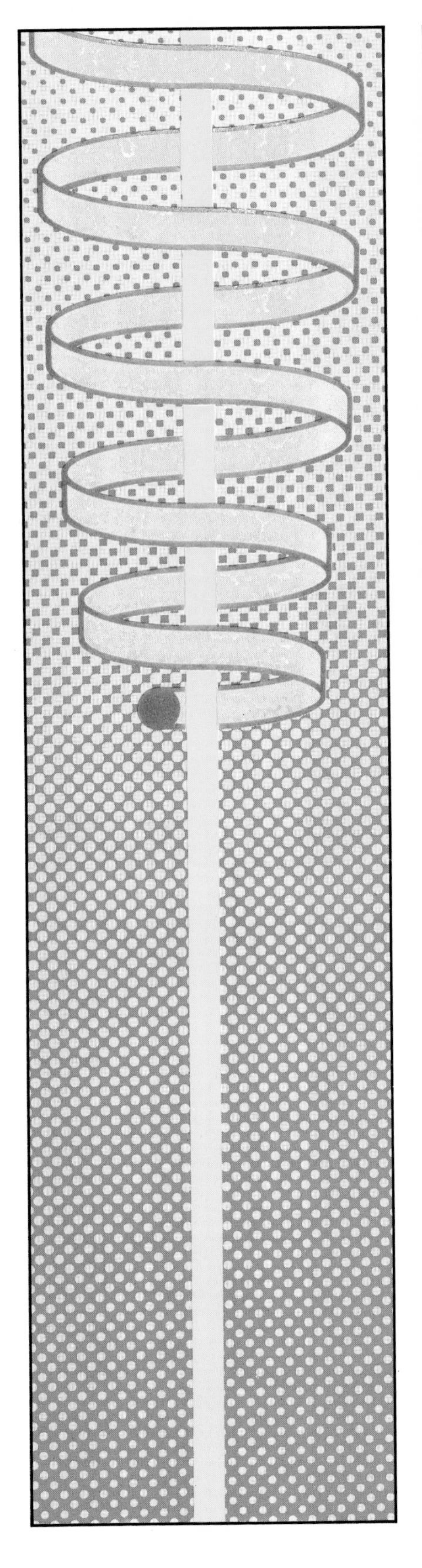

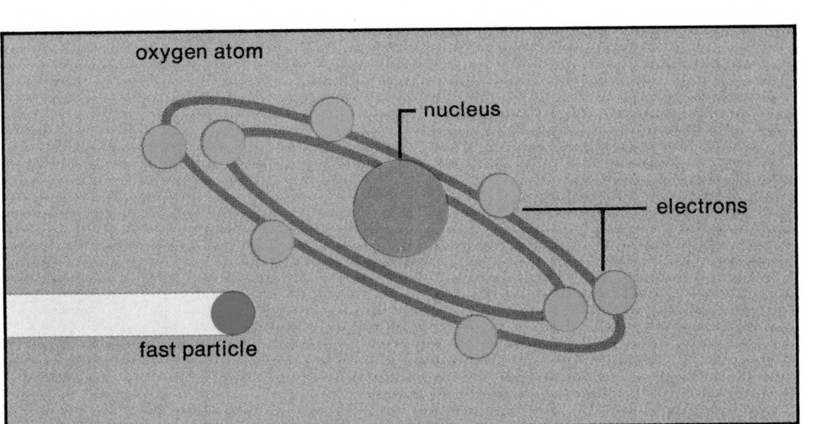

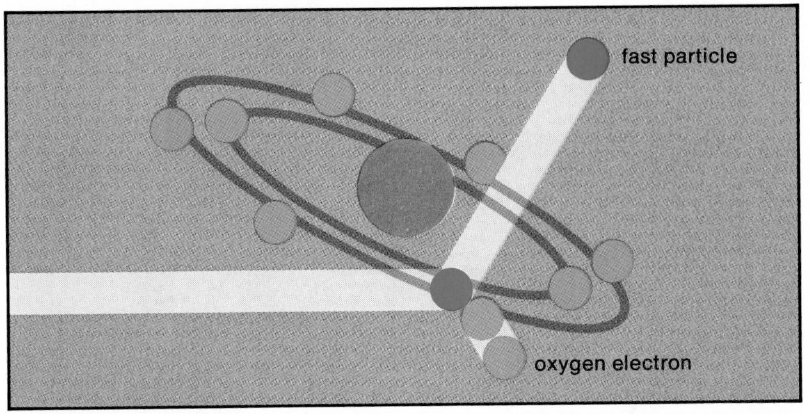

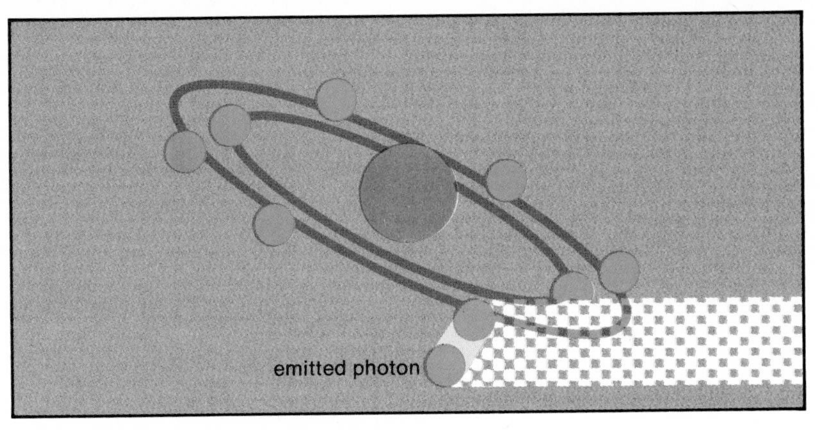

Proton from the solar wind spirals toward the earth counterclockwise around one of the lines of force of the earth's magnetic field (left). When such a proton, or electron, enters the high atmosphere of the earth, it encounters such gases as oxygen (top). The proton or electron knocks one of the outer electrons of the oxygen atom out of its orbit into a higher energy state (center). When the energized electron falls back to its original level, it emits a photon of energy in the form of visible light (bottom).

Courtesy, Geophysical Institute,
University of Alaska

S. -I Akasofu

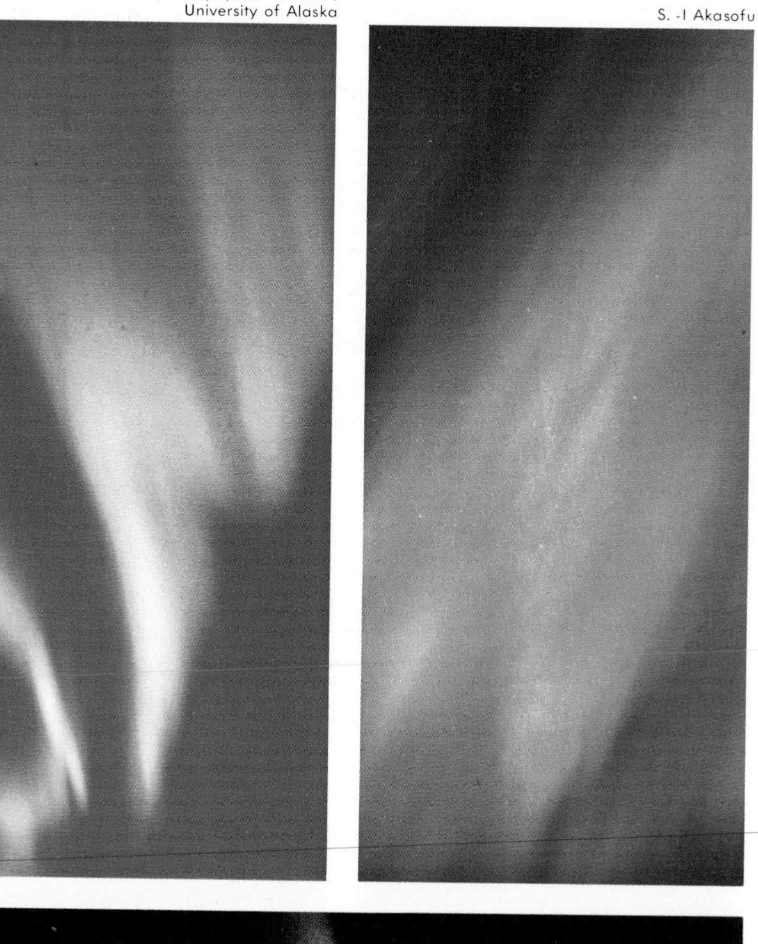

Auroras appear in a variety of forms and colors that depend on the intensity of the solar wind particles and the altitude in the atmosphere at which their interactions with gases take place. Bands (above, left) and rays (above, right) are typical forms. The red aurora (right) resulted from intense fluxes of low-energy particles impinging on the atmosphere at very high altitudes. The green auroral arcs (opposite page, top) were caused by electrons exciting oxygen atoms. Especially energetic electrons penetrating closer than usual to the earth's surface and exciting red-band emissions in nitrogen caused the lower red band in the aurora (opposite page, bottom).

Alfred P. McNeil

Takeshi Ohtake

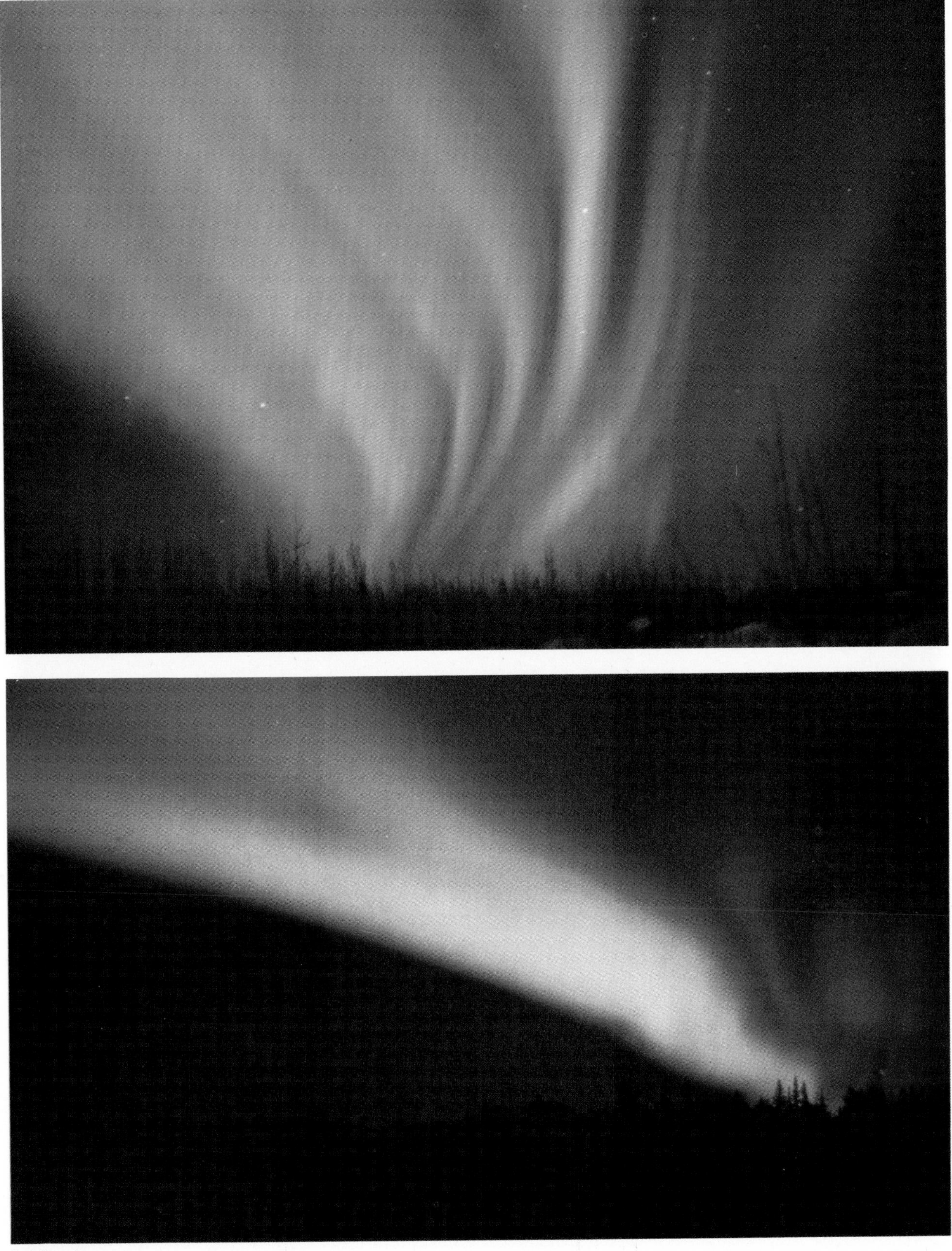

S. -I Akasofu

Courtesy, Geophysical Institute, University of Alaska

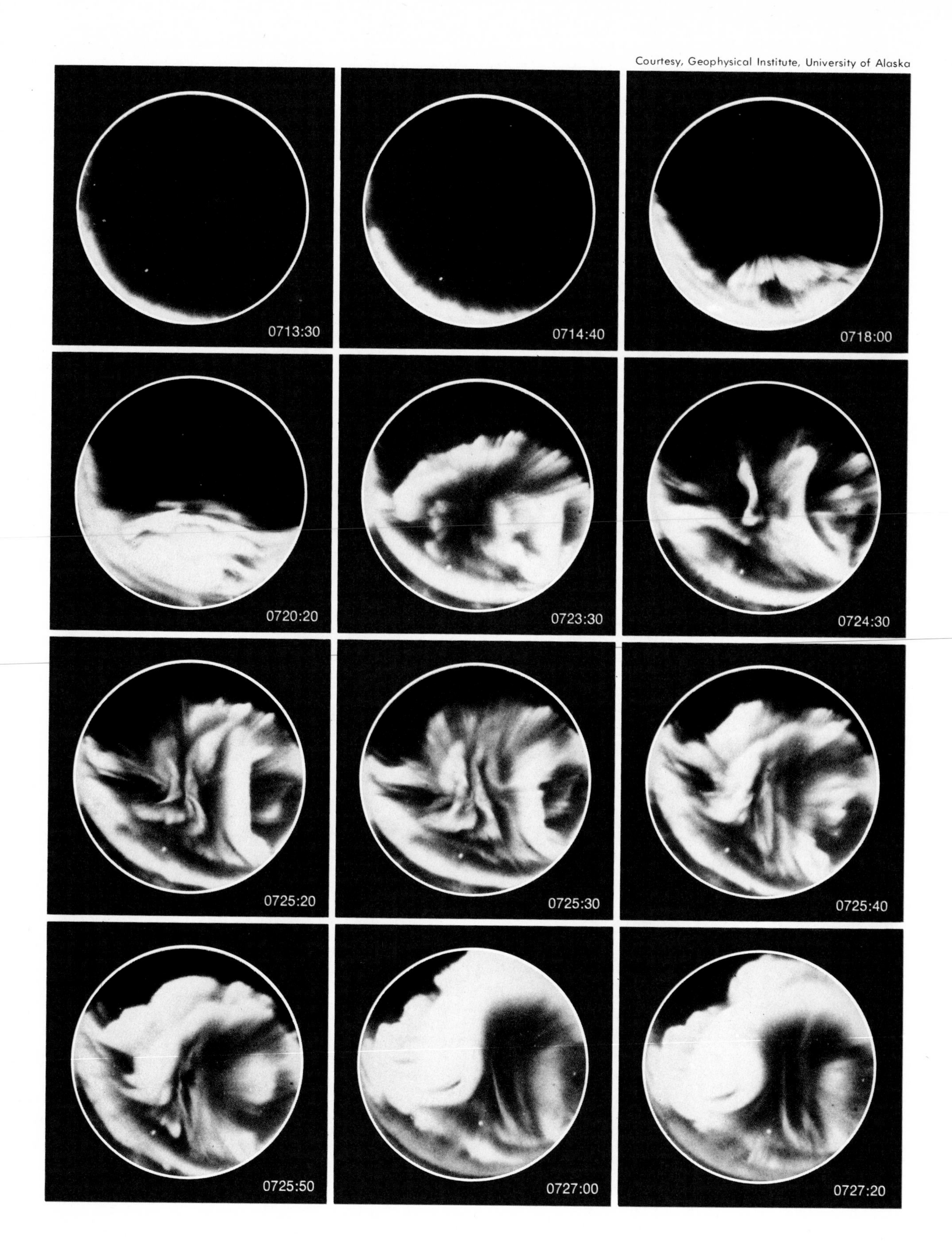

the earth's magnetic field can at times be severely distorted. Occasionally, especially during very large substorms, an aurora is seen in one hemisphere but there is no corresponding aurora in the other.

If a viewer could be suspended high above a polar region so as to look down on the entire auroral zone, that viewer would probably see a continuous ring of aurora encircling the geomagnetic pole. This ring, however, would not lie exactly along the auroral zone except near the local midnight position. The viewer would notice that the auroras on the noon meridian are displaced poleward by a few degrees of latitude from the auroral zone. Thus, the instantaneous auroral pattern is oval-like; in fact, the pattern is called the auroral oval. Its width increases with the level of activity as does its diameter. During large magnetic storms the auroral oval expands, and then auroras are observed at much lower latitudes than usual. Auroras within the dayside portion of the oval are not often seen, partly because daylight obscures them from view except in midwinter and partly because very few people inhabit the near-polar region normally occupied by the dayside portion of the auroral oval.

The magnetosphere and why auroras occur

To understand more about auroras and why they occur it is necessary to look up above the atmosphere to the region from which the primary electrons and protons come. This region, the magnetosphere, was discovered and mapped by the host of increasingly sophisticated scientific satellites launched since the late 1950s.

As the name suggests, the magnetosphere is a region surrounding the earth wherein the earth's magnetic field is the dominant force field. On the earth's surface most matter is in the solid, liquid, or gaseous state; and the dominant force field affecting the matter in those states is gravity. About 100 km above the earth's surface, however, most matter is electrically charged (ionized) and is said to be in a fourth state of matter, the plasma state. Magnetic and electric fields strongly affect the charged particles in plasma; the force of gravity exists but is weaker, since the gravitational force on each particle is weaker than the electromagnetic forces. To matter in the plasma state the earth's magnetosphere acts as a semirigid extension of the earth itself, but it has no effect upon light waves or upon the solid, liquid, and gaseous material as it exists on the earth's surface.

The magnetosphere is a large structure extending toward the sun a distance equal to 10 earth radii (Re) and in the antisolar direction a distance of perhaps 1,000 Re (1 Re = about 4,000 mi). Its boundary acts as a barrier to the tenuous plasma, or solar wind, that streams out from the sun. That plasma consists mostly of electrons and protons moving at a speed such that they require one or two days to travel from the sun to the orbit of the earth. There, the magnetosphere prevents their direct entry to the earth. In fact, to the charged particles of a plasma, the magnetosphere acts as a gigantic magnetic bottle. Not only does it shield the earth from the solar-wind plasma but it

An auroral substorm, the most violent episode in the creation of auroras, is seen forming and breaking up in the series of all-sky photographs on the opposite page. Before the substorm begins the aurora is a quiet arc at the edge of the auroral oval. The arc gradually brightens, and then at 0718:00 the breakup begins. Bands and rays move poleward rapidly, and the sky is soon covered with the swirling formations.

Courtesy, Los Alamos Scientific Laboratory

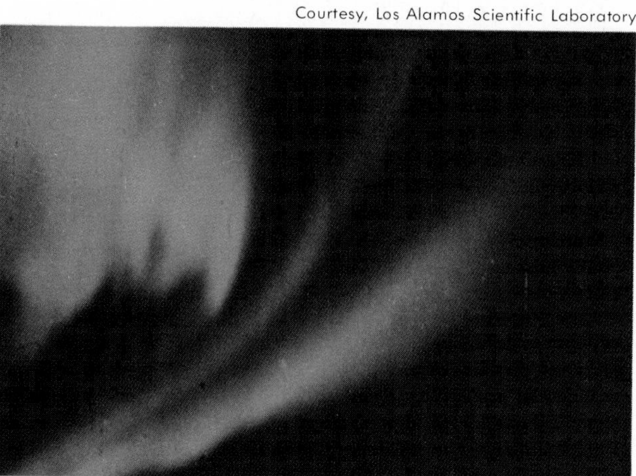

Identical auroras in the northern and southern hemispheres appear in simultaneous photographs taken at conjugate points (points on the same magnetic field line). Passing high above the equator, these magnetic field lines link auroras in the two hemispheres.

also traps plasma that is inside it. The physical agent causing this trapping is the earth's magnetic field. The Van Allen belts, two regions discovered by James Van Allen of the University of Iowa, contain electrons and protons trapped in the magnetospheric bottle. These particles are in constant motion, drifting around the earth in an east-west direction while, at the same time, bouncing in a north-south direction from hemisphere to hemisphere along paths parallel to the direction of the magnetic field.

The Van Allen belts lie in the inner part of the magnetosphere, where the magnetic bottle is particularly effective. Some particles trapped there have been circulating about the earth for years and will be for years to come. Out beyond the Van Allen regions, the magnetospheric bottle is leaky; some particles can enter it, circulate within the bottle for a short time, perhaps only seconds, and then escape. In dealing with the aurora scientists are concerned with this part of the magnetosphere.

The existence of the magnetosphere provides a partial answer to a mystery that has perplexed auroral scientists for 50 years. Large auroral displays tend to occur one or two days after major solar flares. By noting this time lapse, scientists can calculate the speed of particles coming out from a flare. But these calculations result in a speed that is too low to allow the particles to penetrate into the earth's atmosphere to the altitude where the aurora is observed. Therefore, if the electrons and protons causing an aurora come initially from a solar flare, it is obvious that they must be accelerated during or before the time they enter the atmosphere to produce the aurora. Scientists now believe that most such particles do indeed come from the sun and that they leak into the magnetosphere, where they are accelerated before plunging into the polar atmosphere. Just how or where that acceleration occurs is not yet known.

Current studies of the magnetosphere rely on measurements made on satellites and also on observations of auroras. Insight into the magnetosphere and the processes therein is provided by the aurora in a manner illustrated by an analogy that relates the magnetosphere to a television tube. A television tube contains within it a cathode source of electrons. Electrons are ejected from the cathode and then accelerated and guided by electric and magnetic fields within the tube so that they impinge on fluorescent material on the face of the tube. In the analogy, the atmosphere corresponds to the face of the TV tube and the aurora corresponds to the image thereon. Just as the changing image on the TV tube reflects the effect of the varying electric and magnetic fields on the beam from the electron source within the tube, so do the auroras playing across the atmosphere give information about the electric and magnetic fields and particle source regions within the magnetosphere.

Partly as a result of information provided by the aurora and related phenomena, scientists have learned that the magnetosphere is not a static shield encircling the earth. Instead it expands and contracts with changes in the intensity of the solar wind; with these changes there is thought to be a change in the rate at which charged particles can diffuse from the solar wind into the day-side auroral oval and into the outer portions of the dawn, dusk, and night-side magnetosphere. Sooner or later, the magnetosphere becomes unstable; deep within it, actually at a location approximately 30,000 km above the earth's equator on the night side of the earth, a slow but violent plasma explosion occurs. In the process electrons and protons are accelerated. Some of these follow down the magnetic field lines to strike the atmosphere and cause auroras, while others are driven across the magnetic field, toward the earth, where they increase the population of the Van Allen belts. An internal explosion occurs once each substorm; the auroral behavior during the substorm is attributed directly to changes occurring in the magnetosphere just prior to, during, and just after each explosion. During a large magnetic storm, the explosions follow one after another every few hours, and each time the earth is given a spectacular display of auroras.

None of the details of the processes occurring during the substorms are known; scientists are not even certain that all of the primary electrons and protons that cause auroras actually come from the sun. Some may diffuse upward from the atmosphere to be accelerated during the substorm and thereby plunged downward again. It is suspected that the explosive processes that the magnetosphere undergoes are similar to the explosive processes on the sun that cause solar flares. Therefore, once the dynamics of the magnetosphere are understood, the solar flares may be also.

Auroral effects on man's environment

The aurora moved from the realm of scientific curiosity to that of practical importance in the 1920s after communication by means of radio waves bounced off the ionosphere was introduced. The ionization (electric charging of atoms) that accompanies the aurora severely affects that form of radio communication by absorbing the signals and in some cases causing peculiar reflections. Interest in learning more about these effects motivated many of the studies up to the time of the International Geophysical Year (IGY) in 1957–59. During the IGY a massive worldwide effort was undertaken. Many scientific stations were equipped with all-sky cameras to photograph the aurora, magnetometers to measure magnetic variations, radars to detect reflections from the electrons in the aurora, and instruments to measure the absorption of radio waves and other phenomena related to the aurora.

The IGY resulted in the most comprehensive data collection ever obtained on auroras and related phenomena; these data clearly showed the global nature of the phenomena and the importance of their effects. Because of auroral disturbance to the ionosphere, radio communication that uses high-frequency waves bounced off the ionosphere cannot suffice to serve the needs of a civilization that demands increasing communication, reliable at all times and to all places. (Auroral disturbances on Nov. 7, 1972, the date of the United States presidential election, delayed election results from some parts of Alaska.) The demand for more and better communications is being met by increasing the use of satellites in high orbits so that radio signals are sent directly through the ionosphere instead of being mirrored off it. Satellite communication allows the use of higher frequencies that are least affected by auroral disturbances, but the problem still exists to some degree. Also, uncertainty still exists about whether aurorally related disturbances can affect the reliability of the navigational aids used by aircraft; the question was raised in September 1971 when 111 lives were lost at Juneau, Alaska, in the most disastrous single-aircraft crash in U.S. history.

Accompanying auroral displays are transient electric fields in and above the high atmosphere. These cause intense electrical currents in the ionosphere, sometimes measuring hundreds of thousands of amperes, and also induce currents at the earth's surface. These induced currents are strong enough at times to interfere with some forms of geophysical exploration used to find oil and mineral deposits and to disrupt voice and teletype traffic over long landlines used in telephone communications. Occasionally, large enough surges are induced in giant power grids to throw circuit breakers and thereby cause power blackouts. Considerable interest has also arisen in the possibility that the electric fields and currents accompanying auroras may modify high-altitude wind patterns enough to have some effects on the earth's weather.

An interesting and controversial question is whether the aurora has auditory effects upon man. Many persons have reported hearing crackling and rustling sounds during violent auroral displays. It is certain that these persons have not heard sound waves generated directly by the aurora, a fact that has led to the suggestion that the sounds were imaginary. The author has never heard these sounds, but several of his co-workers and acquaintances have. One person who hears the sounds frequently is known to have exceptionally good hearing. The most likely explanation of these reports is that some sound is being generated at the earth-air interface by the electrical and magnetic phenomena accompanying violent auroras. Another possibility is that the sensation of sound is being created within an individual by some physical mechanism not involving sound waves.

The future

Because of its direct association with the magnetosphere and because of its usefulness in studying that entity, auroral physics has recently changed from being a primarily atmospheric endeavor to being a part of the new field of magnetospheric physics. This in turn is part of a larger branch of physics, plasma physics. Since most of the universe is in the plasma state, this branch is taking on increasing importance in cosmology. Plasma physics now also includes the laboratory and theoretical investigations under way to develop means of containing thermonuclear fusion reactions for the production of power (see *1973 Britannica Yearbook of Science and the Future* Feature Article: NUCLEAR FUSION: POWER SOURCE OF THE FUTURE?). Though probably in a roundabout way, it is possible that auroral studies are thus contributing to one of the most serious problems facing man today, the development of clean, cheap, and limitless power.

Beginning in the late 1960s scientists increasingly used the magnetosphere as a full-scale plasma laboratory within which to perform experiments. One of the first of these, in 1969, involved the use of an electron accelerator flown on a rocket from Wallops Island, Va. The accelerator shot pulses of fast (10 KeV) electrons down into the atmosphere and created artificial auroras over Maryland and Virginia. In 1972 the experiment was repeated in Hawaii; this time the accelerator was directed upward. Its ejected electrons were guided along the magnetic field over the Equator a distance of more than 7,000 km to a point south of Samoa where an artificial aurora was created. These experiments clearly demonstrated that charged particles are guided by the magnetic field and that they do produce auroras when they strike the atmosphere. Experiments of this and other types, mounted on both rockets and satellites, will be used more in the future. During 1972 the first useful images of the aurora as seen from above were acquired with satellites. Soon it will be possible to use satellites to obtain photographs showing the entire auroral display over a hemisphere. Such photographs should help eliminate some of the present uncertainties about the global distribution and variations of the aurora.

Science Year in Review Contents

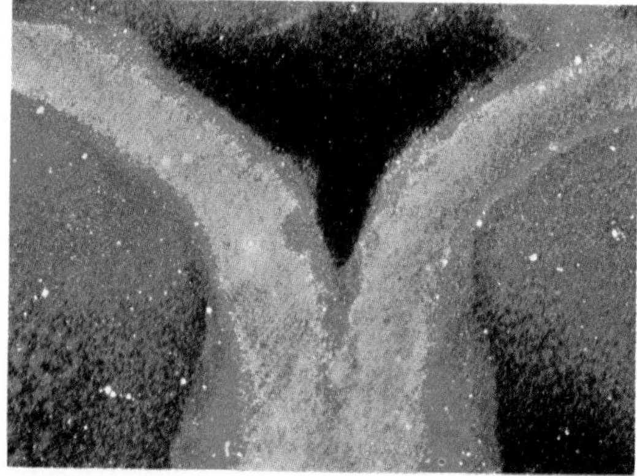

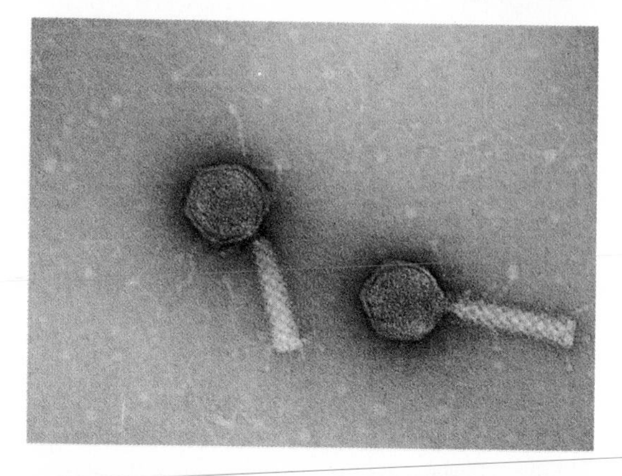

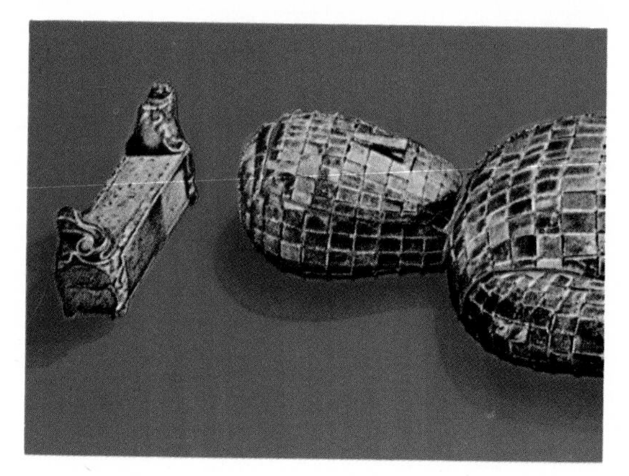

Agriculture

Agricultural science in the 1970s faced a critical test of its ability to serve consumers. Consumer needs had been changing—often dramatically—in recent years, and science was being called upon to develop new products and improve old ones in order to meet those needs.

Toward improved nutrition. Although milk is often called nature's most perfect food, scientists of the U.S. Department of Agriculture's Agricultural Research Service (ARS) were trying to make it better by increasing its content of unsaturated fats. Such "polyunsaturated" milk could significantly reduce the amount of saturated fat in the average diet. ARS scientists found that feeding encapsulated safflower oil to dairy cows increases the proportion of unsaturated fats in their milk. Safflower oil contains a high percentage of linoleic, an unsaturated fatty acid, and encapsulation protects the acid from saturation by the organisms in the cow's stomach.

Whey, a by-product of cheese making, contains proteins and other nutritious elements in a highly dilute form. ARS scientists experimentally fortified several flavors of soft drink with whey and found that the resulting beverages had excellent keeping properties and good taste. If whey proteins can be concentrated commercially at reasonable cost, this process could open new possibilities for the nutritional fortification of widely used snack beverages.

Protein from another unconventional source, cottonseed, was approaching commercial availability as the result of the newly developed Liquid-Cyclone-Process (LCP). Pilot plant runs at the ARS Southern Regional Research Laboratory, New Orleans, La., indicated that 100 tons of cottonseed should yield about 25,000 lb of high-quality edible flour containing about 65% protein. This could play a major role in increasing protein consumption in less developed countries that grow cotton.

Especially timely in view of the rapidly rising cost of meat was the development, by ARS scientists, of attractive, precooked, frozen convenience items made from low-demand lamb carcass parts. Lamb curry, shanks, loaf, riblets, sausage, and shish kebab were among the products. The lamb curry, for example, was a fully prepared and cooked dish made from lamb shoulder cubes, seasonings, fruits, and vegetables, all packed and frozen in a boilable plastic bag. Opportunities for such products were expected to expand as it became more and more necessary to utilize efficiently every bit of raw protein produced. In other research on meat, an ARS study indicated that, in general, young bulls produce 25 to 40% more lean meat than steers on the same diet. A related study showed that Holstein-Friesian dairy steers may become an important source of beef. In tests, these steers produced acceptable carcasses and performed as well, if not better, in the feedlot than steers of the Hereford breed.

An interesting aspect of human nutrition involves the interplay of nutrients. For example, calcium, which is highly important in the formation of bones and teeth and essential for blood clotting, muscle tone, and nerve function, must have the help of vitamin D before it can be absorbed into the bloodstream. ARS-sponsored research indicated that calcium metabolism is also affected by the amount of protein consumed, and that a high intake of protein increases the requirement for calcium. Thus the amount of protein in the diet may be as important to the body's calcium needs as the amount of calcium ingested. Future research would consider the relationship between protein intake and calcium retention and the possible effect of a high-protein diet on the development of certain bone diseases.

Environmental problems. ARS scientists found that severe mosquito infestations originating in poorly drained fields may be curbed by drilling through impermeable soil to permeable layers in small areas that are chronically affected by poor drainage. This procedure should be a useful tool for mosquito-control agencies, farmers with small problem areas, and on land where alternate cropping systems and good cultivation practices are not feasible. The technique could also be useful in a variety of other situations, such as grader ditches, where the seepage of irrigation water creates temporary mosquito habitats.

Fish that dine on mosquito larvae may have sufficient potential as biological control agents to warrant their commercial production and distribution. An ARS field experiment in California showed that topminnows of the species *Gambusia affinis*, released at a rate of 300 per acre per year, achieved good control of mosquito larvae over 2,017 ac of rice fields. The fish are cheaper than insecticides. The cost for 1,000 ac would be about $1,040, compared with $2,200 for chemical agents.

A new lure that gave promise of controlling yellow jackets and related *Vespula* wasps without contributing to environmental pollution was being tested by ARS scientists. More than 500 experimental lures were screened before one—octyl butyrate—was found that brought a rapid response. In the first ten days of testing, 130,000 wasps were trapped in a 22-ac Oregon orchard. Initial field tests of muscalure, a newly identified sex attractant for houseflies, showed that the attractant more than tripled the response of flies to several types of traps and baits. The potential usefulness of muscalure in reducing the need for insecticides was greatly enhanced by its ability to attract both males and females. Unlike some malodorous baits, muscalure is a clear, odorless oil.

The use of insects to control a major plant pest was under investigation. In many parts of the southern U.S. alligatorweed renders waterways unnavigable, interferes with fishing and other aquatic sports, contributes to flooding, and creates a favorable breeding habitat for mosquitoes. It is also highly tolerant to herbicides. Recently a tiny flea beetle, *Agasicles hygrophila*, has been used to curb alligatorweed without any help from herbicides. Originally imported from Argentina, the beetle multiplies rapidly, producing up to 1,000 eggs per female. Two other weed-eating insects from South America were also being used in the biological attack.

Biological control may produce adverse side effects, however, and constant vigilance is necessary. Scientists from the Agricultural Research Service noted that where alligatorweed had been controlled biologically, the equally destructive water hyacinth took over as the dominant weed. They planned to introduce other insects to feed on the water hyacinth. Meanwhile, still another aquatic weed, the European water milfoil, threatened to succeed the other two pests.

Alfalfa plants show the effects of air pollution. Plant at bottom, exposed for four hours to ozone, contrasts markedly with unexposed plant above. Tests were done to develop a pollution-resistant strain.

Courtesy, U.S. Department of Agriculture

New ways of dealing with municipal sewage sludge, the solid matter left after sewage has been treated, were being investigated. One possibility is to incorporate the sludge into cropland, as is done with livestock and poultry manure. A team of scientists was incorporating a variety of industrial and domestic sludge into a 75-ac test site at Beltsville, Md. In one field trial, the sludge was buried in trenches 2 ft wide and 2–4 ft deep, then covered with a 1-ft layer of soil. Other incorporation methods, such as deep disking or rotary tilling, were also being tested.

At the test site, drain lines were installed that emptied into a man-made pond so that groundwater and surface runoff could be captured, controlled, and studied. Test wells enabled the scientists to keep a close check on possible contamination of groundwater with nitrates and other chemicals, heavy metals, and harmful bacteria and viruses. Scientists also were watching for signs of crop damage from volatile materials, salts, and heavy metals. The study would enable them to determine the maximum amount of sludge that can be applied to land to provide the greatest benefit and the least hazard. In addition to solving the disposal problem, incorporation of sludge into cropland may improve nutrition for crops, increase rooting depth for plants, and improve retention and transmission of water in soil.

One of the steps in reclaiming water from industrial effluent involves the removal of heavy metals. A promising method of doing this was discovered by accident. While searching for new durable-press finishes for cotton, ARS scientists found that the compounds they were testing have a high affinity for metal salts and can make cotton into a highly efficient "trap" for water-borne heavy metals. The trap, which can be regenerated for repeated use, is capable of reducing the mercury content of contaminated water below the five parts per billion now permitted in drinking water.

Plant and animal health. The commonly used growth retardant SADH (succinic acid 2,2-dimethyl hydrazide) was found to reduce the sensitivity of petunias to ozone air pollution. Petunias are one of the most sensitive of plant species to this type of air pollution damage. ARS scientists obtained the best results by using the chemical as a foliar spray. They believed that SADH acted by reducing the size of the minute openings or stomata in the plants' epidermis or by causing them to close, thus restricting the entrance of ozone into the leaves.

Hope of curing elm trees infected with Dutch elm disease fungus (DED) was raised by two developments: a pressure technique for injecting mature trees with liquids containing chemicals and a water-soluble form of the fungicide benomyl that inhibits the disease without damaging the tree. With the new injection technique, benomyl is introduced directly into the tree, eliminating the hazard of environmental contamination. Two to four years would be needed before the effectiveness of this procedure against Dutch elm disease could be thoroughly tested.

A second viroid, one of a newly discovered class of infectious particles, was isolated by ARS scientists. The newly identified viroid is the agent of chrysanthemum stunt disease, previously thought to be caused by a virus. ARS scientists isolated the first viroid, the agent of potato spindle tuber disease, in 1972. The viroid is a fragment of ribonucleic acid (RNA), smaller than any known virus. It was believed that viroids might be implicated in many other plant diseases, as well as such animal and human diseases as multiple sclerosis, infectious hepatitis, and some types of cancer. (*See* Year in Review: MOLECULAR BIOLOGY, *Genetics*.)

Through operations on unborn lambs, ARS scientists were clearing up some of the mysteries of fetal immunity. It was hoped that this new research would lead to a better understanding of how vaccines affect the fetus, how fetal disease develops at different ages, and how disease organisms may be influenced by the fetus. An experimental vaccine for bovine viral diarrhea, a major cause of calf deaths, was found to produce an immune response without the risk of spreading the disease to susceptible cattle. ARS-sponsored research in Yugoslavia produced some of the first specific information developed on the nature of the antibody response to higher parasitic forms. The findings suggested the possibility of inducing immunity to liver flukes in livestock by infecting animals with irradiated, immature flukes before they grazed in fluke-infested areas.

—Marcella M. Memolo

Archaeology

The trade in antiquities, and the resulting wholesale destruction of archaeological sites, continued to be a major issue in archaeology. An impending crisis in this situation was feared as a result of the worldwide publicity produced when New York's Metropolitan Museum of Art purchased an ancient Greek vessel at a price purported to be about $1 million, many times the highest price previously paid for such an object. Such publicity inevitably creates a new burst of clandestine digging throughout the world.

Many organizations, such as the Archaeological Institute of America, took strong positions, condemning art museums for purchasing such objects, whether legally imported or not, and urging international action to stop the destruction of sites. Presumably, pressure of this sort will bring prices down in the international market. But the trade in antiquities was expected to continue to affect archaeological excavations in many, if not all, countries. These nations had for some time been expressing a growing resistance to excavation by foreigners. Their objections were based upon feelings of nationalism, a growing sense of ethnic identity, anticolonialism, and, perhaps in many countries, an increasing dislike of Western tutelage. The cry from many less-developed countries that they were being looted of their national artistic heritage by the fantastic antiquities prices paid in Western countries certainly exacerbated antiforeign feelings.

This trend leads to problems for all researchers in archaeology working in countries other than their own, problems that must be resolved if such people are to continue to do significant research. Probably the wave of the future is cooperative research in which nationals and foreigners work in teams with equal responsibilities, expenditures, publications, and direction. Fortunately, by 1973 most countries had well-trained and sophisticated archaeologists who were aware of the need for international cooperation. With some shift in the attitudes of Western scientists, productive research can be done.

Near and Middle East. What is surely one of the most remarkable archaeological discoveries in many years is that of a "proto-industrial urban center," dating from approximately 3000–2000 B.C., in the arid Seistan Basin of eastern Iran. Moreover, in one respect it is unique in the history of Near Eastern excavations. A century ago travelers reported a huge plateau, almost 300 ac in extent, totally covered by a thick layer of pottery fragments. Again in 1916 Sir Aurel Stein visited the site and dug some test pits to find that the layer of pottery fragments was only four inches deep. He concluded that the plateau was an ancient wind-eroded site and of no use to archaeologists. Then, recently, an Iranian/Italian excavation discovered that the hard layer below the fragments, which Stein thought to be a natural sedimentary deposit, was in fact an 8-in.-thick concretion of salt, sand, and clay covering and sealing in a deposit up to 23 ft thick—the remains of a 5,000-year-old city with some buildings standing to roof level.

Because of the hard overlay, containing salt, the buildings of the city contained masses of perishable objects preserved as in no other city site in the Near and Middle East. In addition to the incredible numbers of pottery fragments (about two million) and stone implements (25,000 to date), there were ropes, baskets, dyes, textiles, wooden objects,

Stone tablet below was found at Tall-i-Malyan site in Iran (right), capital of the ancient Elamite civilization. The tablet bears the Proto-Elamite script of about 3000 B.C.

Courtesy, Dr. William Sumner, Ohio State University

bones, and other perishable materials. With such a discovery all of the new scientific techniques for analysis and interpretation were brought to bear with teams of scientists from several countries. Carbon-14, uranium-238, and paleomagnetic techniques were used to date the remains; paleobotanists and zoologists were identifying remains of ancient plants and animals using the flotation process for isolating micro-remains in the debris; and others were performing microscopic analysis of animal dung and other materials to fill in the knowledge of the environment, the climate, the degree of domestication of plants and animals, and manufacturing processes and techniques.

The site was named Shahr-i-Sokhta. As of 1973 it appeared that the city remains covered an area of 173 ac and the cemetery an area of 104 ac (calculations indicated at least 21,000 graves). The excavations showed that it was truly an industrial city with evidence of people working as masons, precious-stone cutters, potters, coppersmiths, weavers, and dyers. Lapis lazuli from Badakshan in Afghanistan demonstrates that the city traded over a large area and perhaps had a monopoly on the transfer of this precious material to other cultural centers in the Near and Middle East. The city may even be the mystical Aratta referred to in Sumerian documents.

Cultivated plants included wheat, barley, cucumber, millet, melons, poppies, and grapes. Domestic animals were cattle, sheep, goats, and, judging from the masses of egg shells, possibly some domesticated birds. There were also remains of many wild birds, animals, and fish. Pottery from the lowest levels indicate some connection with Soviet Armenia, while the latest materials resemble in some respects that of the Indus Valley.

Another major city site in Iran known as Tall-i-Malyan, 29 mi N of Shiraz, appears to be the Elamite capital known from ancient inscriptions as Anshan. Robert Dyson, Jr., and William Sumner, directing the work there for the University Museum, Philadelphia, noted that the site covers an area of 350 ac, which is more than twice the combined area of all 77 sites in the Kur River basin known to have been occupied during that period. The deposit is 40 ft deep with the earliest levels apparently dating to 4000 B.C. This site and Shahr-i-Sokhta alter previous ideas about the genesis of urban civilization occurring exclusively in Mesopotamia.

The rare circumstance of identifying an ancient city with inscribed materials at the site was being demonstrated at Tall-i-Malyan by the comparison of tablets found nearby on the surface that are inscribed with the name Anshan with fragments found in excavation. Texts written in Elamite are the same, and it seemed only a matter of time until the complete texts including the name Anshan are found in controlled cuts in the debris of the city. It now seems clear that Malyan was a great cultural center in Iran during the 2nd and 3rd millennia B.C.

Surprises continued during the Egyptian excavations of 1972. They included a mining district of the Middle Kingdom (20th century B.C.), another ancient site 99 mi SW of Kharpa on the old caravan route between the Nile and Nubia with inscriptions of the 12th dynasty (1991–1778 B.C.), and a cemetery of the 1st century A.D. near the village of Al Qasr in Wahat ad Dakhilah. The most interesting features of this last discovery are the tomb paintings, beautifully preserved and of extraordinary quality—some figures reminding one of the Etruscans. Moreover, on the ceilings are painted circles representing the zodiac, with the 12 signs, 5 planets, the sun, the moon, and other astronomical representations. Demonstrating excellent painting and well preserved, these are the most important zodiacs ever discovered in Egypt.

Asia. Evidence of the world's earliest known agriculture and pottery manufacture, found in South and East Asia by W. G. Solheim, Chester Gorman, and others, apparently can be linked with the earliest known bronze manufacture. The site of Non Nok Tha in northeast Thailand, excavated by the University of Hawaii in the late 1960s, produced evidence of bronze metallurgy as early as the 3rd millennium B.C., using radiocarbon dating methods. Then excavations at Ban Chiang, also in northeast Thailand, undertaken by Thai archaeologists in 1970 and 1971, exposed a vast cemetery with Neolithic, Bronze Age, and Iron Age materials, including a distinctive red-and-white painted pottery resembling the neolithic Yang Shao pottery of China. At Ban Chiang this type of pottery was clearly associated with large numbers of bronze objects.

Early in 1973 Bennet Bronson and Mark Han published in *Antiquity* the results of thermoluminescence analysis of 21 samples of Thai pottery from seven sites ranging in age from about 4500 B.C. to A.D. 1350. Among these were three samples of the red-and-white ware from Ban Chiang, and their dates extended from about 4600 B.C. to about 3500 B.C. with possible errors ranging from 520 to 275 years. The dates are consistent with the archaeological chronology of the sites.

Significantly, the earliest series in the 26 radiocarbon dates from Non Nok Tha represent a span of time from the 3rd to the 2nd millennium B.C., and Solheim concluded from typological analysis that Ban Chiang was contemporary or earlier than the lowest strata at Non Nok Tha. Thermoluminescence dates from the iron-producing sites of Chansen I and Lopburi Army are not as sure and as of late 1973 they had no support from radiocarbon dates, but they did suggest that iron may have appeared in Southeast Asia a half-millennium earlier than in China and India.

Much attention has been focused on recent archaeological excavations in China. Most dramatic was the opening of Han Tombs (roughly contemporary with ancient Rome) containing armorlike jade suits covering the bodies of the king and queen. The two most complete jade suits were made up of 2,690 and 2,156 pieces of jade tied together with golden wire. Nothing quite like them

Archaeologists at the Egyptian oasis of Wahat ad Dakhilah uncovered a 1st-century A.D. depiction of the moon in a zodiac painting (left) and the limestone stela of a local ruler of about 2200 B.C.

Courtesy, Dr. Ahmed Fakhry

Courtesy, UNESCO COURIER

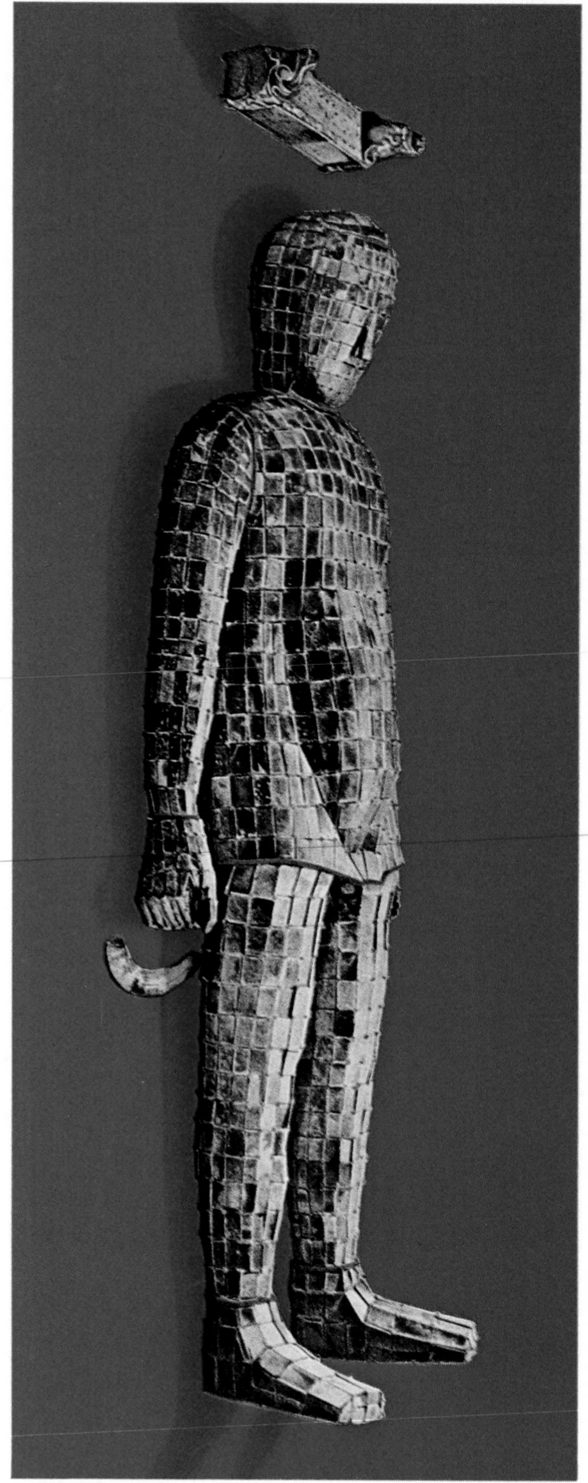

Funeral shroud of a Chinese princess of about 200 B.C. consists of 2,156 domino-sized pieces of jade bound together with gold filament. Discovered in the Han Tombs by Chinese archaeologists in 1968, the shroud was a part of an exhibit of ancient Chinese treasures shown in Paris and London during 1973.

had ever been found although they were mentioned in Chinese documents of 235 B.C., before the Han period. The tombs were also rich in gold, bronze, and silver objects of handsome design (2,800 pieces in the two tombs).

In August 1972 the Chinese news agency reported still another Han Tomb discovery as the most important in recent years. This was a charcoal-sealed tomb containing the remains of a woman preserved in such an extraordinary manner that it was like that of a recently deceased person. The body was so well preserved that it could be embalmed after discovery in the modern way. The woman was found half-submerged in a reddish fluid that may have had some effect upon preservation.

The importance of the find, from an archaeological point of view, lies in the materials found in the tomb. Wrapped in 20 silk garments—brocade, embroideries, damask, and gauze, exquisitely ornamented—the burial produced what the Chinese describe as the most valuable Chinese silk ever found. Moreover, there were 120 small wooden figurines dressed in miniature clothes, 26 of them representing musicians and dancers. On a banquet table were remains of peaches, pears, melons, eggs, pickled vegetables, and rice cakes—all recognizable after 2,000 years.

Mediterranean and Europe. A significant trend in Mediterranean archaeology was the adaptation, with some revising, of a new scientific technology pioneered by U.S. and British archaeologists working with prehistoric remains. These techniques supply new evidence about the ancient environment, the food supply, and technical skills to interpret life in the past. The archaeologists of the classical civilizations on the Mediterranean had reservations about this self-proclaimed "new discipline" with its statistical concerns and the jargon that accompanied it. Hopefully, the new techniques may produce in the Mediterranean a sophisticated synthesis of the methods of the "new archaeology" and the humanistic concerns of traditional classical work.

The most spectacular excavation in the Mediterranean continued to be that of Spyridon Marinatos on the island of Thera (Santorini). New and impressive wall paintings discovered in fragments in 1972 were pieced together and described as "the most important historical document we have so far from the Bronze Age in Greece." Discovered in the late Minoan site of Akrotiri, which was buried under the ash fall from the volcanic explosion of the island about 1500 B.C., the wall painting is a panel extending around three walls of a house. It measures 21 ft in length and 8 to 16 in. in width. Depicted is a fleet of warships, the sacking of a city, a harbor, wooded hills with animals, and other scenes which Marinatos interpreted in part as an Aegean fleet attacking Libyan cities.

In the Soviet Union M. M. Shmagli announced the discovery of a Stone Age town near Kiev that contains the ruins of about 1,500 solidly constructed houses, some of two stories, arranged in a concentric pattern with radially oriented streets. The population was estimated to have been 20,000, making it the largest Stone Age settlement so far discovered in Europe. The town dates from about 3000 B.C., thereby supporting the theory of much earlier urban settlement than was once thought to be possible. It also confirmed the idea that the shift from hunting and gathering to early agriculture very soon led to an urban pattern of living without a long intervening period of village life—at least in certain areas.

America. Two recent papers in *Science* (January 26 and March 9, 1973)—one by W. N. Irving and C. R. Harington and the other by Paul Martin—emphasized that the centuries-old argument about the first settlement of America is by no means solved. Martin saw "an ephemeral or scarcely detectable invasion by or before 15,000 years ago" and a major occupation of America by "big game hunters 11,200 years ago." He also theorized that a rapid spread and population explosion of these hunters caused the extinction of many of the large Late Pleistocene mammals, such as the mammoth and the mastodon, in a relatively short period. Irving and Harington, on the other hand, reported the discovery of three certain man-made bone artifacts, and many probable ones, with large collections of Pleistocene mammal bones along the Old Crow River in Yukon Territory, northern Canada. They gave radiocarbon dates between 25,000 and 29,000 years ago for the three bone artifacts.

For many this conflict recalled the discussions of 40 years ago when the radicals in American archaeology claimed that man and mammoth were contemporaries in America in opposition to such authorities as Ales Hrdlicka. Certainly, the trend since then has been acceptance of earlier and earlier Americans, but there is still no agreement as to how early.

Africa. The discovery of many more fossils of ancient man in East Africa continued to make world news. Richard Leakey of the Kenya National Museum and Glynn Isaac of the University of California at Berkeley, working east of Lake Rudolf in northern Kenya, reported at the end of the 1972 field season 93 new hominid fossils. These included several almost complete skulls, a number of lower jaws, many pieces of crania, and one complete femur—the only one known from the lower Pleistocene period.

Most interesting are the fossils lying below a layer of tuff that can be dated by the potassium-argon method to 2.61 $\pm$ (0.26) million years. Some of these fossils are similar to the robust australopithecines well known from South and East Africa, but one skull, with a cranial capacity of 49 cu in., is approximately 18 cu in. larger than the australopithecines. The complete femur is almost identical with that of modern man. These two fossils add new dimensions to the continuing reevaluation of the early development of man.

At several archaeological sites excavated stone tools, like those from Bed I at Olduvai, were found below the 2.61-million-year tuff layer. Also bones of large mammals, such as the extinct hippopotamus and horse, appeared with the stone tools, indicating that these early hominids were hunters of large game.

—Froelich Rainey

Architecture and building engineering

Outstanding developments in architecture and building engineering during the past year included the start of construction on the world's longest suspension bridge and on a huge new water tunnel, new designs for docking giant supertankers, and the completion of the world's tallest dam. Two crisis situations, the worldwide energy shortage and pollution of the environment, stimulated innovative efforts by architects and engineers to cope with such diverse problems as power overloads caused by air conditioning and pollution-induced corrosion of historic buildings.

New water tunnel. One of the oldest uses of tunneling is for water supply. In the sixth century B.C. a Greek engineer named Eupalinos drove a tunnel through Mt. Castro on the isle of Samos to provide the tyrant Polycrates with fresh water. Twenty-five hundred years later a modern successor to Eupalinos's tunnel is being dug through the rock beneath New York City. Running from Hill View Reservoir, just north of the city, down through the west Bronx, under the Harlem River, through uptown Manhattan, and east under the East River at Welfare Island to Queens on Long Island, City Water Tunnel No. 3 is 13.7 mi long. The difficulties of digging it derive less from its length than from its depth, which ranges up to 780 ft. Water tunnels are placed at great depth to conserve natural pressure resulting from the descent from sources above sea level. Because New York City water drops 280 ft on its way down from the Catskills, it rises 28 stories in Manhattan buildings without pumping.

The new tunnel is not as long as either of its predecessors in New York, City Water Tunnels No. 1 and No. 2, 18 and 20 mi long, respectively. It is, however, much larger in diameter (24 ft, finished and lined). No. 1 took ten years to complete (1907–17) and No. 2 seven years (1929–36). Despite the improvements in tunneling technology, the new tunnel was expected to take at least five years to finish, partly because of its greater diameter, partly because it contains three large and complex chambers for valves and controls, and partly because of problems with shifting rock in the vicinity of the Harlem River. Two tunnel headings were under construction there from a shaft sunk at Highbridge. The two other working shafts were north of Highbridge in the Bronx and on Welfare Island. The three underground chambers served different purposes and were of widely varying design. The most complex was 44 ft by 640 ft in plan, with a 64-ft-high arched roof 250 ft below the surface.

Super-suspension bridges. Work began in 1973 on a new suspension bridge in Great Britain that will take the title of world's record span length away from the United States for the first time in half a century. The new bridge over the Humber River at Hull, a steel box girder deck hung from inclined suspenders from two parallel-wire cables, will have a main span of 4,580 ft, 320 ft longer than the Verrazano-Narrows Bridge in New York.

Britain will not retain the title long, because Japan is planning an even longer suspension span, over the Inland sea between the main island of Honshu and Shikoku. The main span of the Japanese bridge will measure just under 5,000 ft.

Shattering of a granite slab 0.4 in. thick is achieved by subjecting it to a burst of electrons at medium energy. Such "electron guns" might prove a major breakthrough in underground excavation.

Courtesy, Lawrence Berkeley Laboratory, University of California, Berkeley

The same British engineering firm that designed the Hull bridge saw another of its projects draw near completion on the far side of Europe. The Eurasian bridge across the Bosporus at Istanbul represented the fulfillment of a dream at least a thousand years old. With a main span of 3,524 ft, the new bridge places fourth among existing world span lengths behind the Verrazano-Narrows (4,260 ft), the Golden Gate (4,200 ft), and the Mackinac (3,800 ft). Besides the main span the bridge consists of nine structurally independent side spans, five on the European side and four on the Asian side, totaling 1,593 ft.

Berths for giants. The supertanker, which produced a maritime revolution by setting afloat ships several times as large as the "Queen Mary," has like most technological revolutions created as well as solved problems. A 350,000-ton tanker—"very large crude carrier" or VLCC in marine jargon—cannot be berthed at a pier built to handle an 80,000-ton liner. Therefore, several new ports were built in Europe, Japan, and elsewhere to accommodate the new giants. One under construction in 1973 for a British carrier firm was on Grand Bahama Island, designed to serve the U.S. east coast. Since the island is about 50 mi from Florida, the new port was to operate as a transfer point to unload oil from the supertankers and load it into ordinary-size tankers that could dock at North American ports. The port's 1,470-ft steel pier was designed to stand in 100 ft of water about three-quarters of a mile off the Bahama shore. As each carrier docks, it will be girdled with floating booms to contain potential spillage. The main reason for locating in the Bahamas is the environmental opposition to supertanker ports closer to the U.S. shore, for example in Delaware Bay and in the Gulf of Mexico off the Louisiana or Texas coast.

Japan in 1973 completed what was at the time the world's largest tanker dock. Located at Kagoshima, it could handle 477,000-ton ships, the largest under construction. Only 1,300 ft offshore, the new Kagoshima pier was designed to pipeline oil directly from the supertankers to storage facilities. It supplemented an existing pier that could handle vessels of 400,000 tons.

An entirely different approach to the problem of the supertanker was put forward in West Germany, where a Munich engineering firm designed a huge U-shaped concrete floating tube capable of protecting a pair of 500,000-ton tankers during unloading in the open ocean. The tube, a self-propelled vessel of shallow draft, could then move into shore and connect to a pipeline buoy. A boom across the open end of the U would contain any

(Left) Novosti from Sovfoto; (right) APN from Sovfoto

Chambers for the volute pumps of the No. 8 and No. 9 hydropower units at Nurek Dam are installed, below. At the right, construction continues on the dam, located on the Vakhsh River in the Soviet Union.

spillage. Fourteen 3,000-hp engines would power the strange vessel and also supply guidance. Its bulkheads were designed to withstand accidental ramming.

Island in the sea. Another approach to supertanker docking was a project for which a Dutch-British engineering combine began feasibility studies: an artifical island in the North Sea. The primary purpose of the island would be to house industrial-waste-disposal plants for the industry of northwest Europe and Britain, but a second would be to act as a harbor for supertankers. The island would be built on one of the many shallows in the North Sea, perhaps 50 mi off the coast of the Netherlands.

Artificial islands are nothing new. The Chesapeake Bay Bridge-Tunnel crossing employs artificial islands for the tunnel portals in the navigation channel. For more than a hundred years French and British engineers have discussed building an artificial island on the Varne Bank, a shallow in the English Channel, to facilitate construction of the Channel Tunnel.

Big dams. Two giant dams at opposite ends of the earth began contributing toward the solution of the worldwide energy crisis during the year. In Argentina, El Chocón, an earthfill dam 7,700 ft long and 312 ft high, delivered the first 400 of an eventual 1,200 Mw of power. Built by an Italian-Argentine consortium, El Chocón is part of a large hydroelectric complex under construction on Río Limay and Río Neuquén. The Limay was diverted into two giant diversion tunnels during the flood season until the dam was high enough to withstand the torrent. El Chocón was to be supplemented by a smaller facility on the Neuquén at Cerros Colorados, a complex of dikes, control structures, and power plants designed to produce an additional 450 Mw.

In Siberia the huge Nurek Dam, on the Vakhsh River in Tadzhikistan, began delivering power late in 1972. The first three generators were operating at only half their rated 300-Mw capacity because the industrial plants that were under construction as Nurek's customers were only partly complete. Eventually, nine generators will produce 2,700 Mw.

Nurek is an earthfill dam 1,040 ft high with a crest length of 2,624 ft, under construction since 1962. Four diversion tunnels were completed or under construction to fend off the floods of the Vakhsh, an exceptionally turbulent river whose spring-summer torrent runs at a rate more than 20 times the volume of its winter flow. The four tunnels were situated one above the other to correspond with the increasing height of the dam; the first and lowest was plugged before the fourth and highest was built.

Nurek is a dual-purpose dam. In addition to supplying power for industry, it will irrigate 250,000 ac for cotton growing. It will ultimately be linked to the giant nationwide power grid being developed to encompass the entire Soviet Union by 1980.

Another even larger dam was being planned by Soviet engineers for Ragun, farther upstream on the Vakhsh. The Ragun earth and rockfill structure was designed to tower 1,060 ft over the river, taking the title of world's highest from Nurek. The U.S.S.R. also planned a huge three-mile-long dam on Kerch Strait between the Black Sea and the Sea of Azov. The purpose of the dam would be to prevent the saltier water of the Black Sea from entering the Sea of Azov and thereby destroying important fisheries there.

In the U.S. the governors of Oregon, Washington, and Idaho joined environmentalists in trying to prevent construction of two hydroelectric dams in the last wild stretch of the Snake River, in Hells Canyon. "It is time that we recognized that Hells Canyon is truly a unique and manificent national treasure," the governors wrote the Federal Power Commission.

See also Feature Article: THE TROUBLE WITH SUPERDAMS.

Air pollution and Europe's landmarks. Venice, Milan, Rome, and other European cities with rich heritages from the past continued to study ways of saving their cathedrals, castles, outdoor sculpture, and other works of art from the corrosion caused by modern air pollutants. The architect to the Cologne Cathedral reported a failure in the battle against sulfur, the most harmful pollutant for most ancient masonry. Protecting old stones against sulfur's ravages by the use of chemicals, including plastic coatings, proved unsuccessful in tests because sulfur already inside the stone continued to work on it. The only solution in sight was the replacement of affected areas by resistant types of stone.

Even relatively modern structures were not immune from damage. Much of the Cologne Cathedral dates only from the 19th century, while in Paris the wrought-iron Eiffel Tower, dating from 1889, was in trouble. The Eiffel was to be saved by an expensive replacement of the rusted iron elements with steel—a somewhat ironic commentary on the furor following its construction, when many Parisians vehemently demanded that it be removed as an eyesore.

A European landmark affected by a non-pollution problem was the Leaning Tower of Pisa, whose long-threatened collapse grew more imminent. Numerous plans for saving the 185-ft campanile have been advanced from the very beginning of its eight-century life. The first, adopted during its construction, contributed to the problem: the engineer in charge lengthened the columns on the sinking side to compensate esthetically, and thereby added weight. The Italian government specified in its request for engineering bids that the technique for saving the tower must remain completely invisible from the outside and must not alter the present lean by more than one-sixtieth.

Facade and statuary of the cathedral in Milan suffer corrosion from acid pollutants in the air. The iron structures holding the marble sections of the cathedral together are being eaten away by smog.

Mondadoripress from Pictorial Parade

Courtesy, Gulf Oil Co.

Supertankers unload oil at Point Tupper in Nova Scotia, described as the most developed deep-water oil port in North America. Built by Gulf Oil Canada, Ltd., the $105 million facility has a 2,000-ft dock that stands in water 100 ft deep and can service the largest tankers afloat.

The air-conditioning problem. One of the contributing factors to both the environmental and energy crises was high-powered air conditioning. Air conditioning of houses, apartments, factories, stores, and office buildings raised the demand for power higher and higher every year; to meet it either more nuclear power plants had to be built, causing thermal pollution, or more oil had to be burned, causing air pollution. Air conditioning is hardly a new technology, having been pioneered by John Gorrie of Apalachicola, Fla., for his yellow fever patients in 1842. Modern air conditioning, involving electric powered cooling and blowing, was one of the first applications of electric power, and existed in a mature form before World War I. Not until after World War II, however, did it suddenly become "necessary" and spread rapidly, first among commercial buildings, then among residences, and finally into automobiles (where it increased gasoline consumption, also adding to pollution).

Most approaches to the problem were in terms of producing more power while at the same time containing the power-produced emissions. In the past year, signs appeared suggesting another approach: improvements in insulation and in the technology of air conditioning itself to reduce the need for electric power. The University of Delaware built an experimental house with 24 solar panels on the roof. Cadmium sulfide cells in the panels collect the sun's heat and transform it into electrical energy, stored and used for all the house's electrical needs, including air conditioning. Another technique, researched at the University of Pennsylvania, employed a solution that freezes at night and is used for cooling the next day. Because the freezing point of the solution is high, little power is needed for operation.

An even simpler technique was tried out by a California building firm with promising results. Shallow water ponds on a flat roof were covered by panels during the day to keep the water from heating up. The panels then were slid back at night, exposing the water to the cool night air. During the next day the chilled water absorbs the daytime heat of the house.

One of the most interesting experiments was that of a former U.S. Air Force officer, Robert Reines, who designed a house in New Mexico "as a system to conserve heat, water and electricity" so effectively that its heating and electrical needs could be supplied by the sun and the wind. The house is a white hemispherical dome of metal panels insulated by a thick coat of plastic foam sprayed on the inside. A short entrance hallway with two doors helps insulate against winter cold and summer heat. Hot water for radiators is supplied from a storage tank fed by solar heat collectors, black copper tubes containing a water-glycol solution that are run through the water tank. Reines planned to modify his heating system in 1974 so that it could be reversed into a cooling system by running it at night and filling his radiators with chilled water during the day. Skylights and portholes provided enough light to reduce electrical needs to a point where they could be met by power from three windmill-driven generators.

Total jetport. For Montreal's new jetport the Canadian government took the far-sighted step of purchasing not merely the 18,000 ac of land needed for the runways and terminals, but an additional 70,000 ac surrounding the field. By this means the government intended to ensure that the jetport would have only compatible neighbors, thereby avoiding problems that have plagued modern commercial aviation throughout the world.

Initially, the port will have two operational runways, but ultimately, to handle the traffic of the year 2000 and beyond, four more will be added, with support facilities sufficient for a city of 30,000. The passenger terminal will eventually be connected to an as yet unbuilt public transit system, while transport to airplane gates will be by a special vehicle, a "mobile lounge."

—Joseph Gies

Astronautics and space exploration

The Apollo program of manned space flights to the moon ended in December with the successful mission of Apollo 17. A new U.S. manned program began in May with the launching of Skylab I, which, after some difficulties, was repaired by the astronauts while in orbit. Pioneer 10, an unmanned probe en route to Jupiter, successfully navigated the hazardous asteroid belt.

Earth satellites

Earth-oriented satellites that utilize the unique vantage point of space to accomplish a variety of missions are called applications satellites. They are used to obtain useful knowledge of the earth, for economic benefit, and for military security.

There are three general classes: communications, earth survey, and navigation. Satellites in each of these categories continued to demonstrate improved performance and promise for the future. Although the United States and the Soviet Union remained the only countries that launched and operated applications satellites, there were many cooperative international programs.

Communications satellites. This category of satellite operations was dominated by the U.S. Communications Satellite Corporation (Comsat). Comsat operated a global space communications network for an 81-nation consortium, Intelsat. All Comsat satellites were launched into geostationary, or synchronous, orbits about the equator. This meant that at a certain orbital altitude, 22,300 mi, the satellite velocity matched the angular rate of the earth's rotation and thus remained fixed over one portion of the earth's surface.

In 1965 the first commercial communications satellite, Early Bird, was launched. With a capacity equivalent of 240 telephone circuits, it increased transatlantic telecommunications capacity by 50% and made live commercial television transmission across the ocean possible. In 1969 a global system of communications satellites began operating over the Atlantic, Pacific, and Indian oceans. By 1973 these satellites provided 255 communications pathways among 80 antennas in 49 countries. Not only were long-distance telephone and television carried but also telegraph, computer data, and facsimile transmissions. One out of four people on earth could see important events on television screens as they happened, "live via satellite." Costs of long-distance telephone calls dropped dramatically. In 1966, for example, a three-minute, station-to-station telephone call between New York and London cost $12.00. By 1973 it was $5.40 (less at specified hours).

The latest in a series of Comsat satellites, the Intelsat 4, has a capacity of 6,000 telephone circuits or 12 color television channels. Three advanced Intelsat 4s, to be launched beginning in 1975, would each have double that capacity.

Several applications were made by Comsat and other U.S. firms to the U.S. Federal Communications Commission (FCC) for domestic telephone and television satellite systems. There was considerable uncertainty as to how domestic service would evolve in the U.S., because of the number of competitive industrial applicants, the existing massive capital investment in land systems, and federal regulatory complexities. However, U.S. domestic systems were expected to become operative during the 1974–75 period. In January the FCC approved the plan of Western Union Corp. to build and orbit three satellites for handling the company's telegraph, Telex, and private-line services.

An important development in the use of a communications satellite was scheduled for April 1974 when the U.S. National Aeronautics and Space Administration (NASA) would launch Applications Technology Satellite F (ATS-F). This heavy, geostationary satellite was designed to be used in an interagency cooperative program of education and health care. Regional areas in Appalachia, the Rocky Mountains, and Alaska would participate. A year later, the satellite would be moved halfway around the world to allow the Indian government to conduct a one-year instructional television experiment. In this experiment thousands of small remote villages were to receive educational programs broadcast via the satellite by means of low-cost receiving antennas and television equipment.

In 1973 the U.S. Department of Defense continued its development and launching of geostationary satellites to provide a secure communications system for use by the president, Department of State, and other federal agencies. Sixteen nonsynchronous satellites of an earlier limited capability system were still operational. These interim spacecraft were designed to turn off automatically in late 1974. The U.S. Navy/Air Force FleetSatCom communications system, under development in 1973, was intended to provide a worldwide, ultrahigh-frequency communications link between naval aircraft, ships, submarines, and ground installations. Launch of the first FleetSatCom was scheduled for 1975–76.

The Soviet Union operated a communications satellite system that differed from the geostationary one of the U.S. The Soviet satellites flew in a 65° (inclined to the equator), highly elliptical orbit with a perigee (low altitude) in the Southern Hemisphere. Thus, such satellites, called Molniyas, were within view of major Soviet ground terminals for about eight hours a day. An advantage of this orbital path was good reception at near north-polar latitudes, harder to reach from the U.S. equatorial orbit system. By spacing three Molniyas in 12-hour orbits, it was possible to obtain full 24-hour coverage. For the future, the U.S.S.R. planned a geostationary satellite, Statsionar, to become operative in the mid-1970s.

An interesting cooperative satellite communications development began taking place between the U.S. and the U.S.S.R. during the year. The so-called hot line, linking the offices of heads of the two nations, was established in 1963 and used terrestrial cable, wire-telegraph networks. A backup radiotelegraph circuit was established also. To upgrade this service, potentially vital to world peace, two satellite communications circuits were being established between the two capitals. Intelsat was to operate one, and the other would utilize Molniya 2 satellites.

Technicians make final adjustments on Anik 1, a communications satellite built by Hughes Aircraft Co. for Telesat Canada. Launched in November 1972, Anik 1 was designed to provide Canada with 12 channels of color television or up to 5,000 telephone circuits.

Courtesy, Hughes Aircraft Co.

In late 1972 Canada orbited the first of three satellites, Anik 1, to establish nationwide domestic communications. NASA provided the launching service on a reimbursable basis. The Franco-West German Symphonie communications satellite program continued, with an anticipated launch of a prototype in 1974.

Earth-survey satellites. In this category are included meteorological (weather), geodetic, earth resources, and reconnaissance applications. Such satellites are designed to survey the earth with various sensors, photographic and electronic, in order to obtain and transmit data that can not be gained by other means.

Weather satellites. The U.S. and the Soviet Union continued worldwide, daily operations to obtain atmospheric and environmental data. In the U.S., the National Environmental Satellite Service (NESS) was responsible for conducting weather satellite operations. NESS is a major element of the National Oceanographic and Atmospheric Administration (NOAA) of the U.S. Department of Commerce. All NOAA satellites are launched into polar orbit, providing global coverage as the earth turns. Developmental work on advanced design satellites and all launchings are conducted by NASA.

In October 1972 the first of a new class of environmental satellites, NOAA-2, was launched. This was the first operational satellite to utilize scanning radiometers (SRs) only for imagery, replacing the earlier optical television system. In addition, NOAA-2 carried very-high-resolution radiometers (VHRRs) and vertical temperature profile radiometers (VTPRs). These latter sensors obtained vertical temperature profiles of the atmosphere on a near-global basis. The SR subsystem gathered data in the visible and infrared spectra both day and night. Global cloud-cover imagery was stored for later readout to U.S. command and data acquisition stations. Direct readout of local regional cloud-cover imagery was available to all nations.

The operational VTPR marked an important step toward the obtaining of quantitative global measurements needed for detailed weather forecasting. The VTPR measures infrared energy radiated from the earth's surface or tops of clouds. These data are then reduced to calculate the vertical temperature distribution in the atmosphere.

The VHRR obtains data similar to the SR but at a much higher resolution (½ nautical mile as compared with 2–4 nautical miles). Measurements from VHRRs can be used to obtain cloud-top temperatures and, in clear areas, temperatures of the ocean surface.

From NOAA-2 data it was learned that there are large eddies of cool water that break off from the pattern of flow of the Gulf Stream. These eddies represent enormous pools of stored energy and were detected by VHRR visible and infrared imagery. Similar studies of the Beaufort Sea showed some ice areas significantly warmer than others, suggesting consistently thinner ice in those areas. Infrared imagery permits the mapping of ice and floes during the polar night.

NASA's Nimbus 5 environmental satellite placed in orbit advanced-design radiometric and spectrometric sensors. Information obtained from these was expected to be useful in providing improved range and accuracy of weather forecasts, in determining the feasibility of large-scale weather modification, and in assessing the long-term effects of pollutants in the atmosphere. Such research was the aim of the Global Atmospheric Research Program (GARP) of the International World Weather Program.

Planned for launch in late 1973 was NASA's Synchronous Meteorological Satellite (SMS), the first of two. Whereas the lower-orbiting NOAA satellites provided integrated data on the world's weather, a geostationary satellite would provide near-continuous observation of short-duration phenomena such as severe weather features. These phenomena, of which a tornado is an example, are frequently small and have a brief lifetime in comparison to large-scale weather systems. Accordingly, they are not always recognizable in conventional, land-based observational networks. The SMS would be able to take full-earth pictures every half hour, day and night.

The cooperative program of direct exchange of environmental satellite data between the U.S. and U.S.S.R. continued. At the beginning of 1973, NOAA-2 data was being exchanged for data from the Soviet Meteor 10, 11, 12, and 13 satellites. The data exchanged included television and infrared photographs and analyses of meteorological information.

Geodetic satellites. By means of simultaneous observation of a satellite from two or more distant points on the earth's surface, the distance between those points can be determined with great accuracy. In 1972 data processing and analysis for the Worldwide Geometric Satellite Triangulation Program was completed, resulting in world ellipsoid maps of increased geometric accuracy.

Courtesy, NOAA

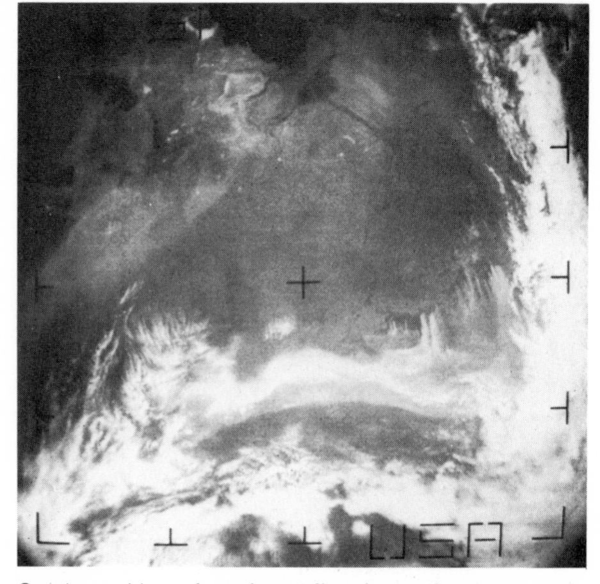

Smoke and haze from forest fires in southwestern Soviet Union are photographed by the U.S. ESSA-8 weather satellite on Sept. 9, 1972. West German scientists at Bochum Observatory processed the picture to increase contrast and definition.

Earth resources satellites. In July 1972 NASA launched the Earth Resources Satellite (ERTS 1) into near-polar orbit. Multispectral sensors on the satellite revealed distinct "signatures" of electromagnetic radiation from different surface and subsurface features of the earth. These data were televised to the earth, where processing yielded multicolored photoimages. By study of these photoimages and comparison to "ground truth" and similar features elsewhere, a new and powerful economic and social tool became available. More than 200 principal investigators in the U.S. and 106 in 37 foreign countries examined and interpreted ERTS imagery.

The quality of the photograph-like images was excellent. Each of the images depicted a square 115 mi on each side (13,000 sq mi). Details revealed exceeded expectations and were equivalent to actual photographs made during earth-orbiting Apollo missions. As the earth turned slowly within the satellite's 565 mile-high orbit, ERTS 1 circled the earth every 103 min and telemetered to three data acquisition stations millions of square miles of imagery daily. Images overlapped for ease in making composite maps. Approximately 1,300 images were acquired each day. Every 18 days the same area was mapped again, providing a comparison with earlier images.

In the U.S., agricultural inventories and forecasts of crop estimates in states and regional areas were expected to be more complete and less costly by using ERTS data than by previous techniques. The satellite was considered able to determine fertility and moisture content of soil and to identify the quality and rates of growth of crops. In forested areas, ERTS detected dead and diseased trees and forest fires. Burned-over areas can be accurately mapped and measured rapidly with little effort.

Accurate geological information about an area is extremely valuable because of its relationship to economical construction of roads, bridges, buildings, waterways, dams, and tunnels. Active fault zones are related to hazards of life and property. Despite previous intensive study and mapping of central California, many important geological features there were revealed by the large-scale images of ERTS 1 that had never before been mapped or detected. Some of these detailed features appear in areas, which, although well mapped, were not previously visible from the ground or from aircraft.

Searching for deposits of concentrated minerals is of obvious economic importance. Furthermore, the location of deposits of lower concentration is difficult to determine by conventional methods. It is believed that subtle variations in tones of ERTS images of large areas may be interpreted to locate promising sites for test drilling of such minerals. As the mineral resources of the U.S. dwindle, such deposits gain increased importance.

Information about the quantity and quality of water available for drinking, industry, recreation, agriculture, and wildlife has been increasingly difficult to obtain. As demand increases in the years ahead, the management of water resources must be improved. ERTS imagery revealed new sources of underground water by noting its effect upon surface vegetation. The amounts of water available for hydroelectric use as well as some early indication of possible danger from flooding are examples of the value of ERTS data.

ERTS 1 was providing high-quality imagery of coastal-zone regions where there is particular concern for the quality of the marine environment. It clearly identified offshore dumping of sewage sludge and acid waste products by noting the differences in the color of water attributed to such dumping. By showing how these colored areas move over the ocean's surface, ERTS was able to yield further insight into the rates of dispersal by tides and currents, and offer a means of assessing and monitoring this aspect of industrial practice.

In other nations investigators were finding ERTS 1 data of value in both developed and undeveloped areas. In Brazil errors in watershed maps had caused bridges to be built on the wrong sites; previously determined locations of villages were discovered to be inaccurate; and some roads had been improperly located by more than 12 mi. Deposits of nickel and copper were almost positively located in South Africa and Pakistan.

The launching by NASA of a second ERTS was postponed until 1975 because of budgetary restrictions. Meanwhile, a great flow of data continued to be generated and disseminated by the U.S. to interested nations. It seemed evident that, as progress is made in the refinement of techniques of processing and analysis of ERTS data, this class

Courtesy, NASA

Nimbus 5, an environmental satellite of the U.S. National Aeronautics and Space Administration (NASA), undergoes final checking prior to its launch on Dec. 12, 1972. The satellite placed in orbit instruments expected to be useful in providing improved weather forecasts and assessing long-term effects of pollutants in the atmosphere.

of satellite would be increasingly important as a tool for discovering and managing the earth's resources.

Reconnaissance satellites. The U.S. and Soviet Union continued their reconnaissance satellite programs, but also continued to release no details. It was presumed, however, that the spacecraft involved in these programs utilized extremely-high-resolution photographic and narrow-band electromagnetic radiation surveillance techniques. Since the early 1960s both countries had regularly launched military observation, low-orbit satellites. Payloads were recovered after a few days in orbit. The Big Bird satellites of the U.S., first launched in 1972, had a lifetime in orbit of more than 50 days, giving them the ability to take a close look at desired targets during an orbit subsequent to detecting them. This was not possible with satellites that had a short orbital lifetime.

The national reconnaissance satellite programs of both the U.S. and U.S.S.R. may be vital to maintaining peace, since each country regularly surveys the other's missile installations. The strategic arms limitation talks (SALT) recognized the existence of reconnaissance satellite programs and provided that, "Each party undertakes not to interfere with national technical means of verification at its disposal" and "Each party undertakes not to use deliberate concealment measures. . . ." As of 1973 the optical resolution of satellite reconnaissance cameras was believed to be such as to permit spotting and identification of objects less than a few feet in size.

Navigation satellites. In the area of navigation satellites, the U.S. Navy Transit system continued to be fully operational and in use by the U.S. fleet throughout the world. Although commercial ships were permitted to use this system, it required a shipborne computer to calculate position based on the Doppler shift of the satellite's signal. Thus, it was too expensive for widespread public use. Future civilian navigation is more likely to depend upon a system wherein a signal is transmitted from a ship via satellite to a shore-based computer; there, position data can be determined and broadcast back.

The multinational European Space Research Organization (ESRO) was negotiating with a number of U.S. firms to select a partner to develop an aeronautical satellite system, aerosat. The purpose of such a satellite would be to determine accurately the position of aircraft in flight at all times. In the crowded North Atlantic air route, commercial aircraft in 1973 were required normally to maintain fixed lateral distances of about 120 mi. This requirement forced many airplanes in the "crowded" sky to travel at altitudes with unfavorable wind velocity and along longer routes than necessary. An aerosat system could maintain constant surveillance and communications, resulting in continuous monitoring of aircraft positions along the most efficient altitude and most desirable route. The first anticipated launch of an aerosat would be in the 1976–77 period.

—F. C. Durant III

Manned space exploration

During the past year, manned space exploration was limited to the sixth and last U.S. manned Apollo lunar landing, the launch of a Soviet unmanned space station, and the flights of Skylab I and II. The United States and the Soviet Union agreed to launch a joint mission to be flown in 1975. Design of the necessary docking module required to join an Apollo command module and a Soviet Soyuz spacecraft was initiated. The U.S. awarded the prime contract for the design, development, and procurement of the space shuttle for the initial flight test phase. The U.S.S.R. continued its basic policy of not announcing to the world its future manned space flight plans.

Apollo. The Apollo lunar landing exploration program initiated by U.S. Pres. John F. Kennedy on May 25, 1961, was completed with the successful flight of Apollo 17. The 11 manned missions were each major steps in man's conquest of space and the exploration of the moon. The Apollo program afforded scientists on earth the first opportunity to examine, in detail, lunar geological samples and to develop a better understanding of the moon's chemical composition and the possible processes that took place in the formation of the moon. The scientific instrumentation placed on the moon provided a network of scientific stations that transmits information on lunar seismic activity, magnetic fields, particle impacts on the moon, lunar gravity, and lunar thermal properties. Other experiments placed on the moon permitted scientists on earth to study, through the use of laser beams, the precise geometric relationship between the moon and earth. These lunar surface experiments were expected to provide continuing information to scientists for many years.

On Dec. 7, 1972, Apollo 17, the last manned lunar landing mission, was launched from Cape Kennedy, Fla. The crew was commanded by Eugene Cernan, a veteran astronaut who had flown in the Gemini program and on Apollo 10. The lunar module pilot, Harrison Schmitt, was the first scientist-astronaut assigned to a U.S. manned space flight mission. Schmitt completed his academic training with a Ph.D. degree in geology prior to his selection as an astronaut in 1965. In the years preceding

Courtesy, NASA

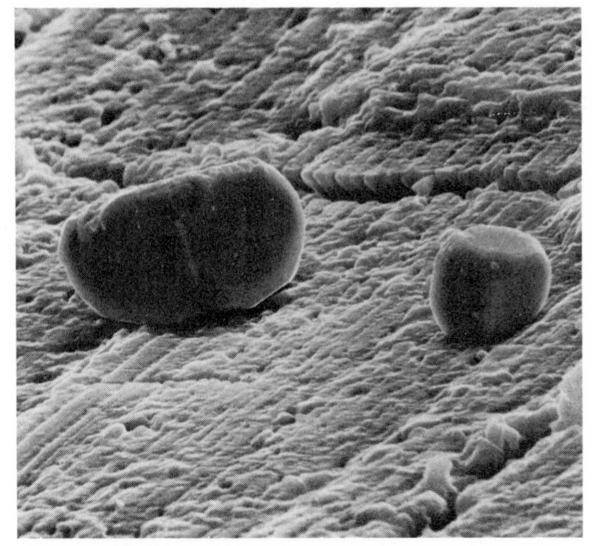

Light scattered by the earth's atmosphere (left) is photographed by an ultraviolet camera aboard Apollo 16. The bright crescent near the center is the geocorona, a halo of low-density hydrogen surrounding the earth. Iron crystals growing on a pyroxene crystal at the Apollo 15 landing site are greatly magnified in photograph taken by a scanning electron microscope (right).

Apollo 17, he completed flight training as a jet aircraft pilot. The command module pilot was Ronald Evans.

The launch of Apollo 17 was delayed 2 hours and 40 minutes due to a failure in one unit of ground launch support equipment, lift-off taking place at 12:33 A.M. EST. After approximately 3 hours in earth orbit, the Saturn IVB engines were used to accelerate the command and lunar modules to escape velocities in order to propel the spacecraft on a path to the moon. Approximately 86 hours after launch, the spacecraft was slowed by rockets to go into a lunar orbit, where it remained for 21½ hours. The lunar module then separated from the command module, and approximately 2½ hours later astronauts Cernan and Schmitt landed on the moon on the southeastern rim of the Sea of Serenity at a site called Taurus-Littrow. This landing site was of importance to lunar scientists because it contained evidence that helped explain the formation and history of the northeastern section of the moon. At the site was a very large landslide that was sampled by the crew. Scientists hoped that the analysis of rocks collected at Taurus-Littrow, coupled with the results of lunar sample analysis from previous missions, might provide important answers to questions about the early geological history of the moon.

Cernan and Schmitt stayed on the lunar surface for approximately 75 hours and made three 7-hour explorations. They used a four-wheeled electrical-powered rover vehicle to move approximately 22 mi on the lunar surface during the three excursions. During the first the crew deployed the rover vehicle and the lunar surface experiment package. The latter was a self-contained lunar surface science station that received electrical power from a nuclear-powered electrical generator; it contained a radio transmitter to send information back to earth and a radio receiver to permit scientists on earth to operate the experiments' equipment. The first experiment required the crew to place a heat probe into the lunar surface to measure the heat flow from the interior of the moon to its surface. It was hoped that this would help to develop an understanding of the moon's core temperature and the processes involved in the formation and activity of the moon. Other experiments included equipment to measure lunar surface gravity, lunar atmospheric composition, lunar ejecta and micrometeorites, and lunar seismic activity. The Apollo 17 lunar science station was designed to send information to earthbound scientists for a minimum of two years.

On the second lunar excursion the crew collected geological samples at various designated areas. This sampling included gathering rocks from the lunar surface and using a coring device to extract subsurface material. While on this excursion, the crew observed and sampled orange and red soil that greatly interested lunar geologists. The third lunar excursion consisted of geological sampling and traverse experiments. Photographic and remote sensor experiments were also conducted from the orbiting command module while the lunar surface was being explored.

Courtesy, NASA

Skylab astronaut checks the retrieval area for special film during a simulation test in an underwater chamber. During missions space walks will be necessary to install and retrieve the film.

On December 14 the crew flew the lunar module off the moon's surface and rendezvoused and docked with the command module. Two days later Apollo 17 started the trip back to earth. Earth landing occurred in the Pacific Ocean south of Hawaii on December 19.

Skylab project. Skylab is the fourth U.S. manned space flight program. Its main component is an experimental earth orbital workshop designed to be used to carry out a comprehensive series of experiments in various scientific and technological fields. Earth resources studies were scheduled to be made by the crewmen to acquire data that might be used in solving problems of the earth's environment. For example, this information was expected to assist in studies of crop and forest inventories, crop health, mineral and water resources, and air and water pollution. Approximately 150 scientists in the U.S. and other countries were to analyze the earth resources data from Skylab and compare them with similar data collected on the earth in order to determine the usefulness of these space measurement techniques.

Extensive medical studies were to be performed by the crewmen to study the effects of prolonged weightless exposure on the major body systems. For example, a series of tests was to be performed at regular intervals to study the effects on the cardiovascular system; measurements would be made on the quality of sleep; and the work capacity of the crewmen was to be measured with a bicycle ergometer.

The Skylab project utilized major components developed for the Apollo program. The orbital workshop was made from the Saturn IVB upper stage of Apollo's Saturn V launch vehicle. This large workshop was 21.7 ft in diameter, 48.5 ft long, and weighed 78,000 lb. It had two floors in the habitable area: the first floor contained living quarters for the crew and medical experiment equipment, while the upper floor housed stowage lockers and science and technology experiment equipment. A large area in the base of the workshop was to be used by the crewmen to dispose of trash and other wastes. Electrical power was to be supplied by a solar array panel. The workshop was to be operated at 5 psi or 1/3 of earth's average atmosphere with a gas mixture of 70% oxygen and 30% nitrogen, and was capable of maintaining crewmen for periods up to 56 days. Scientific airlocks were provided to permit samples and scientific instruments to be deployed into the vacuum of space for special studies. Onboard television cameras provided live coverage of crew activities.

The Apollo command and service module was to be used to transport three-man crews to the orbital workshop and return them to earth. In operation, Skylab I—the orbital workshop, multiple docking adapter, airlock module, and Apollo telescope mount—would be launched into earth orbit by a Saturn V launch vehicle. One day later, Skylab II—an Apollo command module with a three-man crew—was to be launched by a Saturn IB rocket into earth orbit to accomplish rendezvous and docking with the orbital workshop.

Skylabs I and II. On May 14, 1973, the Skylab I unmanned orbital workshop was launched from Cape Kennedy. During the powered phase of the launch a failure occurred with the orbital workshop micrometeoroid shield, causing one of the large solar array wings to be ripped from the orbital workshop and the second solar array not to deploy properly. The failure of the micrometeoroid shield also exposed a large section of the orbital workshop to high solar energy, causing the workshop temperatures to rise above 120° F. This failure resulted in the delay of the Skylab II launch scheduled the next day. For 11 days engineers and scientists on the ground monitored and controlled the overheated orbital workshop as it orbited above the earth. At the same time, a crash program was undertaken to develop a thermal screen that could be deployed by astronauts to provide insulation from the sun's heat. In addition, special tools and other equipment were developed in order to permit crewmen to attempt to repair the orbital workshop. In less than two weeks the new equipment was developed and tested. The Skylab II crew was trained in the new procedures, and the launch vehicle was readied.

Skylab II was launched from Cape Kennedy on May 25. The crew commander for this mission was Charles "Pete" Conrad, a veteran of two Gemini flights and one Apollo lunar landing mission. The science pilot was Joseph P. Kerwin, the first physician flown in a U.S. space flight mission. He was selected for this mission because of his unique training and to provide a trained observer for the medical experiments to be performed. The third crewman was Paul J. Weitz, a veteran naval aviator selected for the astronaut program in 1966.

Almost eight hours after launch the crew rendezvoused with the orbital workshop and transmitted television pictures of the damaged vehicles. These pictures clearly showed that one of the solar array panels was missing and that the second solar panel was only partially deployed. A large rectangular gold-coated area was exposed where the meteoroid shield had been lost. The crew then opened the command module hatch and Weitz stood in his pressurized space suit in the command module hatch area and attempted unsuccessfully to deploy the solar array panel. This maneuver required Conrad to fly within very close range of the workshop. After completing this unprecedented and first attempt of using man in space to repair an unmanned spacecraft, the crew of Skylab II docked with the orbital workshop.

The second day the crew entered the overheated orbital workshop and successfully deployed a 22 ft by 24 ft rectangular thermal shield over the damaged outer skin of the workshop. The device was designed to resemble an umbrella with extendable ribs. Once the ribs were fully extended, they opened like an umbrella to provide the protective thermal shield. Within 24 hours after deployment the orbital workshop temperatures dropped from greater than 120° F to less than 100° F, and within a few days the temperatures dropped to livable levels. On the fourth day of the mission the crew occupied the workshop on a continuing basis and started the medical experiments and other science studies. Conrad and Kerwin on June 7 were able to deploy the solar array panel.

Skylab II was thus saved from an apparent failure through the ingenuity and dedicated efforts of engineers, technicians, and the astronauts. The spacecraft completed its 28-day flight on June 22, 1973, splashing down safely in the Pacific Ocean. Skylab I, the workshop, remained in orbit to be used by later missions. These included Skylab III, scheduled to be launched in August 1973. The crew was to be commanded by Alan L. Bean and include science pilot Owen K. Garriott and pilot Jack R. Lousma. Skylab IV was to be launched in November 1973. The crew was to be under the command of Gerald P. Carr and also include science pilot Edward G. Gibson and pilot William R. Pogue.

Space shuttle. During the past year the U.S. proceeded with the development of a reusable manned space vehicle that will be used as a transportation system for a wide variety of low-earth-orbit missions. The shuttle was designed to be capable of carrying large payloads of up to 65,000 lb into orbit and return to earth. If development of the shuttle is successful, most expendable launch vehicles currently used will be eliminated. The shuttle will be used to deploy earth-orbiting scientific application satellites of all types and is also to have the capability of retrieving satellites from orbit to repair and redeploy them or bring them back to earth for repair and reuse. Development of the space shuttle was planned to be accomplished over the next six years, and full operational activities were scheduled to begin in 1980. Developmental flight testing was to start in 1976–77.

The shuttle was designed to utilize a reusable manned aircraft-type vehicle called the space shuttle orbiter. This vehicle would carry a four-man flight crew and resemble a delta-winged airplane about the size of a DC-9 jet liner, with a wing span of 80 ft and a length of 124 ft. The orbiter would have three liquid-fuel rocket engines and a cargo bay 60 ft long and 15 ft in diameter. This large bay could be used to transport unmanned satellites into space or to carry scientific laboratories. The shuttle would be able to carry up to six science passengers.

In operation, the orbiter would be mounted on a launch vehicle that has two solid-fuel rockets. At launch, the two solid-fuel rockets and the orbiter's three liquid-fuel rocket engines would ignite and burn simultaneously until the vehicle reached an altitude of approximately 25 mi. The solid-rocket stage would then drop away, and the orbiter would continue on its journey into earth orbit. Missions would normally be 7 days in length with a maximum of 30 days. When the mission was completed, the orbiter would return to earth and land like an airplane.

Apollo/Soyuz Test Project. A joint U.S.-Soviet manned space flight program called the Apollo/Soyuz Test Project was planned for flight in 1975. The project was formally approved by U.S. Pres. Richard Nixon and Soviet Premier Aleksei N. Kosygin in May 1972 during Nixon's visit to Moscow. The objective of this joint flight is to exercise the techniques of rendezvous and docking with another spacecraft to demonstrate an international space rescue capability. Scientific and engineering in-flight experiments will be conducted by both countries. This program also affords an opportunity for engineers and technicians from both countries to work closely with one another.

During the last year, joint meetings were held in both countries to develop detailed docking module designs and to start operational planning for the mission. It was agreed that Soviet cosmonauts would train for approximately one month at the Lyndon B. Johnson Space Center in Houston, Tex., and the U.S. astronauts would go to the Soviet Union for training. The U.S. crew for this mission was to be commanded by Thomas P. Stafford, a veteran of both Gemini and Apollo flights. Vance D. Brand was to be the command module pilot and Donald K. Slayton the docking module pilot. This was to be Slayton's initial space flight since his selection as an astronaut in 1959. A medical problem had eliminated him from the second Mercury orbital flight, and not until 1972 was he restored to flight status by medical personnel. The two-man Soviet crew was to consist of A. A. Leonov and V. N. Kubasov.

The two spacecraft used to transport crewmen into earth orbit and return them safely to earth were to be the U.S. Apollo command module and the Soviet Soyuz spacecraft. A docking module was being developed to allow both spacecraft to join in space. The docking module, which is 10 ft long and 56 in. in diameter, was also to serve as a transfer tunnel for crewmen to move from one spacecraft to the other. The U.S. was building the docking module and planned to launch it with the command module on a Saturn IB launch vehicle.

In the operational sequence, the Soviet Union would launch the two-manned Soyuz spacecraft into earth orbit. Shortly thereafter the United States would launch its three-manned command and docking modules into earth orbit. About one day later, the two spacecraft would rendezvous and docking would be accomplished. Two days of docked activities would then take place. This would include the exchange of crews between the spacecraft and the performance of joint science experiments. Upon completion of this period, the spacecraft would undock and proceed with normal reentry and landing procedures.

Soviet manned space flight. The Soviet Union did not launch any cosmonauts into space during the year and had not done so since the death of three crewmen on the 24-day Salyut 1-Soyuz 11 flight on June 30, 1971. On April 3, 1973, the Soviet Union did launch an unmanned Salyut 2 space station into earth orbit. For the next several weeks, the world awaited the launch of a three-manned crew in a Soyuz spacecraft to dock with the Salyut; however, the manned launch did not occur. It would appear that the Salyut 2 malfunctioned and did not permit the mission to proceed as scheduled.

—Richard S. Johnston

Space probes

Photographs made by a Soviet unmanned probe played an important part in the success of the last landing of men on the moon in the decade of the 1970s. Aboard Apollo 17 were two photographs of the lunar surface made by Zond 8, launched on Oct. 20, 1970. The pictures were used by Ronald Evans, command module pilot of Apollo 17, in locating and studying features in the crater Aiken on the far side of the moon. Another Soviet probe also made a significant contribution to the knowledge of lunar geophysics. Luna 19, which ended its mission in October 1972, discovered plasmas on and near the surface of the moon. They were formed by the interaction of cosmic rays with the material of the lunar surface.

Probing Mars. The U.S. spacecraft Mariner 9 ended its scrutiny of Mars on its 698th orbit of the planet on Oct. 27, 1972. The probe had transmitted 7,329 pictures of the red planet and acted upon 46,000 commands from the control center at the Jet Propulsion Laboratory (JPL) in Pasadena, Calif. Engineers at JPL estimated that the now-silent probe would continue to orbit Mars from 50 to 100 years before spiraling onto its surface. The last picture was sent to earth on October 17; by then, the probe had completed its task, mapping the entire surface of Mars.

Among the many new facts revealed by Mariner 9 was that ozone in the Martian atmosphere is inversely related to temperature. The highest reading occurs in the winter, and the lowest shows up during the summer. Furthermore, the probe showed that water-ice crystals make up the clouds that are formed near four large volcanoes on the planet's surface. Other evidence from the pictures returned from Mars indicates that there is a global circulation of the planet's tenuous atmosphere.

Equally significant information on Mars came from two Soviet probes, Mars 2 and 3. They were sent into carefully selected orbits that took them into a southeast to northwest path over the surface of Mars from the dayside to the nightside. Thus, the probes covered areas that were impossible or extremely difficult to observe from earth. Among the discoveries was the fact that the amount of water vapor in the Martian atmosphere was about 2,000 times less than that in the atmosphere of the earth. This finding led L. Ksanfomaliti of the Soviet Institute of Space Research to propose that if all the water vapor in the Martian atmosphere were to be precipitated onto the surface of the planet, the resulting layer of snow would be no thicker than the diameter of a human hair.

Perhaps one of the most important discoveries made by the Soviet probes was that certain spots of the planet are inexplicably warmer than the surrounding areas. A team led by V. I. Moroz of the Soviet Laboratory for Infrared Astronomy discovered such a spot on the nightside of the planet whose temperature was 20° C to 25° C (68° F to 77° F) above that of its surroundings. Moroz suggested that the spot might be a volcano.

A significant meeting took place in Moscow on January 29 to February 2. Members of the Academy of Sciences of the U.S.S.R. met with scientists from the U.S. National Aeronautics and Space Administration (NASA). The Soviet scientists promised to provide data from their Mars 2 and 3 probes to their U.S. counterparts for use in planning the landing zones on Mars for the U.S. Viking probe. In return, the Americans agreed to give their Soviet colleagues maps and photographs of Mars made by Mariner 9. The Soviet scientists also agreed to release any data that they might gain from a soft-landing probe that reached Mars before Viking.

En route to Jupiter. While scientists at the NASA Ames Research Center held their breaths, on July 15, 1972, Pioneer 10 entered the asteroid belt, a region 280 million km (175 million mi) wide between Mars and Jupiter. On Oct. 30, 1972, the probe was at the mid-point in the belt, 434 million

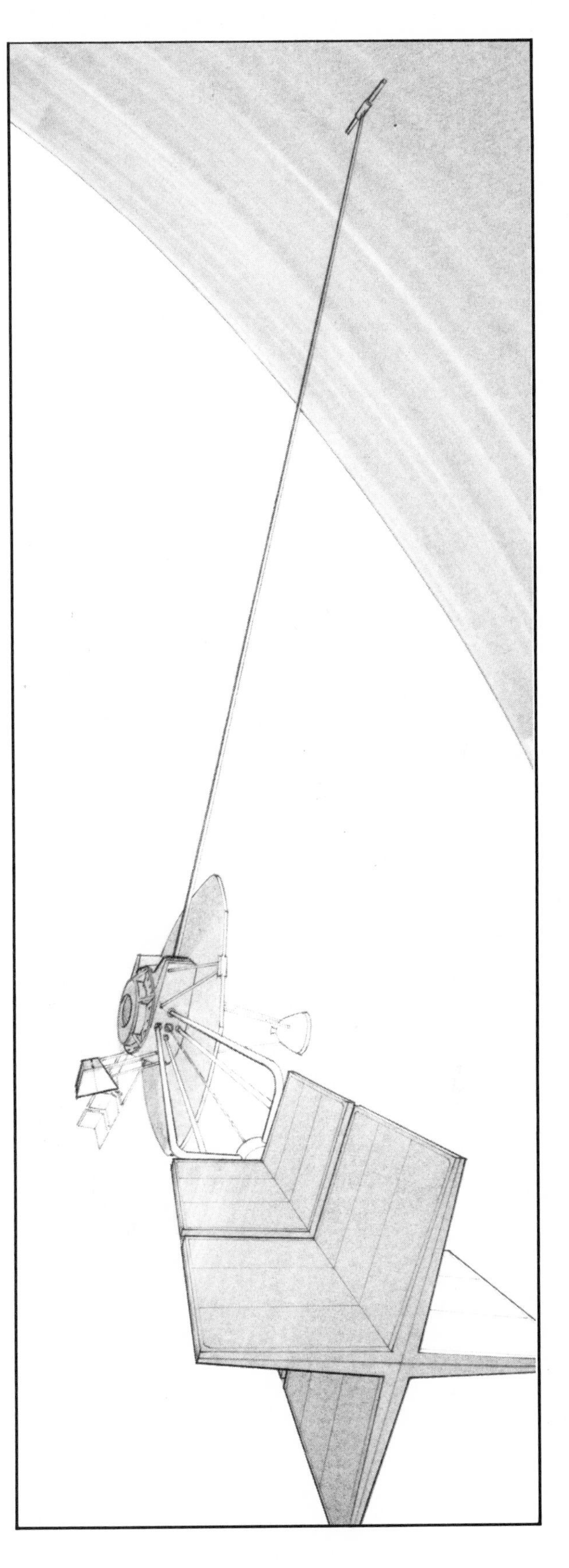

Pioneer 10 passes over the surface of Jupiter as rendered by an artist: the actual event will take place in December 1973. The U.S. probe will return the first measurements of the giant planet's twilight side, never seen from the earth.

Sovfoto

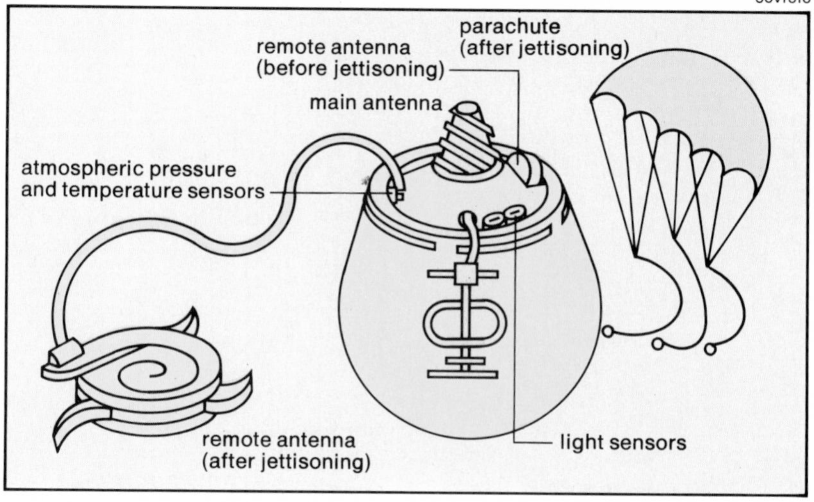

Soviet unmanned space probe Venera 8 soft-landed on Venus July 22, 1972, and transmitted data for about 50 minutes. From this information scientists concluded that the surface of Venus is similar to that of the earth.

km (270 million mi) from the sun. In traversing the asteroid belt, Pioneer 10 received only 83 hits by tiny dust particles. The four small telescopes aboard the craft "saw" between 100 and 200 chunks of debris within several miles of the probe. After seven months in the belt, Pioneer 10 emerged from it operationally unscathed on February 15.

During its journey to Jupiter, Pioneer 10 reported several things of interest. Among those was that the effect of the sun on cosmic rays barely diminished at a distance of three astronomical units from the sun (1 astronomical unit = approximately 93 million mi, the mean distance between the earth and the sun). In addition, the probe reported for the first time the presence of free helium in interplanetary space. Also, free hydrogen was detected entering the solar system from a different angle than had been observed four years earlier. Pioneer 10 also reported that the number of very small meteoroids between the earth and the far side of the asteroid belt was more or less constant. Scientists had previously thought the number would increase toward the middle of the belt.

On Aug. 2, 1972, a rare event happened at a most fortunate time. There was a huge solar storm, the largest ever measured in space, at a time when Pioneers 9 and 10 were positioned in a straight line with the sun. The former was at a distance of 0.78 astronomical units (AU), while the latter was at a distance of 2.2 AU. The activity was measured by these two probes as well as Interplanetary Monitoring Platform 5 (IMP 5), at a distance of 1.0 AU. For about a week, the sun spewed forth bursts of gas containing electrically charged atomic particles (the solar wind).

Previously, the solar wind had been assumed to travel at a constant velocity. However, measurements by instruments aboard the three probes showed that the solar wind slowed down and its temperature increased as it moved farther from the sun. As the plasma swept past Pioneer 9, IMP 5, and Pioneer 10, instruments reported that its velocity dropped from 1,100 km/sec (683 mi/sec) to about 700 km/sec (434 mi/sec). The 1,100 km/sec was the greatest velocity ever recorded. While the velocity decreased with distance from the sun, the temperature rose from approximately 500,000° C to greater than 2,000,000° C.

Jupiter, which usually responds to such solar disturbances, in this case did not. Apparently, time and distance had dissipated the solar wind before it reached the planet. In addition, it is known that comets also are affected by such solar events. However, the comet Schwassmann-Wachmann 1, located near Jupiter at a distance of 5.6 AU, showed no reaction.

On Sept. 19, 1972, commands from the earth "tweaked" the thrusters of Pioneer 10 to increase its velocity by 0.23 m/sec (0.75 ft/sec). The purpose was to insure that the probe would pass behind one of Jupiter's most mysterious moons, Io, which appears distinctly orange to earthbound astronomers. Scientists hoped that the occultation of the probe by the moon would determine whether or not it had an atmosphere. Shifts in the strength of Pioneer 10's telemetry signals as they pass through the atmosphere of Io should confirm or deny the currently held theory that the atmosphere consists of nitrogen and methane. Io appears brighter than usual after it emerges from the shadow of Jupiter. Some scientists believed that this brightness is the result of a layer of ice that forms while Io is in the shadow of Jupiter and melts as soon as the moon is bathed in sunlight.

Based upon how well Pioneer 10 survives its voyage to Jupiter, particularly its journey through the planet's radiation belt, scientists may elect to send Pioneer 11, launched on April 5, onto a five-

Sovfoto

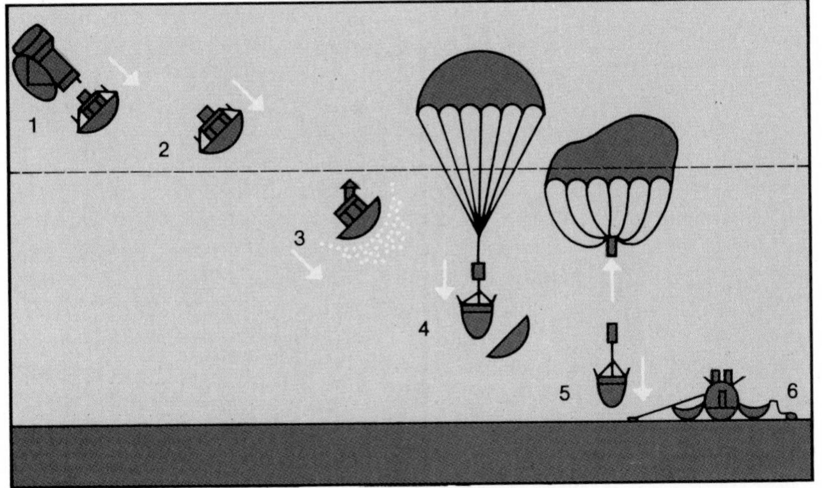

Instrument capsule of Soviet space probe Mars 3 soft-landed onto the surface of Mars Dec. 2, 1972. In sequence, the capsule parted from the spacecraft (1); a motor in the capsule began powering it (2); aerodynamic braking occurred (3) after the capsule entered the planet's atmosphere (dotted line); a parachute was deployed for the descent through the atmosphere and the capsule's heat shield was jettisoned (4). Near the surface the capsule separated from the parachute and a special motor began operating (5), placing the capsule in its working position on Mars (6).

year encounter with Saturn after it visits Jupiter early in 1975. A final decision on the change in trajectory was to be made after information from Pioneer 10 was evaluated. The course correction would be made in space.

Data from Venus. The Soviet probe Venera 8, which soft-landed on Venus, July 22, 1972, transmitted data for about 50 minutes, more than doubling the time of Venera 7, which transmitted for only 23 minutes in December 1970. From this information scientists concluded that the surface of Venus is similar to that of granite rocks on earth despite the higher temperature and far greater atmospheric pressure.

The soft-landing capsule from Venera 8 required 109 minutes to descend from orbit to the surface of Venus. During the descent via parachute, the probe reported that there were wind velocities of nearly 190 kph (112 mph) at an altitude of about 50 km (31 mi) above the surface. However, at a distance of only 10 km (6 mi) the velocity was only about 6 to 7 kph (5 mph). These findings led scientists to conclude that the winds in the upper regions move continuously in the direction of the planet's rotation. The atmosphere itself turned out to be warm and moist. Measurements made by the probe's instruments reported that it was composed of 97% carbon dioxide, 2% nitrogen, less than 0.1% oxygen, and less than 1% water vapor. On the surface, the temperature was 470° C (878° F) and the atmospheric pressure 90 kg/sq cm (1,280 psi). These values were in good agreement with those reported by Venera 7 two years earlier.

Venera 8's gamma-ray spectrometer indicated that the Venusian surface in the landing area contained 4% potassium, 0.0002% uranium, and 0.00065% thorium, about the same ratio as these elements occur in many magmatic rocks on earth. The average density of the material is about 1.5 g/cc (0.05 lb/cu in.) or about half the value for the earth.

The probe's photometer, which measured the sunlight falling on the planetary surface, indicated that approximately two-thirds of the solar radiation does penetrate the thick clouds over Venus. The surface apparently does not undergo severe temperature changes from day to night. Because of the long length of the day—Venus rotates once every 243 earth days—one would expect drastic temperature changes. However, the large amount of carbon dioxide in the atmosphere prevents the escape of the heat during the night.

Exploring the moon. Another Lunokhod returned to the moon to continue the unmanned exploration of it by the U.S.S.R. On January 8, Luna 21 was launched from Tyuratam. The flight to the moon was uneventful except for a momentary fear that one of the landing legs of the craft had failed to deploy. However, engineers at the mission control center near Moscow soon found that the trouble lay in a faulty telemetry system and not in the landing gear itself.

Luna 21 went into lunar orbit on January 12 with an initial apolune of 109.9 km (68.3 mi) and a perilune of 89.9 km (55.9 mi). It completed one revolution every 118 minutes. The orbit was later adjusted to permit a perilune of 16 km (10 mi), and the craft descended to a target area inside the crater Le Monnier on January 16. The landing area was on the eastern edge of the Sea of Serenity about 160 km (100 mi) north of the landing site of Apollo 17. Thus, data retrieved by the Soviet probe was expected to enhance greatly the similar data reported by the instruments of Apollo 17.

Lunokhod 2 weighed 838.9 kg (1,848 lb), 85 kg (185 lb) more than Lunokhod 1. It remained on its Luna descent stage for almost three hours before driving down onto the lunar surface. After an hour of checkout, the probe was parked for three days to charge up its batteries. The probe was an improved model of Lunokhod 1. It carried a magnetometer, extending 1.2 m (4 ft) in front of the vehicle. Nicknamed "the cobra" by engineers at mission control center, it was designed to measure the local magnetic fields of the crater as well as the magnetic properties of the rocks and lunar material over which the probe traveled. The robot also had a much improved steering system. Lunokhod 1 was designed so that one set of wheels could be driven backward while the other turned forward. Its successor had a braking system that permitted the operator to slow either of the four wheels on a side while the others continued to move. Thus, Lunokhod 2 could "turn on a kopeck."

Also on board Lunokhod 2 was an improved television system. Whereas the older model could transmit only one picture every 20 seconds, the new instrument could send one every 3 seconds. This faster rate made it much easier for the crew on earth to operate the vehicle. In addition, the lens for the steering television was mounted much higher on the robot and was shielded. This change permitted the "driver" on earth to view the surface at eye level, allowing him to operate the vehicle during the lunar noon, when the lack of contrast of surface features had required the Lunokhod 1 to be halted.

During the first 100 hours of operation, Lunokhod 2 traveled approximately 1,159 m (3,800 ft). On January 23, the robot closed its solar lid and prepared for its first lunar night. Sensors within the craft monitored its internal environment and reported that all was well to mission control center. On February 8 a command from them opened the solar lid to charge the batteries. On February 9 the vehicle began its second lunar day. During that period it traveled more than 11,067 m (36,299 ft) toward the southern end of the crater. The third "day" began on March 9, with the vehicle headed still farther south toward a fracture some 300 m (990 ft) wide and 16 km (10 mi) long. It ended on March 23, about 2.5 km (1.5 mi) short of the fracture at the end of the third day. During its trip to the fracture, the vehicle's instruments reported that the iron and aluminum content of the soil was decreasing from previous readings closer to the center of the crater.

Future outlook. Preparations continued for launching the first U.S. probe to Venus and Mercury in October. The Mariner 10 (as it would be designated if successful) was designed to weigh 454 kg (1,000 lb) and pass within about 5,300 km (3,300 mi) of Venus in February 1974, and within 1,000 km (621 mi) of Mercury in March. The craft was scheduled to take about 8,000 pictures of each planet. It would utilize the gravity of Venus to swing

Viking unmanned space probe now being designed for NASA is scheduled to soft-land on Mars in 1976. NASA scientists planned to land the first of the two Vikings near the planet's equator; the second would probably descend between lat 50°–55° N, where there was a probability of finding water vapor.

itself on for the encounter with Mercury, the innermost of the sun's family of planets.

After delaying the project for a year, NASA hoped to revive the plan for launching two probes to Venus, one in January 1977, and the other in the spring or summer of the following year. The former would carry four payloads to eject into the Venusian atmosphere, while the latter would be a probe that stayed in orbit around the planet for at least a year.

Planning also continued on Mariner 77, a mission destined to visit Jupiter and Saturn after being launched in 1977. Two of the probes would be launched in that year on "swing by" missions, using the gravitational field of Jupiter to speed them on past Saturn. More than 200 scientists throughout the world responded to invitations from NASA to participate in designing the instrumentation for the two craft, which was to be identical. Ninety scientists from the U.S. and four from foreign nations were selected. The experiments were to be in 11 categories: imaging, radio science, infrared spectroscopy and radiometry, ultraviolet spectroscopy, magnetometry, plasma research, low-energy charged particles, interstellar cosmic rays and planetary magnetospheres, interplanetary interstellar particulate matter, photopolarimetry, and planetary radio astronomy.

The possibility existed that one of the two Mariners could pass within 160 km (100 mi) of Titan, the largest of the 10 moons of Saturn. The prospect excited astronomers and exobiologists because Titan is known to have an atmosphere. Particularly interested in the possibility was Carl Sagan, professor of astronomy and director of Cornell University's Laboratory for Planetary Studies, who stated "Our research has shown that at the least Titan should be littered with the kind of organic molecules which, in the early history of the earth, led to the origin of life."

The two probes, if launched on schedule in 1977, would pass Jupiter in 1979 and Saturn in 1981. If the schedule is missed, the proper alignment for another "swing by" mission will not occur until 1997.

Work continued on the two Viking probes to be soft-landed on Mars in July and August 1976, after launchings in July 1975. A tentative landing site for the first of them had been selected by March. NASA chose a low, smooth area near the planet's equator. A decision on the landing site for the second Viking was expected to come after much greater study of the region between latitudes 50° and 55° N, where scientists believed that there was a great probability of finding water vapor.

The success of the geochemical experiments in the Apollo program caused NASA scientists to revise their thinking about instrumentation for Viking. It was announced that the probe would carry an X-ray fluorescence spectrometer when it soft-landed on Mars in 1976. This device was designed to provide data on the mineralogical characteristics of the Martian soil in the region where the probes land.

—Mitchell R. Sharpe

Astronomy

In 1973, when the 500th anniversary of the birth of Copernicus was celebrated throughout the world, astronomy was characterized by major advances in observational facilities, and by a flowering of the new fields of X-ray and gamma-ray astronomy.

Large telescopes. At Kitt Peak National Observatory, 52 mi west of Tucson, Ariz., construction of a 158-in. reflecting telescope was completed in the spring of 1973. After tests and adjustments are finished, it will briefly be the second largest optical telescope in operation, until the 236-in. Soviet reflector begins working near Zelenchukskaya in the Caucasus Mountains. (Currently, the world's largest operating telescope is the 200-in. Hale reflector on Palomar Mountain in California.)

Known as the Mayall telescope (after former Kitt Peak director Nicholas U. Mayall), the $10 million instrument was funded by the U.S. National Science Foundation. The primary mirror, 158 in. in diameter and 24 in. thick, consists of fused quartz and weighs 15 tons. This versatile telescope can be used in any of three optical arrangements: prime focus, coudé, and Cassegrain. The total moving weight of the Mayall telescope is 300 tons, and its dome is 105 ft in diameter. The telescope building is 185 ft high, so that the instrument is raised above most of the atmospheric turbulence caused by hot air currents rising from the ground. Observing programs for the telescope are scheduled to include optical studies of radio and X-ray sources, faint quasars and pulsars, and, especially, galaxies.

A second 158-in. reflector, described as nearly a twin of the Mayall telescope, was under construction at Cerro Tololo Inter-American Observatory in Chile. It was expected to go into operation in 1975.

Completion in 1973 was expected for the 150-in. Anglo-Australian telescope for the Siding Spring Observatory in New South Wales, Austr. This site is the field station for Mount Stromlo Observatory near Canberra. The 16-ton mirror for the 150-in. reflector was ground and polished by the British firm of Grubb Parsons.

Siding Spring Observatory also gained a new 48-in. Schmidt telescope. This very fast, highly corrected wide-field photographic instrument closely resembled the 48-in. Schmidt in use on Palomar

Courtesy, Sacramento Peak Observatory, Air Force Cambridge Research Laboratories

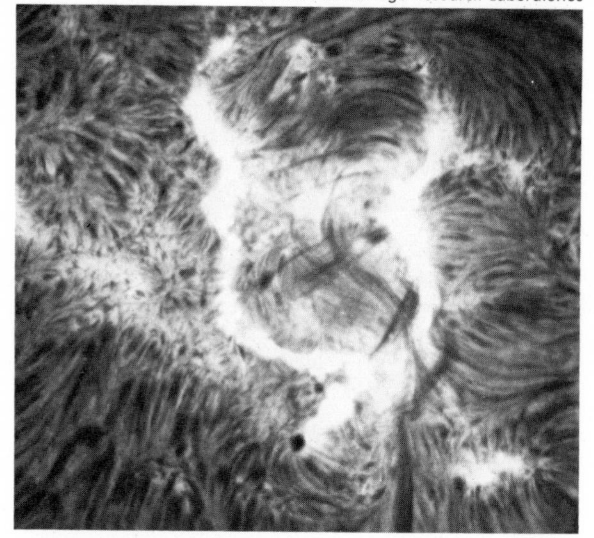

Solar flare of Aug. 2, 1972, ejects a stream of ionized gas into interplanetary space. Instruments aboard Pioneer 10 measured the speed of the gas at 700 km/sec at 320 million miles from the sun.

Mountain. It could photograph a sky area 6.6° wide on glass plates 14 in. square. The Siding Spring telescope was expected to be used for mapping the southern sky in red and blue light in order to extend the National Geographic Society–Palomar Observatory sky atlas to the south celestial pole. This program was being shared with the 39.4-in. Schmidt telescope of the European Southern Observatory at La Silla, Chile.

Record solar activity. During early August 1972, a remarkable series of disturbances occurred on the sun, associated with a large and complex sunspot group that appeared at the eastern edge of the solar disk on July 29. This solar active region, marked by strong and complex magnetic fields, was the seat of a succession of brilliant solar flares. The chief of these appeared on August 2, 4, 7, and 11. Each flare was characterized by bursts of solar X rays and radio wave emission, and was followed by bright auroral displays and widespread geomagnetic disturbances.

The U.S. spacecraft OSO-7 (Orbiting Solar Observatory 7) measured gamma-ray emission from the sun during the flares of August 4 and 7. For the first time, gamma-ray spectral lines were detected, consisting of photons with energies of 0.51 and 2.22 million electron volts (MeV). According to a team of University of New Hampshire researchers, these lines are produced in the sun by the interaction of fast downward-moving protons with material in the solar photosphere.

When a flare erupts, it ejects a fast stream of ionized gas into interplanetary space, adding to the ever-present outward-moving solar wind of charged particles. On August 4 such a stream (believed caused by the flare two days earlier) was encountered by the Jupiter-bound Pioneer 10 spacecraft, then about 320 million mi from the sun. As measured by Pioneer's instruments, this gas stream was moving 434 mi/sec. It was preceded by a shock front in which a strong magnetic field was detected.

Total eclipse. The total eclipse of the sun on June 30, 1973, was remarkable for the long duration of totality, which at one point in the Sahara Desert reached 7 min 4 sec. This is almost the longest duration possible, and will not be surpassed until the eclipse of June 25, 2150, when totality will last 7 min 14 sec.

During the 1973 eclipse, the long, narrow path of totality started at sunrise in South America near the Guyana-Brazil border. It crossed Surinam and the South Atlantic Ocean; extended eastward through the African republics of Mauritania, Mali, Niger, Chad, Sudan and Uganda; continued southeastward across Kenya and Somalia; and ended at sunset in the Indian Ocean.

Most expeditions to observe the eclipse chose the vicinity of Lake Rudolf in Kenya. Despite the relative freedom of the Sahara Desert from clouds, the desert air is often too dust-laden for delicate eclipse observations. Some eclipse studies were made from high jet aircraft, flying eastward along the central line of the eclipse at a speed approaching that of the moon's shadow and thereby gaining a considerably longer duration of totality than a fixed station on the ground would have.

However, eclipses have lost much of their scientific value for investigation of the solar corona. At one time, the corona could be studied only when the dazzling solar disk was blacked out by the moon. By the 1970s, however, man's understanding of the physics of the corona came increasingly from satellite observations at ultraviolet and X-ray wavelengths, from radio observations, and from observations with coronagraphs.

Two notable fireballs. A great meteor was seen in full daylight on the afternoon of Aug. 10, 1972, attracting widespread attention. Intermediate in brightness between the sun and full moon, this fireball entered the earth's atmosphere on a nearly horizontal path, moving northward from over Utah to over Alberta. It left a smokelike trail that remained visible for nearly an hour. Remarkably, the trajectory of this meteor did not meet the earth's surface. Instead, the massive body is believed to have swung around the earth in a hyperbolic orbit, escaping into interplanetary space after having come within about 25 mi or less of the ground. This unique behavior of a fireball leaving the earth's atmosphere was first pointed out by Luigi Jacchia of the Smithsonian Astrophysical Observatory in Cambridge, Mass., a leading meteor expert who happened to be an eyewitness.

The Czechoslovakian meteor of Jan. 2, 1973, was among the most thoroughly observed on record. Because it was photographed from no fewer than five meteor-camera stations, the meteor could have its path through the earth's atmosphere determined with unusual accuracy. It appeared at a height of 52 mi over a point very close to Prague, and became as bright as the full moon as it traveled east-northeast before being completely consumed at a height of 26 mi. The fireball's spectrum was photographed at Ondrejov Observatory, showing bright lines of iron and other metals. Before entering the earth's atmosphere, this body was traveling around the sun in a highly elongated elliptical orbit, with a perihelion point near the orbit of Mercury.

Saturn. In December 1972 and January 1973 radar astronomers at the Jet Propulsion Laboratory in Pasadena, Calif., succeeded in detecting echoes of radar signals they had beamed at the planet Saturn. Thus, Saturn became the most remote celestial body with which radar contact was made. For this experiment, the 210-ft-diameter steerable antenna of JPL's Goldstone Station in California was used both to transmit the 400-kw beam toward Saturn and to receive the enormously weakened echo. The microwave beam, of 12.5-cm wavelength, required 2 hours 15 minutes to travel the 1,500,000,000-mi round trip to Saturn and back.

The radar reflection from Saturn's rings was considerably stronger than expected. Results indicate that the rings cannot be made up of tiny ice crystals or dust, as many astronomers had believed, but probably consist of rough, jagged, solid fragments at least a yard in diameter and possibly much larger. Hence, the ring system must be considered an extreme hazard to any spacecraft sent through it.

Although the outer diameter of Saturn's ring system is about 170,000 mi, the thickness is less than a mile. The radar experiment succeeded because it was conducted at a time when the rings were at their maximum tilt to the line of sight. It would be impossible in 1980 when the rings will be edgewise. No reflections could be detected from the globe of Saturn itself.

Planet X. Considerable public interest was aroused by press announcements in mid-1972 of a new major planet beyond Pluto. Its existence was predicted by Joseph L. Brady of the University of California as a byproduct of extensive computations by him and Edna Carpenter on the motion of Halley's comet from A.D. 295 until its next return in 1986.

continued on page 188

Observatory near Zelenchukskaya in the Caucasus Mountains of the Soviet Union will house a 236-in. optical reflecting telescope, the largest instrument of its kind in the world. Expected to become operational in 1974, the telescope surpasses the previous recordholder, the 200-in. Hale reflector at Palomar Mountain.

Tass from Sovfoto

The Mayall telescope

On the evening of June 19, 1973, 500 years after the birth of Nicolaus Copernicus, a hushed audience in the auditorium of the University of Arizona heard astrophysicist Jesse Greenstein describe his "Exploration of a Strange and Beautiful World," a domain in which he is well traveled. (*See* Feature Article: ASTRONOMY FOR THE '70S.) On the following morning, about 52 mi to the west and at an altitude of 6,900 ft, Leo Goldberg, distinguished director of the Kitt Peak (Ariz.) National Observatory, dedicated the second largest telescope in the world to his predecessor, Nicholas U. Mayall.

There was much to celebrate. With a mirror 158 in. in diameter, the Mayall telescope was the largest optical instrument to be built since the construction in 1948 of the 200-in. Hale telescope at Palomar Mountain in California. The new instrument, which was not scheduled to become fully operational until October 1973, would have a wider field of view than any other existing large reflecting telescope—six times wider than that of the largest optical instrument. One photographic plate taken through it would give astronomers over 40 times more sky coverage than any other existing large optical telescope.

Although smaller than the Hale (which was itself due to be surpassed shortly by a 236-in. giant under construction in the Soviet Union), the Mayall telescope benefited from a cornucopia of technological improvements not available to the builders at Palomar. One of the most central was the development during the 1960s of a technology for fusing together large chunks of quartz into a homogeneous mass. The Mayall instrument is the first large reflecting telescope to utilize a fused-quartz mirror.

The goal of the mirror-maker is to reduce to its absolute minimum the distortion that occurs when the surface of the mirror is warped by sudden temperature changes. Because Pyrex glass shows much less thermal distortion than ordinary glass, it was used in every large reflecting telescope made between 1925 and 1960. Fused quartz, however, has a yet much lower expansion coefficient than Pyrex; its use in the Mayall will tremendously increase its efficiency.

The 15-ton mirror will also benefit from the fact that it was the first such to receive its final "figuring" in a fully equipped optical shop rather than in the telescope itself. The construction of a duplicate support system—a pneumatic cushion on which the mirror rides and is positioned for obser-

At the prime focus position the Mayall telescope obtains a detailed optical image of the Trifid Nebula (right) in 18 minutes as compared with the 60 minutes required by an 84-in. telescope (left).

Courtesy, Kitt Peak National Observatory

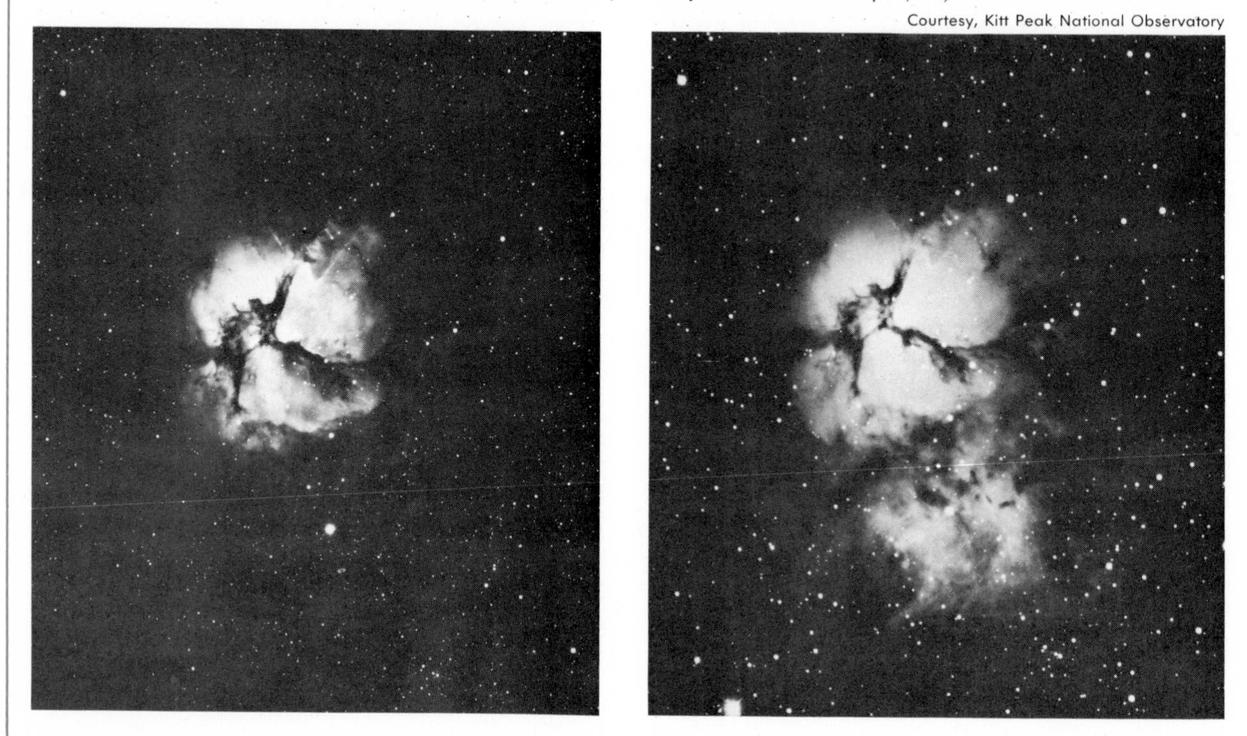

vations—in the Kitt Peak optical shop in Tucson made possible the polishing of the mirror there to accuracies of 1/10 millionth of an inch. Three years of grinding and polishing were required to prepare the mirror for installation in its ultimate bed.

Temperature fluctuations in the air surrounding the observatory can also distort viewing. Experiments conducted at Kitt Peak before the new instrument was designed showed that a smaller star image—hence, better seeing—results if a telescope is elevated at least 100 ft from the ground. Consequently, the Mayall telescope, all 375 tons of it, is perched on a hollow cylindrical pier, 37 ft in diameter and 92 ft high. The building in which the telescope is housed is 185 ft high overall, about as tall as a 19-story building.

The hemispheric shell that protects the telescope from the elements has a diameter of 105 ft and weighs 500 tons. To enable the observer to direct the instrument at any portion of the night sky, the dome is mounted on 32 tracks that permit it to be moved a full 360°. Viewing is permitted through a 28-ft slit in the dome, which opens by rolling upward and backward on tracks.

The telescope itself is aimed by precise large gear drives that can direct the instrument to any portion of the sky and compensate for the effects of the earth's rotation. The movable portion of the telescope weighs 300 tons. So accurately is it balanced and so delicately mounted—on eight oil pressure pads at 600 psi—the electric motor required to drive the tracking mechanism is rated at one-half horsepower. Actually, the 300 tons could be shifted quite easily by hand.

When the telescope is in full operation, it will have three main observing positions: the prime focus, the coudé focus, and the Cassegrain focus. At the prime focus (with a focal length/aperture ratio of 2.8), the incoming light from distant objects is converged by the aluminized surface of the 158-in. primary mirror (which has the light-gathering capacity of a million human eyes) and brought to the top of the telescope tube, 40 ft from the mirror. There, the astronomer, perched inside an observing cage, records the image.

At the Cassegrain focus, a 52-in.-diameter convex secondary mirror located near the top of the telescope tube intercepts light coming from the primary mirror and reflects it back down the tube and through a central, 50-in. hole in the primary mirror to a focus slightly behind it. Because of its increased light-gathering power, the Cassegrain focus (with a focal ratio of 8) was expected to be the observational mode most used. The Mayall instrument will be especially useful at this focus because it is one of the first instruments to be built so that it is strong enough to carry a great weight of instrumentation at the Cassegrain level.

When the telescope is used at the coudé focus, several other mirrors are used to reflect the light to a focus in the observatory building where even larger and more sophisticated equipment can be used to analyze the incoming light. A further advantage of the Mayall telescope is that through a "coudé feed" it will be possible to make use of the coudé focus at the same time the instrument is being used at another of the positions, thereby doubling the efficiency of the facility. The coudé focus will have alternate focal ratios of 30 and more than 100.

At one or another of the three main observing positions, astronomers will be able to employ an almost endless variety of observational systems, excepting one. To the disappointment of the lay visitor, the historic image of the astronomer with his eye glued to an eyepiece is no more. The more traditional of the astronomers on Kitt Peak make use of photographic film, up to 14 in. square. The enormous light-gathering power of the 158-in. mirror makes possible sharp exposure of short duration.

In other circumstances, electronic light-measuring devices will be used instead of photographic plates and will be, in effect, counting photons, the smallest imaginable quanta of light. Other devices being developed for the Mayall telescope include imaginative combinations of computers and image detectors in which as many as 150 exposures of a single dimly visible object can be studied against each other so that the light from real objects can be separated from the random "noise" of the background sky, with a certainty ten times as great as now possible.

One of the special features of the Mayall telescope and unique to any telescope this size is a "flip secondary," a massive ring supporting a camera on one side and secondary mirrors on the other. This device will enable astronomers to switch the instrument from one operating position to another with great efficiency.

There is one other unique aspect to the Mayall telescope—it will not be unique for long. In 1974 the earth will be wearing, in effect, its first set of binoculars. Another 158-in. telescope, an almost identical twin to the one at Kitt Peak, is under construction on 7,000-ft Cerro Tololo in Chile. It will also be operated by the Association of Universities for Research in Astronomy, in this case in collaboration with the leading university in Chile.

—Howard J. Lewis

continued from page 185

When the calculated motion of the comet (taking all known gravitational effects into account) was compared with early Chinese and European observations up to 1682, Brady found small discrepancies between the comet's observed and calculated dates of perihelion. He noted a rough cycle of about 500 years in these discrepancies, and interpreted this as the first approximation to the period of an unknown planet disturbing the comet's motion. On this assumption, he analyzed anew the numerous observations of Halley's comet at the last seven returns (1456–1910) in order to find by trial and error the orbital elements of the hypothetical disturbing planet.

Thus, Brady deduced that a new planet, nearly as massive as Jupiter, was traveling around the sun in a nearly circular 464-year orbit, about 5,800,000,000 mi from the sun. This orbit, he inferred, was nearly perpendicular to the orbits of the known planets. The present location of the planet should be in Cassiopeia. Brady believed that it would resemble in appearance a faint star of magnitude 13 or 14, moving about 0.8° per year among the background stars.

This much-publicized prediction remained unverified, and there was mounting evidence that it was ill-founded. Photographic searches were conducted at the Royal Greenwich Observatory in Great Britain, Lick Observatory in California, and Lowell Observatory in Arizona; none revealed any such moving object near the predicted position. Moreover, other astronomers pointed out that reasonable assumptions as to the density and reflectivity of planet X would make it about magnitude 11—so bright that, if it existed, it should have been picked up years ago in photographic sky surveys.

Computations were made independently by astronomers at the U.S. Naval Observatory in Washington, D.C., and at California Institute of Technology to ascertain what gravitational effects planet X would have on the orbits of the known planets. Both studies indicated that a body nearly as massive as Jupiter and moving as calculated by Brady would produce significant perturbations in the motions of Jupiter and Saturn, and major perturbations in Uranus and Neptune. Disturbances so gross were not revealed by the observations of these four planets, providing a strong argument against the existence of planet X.

X-ray pulsars. By the spring of 1973 at least 125 sources of celestial X rays had been discovered with detectors carried aboard two artificial satellites, *Uhuru* and OSO-7. The latter contained a small X-ray telescope as one of its nine experiments. Approximately 80 of the 125 sources were located close to the central line of the Milky Way, indicating that most of them lie within our galaxy. These galactic X-ray sources are of many kinds, but the most remarkable are several that emit very rapid regular pulses of X rays.

The best-studied example of this handful is Her (Hercules) X-1, discovered from *Uhuru* data by Riccardo Giacconi and his associates at American Science & Engineering Inc. in Cambridge, Mass. Every 1.24 sec it gives a flash of X radiation, but at regular intervals of 1.7 days the sequence is interrupted for about one-fourth of a day.

These recurring interruptions suggested that Her X-1 is a member of a binary system with an orbital period of 1.7 days, the pulses ceasing when the X-ray source is hidden behind its companion star. This explanation was verified by accurate timings of the individual pulses, which were found to run slightly ahead of or behind schedule, depending on whether the X-ray source was traversing the nearer or the farther half of its orbit. (*See* Feature Article: UHURU: THE FIRST ORBITING X-RAY LABORATORY.)

Other observations of Her X-1 were made with the OSO-7 satellite, and from these George Clark and his co-workers at Massachusetts Institute of Technology were able to pinpoint the position of the source to within 0.3°. A Harvard University astronomer, William Liller, pointed out that within that area was located a faint, little-studied variable star, HZ Herculis, noteworthy for its very blue color. Liller therefore suggested that this optical object might be the stellar companion of Her X-1. The correctness of this identification was established by John and Neta Bahcall at Wise Observatory in Israel. Their photographic measurements of the brightness of HZ Herculis showed that it varied within the same 1.7-day period and that the star was faintest at just the times when the X-ray emission was absent. The optical period of HZ Herculis was also determined by N. E. Kurochkin in Moscow and by Liller at Harvard, who ascertained it to be 1.70017 days.

Thanks to this cooperative effort by scientists in many countries, by early 1973 a clear picture of the binary system could be visualized. The X-ray pulsar itself is doubtless an extremely dense, collapsed object—a neutron star only about five miles in diameter but with about 0.6 as much mass as the sun. It is rotating on its axis once in 1.24 sec, its pulsation period, while it revolves around the large, hot, blue star. In this binary, the side of the blue star that is turned toward the X-ray source becomes much hotter and brighter than the other side, because of the intense X-ray flux. Therefore, the hotter and the cooler faces of HZ Herculis are alternately presented, thus producing this star's 1.7-day oscillation in brightness.

These X-ray pulsars are fundamentally different from the ordinary radio pulsars, such as that in the Crab Nebula. No radio pulsar is known to belong to a binary system. Also, while the pulses of Her X-1 have been observed at both X-ray and optical wavelengths, they have not been detected by radio telescopes.

Furthermore, the periods of all sufficiently observed radio pulsars are very gradually lengthening, because these neutron stars are expending rotational energy in the form of radiation. However, the pulse periods of Her X-1 and Centaurus X-3 are slowly becoming shorter, indicating that in these cases the rotating neutron stars are gaining instead of losing rotational energy. In the Her X-1 binary system, the added energy probably is furnished by gas streaming from the hot primary star to the neutron star. The intense gravitational field of the neutron star would accelerate the gas to about half the speed of light, and the impact of the atoms on the rigid surface of the neutron star would cause the emission of X rays.

Gamma-ray astronomy. Electromagnetic radiation of wavelengths shorter than about 0.1 angstrom (one angstrom = 0.0000001 mm) is known as gamma radiation; it consists of photons with energies of more than 1 MeV (the energy of a photon of visible light is only about two electron volts). For about 15 years, astronomers have sought to detect celestial gamma rays, realizing that these should be produced in many regions of the universe where very energetic particles interact with matter. Because the earth's atmosphere strongly absorbs all but the highest-energy gamma rays, the measurements must be made from high altitudes.

Until late 1972 progress in gamma-ray astronomy had been slow. Astronomers now realize that most of the early reports of celestial gamma-ray sources (other than the sun) were erroneous. A critical summary in October 1972 by L. O'Mongain indicated that at energies of 100 MeV only four sources in our galaxy had been detected with reasonable certainty: the pulsar in the Crab Nebula, two sources near the galactic center in Sagittarius, and a fourth in the constellation Libra. The strongest and best observed of these is the Crab Nebula's pulsar. In January 1973 Cornell University scientists reported that they had measured it at energies above 800 MeV, with a detector carried by a giant balloon to 110,000 ft above Texas. These observations showed clearly that the gamma rays from the Crab pulsar are peaked sharply every 0.033 sec, demonstrating the same periodicity as the radio and optical pulses from this object. Thus, the pulsar in the Crab Nebula has been observed over a 10^{14}-to-1 range in frequencies: gamma-ray, X-ray, ultraviolet, visible, infrared, and radio.

"...it is easier to believe
that the Sun is at rest than
to confuse the issues by assuming
a vast number of spheres which
those who keep the Earth
at the center must do."
1473
1973
Sol.

A major breakthrough occurred with the successful launching on Nov. 16, 1972, of the Explorer 48 satellite, expressly instrumented for gamma-ray astronomy. Also known as SAS-B (Small Astronomy Satellite B), this NASA spacecraft was sent aloft by Italian scientists from the San Marco launch platform in the Indian Ocean off the coast of Kenya. A Scout four-stage solid-propellant rocket carried the payload into a nearly circular 96-min orbit, 342 mi in altitude and inclined 1.8° to the earth's equator. Explorer 48, weighing 410 lb and 51 in. long, carries a single experiment: a spark-chamber gamma-ray telescope, far more sensitive than any flown before.

The spark chamber measures the three-dimensional trajectory of the electron pair produced when an incoming gamma ray interacts with one of the thin metal plates between the 32 spark-chamber decks. The instrument can discriminate against high-energy charged particles. This arrangement is able to determine the arrival directions of individual gamma rays to an accuracy of about 1°. Counts are made of photons in subdivisions of the energy range from about 25 to 200 MeV; the total flux above 200 MeV is also measured.

After more than two months of operation, Explorer 48 and its gamma-ray telescope continued to perform flawlessly. It had already observed gamma-ray sources at the galactic center, along the galactic plane, in the Crab Nebula, and elsewhere. The primary goal of Explorer 48 was to make the first detailed gamma-ray map of the sky. It was also intended to search for the short-lived burst of gamma radiation that should accompany a supernova explosion and to look for pulsed gamma radiation from pulsars.

This sky survey, conducted with about ten times the sensitivity of other current experiments, was expected to produce in 1973 a sharp increase in the number of known gamma-ray sources. Fundamental questions concerning stars, interstellar matter, galactic magnetic fields, and cosmic radiation may be answered when gamma-ray sources are understood.

—Joseph Ashbrook

Atmospheric sciences

In many respects the year was one of steady progress in the atmospheric sciences. Early in 1973, however, decisions by the U.S. government on support for science introduced some serious uncertainties for the future.

GARP and GATE. An old acronym and a new one were in the news: GARP (Global Atmospheric Research Program) and GATE (GARP Atlantic Tropical Experiment). The first referred to a vast international endeavor to develop a high level of understanding of the earth's atmosphere up to an altitude of 30 km (19 mi). It included detailed observations of the entire atmosphere and the upper layers of the oceans on a scale never before accomplished. Weather instruments on land, ships, airplanes, and buoys were involved, but GARP depended chiefly on the widespread use of satellites to measure the properties of the atmosphere and underlying surface. The orbiting vehicles would also serve as means of transmitting data to giant computers and allowing rapid worldwide communications.

Such global data were needed in order to perfect mathematical models of the entire atmosphere. The models would serve as the means for producing improved weather forecasts for periods up to about two weeks in advance and would also be of value in studying the effects of environmental abnormalities on the weather and climate. For example, a valid mathematical model makes it possible to evaluate the effects of abnormally warm or cold bodies of ocean water on the weather.

Before the first completely global program could be mounted—probably toward the end of the 1970s—a number of smaller experiments such as GATE were needed. GATE was scheduled for the period between June 15 and Sept. 30, 1974. It would involve a detailed study of the tropical area from latitude 10° S to 20° N between Central and South America and the east coast of Africa. Some 25 nations planned to operate 25 to 30 specially instrumented ships, at least 13 airplanes, and a number of data-collecting buoys in the GATE area, while the U.S. and the U.S.S.R. would launch weather satellites for observing the region.

Weather modification. Notwithstanding the U.S. government's decision to emphasize research on natural disasters caused by earthquakes, floods, and the weather, there was a major reduction in support for research on techniques to increase precipitation, decrease lightning, and weaken hurricanes. This occurred despite the recommendation for increased support, made in a lengthy report in 1973 by the National Academy of Sciences.

Early in 1973 it was announced that the executive department was impounding funds appropriated for the National Oceanic and Atmospheric Administration (NOAA) and the Bureau of Reclamation for the purpose of conducting weather modification research. The bureau's budget for such activities was reduced from $6.6 million to $3.2 million and, as a result, many university programs dealing with the study of techniques for increasing rain or snow were being discontinued. A NOAA program on lightning suppression was being stopped.

Courtesy, Roger J. Cheng, Atmospheric Sciences Research Center, State University of New York at Albany

Sodium iodide-silver iodide crystals on a microscope slide in high humidity absorb water from the moist atmosphere, simulating the cloud-seeding mechanism. The darker crystals are silver iodide.

More important, Project Stormfury was shut down for a period of at least three years. Stormfury was a joint NOAA-U.S. Navy endeavor aimed at developing a procedure for reducing the destructiveness of hurricanes. In August 1969, Stormfury scientists had seeded Hurricane Debbie with silver iodide particles on two days and observed substantial decreases in wind velocity. Although this effect could not be attributed unequivocally to the seedings, the evidence indicated that such might have been the case. Theoretical studies also indicate that, in some circumstances, peak wind reductions can be caused by seeding with ice nuclei. Some prominent scientists argued that hurricanes approaching land should be seeded, but most meteorologists took the position that more experiments were needed. The curtailment of Project Stormfury would make this impossible, at least for several years.

Several major weather modification research projects continued to move forward, however. Joanne Simpson of NOAA extended her experiments on the effects of heavy ice-nuclei seeding on cumuliform clouds in Florida. Results to date indicated that rainfall from some clouds could be increased substantially. In Colorado, a Bureau of Reclamation project, under the overall direction of Archie Kahan, was in its third year. The program was designed to test earlier ideas on the degree to which snowfall from winter storms could be increased over an area of about 3,400 sq km (about 1,300 sq mi). Also in Colorado, the National Hail Research Experiment, directed by William Swinbank, was in its second year of testing techniques for reducing the fall of damaging hail.

During the winter of 1972–73, a ground-based system for clearing away supercooled fog was installed at Elmendorf Air Force Base near Anchorage, Alaska. It was similar to one at Orly International Airport in France in that it used expanding propane gas to transform the water droplets into ice crystals. In more than two dozen fog situations at Elmendorf AFB, the system cleared the fog enough to permit more than 96% of the scheduled flights to proceed.

One reason for the lack of rapid development in the field of weather modification was that knowledge of the chemistry and physics of clouds was inadequate. Basic research in this subject had suffered in recent years, and the reduction of support for weather modification research would inevitably reduce the amount of cloud-physics research done in university and government laboratories. Interestingly, the year saw considerable progress in plans to conduct cloud-physics experiments in an orbiting satellite. This came about largely at the instigation of Charles Hosler at Pennsylvania State University, working with scientists at the McDonnell Douglas Astronautics Co. and the National Aeronautics and Space Administration (NASA). Gravity in a satellite is almost at zero, so the air in a test chamber undergoes little up-and-down convective motion and particles do not "fall" to the bottom. This condition facilitates study of the behavior of small individual water and ice particles as they evaporate, grow, or break up.

Long-term weather prediction. Climatological observations yield much useful information about conditions far into the future. For example, they can be used to indicate the normal rainfall that can be expected during the growing season in Kansas, the Ukraine, or other places where there are adequate data. They also reveal the variability observed in the past and make it possible to calculate the

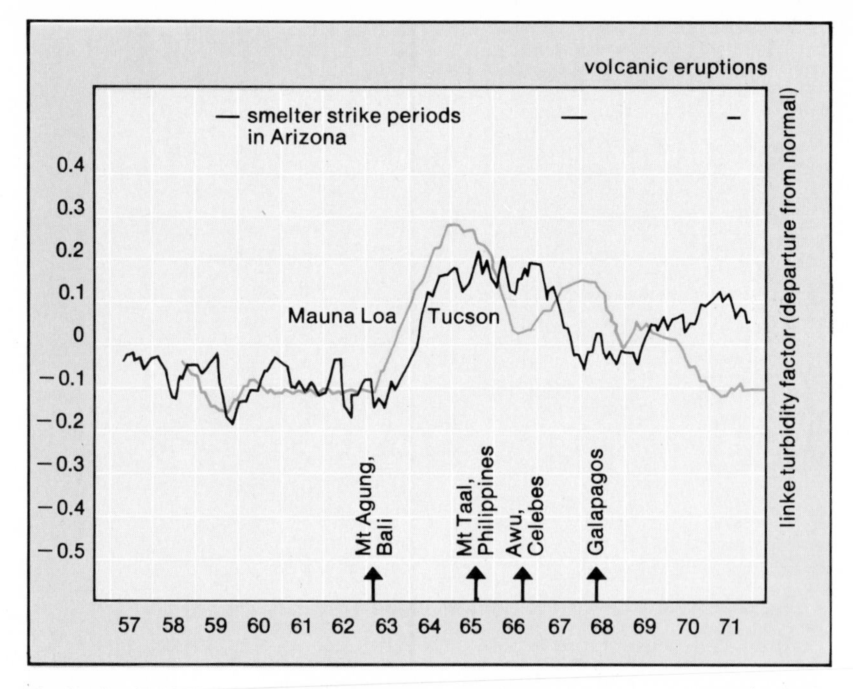

Graph shows the departure from normal of atmospheric turbidity measurements over Mauna Loa, Hawaii, and Tucson, Ariz., from 1956 to 1971 and the correlation between changes in turbidity and volcanic eruptions. Curves are based on average yearly values.

probability of a season occurring with, for instance, only 50% of the normal rainfall. Unfortunately, such information is not enough to provide accurate predictions of future wheat yields. Climatological data alone do not make it possible to predict the extreme abnormalities associated with droughts or floods.

During the past year there was a resurgence of interest in the use of correlations between present and future weather conditions and solar or magnetic phenomena. Interest in this subject was particularly intense in the U.S.S.R., which, in 1972, suffered a disastrous drought and drastic reductions in agricultural output. Not surprisingly, the drought was not forecast by the Soviet Hydrometeorological Service. On the other hand, it was claimed that a Siberian "heliometeorologist," A. D'yakov, predicted the drought by means of techniques involving solar-terrestrial relations. Specific information about his procedures was not available outside the Soviet Union, but apparently the Soviet Hydrometeorological Service was attempting to evaluate them. During a conference held in late October 1972, D'yakov was asked to make a long-range forecast for the period 1973–74 so that his forecasting methods could be evaluated objectively.

Satellites. In order to describe the properties of the earth's atmosphere adequately and to make predictions one to two weeks in the future by means of available mathematical models, it is necessary to observe the entire atmosphere. Traditionally such measurements were made by instruments near the ground and balloon-borne radiosondes. The network of stations was adequate in North America and Europe, but over the oceans, and particularly in the Southern Hemisphere, there were large unobserved regions. There was little doubt that the observational data essential to the success of programs such as GARP could be collected only through the use of advanced weather satellites.

Several satellites of geophysical interest were launched during the year. On July 23, 1972, NASA placed its first Earth Resources Technology Satellite (ERTS 1) in a near-polar orbit at an altitude of about 920 km (565 mi). It was designed to have a one-year lifetime, during which it would make observations of interest to atmospheric scientists and weather forecasters. By means of various radiometers, ERTS 1 measured such quantities as the surface characteristics of oceans, lakes, and bays, sea-ice distribution, snow cover over the continents, and details of the clouds producing severe storms.

On Oct. 15, 1972, an operational environmental satellite, NOAA-2, was placed in orbit at an altitude of about 1,500 km (932 mi). It provided global coverage of the earth's atmosphere and oceans twice a day. Unlike earlier operational satellites, it carried no conventional cameras, relying instead on scanning radiometers to produce images of clouds and other phenomena. Equipment on NOAA-2 made it possible to obtain routine vertical temperature profiles of the atmosphere over most of the globe.

For a number of years, meteorological data have been collected by means of synchronous satellites having spin-scan cameras that can yield daytime cloud patterns over half of the earth at intervals of about 20 minutes. Work was in progress on a Synchronous Meteorological Satellite (SMS-A) equipped with visible and infrared spin-scan radiometers mounted on a 16-in. telescope. Such a system would provide day and night viewing and would permit objective determinations of cloud types, temperatures, and heights, as well as wind fields determined from the movement of cloud elements.

The value of sensitive detection radiometers on satellites was indicated by the recently declassified military satellite system known as the Data Acquisition and Processing Program (DAPP). It provided visual, infrared, and other remote sensing data having meteorological significance. Its measurements included global coverage of visual and infrared imagery having a spatial resolution of about 3.5 km (2.2 mi) and, in some cases, as fine as about 0.5 km (0.3 mi); remote soundings of vertical temperature profiles of the atmosphere; and concentrations of electrons associated with auroral events. On the night side of the earth the visual sensor could detect clouds as long as there was at least a half moon. Spectacular photographs were obtained showing cloud patterns and auroral displays (*see below*).

See also Year in Review: ASTRONAUTICS AND SPACE EXPLORATION, *Earth Satellites.*

Radar observations of the ionosphere. Satellite measurements of the properties of the atmosphere were being augmented by the so-called incoherent backscatter radar technique. One radar for this purpose, operated by Cornell University, was located at Arecibo, P.R., and employed a circular antenna about 305 m (1,000 ft) in diameter. In recent years, incoherent radar scatter had become an exceedingly powerful method for measuring such properties of the ionosphere as electron density, temperature of the atmosphere, ion temperature, and mass and drift speed over a height range of hundreds of kilometers.

Aeronomists were attempting to obtain funding for the construction of an advanced incoherent radar to be located near the U.S.-Canadian border and used by scientists of both countries. Such an installation, coupled with optical, rocket, and satellite observations, should lead to major advances in understanding of the upper layers of the atmosphere.

Atmospheric composition and climate. Interest in the earth's climate and the factors governing it continued at a high level. At the UN Conference on the Human Environment, held in Stockholm in June 1972, a decision was reached calling for the establishment of a worldwide network of stations for monitoring various constituents of the atmosphere.

Since 1956 the U.S. government had maintained an observatory at an altitude of 11,200 ft (3,416 m) on Mauna Loa in Hawaii for the purpose of measuring atmospheric properties. This was to be the central station of a string extending between the North and South poles. Measurements were to be made of many quantities, including concentrations of carbon dioxide and particulate matter in the air and of incoming solar radiation, which serves as an indication of overall atmospheric turbidity. Ten long-term air-quality stations also had been put into operation in different regions of the U.S. They would measure atmospheric turbidity, the chemical

Nighttime view of a part of the Northern Hemisphere is produced by sensitive optical detectors on a satellite of the U.S. Air Force. Bright arcs are the aurora borealis, while light images indicate clouds and bright spots denote cities in Europe.

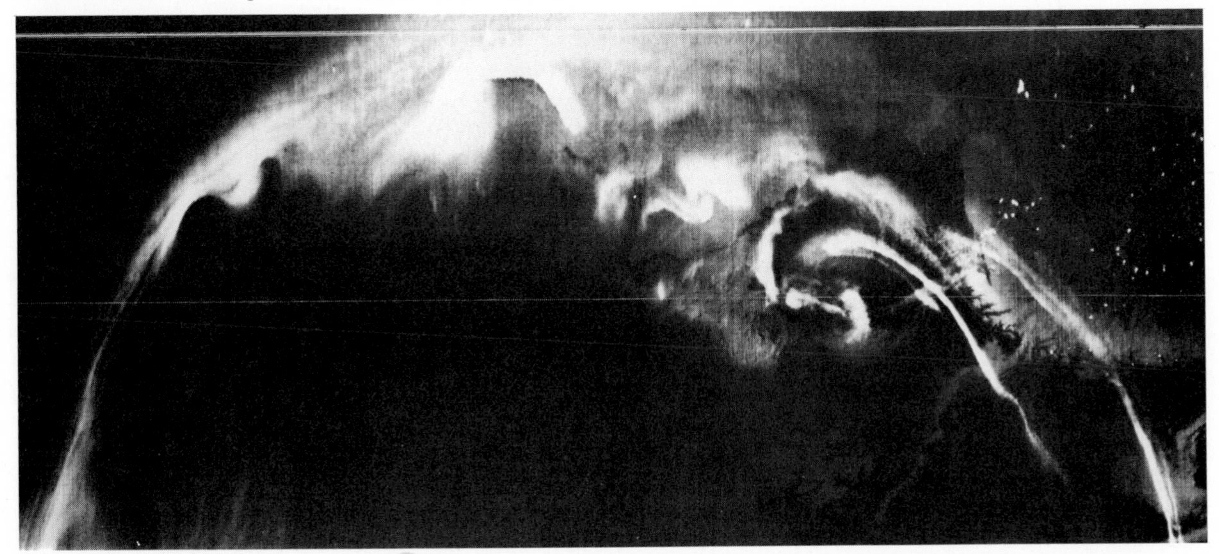

Courtesy, U.S. Air Weather Service

composition of rain and snow, and other atmospheric parameters.

The measurements of atmospheric turbidity at Mauna Loa showed that the major changes in particulate loading of the atmosphere since 1956 had been caused by volcanic eruptions. The major contributor of atmospheric particles was Mt. Agung in Indonesia, which erupted in 1963. Various smaller volcanoes added aerosols to the atmosphere between 1965 and 1969, but by 1971 the particulate concentration over the Hawaiian mountain station had returned to the level prevailing before 1963. The effects of the volcanoes also were evident in the atmosphere over Tucson, Ariz., but in 1971 the dust loading there was higher than in the early 1960s. This suggested an increase in the particulate pollution level resulting from human activity.

The U.S. Department of Transportation was involved in a program to assess the environmental and meteorological effects of high-altitude aircraft, including both subsonic and supersonic types. This program, known as the Climate Impact Assessment Program (CIAP), was sponsoring a series of working and writing conferences. Scientists attending them prepared a series of lengthy reports dealing with the present state of and dynamic processes in the stratosphere, as well as the complex interactions of aircraft emissions there.

The CIAP reports would also include projections of possible future states of the stratosphere and analyses of possible effects on climate close to the earth's surface. These studies should contribute to the resolution of some important questions troubling environmentalists. For example, various scientists had concluded that nitrogen oxide emissions from the engines of a fleet of supersonic aircraft would decrease the ozone concentration in the upper atmosphere. If this occurred, the amount of ultraviolet solar radiation reaching the ground would rise and more cases of skin cancer could be expected. For a number of reasons, questions were raised about the validity of these assertions. The amount of nitrogen oxide likely to be emitted by the projected fleet of airplanes was still not known, and the factors governing formation and destruction of the ozone layer were still being debated.

Although the problem was far from resolved, important early steps were taken in the formulation of a mathematical model of the earth's climate. Such models take into account many of the pertinent quantities, such as incoming solar radiation, configuration of land and water, and various energy-transfer mechanisms. One of them, developed by William D. Sellers, calculated conditions over the earth over time steps equivalent to one month, with each time step requiring less than two seconds of computing time. Sellers had extended the calculations over a period equivalent to about 400 years. His results indicated that the mathematical approach to the study of climate was likely to be a fruitful one.

—Louis J. Battan

Behavioral sciences

Anthropology continued to grow during 1972–73, but that growth took place against a background of new opportunities and challenges to the discipline. New developments included an increasing awareness of the implications of protein deficiency for many of the peoples studied by anthropologists. The old issue of the relationship between "race" and intelligence was revived, as was speculation on the origin and evolution of human language. In psychology, many years of work in linguistics and artificial intelligence began to crystallize into a new and rigorous science of the mind.

Anthropology

In a speech before the American Anthropological Association, the incoming president for 1972, Anthony F. C. Wallace, reflected on the pressures for change confronting anthropology in the 1970s. Increasingly, foreign populations objected to being studied under conditions that favored U.S. researchers over their own social scientists. There was concern to direct anthropology (and other fields) toward such moral and practical problems as war, poverty, and the environment. Centralization of federal decision-making and "cost-benefit" approach to scientific research were leading to a greater insistence that federally funded projects be relevant and useful. In short, anthropology was being challenged in many ways to rethink its aims and justify its existence.

One of the more interesting results of this reassessment was a book, edited by Dell Hymes, entitled *Reinventing Anthropology* (1972). In an opening chapter, Hymes defined anthropology as "a source of knowledge about the exploited by those who exploit them" and asked if such a phenomenon ought to continue. Answering his own question in the negative, he suggested that, at the very least, the departmental structure of the discipline be loosened to permit the "reinvention" of anthropology by its practitioners, "following personal interest and talent where they best lead." Hymes saw this process as a merging of many subdisciplines into related but nonanthropological fields—for example, economic anthropology would be absorbed by economics—with perhaps only archaeology and prehistory remaining as distinct

Dr. Gerald S. Hawkins, Smithsonian Astrophysical Observatory

Immense geometric patterns were marked on the floor of the desert near Nazca in southern Peru at least 2,000 years ago. Some archaeologists believe they have an astronomical significance, although no correspondence with the movements of heavenly bodies has yet been discovered. Many so-called primitive peoples had an astounding knowledge of astronomy and engineering.

disciplines. Unfortunately, the book did not discuss how or whether other disciplines would absorb either the anthropologists or anthropological insights and approaches.

As an alternative, Hymes suggested that the rigid compartmentalization into physical anthropology, linguistics, archaeology, and social anthropology be ended, along with the insistence that every student master all four subfields. In its place he envisioned a discipline "integrated at some levels, but comprising much diversity, part maintained from the past, part emergent, as sufficiently autonomous groups shaped their own cultural worlds."

A particularly encouraging sign was the fact that two of the more important recent critiques of anthropology had emerged from the third world. One was *De eso que llaman antropología mexicana* ("On Mexican So-called Anthropology"), a polemical, angry book written in the aftermath of the 1968 student demonstrations by a group of young anthropologists on the point of completing their advanced degrees. Though published in 1970, it did not come to the attention of most U.S. anthropologists until it was reviewed in the *American Anthropologist* in 1972 and 1973. The authors argued that Mexican anthropology had not reflected the needs of the people it studied and had been too bound to the "neocolonialist orientation" of the Mexican power structure. It appeared that the work had already had an important influence on anthropology in Mexico.

The other work, *Review of Social and Cultural Anthropology in India* (1972), edited by M. S. A. Rao, was part of an ambitious and remarkably well organized attempt by leading Indian scholars to assess the development of all the social sciences in India. It critically analyzed the historical development and present condition of each subdiscipline of anthropology, assessed the work of all scholars who had worked in India, with particular attention to Indians, attempted to delineate areas where reseach and the development of teaching facilities were particularly needed, and outlined priorities for the future. The Indian Council of Social Science Research would use these priorities as guidelines for assessing research proposals submitted to the Indian government, as well as research in India proposed by foreigners.

Race and IQ. With rather monotonous regularity, studies had appeared during recent years purporting to show that certain racial, ethnic, or social groups are poorer learners and make significantly lower scores on IQ tests than members of the group to which the researcher happened to belong. From this, it was assumed that lower test scores reflect genetic inferiority. Although the evidence was overwhelmingly against this interpretation, such studies nonetheless enjoyed a brief fame before being consigned to obscurity.

During the year the ire of anthropologists was focused on the latest work of this kind by Arthur Jensen, professor of educational psychology at the University of California at Berkeley. In particular, the attack was directed against his article "How Much Can We Boost IQ and Scholastic Achievement?" (*Harvard Educational Review*, Winter 1969). At its meeting in November 1971 the American Anthropological Association had unanimously passed a resolution condemning the views of Jensen and others like him as "dangerous and unscientific," and the subject continued to be discussed among anthropologists throughout 1972–73.

Probably the best short summary of the evidence against Jensen's views was an article by Ashley Montagu, "Sociogenic Brain Damage," which appeared in the October 1972 issue of the *American Anthropologist.* Without denying that a significant genetic element contributes to basic intellectual potential, Montagu reviewed the studies indicating that intelligence, perhaps more than any other characteristic, is dependent on the kind of environmental stimulation to which it is exposed. Jensen conspicuously failed to investigate the complexity of environmental factors that contribute to intelligence and to performance in a test situation (which is not the same thing). He assumed that identical incomes produce identical environments. However, as Montagu observed,

> if nutrition is poor, health care deficient, housing debasing, family income low, family disorganization prevalent, discipline anarchic, ghettoization more or less complete, personal worth consistently diminished, expectations low, and aspirations frustrated, as well as numerous other environmental handicaps, then one may expect the kind of failures in intellectual development that are so often gratuitously attributed to genetic factors.

Protein deficiency and mental retardation. Montagu's article included a summary of recent research on protein deficiency, which has a critically important bearing on many of the peoples studied by anthropologists. The link between protein malnutrition in children and permanent mental retardation had been amply demonstrated. Certainly it is extremely difficult for a society to attain a high level of socioeconomic development if its human resources are seriously handicapped by the biological consequences of malnutrition. It is also a serious error to describe the sociodynamics or "world view" or to speculate about the future economic and social development of such a community without taking the effects of malnutrition into account.

Unfortunately, hard data on malnutrition were scarce, and estimates of its extent varied widely. For example, the April 1972 issue of the *WHO Chronicle* (World Health Organization) gave a much more conservative estimate of the worldwide incidence of protein malnutrition than the studies cited by Montagu. However, even the lowest estimate—*i.e.*, that 20% of the world's children five years old or under were not receiving enough protein to ensure "normal" brain development—represented a major world problem. If the higher estimates of 50% or more were correct, the world faced a crisis of major proportions.

Physical anthropology. Oct. 1, 1972, marked the death of Louis S. B. Leakey (*see* OBITUARIES), who since 1931 had explored and excavated in Olduvai

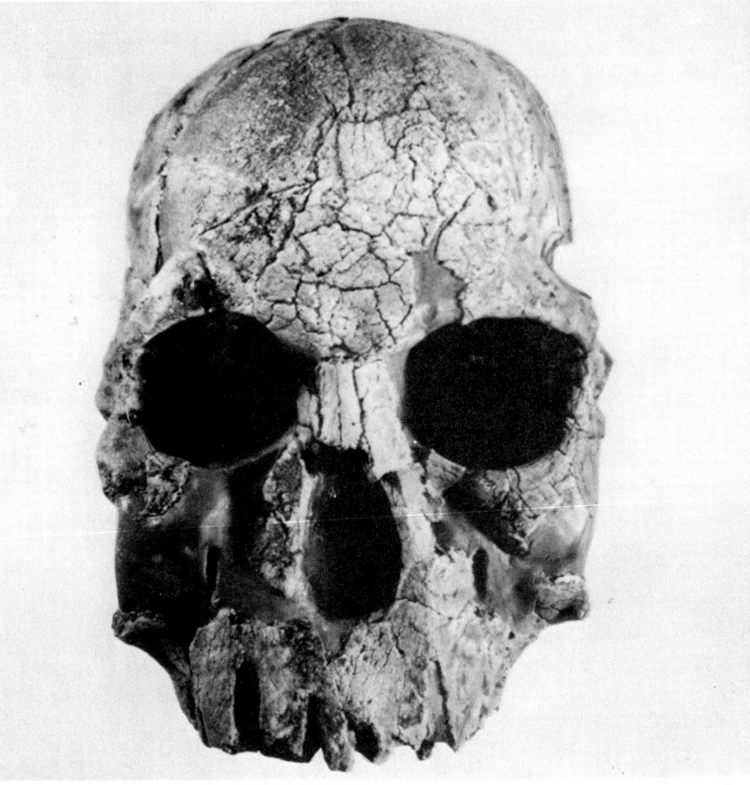

Controversial fossil skull found by Richard Leakey near Lake Rudolf, Kenya, presents a surprisingly modern appearance. Gaps in the skull have been filled with plastic. Leakey believes the fossil is some 2.8 million years old and belongs to the genus Homo. *This would make it the earliest complete skull of early man ever found and would push man's appearance even farther back in time than had been believed.*

UPI Compix

Gorge, Tanzania, an East African site of primate finds that moved the date of man's emergence back thousands of years. Leakey's interpretations and his revolutionary discoveries were often the subject of spirited controversy, and controversy continued to surround the discovery, shortly before his death, of a 2.8 million-year-old fossil.

Leakey's son, Richard E. Leakey, tentatively identified the fossil as belonging to the genus *Homo*, which would call for a considerable revision of the generally accepted theories of early man. The fossil had a surprisingly large brain case for the period (about 800 cc) and lacked brow ridges. According to Richard Leakey, its discovery "leaves in ruins the notion that all early fossils can be arranged in an orderly sequence of evolutionary change. It appears that there were several different kinds of early man, some of whom developed larger brains earlier than had been supposed" (*National Geographic*, June 1973). The additional discovery, in a closely related stratum, of leg bones remarkably like those of modern man and unlike those of the australopithecines seemed to give further support to Richard Leakey's statement, but lively discussion on the subject was expected to continue for some time to come.

The origin of language. One of the most recent developments in linguistics was the revival of interest in the evolution of human language, an issue that greatly concerned anthropologists when the discipline was taking shape in the 19th century. The most important stimulus for this was research on chimpanzees, especially by R. Allen Gardner and Beatrice Gardner, who taught a chimpanzee named Washoe to use over 300 elementary forms of the American Sign Language for the Deaf.

Data from these and other studies led Gordon W. Hewes to postulate that man's first language was gestural. His provocative arguments were presented and then reviewed extensively by 16 other anthropologists, with a final comment by himself, in "Primate Communication and the Gestural Origin of Language" (*Current Anthropology*, February–April 1973). The discussion was particularly interesting because it involved, as a unit, the distinct approaches of the often disparate subfields of prehistory, linguistics, and primatology.

Growth of the field. According to the *Guide to Departments of Anthropology 1972–73*, the number of anthropologists in the U.S. had risen more than 10% over the previous year and there was a net gain of 26 departments in the same period. Of the total of 258 departments, 84 offered doctoral degrees, more than double the number ten years earlier. Unfortunately, the number of new Ph.D.s in anthropology was increasing more rapidly than jobs in the field. A growing concern of anthropologists, therefore, was to cut down graduate admissions and to find new opportunities for those who completed graduate programs. Some new job opportunities were opening up in federal, state, and community organizations, but most new positions were being generated by the continued expansion of undergraduate and high school teaching.

The continued expansion of anthropology and the difficulty experienced by professionals in keeping pace with the literature was reflected in the replacement of the *Biennial Review of Anthropology* (last issue was for 1971) with the *Annual Review of Anthropology* (first issued in 1972). Anthropology thus became the second behavioral science (psychology was the first) to be included among the fields covered by the prestigious Annual Reviews Inc., a publishing venture devoted solely to the communication of scientific information. A number of new journals, monograph series, and newsletters also emerged. These included six linguistics journals and two newsletters dealing with Indian languages, a journal for physical anthropology, three new archaeology journals, and two general monograph series, as well as a number of journals of peripheral interest to anthropologists.

—Raymond Lee Owens

Psychology

By 1973 it had become clear that psychology was not limited to being a science of behavior. It was also—perhaps primarily—a science of the mind. The development of genuine scientific knowledge and theory about mental processes was a significant event, and it took some effort to adjust to it. Modern man had grown reasonably comfortable in thinking about himself in biological terms; psychoanalytic concepts had become commonplace, and the analytical style of behaviorists had become familiar. But scientific ideas are not built to last, and in 1973 it was possible to judge with some confidence the new content of scientific psychology.

Knowledge structures. One major recent achievement in scientific psychology concerned new theoretical analyses of knowledge structures. A person knows that trees and flowers are plants, that roses and daisies are flowers, and that squirrels are not trees. No psychologist discovered that persons have knowledge of this kind, but it is far from trivial to develop a workable theory about it. In a parallel case from physics, everyone has always known that objects fall when they are dropped, but a theory that describes and explains such events represents an important scientific contribution. Galileo, Newton, and Einstein changed our way of thinking about the universe partly because of their theories of gravitation.

There was as yet no Einsteinian or Newtonian theory of human conceptual knowledge, but there was a Galilean theory; that is, a systematic framework of ideas describing the main characteristics of human knowledge about concepts and the general kinds of things to which concepts refer. There were two main versions of this theory. One was closely related to earlier work on artificial intelligence and described conceptual knowledge as a network in which concepts are connected by the relationships they have with each other; thus: squirrel → mammal → animal. The other related more closely to ideas worked out initially in linguistics and described conceptual knowledge as a structure of propositions. One kind of proposition states a division of some conceptual category into subclasses, such as "Plants: (trees, flowers, shrubs, grass, weeds)" or "Animals: (mammals, birds, reptiles, fish, worms)" or "Animals: (pets, farm animals, wild animals)." Other propositions state relationships, such as "(eat: squirrels, nuts)" and "(place to live: squirrels, trees)."

Chimpanzee in the wild is eating a small monkey. Research has shown that chimpanzees, once thought to be exclusively vegetarian, not only hunt and eat prey but have well-developed social patterns centering on these activities.

Dr. Geza Teleki, University of Georgia

Whether conceptual knowledge was represented as a network or as a structure of propositions probably made little difference to the substance of the theory. Some things, such as knowledge about events that occur, seemed easier to represent as a network. Others, such as the ambiguities of what is meant by the concept "animal" (the general biological class or the more informal concept including pets, farm animals, and wild animals) seemed more easily represented as a propositional structure. With either notation, the psychological theory of conceptual and factual knowledge had moved forward in a new and significant way.

Comprehension of language. A second area in which major advances were made was the theory of comprehension of language. The theory concerns the process in a person's mind when he reads or hears someone speak a sentence, such as, "Mother gave the dog a bone." One thing was certain: the mental response is not like writing a copy of the sentence or recording the sentence on audio tape. In many situations, the reader or listener does not store information about the exact words used in a sentence. The information stored in memory when sentences are understood is some representation of the meaning of the message.

The psychological theory of comprehension is a hypothetical description of the kinds of mental structures that are built when a person understands a message, and of the process involved in the building of those structures. Three general factors are considered. One is the set of relationships expressed within the message itself. The sentence "Mother gave the dog a bone" is put together in such a way that the reader or listener knows that the action involved is giving, that the action occurred in the past, and that Mother is the agent. It is also understood that the thing given was a bone and the dog was the recipient. The process of understanding the sentence involves constructing a pattern of relationships. This pattern would be the same if the sentence had been "The dog was given a bone by Mother." Experiments showed that semantic information is remembered much better than syntactic information, such as whether the sentence is in the active or passive voice.

A second important factor is the reader's or listener's use of information given earlier and then used to interpret later sentences. If one reads, "Katie is riding very well today. Did you see the way she took that last jump?" there is no difficulty in understanding that, in the second sentence, "she" refers to Katie. That understanding occurs because a reader holds in memory a structure representing the information in the first sentence.

A third basic factor involves the listener's or reader's use of general knowledge present in long-term memory. To understand the sentence, "John dribbled the ball down the court and scored a basket," the reader must know what dribbling is. A more subtle use is also made of knowledge stored in the reader's memory. A reader acquainted with basketball knows that the court is flat. Therefore, "down the court" is understood to mean "from one end to the other" rather than "toward the lower end."

Theories about the process of language comprehension were developed that incorporated ideas about the way people may use information stored in memory as a basis for understanding messages. One problem for such theories is to describe a way in which general conceptual knowledge influences the process of constructing a mental representation of a message. One suggestion was called conceptual dependency. We know that certain kinds of things can be related to certain other kinds of things, and we know some of the relations that can occur. For example, we know that horses are among the things people can ride. So when we hear someone say, "Katie is riding a spirited horse," we are able to construct a representation of that message consistent with our knowledge of possible relations among concepts.

Another idea was that much of our knowledge, especially in areas about which we know a great deal, corresponds to procedures either for carrying out actions or for evaluating the truth or falsehood of statements, and that there is a close formal correspondence between the mental processes used in understanding language and the processes used to carry out actions. Thus, this idea provided a connection between the theory of comprehension and theories of complex action of the kind incorporated in computer-controlled robots.

Human problem solving. A third area in which there were important developments during 1972–73 was the theory of human problem solving. A variety of problem-solving situations have been studied by psychologists, from simple puzzles to complex intellectual tasks such as playing chess. When a person works on a problem, the nature of the task determines an environment—a goal to be accomplished and rules for operating on the objects and situations involved in the problem. Operating within this task environment, the person can make changes in the situation that can lead eventually to a solution.

Each change in the problem situation creates a new state, and the solution consists of a sequence of these states leading to the goal. The problem solver is thus considered as seeking a successful path through a space of problem states. There are various general approaches that can be used: one may work forward through the space, starting with the given materials of the problem, or work backward from the goal state trying to see a way to reach it from the given materials. These approaches have been called methods of problem solving.

Computer systems for problem solving based on these ideas had been available for several years. In recently published work, theoretical systems were discussed in relation to known limitations in human mental processes, such as the rate at which human individuals can deal with items of information and the amount of information that can be considered simultaneously in human short-term memory. In addition, detailed records were obtained, with individuals reporting their thoughts during the solution of problems. These records were used to develop theories about the psychological process of solving problems, including detailed hypotheses about the kinds of processes used at each stage.

Computers and human thought. These major theoretical advances all evolved out of the use of computer simulation and artificial intelligence. The year's work appeared to mark a kind of turning point in the development of simulation of cognitive processes. It was now clear that theories developed as computer systems made genuine contact with the kinds of questions about psychological processes that concern students of human mental activity. Theories embodied in computer simulations thus provided genuine psychological understanding.

The development of computer simulation as a theoretical method of established substantive significance came at an appropriate time. During the 1960s a great deal of vigorous scientific work had extended the field of human experimental and theoretical psychology in many directions. Much careful and innovative experimentation had given rise to a growing body of empirical literature regarding many psychological processes, including problem solving, comprehension, memory, decision processes, and perception. At the same time, many new theoretical analyses were formulated increasing the ability of psychologists to express their understanding of those processes with rigor and clarity.

Thus, the addition of computer simulation to the psychologist's theoretical armamentarium came at a time when rigorous theorizing about mental processes had become standard practice in many areas of psychological inquiry. The development of a computer program that does something is not the same thing as developing a psychological theory about the way that people do that thing. However, computer languages provide important formal systems within which psychological hypotheses about complex mental processes can be evolved. During the next several years, much important work could be expected on experimental methods for testing and in extending the hypotheses that arise from the use of computer programs to express theories.

Learning theory. One important achievement that could be expected in the near future was the development of a genuine theory of learning, providing understanding of the psychological processes involved when children or adults come to understand new concepts or acquire new intellectual skills. In the first half of the 20th century, the psychology of learning was restricted to principles regarding changes in the frequency of observable behaviors.

The principles of behavioristic psychology—reinforcement, punishment, extinction, and other important ideas regarding the conditioning of responses—have extremely important applications. Many important advances in clinical and educational psychology were made possible by their effective use. However, more precise concepts were apparently needed for a satisfactory analysis of complex cognitive processes and significant conceptual learning.

Based on current understanding, it seemed quite reasonable to expect that within the next few years a truly usable theory of learning would be formulated. Such a theory would provide an analysis of the processes whereby individuals change important components of their knowledge structures when they learn new ideas and experience new relationships among familiar ideas or things. The theories of learning presently available included useful principles but, because the principles were quite general, it had been difficult to develop specific analyses of a kind that could guide day-to-day teaching of complex material. As theoretical systems were developed describing complex cognitive structures and processes in definite and rigorous terms, the conceptual resources for an educational technology of a new and genuinely systematic kind would be available.

—James G. Greeno

Botany

Most developments in botany in recent months resulted from steady work over a number of years. Some major areas of research activity centered on the morphology and physiology of plant development, plant origins and geography, plant population dynamics, and plant ecology.

Somatic hybridization. A most exciting advance in developmental botany was reported in late 1972 by Peter Carlson, Harold Smith, and Rosemarie Dearing at Brookhaven National Laboratory, Upton, N.Y. They were able to produce plant hybrids by fusing cells of two different species of *Nicotiana* (tobacco genus), and thereby to circumvent the normal reproductive process.

Recent developments in plant cell culture enabled researchers to take cells from the mesophyll layers of leaves and incubate them in nutrient medium. Under the proper conditions such cell cultures regenerated into entire plants. Cells from similar cultures had been stimulated to fuse by experimental manipulation, including digesting off the cell walls. Such fusion was expected to yield cells with the sum of the chromosome numbers of both parent plants. Until the successful experiments of the Brookhaven group, however, no one had ever been able to cause fused cells to divide and produce a new plant.

Carlson and his team secured cells from strips of leaves from *Nicotiana glauca,* chromosome number 24, and *N. Langsdorffii,* chromosome number 18. Cell walls were digested by enzymes, the protoplasts (wall-less cells) harvested by centrifuge, and a mixture of them from both species was stimulated to fuse in a solution of sodium nitrate. At this point it became necessary to determine which fused cells were hybrids and so, following another centrifuging to secure a pellet of cells, they were placed in a medium that favored the growth of hybrid cells alone.

Some hybrid cells produced colonies called calli. Removal of calli to another nutrient medium was followed by the formation of rudimentary shoots and stems but no roots. This was a phenomenon that had never been witnessed previously from hybrid plant cells. Since no roots were produced, it became necessary to graft shoots onto one of the parent plants, *N. glauca,* to witness further development. Careful study of the resulting hybrid plants for chromosome number, morphology, and some physiological characteristics proved them to be identical to hybrids produced by normal sexual means, which was possible in this genus.

Somatic hybridization offered botanists and geneticists the opportunity to study plant hybrids even in groups unable to produce them by normal sexual means. This process, called parasexual hybridization by its developers, might also lead to the production of new agricultural crops heretofore impossible because of hybridization barriers.

Corn ancestry. Hybridization also captured the interest of those seeking to understand the origins of plants. Experiments in crossing plants might give clues as to how closely they are related to suspected ancestors. George Beadle, University of Chicago Nobel prize winner, and a member of the Editorial Advisory Board of the *Britannica Yearbook of Science and the Future,* thought his evidence showed that the ancestor of corn, *Zea mays,* was a corn-like plant called teosinte, descendants of which still grow in Central America. This plant, *Zea mexicana,* has many characteristics in common with corn, including the ability of the two to hybridize in nature where teosinte grows along the edges of and within cornfields of Central Mexico and Guatemala.

The outstanding difference between corn and teosinte is their fruiting structure. The well-known cob of corn bears numerous rows of kernels and depends on the transport of the whole cob for dispersal. This characteristic is not a disadvantage as long as man harvests and plants the crop; however, it is a disadvantage in the wild, since all seeds on a cob would be dispersed to the same place where germination would lead to severe crowding of the seedlings. Teosinte bears its kernels on a spike that fragments when mature, thus dispersing individual seeds necessary to survival in the wild. This difference was being exploited in hybridization experiments with the objective of measuring genetic difference.

Botanists at Brookhaven National Laboratory produced the first viable hybrid plants by means of cell fusion. Leaf cells from two species of tobacco plants were treated, cultured, and grafted to parent plants. Their offspring were exact copies of the original hybrid.

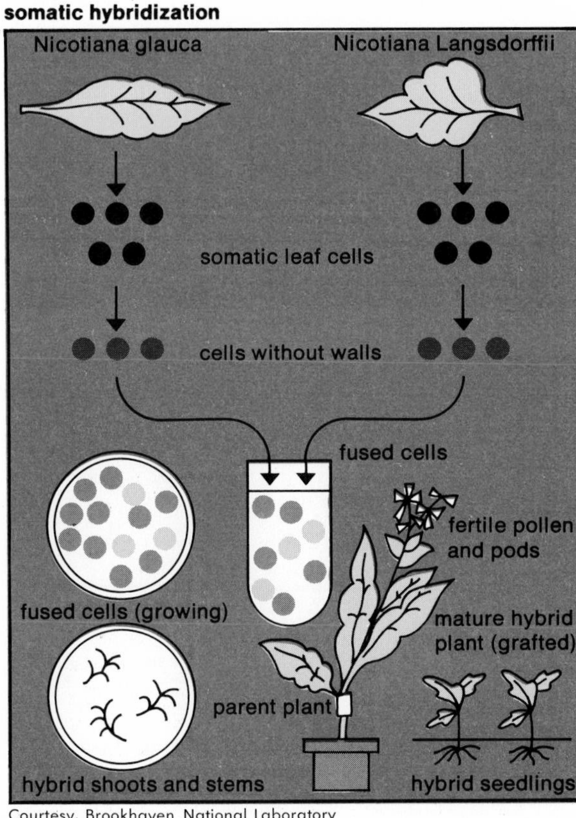

Courtesy, Brookhaven National Laboratory

If corn and teosinte cross, the resulting plants (F_1 hybrids) produce an intermediate maize-like fruiting structure looking like a miniature cob with as few as four rows of kernels. Crosses of these F_1 plants produce F_2 plants with a range of fruiting structures from the cob of corn to the spike of teosinte. According to current genetics, the proportion of these F_2 plants that resemble their grandparents for a certain characteristic would indicate how many genes control the difference in that characteristic. A difference of one gene should produce offspring one-fourth of which are like teosinte and one-fourth like corn. Two gene differences should produce offspring in which one-sixteenth are like teosinte and one-sixteenth like corn, and so on in a geometric fashion.

Beadle crossed the most primitive corn, Chapalote, with the most corn-like teosinte, Chalco. The frequency of grandparental reappearance was about one out of 500. This figure fell between the expected frequency for four and five gene differences, few genes indeed for such an apparent morphological difference. On this basis corn and teosinte must be considered to be closely related.

Whether corn arose from a teosinte-like ancestor or teosinte from a corn-like ancestor was still speculation. Beadle believed that the former was the case, since teosinte maintains its existence without man's intervention. It had also been shown that some of the quality of cultivated corn might, in fact, be due to hybridization occurring normally between teosinte and corn in the field. Some teosinte genes are constantly being incorporated into the corn gene pool. While the reverse may also happen, the former was of more concern since teosinte was being eliminated by increasing cultivation and grazing practices and, hence, the gene flow from teosinte to corn might stop.

Plant geography. Biogeographic relationships of plants both interested and confused botanists. Certain groups of plants as well as of animals are common to widely separated regions of the earth. If it is assumed that these groups had a common origin, then theories must be developed to account for their present discontinuous or disjunct distribution. For many years biogeographers had suggested that large land masses such as continents had to have been connected by land bridges at one time or another to account for plant and animal migration between them. Subsequent submergence of such bridges would have left similar organisms on both sides of a sizable water barrier.

Insurmountable problems contradicted this theory, especially when applied to the Southern Hemisphere. For instance, there was no evidence of a vast land bridge between South America and Australia, New Zealand, and Tasmania, where representatives of the same taxa were found.

Rapid acceptance in the last few years of the theory of continental drift, quite well supported by 1973 by the theory of plate tectonics, provided a new explanation for some problematic plant distributions. It is now thought that about 100 million years ago South America, Antarctica, Australia, New Zealand, and Africa, which then included India and Madagascar, were all connected in one land mass, called Gondwanaland. Subsequent movements of the earth's crust forced all of these land masses apart until they reached their present positions relative to each other.

Both fossil and present-day plant distributions reflected such a process. It was known that the forests of temperate portions of Gondwanaland were characterized by members of the Podocarpaceae, Araucariaceae, Fagaceae, and many other gymnosperms (primitive, nonflowering seed plants) and evergreen angiosperms (flowering plants). The present temperate forests in South America and Australia-New Zealand were characterized by some descendant species of the original families of the common Southern Hemisphere land mass. Since the present land masses had been separated for some time, such descendant species had changed differentially; however, in many cases they were recognizably members of the same genus.

A notable example of such a process was the genus *Nathofagus*, found in both South America and New Zealand but not in Northern Hemisphere paleobotany more recent than 140 million years. What was more, the more recent separation of Australia and South America than that of other southern lands was reflected in the greater similarity in their plants. Continental drift was proving to be a very adequate explanation for major worldwide trends in plant distribution previously noted in the fossil record and in certain contemporary Southern Hemisphere groups.

Fossil plants. The Precambrian Period of geological history was long considered to be without record of life. However, with the discovery of unmistakable traces of primitive plants in the Gunflint Chert of northern Minnesota and southern Ontario around Lake Superior, as well as in other places, researchers concluded that life existed at least 600 million years ago, at the end of the Precambrian. The primitive plants identified were blue-green algae (Cyanophytes), and they formed hemispherical masses of calcium carbonate or silica called stromatolites.

Some researchers studying the growth of filamentous bacteria, called flexibacteria, in Yellowstone National Park geyser pools, believed that the formation of stromatolites by flexibacteria may have taken place in the Precambrian. Such *Conophyton* stromatolites had previously been thought to have become extinct near the Precambrian-Cambrian time border. The Yellowstone finding indicated that bacteria may produce stromatolites and what previously may have been identified as very fine blue-green algae filaments in Precambrian formations might have really been flexibacteria.

Far-reaching implications arose from the fact that photosynthetic bacteria do not yield oxygen during photosynthesis (sugar production). Previous correlation between Precambrian life and the oxygen revolution, assumed to have taken place in the atmosphere because of photosynthesis, might be false. Evidence for Precambrian stromatolites, especially the Archean Bulawayan stromatolites of Rhodesia, which were the oldest life-bearing deposits known, equally supported flexibacterial origins. In addition, the most abundant components of the Gunflint microbiota (microscopic life) closely resembled hot spring microbiota and the resemblance of these organisms to flexibacteria and *Phormidium*-like cyanophytes was striking.

Nitrogen and succession. Research on the nitrogen cycle continued to uncover very interesting ecological phenomena. Plants are known to be

The delicate white hairs that top the long beaks of dandelion seeds (Taraxacum officinale) *permit their ready dispersal by the wind. Interest in questions about seed ecology grew in recent months.*

Charlie Ott from National Audubon Society

Howard A. Miller from National Audubon Society

A sundew, Drosera brevifolia, *has successfully trapped a spider in its sticky tentacles. Like all carnivorous plants, sundews trap insects and digest them to obtain nitrogen, which most plants get from the soil. It was thought that this adaptation helps carnivorous plants survive in nitrogen-poor, marginal environments. Recent studies indicated that plant communities can suppress the nitrification process to conserve energy.*

able to utilize both nitrates and ammonium ion from the soil. These chemicals result from the decay of organic materials containing nitrogen, mainly protein. Ammonia first results from decay and, if it is not taken up by plants or does not escape into the atmosphere, is converted into nitrates by at least two oxidative steps involving bacteria. The bacterial genera *Nitrosomonas* and *Nitrobacter* are known to be involved.

Experiments with many kinds of plant communities now showed that conversion of ammonium ion to nitrate does not occur to the same extent in all communities nor does the frequency of *Nitrosomonas* and *Nitrobacter.* A research team at the University of Oklahoma demonstrated that oxidation of ammonium ion and the frequency of nitrifying bacteria were not only positively correlated but that they both dropped off with advancing community succession. The soils under plant communities judged to be at climax were low in nitrate and the associated bacteria.

Climax is the final, stable stage in the development of an ecosystem. It occurs when a plant community has successfully adjusted to its environment. The drop-off noticed in conversion and in bacterial frequency might well be one of the factors contributing to climax. Plants utilize reduced nitrogen as in ammonium ion in the synthesis of protein. If nitrates are produced in the soil, this not only takes the energy necessary for the oxidation but, even more, requires that plants expend a great deal of energy reducing the nitrogen to produce protein. By suppressing the nitrification process, climax communities conserve energy, which contributes to their efficiency. Added to this is the demonstrated fact that ammonium ion is retained in the soil because of its positive charge while nitrate leaches out very readily because of its negative charge.

Seeds. In the 1880s W. J. Beal set out to discover how long dormant seeds could remain viable. He placed seeds of 23 locally common species of plants in open bottles and buried them in moist sandy soil. Periodically, one of these bottles has been dug up and its contents subjected to germination tests. Tests of seeds held in this way for 90 years were reported in recent months. The seeds of one species, a moth mullein, *Verbascum blattaria,* remained capable of germination. Apparently, as tests indicated, these seeds remained dormant in moist soil; they had to be dried and remoistened in the light to germinate.

Seed dormancy has ecological significance. It occupies a period of time when the embryo undergoes further development, when conditions unfavorable to seedling survival would be encountered, or both. A group of researchers at the State University of Rutgers, New Brunswick, N.J., was able to show that seeds from populations of the same species growing in different habitats had germination requirements that corresponded to the nature of the habitat. This represented definite physiological adaptation to the environment.

Populations of *Danthonia sericea*, a perennial grass, grow in two generally different habitats in the eastern United States. Variety *sericea* populates well-drained sandy soils from New Jersey to the Gulf states. Variety *epilis* populates wet places, such as bogs and stream beds, over the same range. With this combination of factors, effects of differential temperature regimes, as well as of moisture variations, could be studied as they affect seed dormancy and germination of seeds from the same species. It was found that *D. epilis* germinated far better under wet conditions than did *D. sericea*, thus matching its habitat. Seeds from southern plants of both varieties required less prechilling for germination, a process usually provided by winter weather in the north, than did northern plants, also correlating with the habitat conditions.

Much was known about seed germination but relatively little about its ecological implications. Since seed germination was one of the important determinants of the survival and distribution of species, its ecology was of great interest.

Another interesting study on seed ecology showed that not only light may be necessary for germination but also the development in a species of the ability to expose its seeds to light when it is available, and not necessarily when germination may take place. *Draba verna* is an interesting little plant of the mustard family that germinates in the fall and is, therefore, a winter annual. J. M. Baskin and C. C. Baskin showed that even if light is necessary for germination in the fall, it may be given to the seeds as much as four months earlier. Such a phenomenon might be adaptive for such plants whose seeds may become buried during the summer and would not have necessary light for germination available at the time the event itself should occur.

Continuing research in botany promised to give much more information on molecular control of plant development and on the intricate controls that determine the structure and function of plant communities. These would be tied somewhat with the current general interest in environmental problems.

—Albert Smith

Chemistry

Major developments in chemistry included the determination of the complete nucleotide sequence of an RNA molecule capable of replicating itself and new insights into the energy states of molecules. New ways of increasing the world supply of protein were under study, and chemical synthesists worked to develop an effective antitumor compound.

Applied chemistry

During the past year the science of chemistry again demonstrated its utility as a tool for a variety of other disciplines, as well as underlining its self-sufficiency as a field of its own. In materials science, for instance, chemists were helping develop new high-temperature materials that promised to be useful in turbines. In meeting the problem of adequate food supply, chemistry in partnership with plant genetics was unraveling the nitrogen-fixation process, possibly pointing the way to increased protein production. A uniquely chemical contribution was a new set of catalysts that might be of significant commercial value through the production of major industrial chemicals.

A silver route. When a match is put to the common hydrocarbon ethylene ($CH_2{=}CH_2$), the result is a flame and, mostly, two products: carbon dioxide and water. But if the burning is controlled in some way, there is a substantial yield of a third product, ethylene oxide $H_2C\overset{O}{—}CH_2$, a chemical that is not only industrially valuable in its own right—for use in antifreeze, for instance—but is a critical intermediate for the making of surface-active agents (such as detergents) and polymers (plastics, synthetic rubber). Since several billion pounds of ethylene oxide are made each year, improvements in its manufacture can mean considerable economic gains for the industry, gains hopefully to be passed on to the consumer. Such gains may be coming out of some basic work done by British chemists, led by Derek Bryce-Smith, at the University of Reading. Simply put, Bryce-Smith developed a catalyst that not only improves the yield of ethylene oxide in the reaction but also considerably improves the rate at which the reaction proceeds; that is, more ethylene oxide is made with each pass.

The catalyst is a silver compound, silver ketenide $(Ag)_2C{=}C{=}O$, which apparently forms a structure in which the silver atoms Ag are arranged in monoatomic layers that allow the ethylene—the material undergoing the reaction—to slip in between and come into intimate contact with the silver. Chemists have long known that silver can be an effective catalyst for such reactions as transforming ethylene to its oxide. The problem, however, was to find a way to do this so that the yields from the reaction were large enough to be commercially attractive.

The structure of silver ketenide, as discerned by X-ray crystallography, is a layer of silver atoms with the ketenide, or organic, portion sticking up between the silver layers. The structure is nicely regular: the "sticks" of ketenide all point in the same direction, suspended from the array of silver atoms, while the arrays lie in the same direction, parallel to one another. There is no chemical connection between the different layers, and so they can slide past each other, much as the carbon layers of graphite can slide past each other to give graphite its valuable lubrication abilities. Bryce-Smith found that once the layers of silver atoms are formed, the ketenide sticks can be removed (forming incidentally a compound, carbon suboxide, that is itself much sought after) and the catalytic properties are still partly retained. Apparently, even though heating destroys most of the layered or lamellar arrangement of silver atoms, various islands or pockets of silver layers are retained, enough to enable the material to retain sufficient catalytic powers. The removal of the ketenide portion enables the catalyst to be heated without destroying its catalytic powers, a distinct commercial advantage since reaction rates increase with higher temperatures.

High-temperature ceramics. Also in Great Britain, at the University of Newcastle upon Tyne, two chemists, K. H. Jack and W. I. Wilson, provided details on new high-temperature materials based on silicon nitride that are being investigated actively in the U.S. by several companies, principally for use in gas turbines to power cars and produce electricity. The value of synthetic nitrides as revealed by the British team is that they can be formulated into variable compositions with differing properties. This can be done, for example, by replacing silicon and nitrogen atoms in the crystals of silicon nitride—the material is a crystalline, synthetic ceramic—with aluminum and oxygen atoms. The products are solid solutions of aluminum oxide in silicon nitride.

The particular value of these groups of ceramics is that they are readily workable into various engineering formulations that can withstand high temperatures—up to 2,500° F—that destroy other materials, including some nominally high-temperature metal alloys. Thermodynamics states that, for any process involving temperature changes, the efficiency is increased the higher the initial temperature.

Thus, for the operation of turbines, the higher the inlet temperatures, the greater the efficiency. For this reason the Ford Motor Co. became interested in these ceramics as structural materials for gas turbines in cars, and Westinghouse Electric Corp. was investigating them as materials for gas turbines that generate electricity.

More proteins. As the meat boycott of 1973 demonstrated, a shortage of protein—when translated into expensive meat prices—is quickly felt. The high meat prices are an obvious sign that the world is still coping with the problem of supplying adequate proteins, not only to those now living at or below a subsistence level, but even to the more affluent nations whose dietary patterns, particularly in protein, are approaching that of the United States. There were other harbingers of a protein shortage: the world fish catch had apparently slowed its steady upward progression, and the production of soybeans, a major world source of high-quality protein, leveled off, in good part because the yield

Hypothetical scheme shows how root nodule bacteria, Rhizobium, *of leguminous plants might be made to fix nitrogen better by using techniques of molecular biology and plant breeding. Nitrogen fixation is the process by which atmospheric nitrogen is incorporated into plants to make protein.*

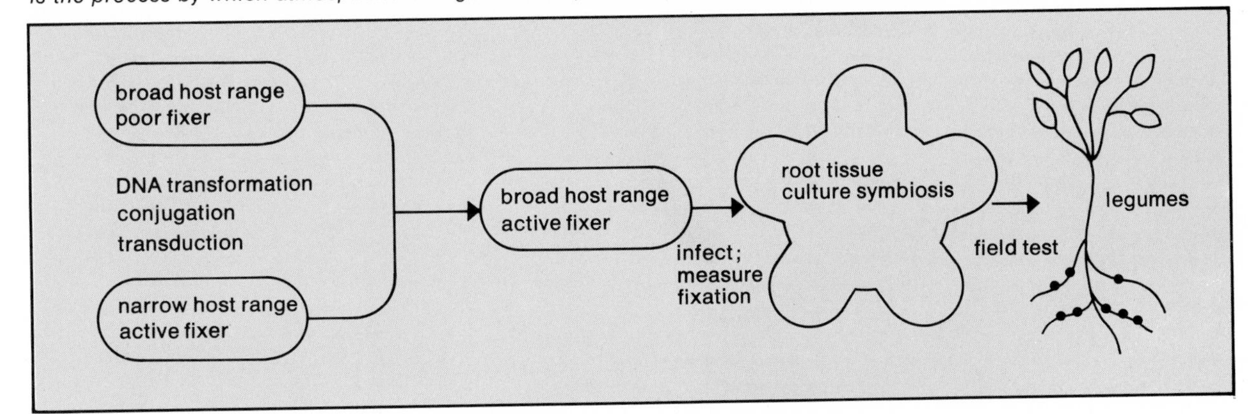

of soybeans per acre had not been increased and also because there were limitations on acreage available.

With that as background, the work during the past year on nitrogen fixation was heartening. Nitrogen fixation is a process mediated by microorganisms (either individually or in symbiotic concert with certain plants) and blue-green algae, by means of which atmospheric nitrogen is incorporated into plants to make protein. There is no semblance yet of a man-made system that imitates nature's way for converting nitrogen to ammonium ions, which in turn go through various transformations to become protein. When plants cannot make protein themselves, they must be fed fertilizers, a costly method that is under increasing attack from environmentalists.

As researchers viewed the problem, the protein supply of the world could be increased if plants that now fix nitrogen, such as soybeans, could do it better, producing better yields; or if plants that do not now fix nitrogen could be "taught" how to do it. In the past year considerable progress was made in understanding the genetic basis of nitrogen fixation and, through an amalgam of chemistry and genetics, in transferring nitrogen-fixation abilities to strains of bacteria that were previously powerless to do anything at all with atmospheric nitrogen. At the University of Sussex in England, John R. Postgate and R. A. Dixon successfully transferred the ability to fix nitrogen from one nitrogen-fixing bacteria to a strain of *Escherichia coli,* a non-nitrogen fixer that lives in human intestines. While the strain receiving the nitrogen-fixing ability was well chosen—it does not have the usual equipment for destroying foreign DNA, for example—the work did show that it could be done.

In a related work, three biochemists from the Berkeley campus of the University of California—Stanley L. Streicher, Elizabeth G. Gurney, and Raymond C. Valentine—reported on a detailed analysis of the genetic basis of the nitrogen-fixation apparatus of one free-living nitrogen fixer, *Klebsiella pneumoniae.* They found strains of these bacteria that could be transformed by infection with a virus from non-fixers to active fixers. With that tool in hand, they were able to begin exploring in detail the genetic apparatus required for a particular organism to be a nitrogen fixer. It is apparently more complex than was previously thought. For example, different sets of genetic directions are required for making the enzyme, nitrogenase, that makes the nitrogen-fixation reaction possible, and for controlling and regulating the process.

The genetic blueprinting is probably even more complex for the symbiotic bacteria, those bacteria that can only fix nitrogen in combination with a plant such as the soybean. (Such plants apparently supply the nutrients that make it possible for the nitrogen fixer to carry on.) Relatively little is known

Window photographed after an explosion demonstrates the effectiveness of Scotchtint polyester film. Developed by the 3M Co., the tough, transparent film is applied to the inside surface of glass panes. Glass in the untreated pane at left shattered and blew across the room, while the glass treated with Scotchtint (center and right), though cracked, was held harmlessly in place.

Courtesy, 3M Company

Courtesy, General Electric Research and Development Center

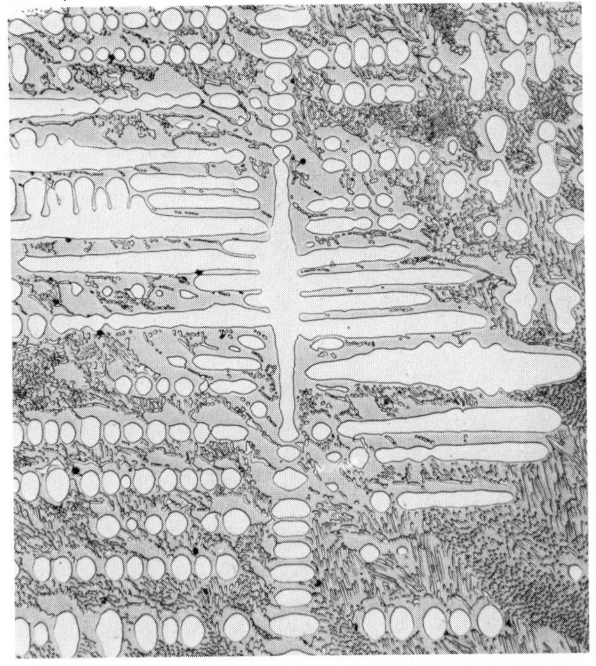

Microstructure of a cobalt-ytterbium alloy, as magnified 1,000 times, reveals large cobalt-rich dendrites (branching treelike shapes). The alloy was prepared as part of a study of cobalt rare earth magnetic compounds.

about this symbiotic process, although in 1972 a group of chemists at Du Pont & Co., led by R. D. Holsten, did in the test tube successfully marry and propagate a symbiotic culture of soybean cells and their nitrogen-fixing partners, a bacterial species called *Rhizobium*.

The work of nitrogen fixation has classically been chemists' territory, in good part because the process is so industrially important. Within the field so-called "pure chemistry" is under way, particularly in trying to determine the very complex structure of the two-part nitrogen-fixing enzyme, nitrogenase.

Meanwhile, both progress and failure took place during the year in efforts to derive proteins from another source, yeasts grown on paraffins such as crude oil. In Japan the cancellation of a proposed commercial protein plant apparently resulted from faulty quality control. In Europe there was progress, exemplified by construction of a 100,000-ton-per-year plant in Sardinia. This protein, from oil-fed bacteria, to be marketed by British Petroleum Co. Ltd. under the trade name of Toprina, was to be used as animal feed by Western European farmers. Although the process was proclaimed as a way to solve the "protein gap," as of 1973 there had been no substantive effort to prepare the protein for human consumption.

—Norman Metzger

Chemical dynamics

During the past year, activity in the area of chemical dynamics proceeded at a brisk pace. Although no spectacular breakthrough occurred, significant contributions were made. Activity in the field can be divided into three broad categories: (1) creating reactions that are chemiluminescent so that the light emitted can be spectroscopically analyzed to determine the internal-state distribution of the reaction products; (2) production of ion beams of increasingly narrower energy spread, permitting the study of transfer of translational energy into internal degrees of freedom such as vibrational and rotational motions of molecules (translational energy is the kinetic energy associated with the translational motion of a molecule, in which every part of the molecule moves parallel to and in the same direction as every other part); (3) use of lasers and sophisticated optical techniques in order to permit the identification of the quantum states of reaction products in some systems. (Quantum states of a molecule are states characterized by discrete energies.)

Chemiluminescence. Two examples of the chemiluminescence technique that appeared during the past year may be cited. The first was a non-beam technique reported in a series of publications by a group of scientists at the University of Toronto. The reactions studied were $X + HY \rightarrow HX + Y$ ($X = Cl$ or Br and $Y = Br$ or Cl), $H + X_2 \rightarrow HX + X$, and $F + H_2$ (or D_2) $\rightarrow HF + H$. The reactions were carried out in a chamber designed so that the reaction zone was surrounded by a wall containing cryogenic (extremely cold) fluid (liquid nitrogen or cold helium gas). The atoms were generated by a microwave discharge, or, in the case of H or D, by a Wood's high-voltage discharge tube. When the molecules react with the atoms, infrared emission results from the vibrationally and rotationally excited products. This radiation was examined either by conventional grating spectroscopy or by Fourier transform spectroscopy. It was found that members of this family of reactions ($X + HY$) diverted a substantial fraction of the energy released by the chemical reaction into vibrational and rotational excitation and only a small fraction into relative translational energy. For $Cl + HI$ and for $Cl + DI$, 84% of the available energy went into vibrational and rotational excitation and only 16% into translational energy.

The reactions $H + X_2$ were investigated to study the effect of the mass of the reacting atom on the conversion of reaction energy into internal and translational energies of the product species. Some earlier preliminary studies indicated that in such reactions most of the total available energy

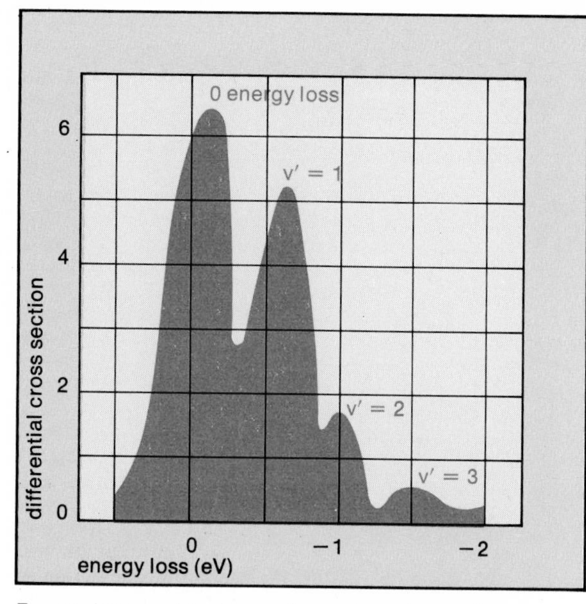

Energy losses of hydrogen ions when they collide with hydrogen molecules are plotted on graph. During the collisions some of the kinetic energy of the ions is converted into internal (vibrational and rotational) energy in the molecules.

went into translational energy of the products. This appeared to be supported by classical theoretical calculations on these systems. In the reaction H + Cl_2, 46% of the energy went into product rotation and vibration and 54% into translation. Thus, the conversion of total available energy into translational energy was considerably greater than for the case cited above, where the mass of the reacting atom was considerably greater. It was also found that translational energy of the products was significantly greater for successively lower vibrational product states.

A second example of the use of chemiluminescence to study the dynamics of chemical reactions was a recently reported experiment by a Columbia University group in which a thermal beam of aluminum atoms intersected an uncollimated (nonparallel) thermal beam of ozone molecules. The resulting visible chemiluminescence was then recorded with a scanning monochromator. The reaction is $Al + O_3 \rightarrow AlO + O_2$. In this example, one is observing light emission in the visible region of the spectrum that results from an electronic transition, and the vibrational and rotational transitions appear as superimposed fine structure. Although from the point of view of reaction dynamics this reaction was not studied in as detailed a manner as were the above infrared chemiluminescent reactions, it was clear that it too could provide information on the distribution of the produced internal (vibration and rotation) and translational energy.

Ion-molecule reactions. Another area of chemical dynamics in which scientists in the United States and in other countries were intensively studying the conversion of translational energy into internal energy was ion-molecule reactions. Although no ion-molecule reaction was studied in such detail that all of the quantum states of both products and reactants could be experimentally established, conversion of translational energy into vibrational and rotational energies during inelastic collisions between ions and molecules was studied successfully during the past year. (An inelastic collision is one in which part of the kinetic energy of the colliding particles changes into another kind of energy.) This type of reaction can be represented by the equation $A^+ + BC \rightarrow BC^* + A^+$.

The A^+ ions are generally produced by electron bombardment of a suitable gas. They are then mass analyzed, a process in which all undesired ions are excluded, and passed through an electrostatic energy analyzer that only transmits an A^+ beam with a very narrow energy spread. The translational energy of A^+ is accurately determined. On collision with molecule BC, some of the kinetic energy of A^+ is converted into internal energy (rotational and vibrational) in BC^*. The energy of A^+ is then measured with an electrostatic energy analyzer, and the difference between the initial and final energy of A^+ is the energy loss. The A^+ ions after suffering an inelastic collision are detected with a suitable detector (electron multiplier), which responds to individual particles. A plot of the number of A^+s arriving at the detector versus the energy loss of the ions gives a curve with peaks corresponding to the spacings of the vibrational or rotational energy levels in the molecule BC.

A somewhat idealized plot of such an inelastic collision as represented by the equation $H^+ + H_2(v) \rightarrow H_2(v') + H^+$ is given in the above figure. The peak designated by $v' = 1$ corresponds to a transition in H_2 from vibrational level $v = 0$ to $v' = 1$ induced by the inelastic collision. Peaks corresponding to $v' = 2, 3$, etc., have corresponding significance. These inelastic collisions were reported for a number of different ions and target molecules, and they have been studied as a function of bombarding-ion energy and scattering angle. With careful attention to resolution and other experimental details, this technique has been extended to the measurement of the differential cross section for individual rotational quantum transitions in the scattering of Li^+ by H_2.

Laser-induced fluorescence. Mention should be made of a new promising detection scheme that first appeared in the scientific literature during the past year. When one is using crossed beams of neutral reactants to study the dynamics of a chemical reaction, one is confronted, in general, with the problem of detecting a neutral product. In the early days of beam work, the experiments mainly concentrated on alkali metals and halogens since the materials and compounds containing them could be detected efficiently and simply. In more recent times, a universal detector that consists of an ionizer and some form of mass analyzer such as a quadrupole mass filter has been used. The beam of material to be detected is passed into the ionizer, and the resulting ions are then identified from the mass analysis. Because only about 0.1% of the beam is ionized it is highly desirable to have an extremely efficient universal detector.

The recently announced laser-induced fluorescence technique promises to be a big step in this direction since not only is it highly sensitive but also seems to be well suited for determining the vibrational-rotational distribution of the products of a wide class of reactions. This technique was applied to the reaction of a beam of Ba atoms with a beam of O_2 molecules to form BaO by a group at Columbia University. At the intersection of the beams, a reaction takes place and, simultaneously, light from a pulsed, tunable laser passes through the reaction zone. As the wavelength of the laser is varied, the BaO molecules are excited to the next permitted energy level whenever the laser wavelength coincides with such an allowed transition. Following the short excitation pulse, the molecule loses its excitation energy by emitting light (fluorescence). The fluorescence is detected by a photomultiplier tube that is electronically switched on during the time the laser pulse is applied and during the fluorescence emission. The time during which the photomultiplier is recording the emitted light depends on the length of the fluorescence lifetime of the excited state. In the case reported, this time was 100 nanoseconds (one nanosecond = one one-billionth second). This process is repeated many times, and the signal associated with many laser pulses is averaged electronically. One of the principal advantages of this technique is its specificity, since it is possible to identify the absorbing molecule and to assign the quantum numbers (integers that indicate energy levels) if the emission spectra of the molecule is known. It also can give information on the structure of the emitting species.

Future outlook. In the near future, studies of the basic principles underlying chemical dynamics can be expected to proceed at a vigorous pace throughout the world. On the experimental side, one can expect more work in which the detailed rates of the reactions are studied as a function of the quantum states of the reactants. Also, as more sophisticated optical techniques are brought to bear on the problem, beam-scattering experiments will reflect the roles played by the various internal quantum states of the reactants. The "ideal" experiment will be more closely approached—the scattering experiment in which all of the quantum states of reactants and products are identified as a function of the translational energy and angular distribution.

From the theoretical point of view, chemists expect to see steady progress that will enable them to understand how and why the translational and internal energies of the reactants determine the energy distribution in the products of a chemical reaction. With the advance of theoretical and computational knowledge, it seems more likely that simple three-atom systems will be calculated with sufficient accuracy so that they can be used as input in scattering calculations in which the cross sections and rate constants for simple reactions will be determined and compared with experimental measurements.

—Walter S. Koski

Chemical synthesis

The ability to synthesize new substances from raw materials is an important aspect of chemistry. This article describes several achievements in synthesis during the past year that illustrate the impact of the synthetic chemist on other areas of the science.

Enzyme analogs. Enzymes are exceedingly complex molecules that act as catalysts in biological systems. Their function is to speed up reactions that would normally occur slowly under biological conditions. It is estimated that they allow reactions to occur as much as 10^{10} (ten billion) times faster than they would otherwise. They are usually very large molecules with molecular weights from about 10,000 up to several million. Mainly protein in nature, they sometimes require the cooperation of a cofactor (a smaller organic molecule) and/or various metal ions for their action. Even though enzymes are such large molecules, the part where reactions are catalyzed is usually only a small region, called the active site. The remainder of the enzyme molecule provides the substance with the proper shape and geometry that will allow it to accomplish its function.

Some enzymes have a prosthetic group associated with the active site. A prosthetic group is a non-protein subunit which is firmly bound to the protein framework and is integral in the mechanism of action of the enzymes.

A related class of biological regulators is the carrier proteins. These important molecules are also catalysts, but in a somewhat different sense. A well-known example is hemoglobin, which carries

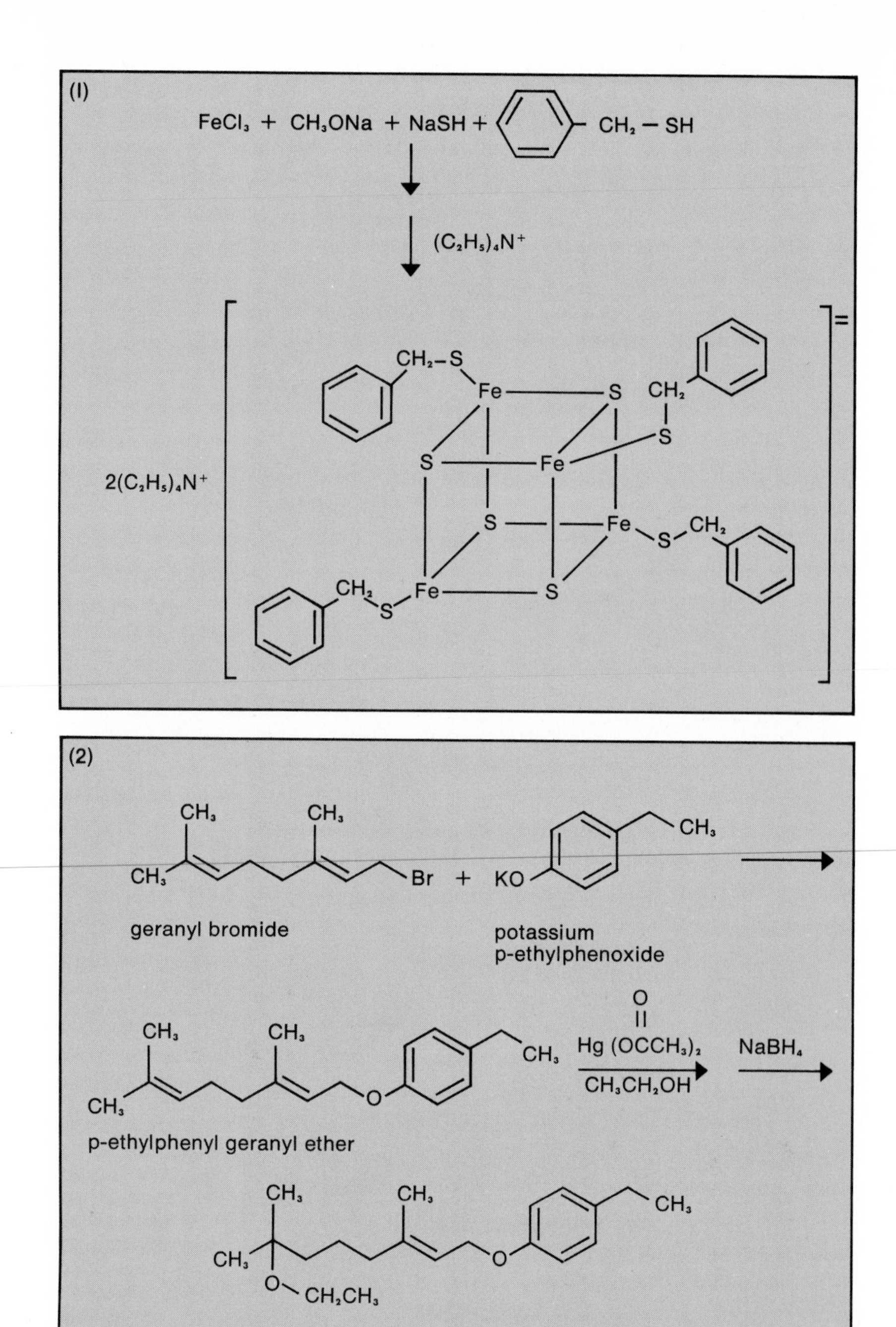

oxygen in the respiratory system of animals. An important group of carriers is the proteins that function in the electron-transport chain. They play a crucial role in the transfer and storage of energy in organisms.

A great deal of research was directed toward elucidating the mechanism of enzyme action, and most of this work was naturally centered on the active sites. Chemical synthesis was playing an increasingly important role in this area. If scientists could synthesize analogs, simple molecules that mimic the properties of enzymes, this would help to unravel the puzzle of their extraordinary catalytic functions. A major advance in this area came in September 1972, when a team of scientists from the Massachusetts Institute of Technology, Northwestern University, and Du Pont & Co. announced that they had prepared a simple analog of the prosthetic group that occurs in the high-potential iron protein (HiPIP) and in ferredoxin from two bacterial strains. The analog was prepared from a mixture of ferric chloride, sodium methoxide, sodium hy-

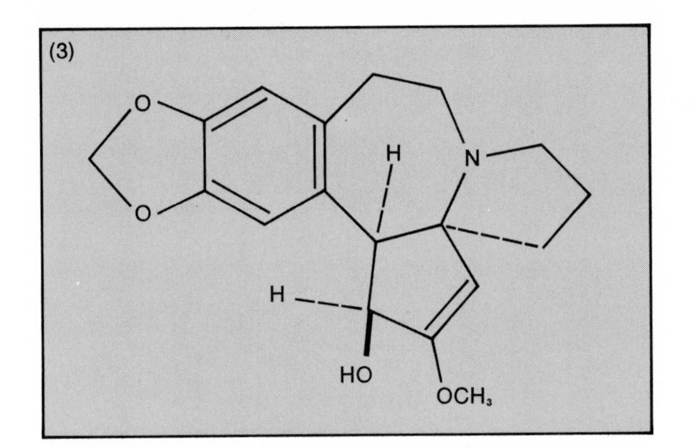

drosulfide, and benzyl mercaptan in methanol solution. It was isolated in crystalline form as its tetraethylammonium salt (1).

The synthetic compound has structural and chemical properties that bear a striking similarity to those of the prosthetic group in the natural iron-sulfur electron-transport proteins. With this simple analog available, the properties of the prosthetic group could be studied in the absence of the accompanying protein. In this way, scientists might be able to learn to what extent the properties of an enzyme are a function of the prosthetic group alone. Syntheses such as this also open up the possibility of using such simple molecules as chemical reagents to accomplish the same sort of tasks that the corresponding enzymes serve for living systems.

Insect juvenile hormones. Growing public concern over the quality of the environment stimulated scientists to search for insect-control methods that are safer and more selective than the methods currently in use. The broad-spectrum insecticides such as DDT are becoming unacceptable because of their nonselective nature and because they leave residues that can build up to a point where they damage higher organisms. A potential method of insect control which has excited scientists for the past decade is the use of natural insect hormones, substances produced by the organism to regulate its own growth.

In March 1973, a team of entomologists at the U.S. Department of Agriculture's Agricultural Research Center in Beltsville, Md., announced that they had synthesized the most potent juvenile hormone analog yet. The analog was prepared from the potassium salt of p-ethylphenol and geranyl bromide, both of which are inexpensive and readily available. These two building blocks were coupled to give an intermediate, p-ethylphenyl geranyl ether, which itself has juvenile hormone activity. The intermediate ether was first treated with mercuric acetate in ethyl alcohol solution, and then with sodium borohydride (2). The product, dubbed JH-25, was more than 100,000 times as effective as the previously isolated cecropia juvenile hormone in controlling growth in the yellow mealworm, *Tenebrio molitor* L., an insect widely used for assaying for juvenile hormone activity.

Even with such a potent product in hand, much remained to be done. Researchers expected to synthesize more new analogs as the search goes on for even more potent and more selective materials.

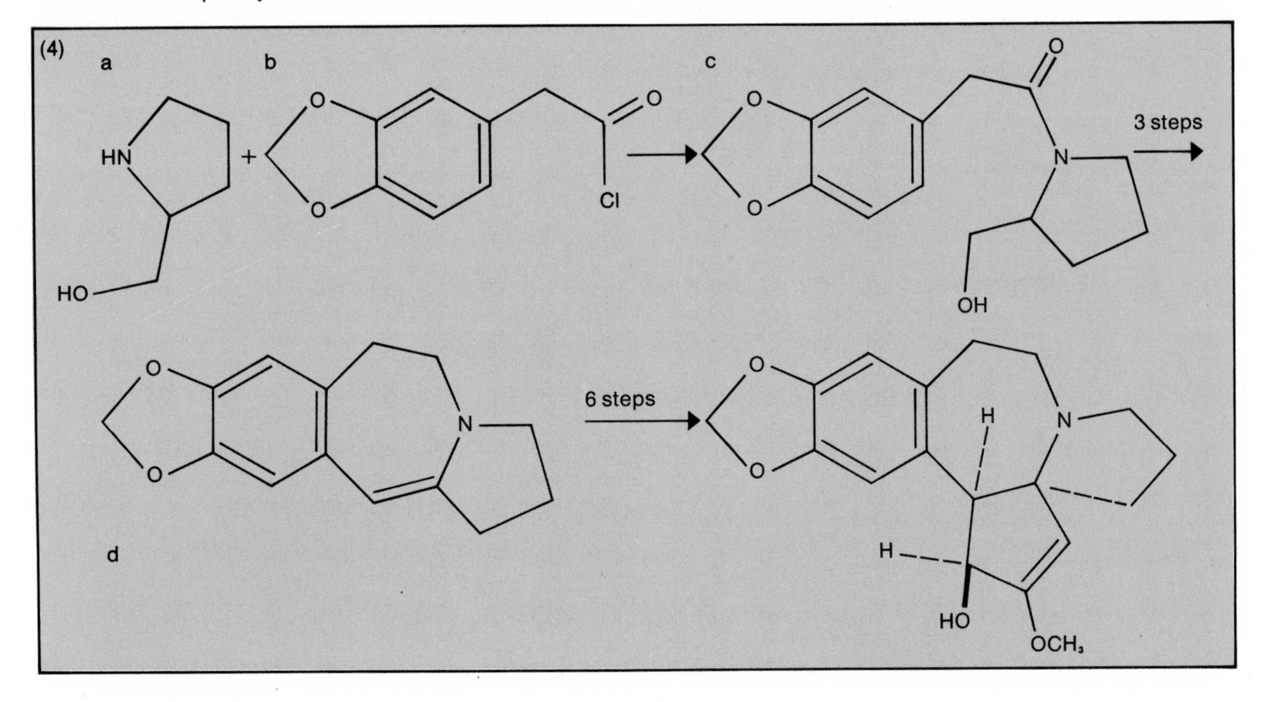

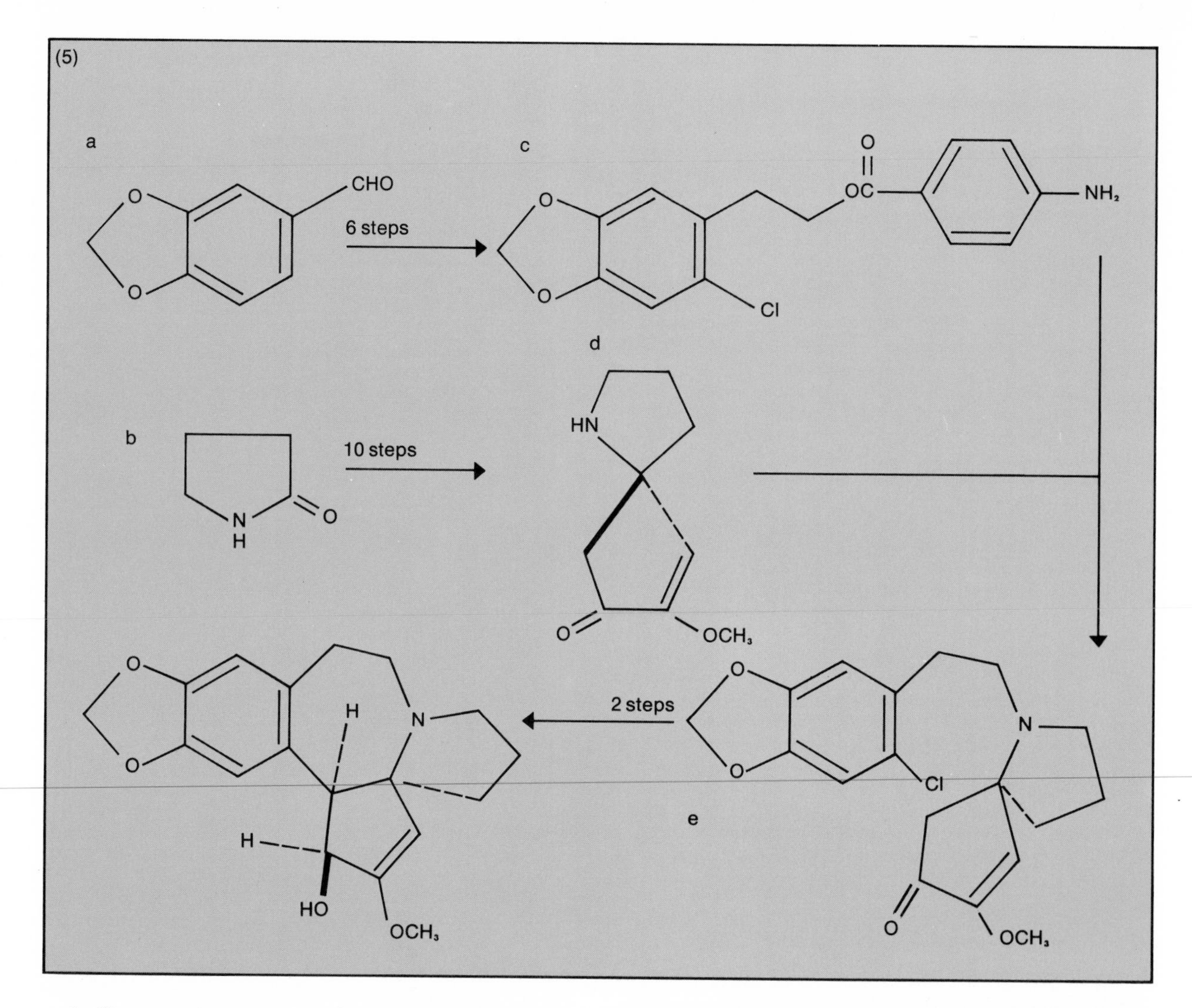

Antitumor compounds. A major effort by organic chemists was the attempt to develop methods for the laboratory synthesis of interesting natural products. A large amount of this synthetic work was concerned with compounds that show antitumor activity. Often the target molecules have promising properties that warrant clinical study, but they may be in short supply. In other cases, it is known at the outset that the actual natural materials are either not sufficiently active or are too toxic for use in chemotherapy. However, there is always the chance that analogous structures will have more desirable properties. If a reliable laboratory synthesis of such a compound can be developed, then it is possible to alter the synthesis in such a way as to produce modified structures for physiological tests.

Cephalotaxine (3) is the parent member of a group of minor alkaloids that occur in the Japanese plum yew (*Cephalotaxus harringtonia*). The actual natural products are simple esters of cephalotaxine, and they have been found to be effective inhibitors of experimental lymphoid leukemia in mice.

In the fall of 1972, two different research teams completed syntheses of cephalotaxine. The first synthesis, reported in October by Steven Weinreb

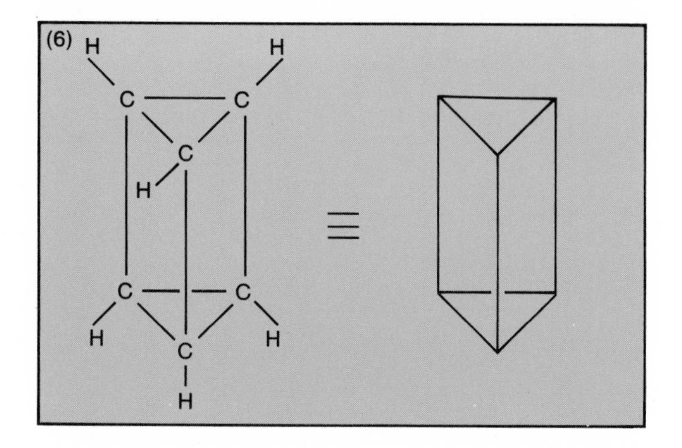

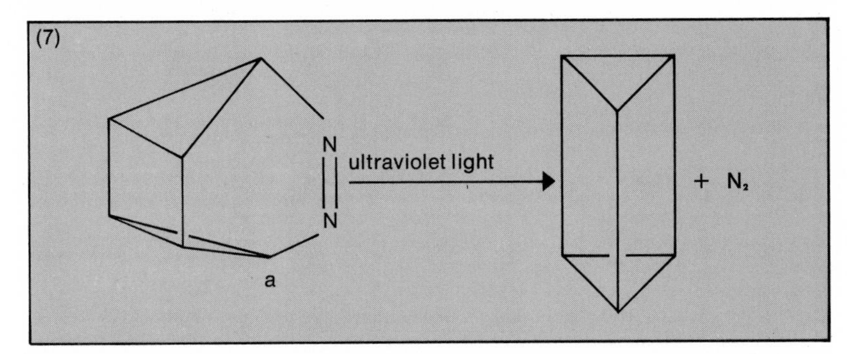

and Joseph Auerbach of Fordham University, Bronx, N.Y., started with the amino alcohol prolinol (4a) and 3,4-methylenedioxyphenylacetyl chloride (4b). These two building blocks were condensed to give the intermediate 4c, which was converted by a sequence of three reactions into the tetracyclic compound 4d. The last ring was added by six more reactions that gave racemic cephalotaxine (a 50:50 mixture of the natural product and its mirror-image isomer).

The second synthesis of the alkaloid was communicated in November by M. F. Semmelhack, B. P. Chong, and L. D. Jones of Cornell University. The Cornell synthesis (5) proceeded along completely different lines than did the Fordham one. The starting materials in this case are piperonal (a) and pyrrolidone (b). Compound a was converted into intermediate c by a six-stage sequence of reactions. At the same time, compound b was converted into the bicyclic intermediate d by a ten-stage route. Intermediates c and d were then coupled to give the intermediate e. The final ring was then formed to obtain racemic cephalotaxine.

Weinreb and Semmelhack both reported that their syntheses could be used to prepare substantial amounts of materials for physiological evaluation. In addition, both synthetic routes were amenable to modification so that further analogs might be prepared.

Prismane. In April 1973, Thomas Katz and Nancy Acton of Columbia University succeeded in synthesizing an interesting organic compound that had long eluded other chemists. The compound is tetracyclo[$2.2.0.0^{2,6}.0^{3,5}$]hexane (6), which was given the name prismane, since its carbon skeleton has the geometry of a prism.

Although prismane has little or no practical importance, a study of its chemical and physical properties is useful in understanding the fundamental nature of chemical bonding. Prismane has already yielded one surprising piece of information. Chemists who had previously attempted to synthesize the molecule had assumed that the large amount of bond strain in it would cause it to be very unstable. Consequently, the previous synthetic attempts had mostly been done at low temperatures. Katz and Acton prepared the hydrocarbon by irradiating intermediate 7a with ultraviolet light. The intermediate decomposes to give prismane and gaseous nitrogen (7). When the reaction was carried out at –65° C, no prismane was produced. Instead, intermediate 7a decomposed to give other products. When the irradiation was carried out at 30° C, however, prismane was produced in 8% yield. In fact, the molecule is surprisingly stable. At 90° C it decomposes to the extent of only 50% in 11 hours. Chemists planned to undertake new studies to find out why prismane is more stable than previous workers had thought it would be.

The future. Over the years the level of sophistication of chemical synthesis has gradually risen. Synthetic objectives unthinkable in 1950 may in the 1970s be attacked with some confidence. No one date or single accomplishment can be pointed to as being the factor responsible for bringing the capability of chemical synthesis to the relatively high level where it now stands. Rather, the progress has been a gradual, almost imperceptible, one. New reagents have been discovered, often in answer to a specific synthetic problem. New apparatus has been invented. New methods of analysis and identification have spurred the progress. And, finally, new concepts of synthesis have evolved, usually subtle innovations, sometimes traceable to a given individual but often not. These conceptual methods often have a profound effect upon the ease with which synthesis may be accomplished.

In the coming years, synthetic chemists will continue to refine their methods and concepts and also to add new ones. Progress in the search for new and more efficient synthetic methods is often painfully slow, but advances are occurring more and more rapidly.

—Clayton H. Heathcock

Structural chemistry

A major breakthrough in structural chemistry occurred during the year when a group of researchers in the College of Physicians and Surgeons of

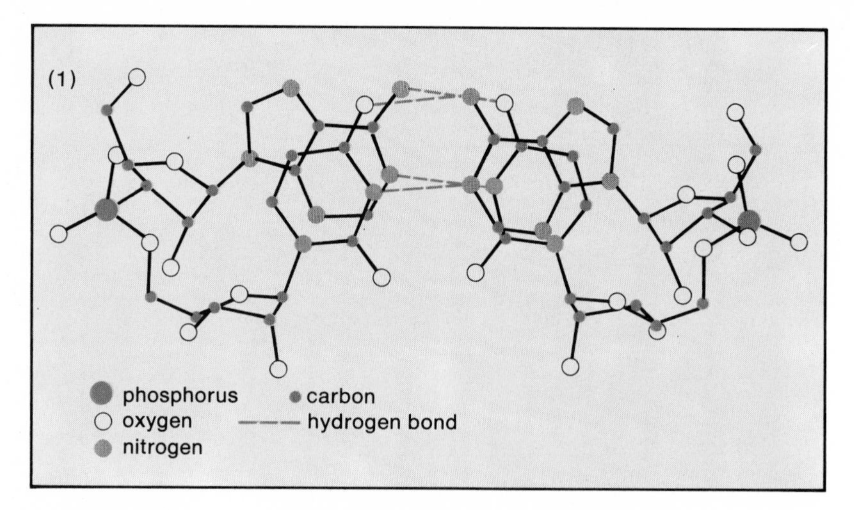

Columbia University headed by Sol Spiegelman determined the complete nucleotide sequence of an RNA (ribonucleic acid) molecule capable of replicating (duplicating) itself. Previously, the structures of nucleic acids, which contain all the information necessary for the reproduction of a living system, were so complex and their size so large as to resist complete structure determination, although much was known about their composition and properties. DNAs (deoxyribonucleic acids) have molecular weights of 1 million up to 100 million, while ribosomal RNAs have molecular weights of about 1 million. (Some nonreplicating transfer RNAs [t-RNAs] are known to have molecular weights of only about 25,000–30,000, corresponding to 75–90 nucleotide units.)

This group of research workers had previously prepared a replicating RNA molecule containing only 218 nucleotides (compounds that are the building blocks of nucleic acids). For this task they first prepared replications of high-molecular-weight RNA molecules, using a replication-mediating enzyme (replicase) isolated from RNA-virus-infected bacteria. Subsequent replication experiments were arranged so that continued survival of the molecules depended upon the speed with which they were able to complete replication. Since smaller molecules would complete replication more rapidly, adaptation gradually occurred in which dispensable sequences of the RNA were eliminated, while only those sequences that were required by the enzyme for recognition and replication were retained. By such a process the researchers were able to isolate a replicating RNA molecule (MDV-1) that contained only 218 nucleotides and was capable of replicating itself. In a test tube one such molecule will autocatalytically generate one trillion copies in 20 minutes. These molecules represent the smallest known nucleic acid capable of self-replication.

Using ribonuclease A, ribonuclease T_1, or alkali, the researchers broke down the MDV-1 molecule to form a series of fragments of up to 12 nucleotide units called oligonucleotides. The complex problem of fitting these oligonucleotides together in a logical sequence for the structure of MDV-1 was aided by the fact that the two complementary strands of the RNA could be isolated and studied separately.

With the primary sequence of this relatively simple replicating RNA known, it should be possible to determine the chemistry of the recognition device used by the enzyme in selecting molecules for replication. Only those nucleotides that are correctly adsorbed on one strand of the nucleic acid are polymerized to form the complementary strand in an extremely fast reaction. It should also be possible to carry out extracellular experiments that will help in determining the chemical evolution of nucleic acid.

Gene structure. In the summer of 1972, Walter Fiers at the State University of Ghent in Belgium determined for the first time the complete chemical structure of a gene. The material that he studied was part of the RNA that forms the core of a tiny virus, bacteriophage MS2, which attacks bacteria. The MS2 virus (as well as some other very small viruses) contains no DNA, where genetic information is normally stored. Fiers and his co-workers were able to establish the identity and location of each of the 387 nucleotides strung end-to-end to form one of the three genes of the MS2 virus, the one that is responsible for the synthesis of the protein that serves as a protective coating for the virus.

The gene was shown to contain 49 codons (triplets of nucleotides) that regulate the incorporation of specific amino acids into the protective protein coating during the synthesis. Although this work represents the determination of the structure of a very small gene, researchers believed that it could lead to elucidation of the structures of more complex genes and to the mechanism by which they control cell functions.

Viewing the double helix. Although the original James Watson-Francis Crick theory for the double helical structure for nucleic acid polymers was first published in 1953, not until 1973 were scientists able to view parts of the double helix at a resolution of one atom. Since very large nucleic acids do not form well-defined single crystals, only approximate data could be obtained by X-ray studies on fibers in which the location of the oxygen, nitrogen, and carbon atoms were known with reasonable precision but the location of the very small hydrogen atoms were not determined with any certainty.

Recently a group of scientists at the Massachusetts Institute of Technology under the direction of Alexander Rich were able to verify the original Watson-Crick postulate of the two helices bound together by specific hydrogen bonds between adenine and uracil (or thymine) groups and between guanine and cytosine bases to form the flat connecting rungs between the long sugar-phosphate chains in the ladderlike molecule.

The MIT group arrived at its conclusion through the use of high-resolution X-ray diffraction analysis of single crystals of two nucleic acid constituents, adenylyl-3′,5′-uridine and guanylyl-3′,5′-cytidine. These are dinucleosides, each of which forms antiparallel double helices in the crystal. The researchers found that a number of water molecules are bound to the double helix. The former structure contains 12 water molecules per double helical fragment while the latter contains 18. The exact structure of the hydrogen bonds is given in (1).

Leucogenenol. Frederick A. H. Rice of American University discovered a new type of hormone, leucogenenol (2), which stimulates the production of those human cells that are the precursors of the peripheral blood cells, such as the lymphocytes and myeloid cells, and in general stimulates the regeneration of myeloid and lymphoid tissues. Strangely, however, he isolated this new hormone not from body tissues but from the metabolic products of a microorganism, *Penicillium gilmannii.*

Later, however, Rice was able to show that this hormone occurs in many body organs, including the human liver, and probably plays an important role in the regulation of the number and type of blood cells in the body. He was able to establish the structure of leucogenenol as a relatively simple tricyclic material unrelated in structure to any of the common steroidal or polypeptide hormones.

Antimatter. Since 1956 scientists have observed

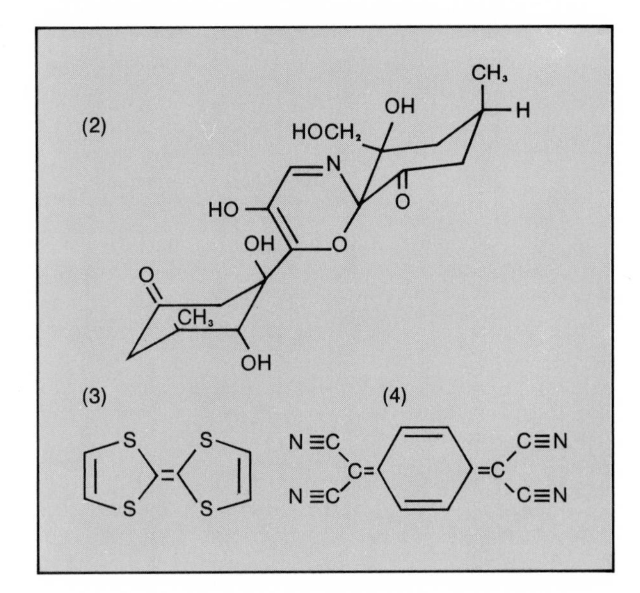

experimentally the existence of antimatter in the form of antielectrons, antiprotons, and antineutrons. When matter and antimatter collide, there is a mutual annihilation to produce a very large amount of energy. Because of the current energy crisis, a number of scientists have been looking for more sophisticated forms of antimatter as a possible way to yield a new form of useful energy.

In 1965 the first sophisticated form of antimatter was discovered in the United States with the discovery of the antideuteron. This is the counterpart of the deuteron, the nucleus of the deuterium, or heavy hydrogen atom. More recently, a group of scientists in the Soviet Union working with their 70 GeV (billion electron volt) accelerator reported the creation of an antihelium isotope nucleus consisting of two antiprotons and an antineutron. They also reported the creation of the nucleus of antitritium, which contains one antiproton and two antineutrons and is the counterpart of tritium, a heavy radioactive isotope of hydrogen.

Metallic hydrogen? From its position in the periodic table of elements, hydrogen would be expected to possess some metallic properties. This postulation initiated a race between scientists in the U.S. and the Soviet Union to produce a metallic form of the element. At the present time, most of the evidence that hydrogen can exist in some form of a solid metal comes indirectly from astronomical studies of the planets. For example, Jupiter, although over 300 times the mass of the earth, has an extremely low density, and 14 out of every 15 atoms on Jupiter are of hydrogen. The tremendous mass of this planet, composed as it is of the lightest element, can be explained only if hydrogen exists in some very dense state. Calculations suggest that metallic hydrogen must be at least ten times as dense as the molecular form of solid hydrogen that exists close to absolute zero and that this metallic form can only be produced under extreme pressures, in excess of 15 million psi.

In 1972 a Soviet team claimed to have detected evidence of metallic hydrogen in shadowgrams taken a few millionths of a second after an explosion designed to convert the molecular form of hydrogen to a metal. A group of scientists at Cornell University were attempting to "squeeze" molecular hydrogen intensively but slowly to produce the metallic hydrogen for a period long enough to carry out experiments on its nature. Metallic hydrogen could be useful as a weight-saving fuel for spacecraft propulsion as well as for producing controlled fusion in nuclear power reactors.

Superconductors. Superconducting metals attracted interest for use in such devices as large electromagnets. Although many metals cooled to within a few degrees of absolute zero (0° K, −273° C) become superconductors, the highest temperature at which superconductivity was found for compounds of several metals is 21° K. Alan J. Heeger and Anthony F. Garito of the University of Pennsylvania reported that the conductivity of crystals of an organic charge-transfer salt of tetrathiofulvalene (3) and tetracyano-*p*-quinodimethan (4) has a conductivity at 58° K that is at least 500 times its conductivity at room temperature. (A charge-transfer salt is formed by the transfer of an electron from one molecule to another, so that the tetrathiofulvalene is converted to a positive ion and the tetracyano-*p*-quinodimethan is converted to a negative ion.) Although normal conductivities of organic salts are from 100 to 200 $ohms^{-1}$ cm^{-1}, special crystals of these charge-transfer complexes have a conductivity that is greater than 10^6 $ohms^{-1}$ cm^{-1}. These observations were expected to touch off an intensive search for superconductors that will operate at the temperature of liquid nitrogen (76° K) and above for many uses, including long-distance transmission of electricity.

Silicon compounds. One of the reasons that carbon compounds are so important in life processes, and also the reason that there are more than two and one-half million known organic com-

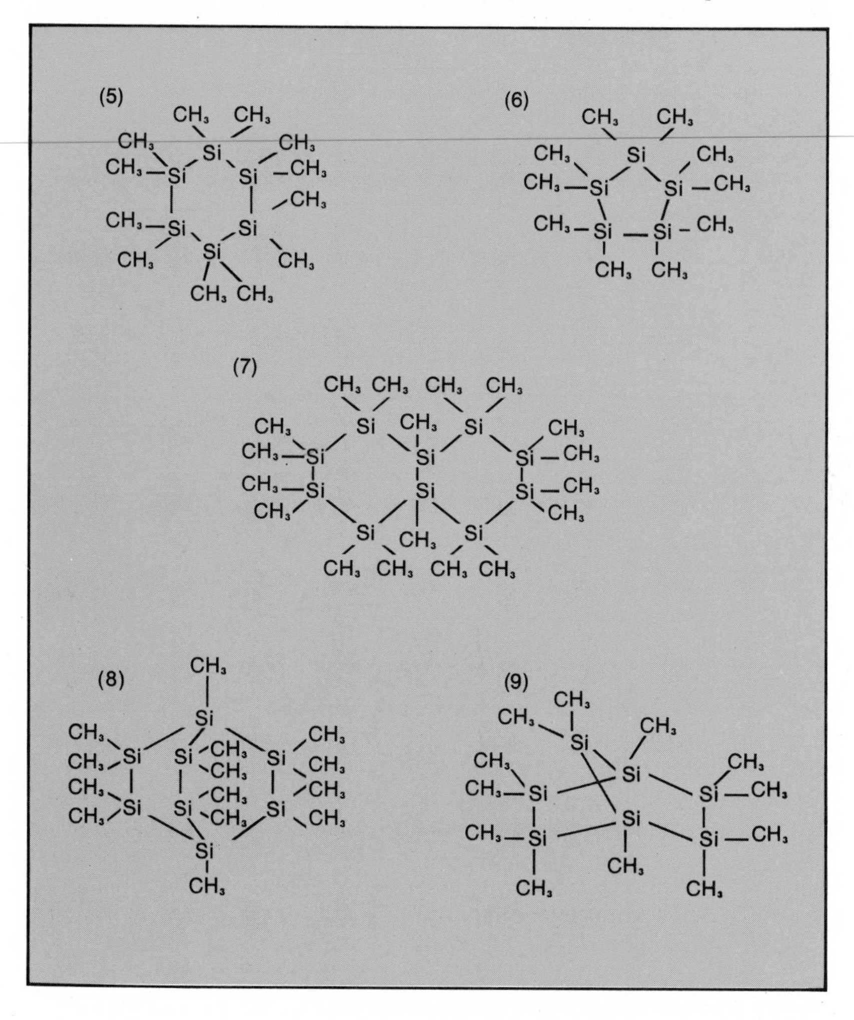

pounds, is the fact that carbon atoms can form long-chain and ring compounds containing bonds with other carbon atoms. Only one other element in the periodic chart appears to share this property with carbon to any large extent; that element is silicon. Recently, Robert West of the University of Wisconsin produced a series of cyclic and bridged silicon compounds that have many features in common with their carbon analogs. For instance, he was able to make $Si_6(CH_3)_{12}$, dodecamethylcyclohexasilane (5), which contains a six-membered ring composed entirely of silicon atoms. Although there are no double bonds in this ring system, some of the properties resemble those of benzene.

Generally, West was able to make analogs of many common ring systems, including methylated derivatives of cyclopentasilane, which contains five silicons in a ring (6), and of cycloheptasilane, which contains seven silicon atoms in a ring. He also isolated (7) a $Si_{10}(CH_3)_{16}$, for which a structure analogous in chemical properties to naphthalene was postulated. Just recently, he made a series of cage structures; these included $Si_8(CH_3)_{14}$, tetradecamethylbicyclo[2,2,2]octasilane (8), which contains two fused six-membered rings with a two silicon-atom bridge. Another caged structure (9) was postulated for $Si_7(CH_3)_{12}$, and very complicated structures for $Si_9(CH_3)_{16}$, $Si_{11}(CH_3)_{18}$, and $Si_{13}(CH_3)_{22}$. This research opened up the possibility that an extensive system of compounds based on silicon can be prepared in the laboratory as analogs of all the carbon compounds that have presently been found to exist.

—William J. Bailey

Communications

The past year was an extremely active one in the field of communications with progress on many fronts. Tiny gas lasers were demonstrated using thin glass tubes as waveguides. Such lasers may amplify light signals in optical communication networks of the future. Work in the United States progressed on the optimum method for correctly receiving sequences of digital signals that have been transmitted over noisy, dispersive communication channels.

The transistor, which has had a truly dramatic effect on communications, celebrated its 25th anniversary. This device, plus its multitudinous semiconductor offspring, can amplify, rectify, memorize, oscillate, switch, emit light, detect light, sense temperature, and more. Miniature radios are the most visible, or audible, application, but all of today's communication arteries—whether under the ocean, by satellite, via microwave towers or coaxial cable—depend on this technology. A truly impressive outgrowth of the transistor occurred: large-scale integration (LSI), which combines thousands of solid-state devices on a single silicon "chip," typically less than $^1/_4$ inch square. The use of LSI came into its own particularly in small hand-held calculators, as well as in some electronic watches and communication equipment for data transmission.

Notable developments took place in regard to automatic remote reading of gas, water, and electric meters within and outside homes. Two major utility companies in Houston and Chicago collaborated with various meter manufacturers and with the Bell System in extensive year-long trials. More than 500 meters in two major cities were read periodically through the existing telephone lines—usually in the early morning, without ringing the phone, and in just a few seconds. The trials were highly successful from a technical point of view.

The growth in stereophonic FM broadcasting continued in the U.S. An event of particular note was the first live quadraphonic (four-channel) broadcast over a leading FM station—WQXR-FM in New York—implemented in collaboration with Columbia Records on March 3, 1973. Spurred by technical developments and by the greatly heightened feeling of presence imparted to the listener when he is completely surrounded by sound, four-channel reproduction was receiving increased attention.

Telephones. Telephones increased at an annual rate of nearly 7% to a worldwide total of 291 million as of the beginning of 1972 and were expected to pass the 300-million mark by the end of that year. Americans not only made more calls during the past year, but the average duration and distance of the calls grew measurably. The greatest increase in calling volume took place for long distances—thousands of miles—reflecting the relative cost improvements for long-haul transmission. These increases were matched by a better than 10% growth in equipped circuit miles in just one year. In preparation for further growth, a successful trial was concluded late in 1972: it was expected to lead to a 25% increase in the capacity of the microwave beams that carry the majority of all nonlocal voice and video traffic, a rise from 1,200 to 1,500 voice channels. The use of solid-state microwave generators and amplifiers instead of tubes allowed this increase without using additional bandwidth.

Plans to accommodate the greater volume expected in the decades ahead included the deployment during the 1980s of a waveguide only millimeters in diameter, a buried "pipe" capable of carrying approximately 250,000 two-way conversations per second in each direction. Even greater capacities may someday be achieved by the use of hairbreadth transparent fibers, carrying information in the form of laser-generated light pulses. Dramatic progress was made on such fibers so that they were able to demonstrate transmission losses as low as approximately two decibels per kilometer, an achievement undreamt of only a few years ago. However, practical ways for making connections and repairing or "splicing" breaks remained to be found.

Overseas telephony continued its explosive growth, with a 24% increase in the number of calls for the most recent year tabulated. Anticipating further growth, the U.S. Federal Communications Commission (FCC) during 1972 approved the installation of a 3,500 two-way channel transatlantic cable, more than quadruple the capacity of its predecessors. Also, early in 1973, Intelsat (International Telecommunications Satellite Consortium) announced plans for doubling the traffic capacity of its communications satellite system. Meanwhile, Western Union received permission for WESTAR, a domestic three-satellite system similar to Telesat Canada's system, which became operational during the past year. WESTAR was designed to have two satellites in orbit and one spare on the ground, each having 12 microwave transponders to handle a total of 7,200 two-way voice channels, 12 television channels, or 600 million bits (binary digits) per second of data.

Microwave transmission. A satellite of a different sort made news during 1972. After circling Mars in 698 picture-taking orbits, Mariner 9 was "turned off" by its 45,960th command from the Jet Propulsion Laboratory, Pasadena, Calif. For nearly a year, the satellite's video camera gear sent back to the earth more than 7,000 pictures by way of its microwave transmitter. From these, the U.S. Geological Survey's Center of Astrogeology in Flagstaff, Ariz., recently produced a detailed map of Mars, the first such achievement for any planet besides the earth.

Terrestrial microwave communication also received increasing attention during the year, as radio spectrum space continued to become scarcer. New insights were gained on potentially good future utilization of the spectrum involving the use of high-speed digital modulation combined with improved antennas. Digital modulation is highly resistant to interference from other microwave beams, and the improved antennas generate sharply focused beams with a minimum of side scatter. This combination thus permits relatively high-density communication networks. Since microwave beams are subject to fading, caused principally by rain, overall systems design also involves the use of space diversity, in which an area is transversed by alternate microwave paths. In this arrangement one such path can automatically take over for another.

Television. Television broadcasting of a noncommercial nature continued to gain in the U.S., as the American Telephone and Telegraph Company (AT & T) made available to the Corporation for Public Broadcasting the largest video network

Fine optical-glass fibers (right) that transmit light waves show promise of being a major communications medium. A coil of such fibers (left) is used in measuring the light lost when a laser is transmitted through them. Because of its shorter wavelength light can carry more information than longer waves.

Courtesy, Bell Telephone Laboratories

ever put into service. On the other hand, opinions diverged on the method of funding and on programming. Substantial progress continued in the field of cable television with a steadily growing number of people within reach of such service. Peter Goldmark, a vigorous proponent of using communications to improve the quality of urban life, revealed a system designed to provide programming for an entire cable television network by using four video cassette players to supply movies, other programs, and special messages by means of automatically preset timing. Meanwhile, the see-while-you-talk PICTUREPHONE® service introduced in Pittsburgh and Chicago during 1970 and 1971, respectively, entered its third year.

From a technical point of view, definite progress was reported during the past year on an old problem, that of exploiting the redundancy in television transmission to cut down on the required bandwidth. This is of particular interest for long-distance digitally encoded video transmission. It was shown that each picture element, one of the roughly quarter-million points required to define fully a television picture of commercial quality, could be rendered with only one bit of information instead of the more usual nine bits. In demonstrating this potential saving, researchers took advantage of the similarity of successive video frames—resulting from the confinement of motion to relatively small areas—as well as certain psycho-visual phenomena. The rather complex instrumentation required to implement such a system has, however, discouraged its application.

Television set sales increased substantially abroad, particularly in Western Europe. Color broadcasting became commonplace in most of northern Europe, with two different systems, PAL and SECAM, in use. The latter was used primarily by France and the U.S.S.R. and is the more complicated; the former is just slightly more complex than the NTSC system used in the U.S. but has the advantage of better hue stability, without hue control on the television set. China, perhaps stimulated by increased contact with the West, began serious study with a view to adopting and implementing a system of color television broadcasting.

Data transmission. One of the fastest growing types of communication during the past year was that between computers and other digital business machines. Data communication expanded on several fronts. The prevalent mode of operation makes use of the voice telephone network, even though this provides analog (continuous signals) rather than digital (discrete pulse) transmission. A so-called "modem" (modulator-demodulator) or "data set" makes the necessary conversion to the digital mode. Such modems are made by many

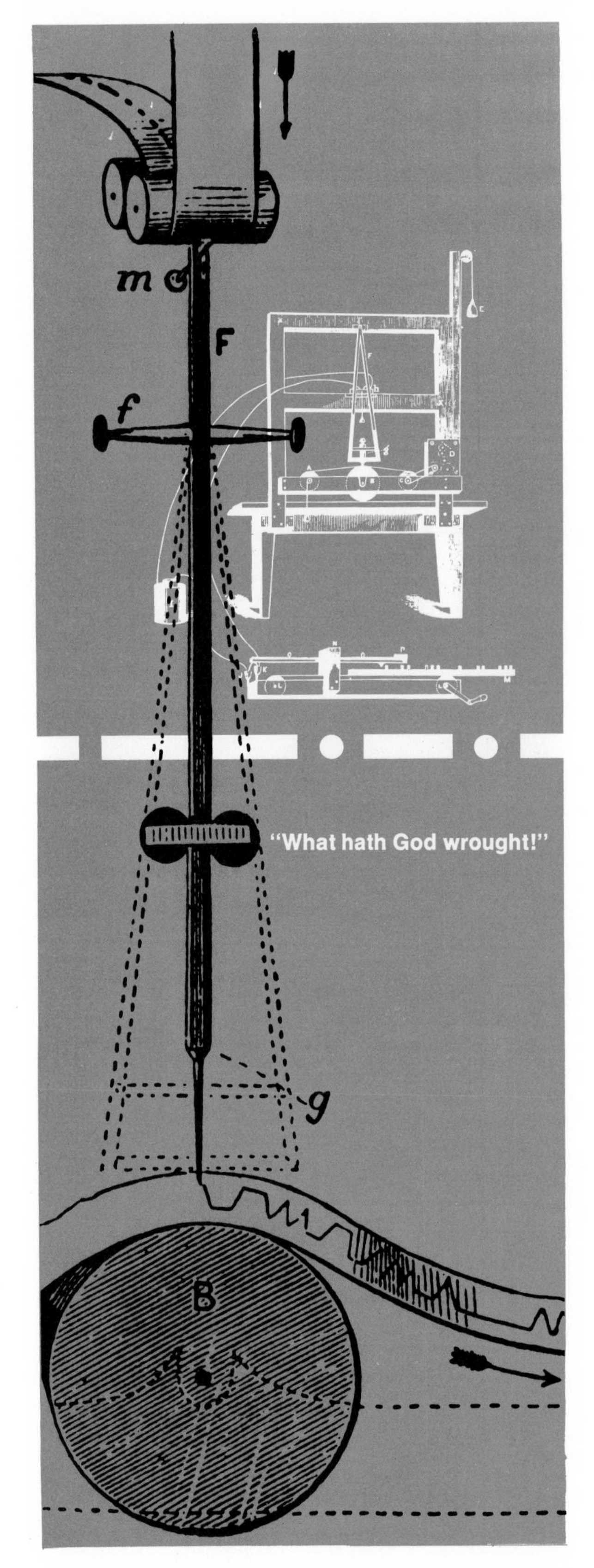

manufacturers; there were important new entries into an already crowded field, including a rapidly self-adapting design operating at 4,800 bits per second. It contains an electronic compensator, which in only 0.0050 sec automatically adjusts itself to cancel out most of the distortion characteristic of the particular voice line being used.

The use of 4,800 bits per second continued to become more commonplace. It is used, for example, in data communication networks interconnecting various branch locations of an organization with its centralized computer installation. Such uses occur in airline reservation systems, stock brokerages, and banks. The highest speed or bit rates transmitted over voice bandwidth channels continued at or around 9,600 bits per second, but several manufacturers made progress to compress the rather complex modems required for such conversions into much smaller space.

By far the greatest attention in the world of data communication during the past year was evoked by the emerging new networks designed exclusively for digital data communication service. The Bell System filed an application to construct equipment required to utilize its already existing and rapidly growing digital transmission facilities for a new data service named the Digital Data System (DDS). It was to operate at speeds of 2.4, 4.8, 9.6, and 56 kilobits per second. A Canadian digital system named Dataroute, offering speeds as low as 110 bits/sec, began service in Canada during the spring of 1973.

Digital transmission facilities continued their remarkable growth. They continued to be used primarily for voice signals, which are "sampled" at the rate of 8,000 times per second. Each sample has a numerical value that is encoded into a group of binary pulses, thus giving rise to the name "pulse code modulation." Future transmission media such as millimeter waveguide and optical fibers are well suited to binary pulse transmission.

Specialized carriers. The past year also witnessed intensive activity by the so-called specialized carriers, engaged in the business of setting up separate nationwide communications networks, primarily for the efficient transmission of digital data. During 1972 Microwave Communications, Inc. (MCI) began service between St. Louis and Chicago, while the Data Transmission Co. (Datran) continued to proceed with construction of its nationwide network. A 1973 entry was Packet Communications, named for its plan to convey packets of digital communication—for example, a block of a few thousand bits—from source to destination. Rapid routing of the message from source to destination is the essential feature of this system, which is an outgrowth of the U.S. government-sponsored Advanced Research Projects Agency (ARPA) network, interconnecting government and major university computing centers. The British independently developed the concept of packet transmission; following proposals by Donald W. Davies of the National Physical Laboratory, a demonstration network was constructed and tests continued during the past year.

Such networks have the potential for quick delivery of "message bursts" to any of a large number of destinations. A far-reaching proposal along these lines was advocated by John R. Pierce, the early proponent of communication satellites. He envisioned national, regional, and local round-robin loops, similar to conveyer belts, carrying high-speed data, with each message finding its own way from end to end, transferring from one loop to another as required.

Policy questions relating to communications in the U.S. received much attention from the FCC. An area of particular concern was the relationship of the new specialized carriers to the established common carriers (such as the Bell System). This issue could be traced back to the 1968 Carterfone decision, which allowed equipment sometimes referred to as "foreign attachments" to be interconnected with the far-flung dialed telephone network. Much attention centered on whether this equipment could be simply "certified" for safety, as its owners desired, or whether protective coupling devices, asked for by the common carriers and supplied by them, were necessary. Task forces dedicated to resolving this question were at work throughout the year.

International activities. On the international scene, the Fifth Plenary Assembly of the Comité consultatif international télégraphique et téléphonique (CCITT) took place in Geneva, Switz., at the end of 1972. It concluded the 1968–72 epoch of this international body's work, the main objective of which is to achieve compatibility among the communication systems of various nations with respect to, for example, telephone numbering plans, equipment design, and operating plans. Approximately 400 delegates registered for the Plenary Assembly, representing 85 countries and about 30 recognized private operating agencies. With the 1968–72 epoch behind it, the CCITT began reorganizing for its 1972–76 epoch.

Another event with international implications was the recent trial of Common Channel Interoffice Signaling. Slated for future use by many nations, the system assigns one transmission channel to handle the signaling for a large group of voice channels. At present, this signaling occurs on each individual voice channel.

—Ernest R. Kretzmer

Computers

The computer business continued to show enormous vitality during 1972–73. Industrial applications were so widespread that in most cases, even where it was of central importance to an undertaking, the computer was no longer in the spotlight. When one read that a computer accomplished (or failed to accomplish) a task, this now generally meant that an interconnected group of men, programs, computers, and other devices had done or failed to achieve some goal; it had become pointless to praise or blame the individual components separately.

Engineering and design. Most computers built during the year were "microprogrammed"; that is, their electronic capabilities were built in as programs and thus were changeable. This generality at the micro-level made them appear at the programming level as almost being special-purpose machines. This was exemplified by the flexibility with which a computer manufacturer issued a calculator specifically oriented to business calculations but internally almost no different from his scientific calculator issued about a year earlier.

Visual displays were becoming indispensable in computer technology. The standard output devices remained the typewriter, the teleprinter, and the high-speed impact line printer. But these devices, with their fixed and limited alphabets and fonts and lack of drawing capability, were forcing the computer to depend more on displays. Although these most commonly showed only characters from a fixed alphabet, new varieties became available that permitted variable alphabets and fonts and drawings. Putting these displays into time-sharing environments or attaching small computers directly to them was opening the entire field of communications media to exploitation by the computer. *See* ELECTRONICS.

Courtesy, Dr. Alan J. Perlis, Yale University

Face was "drawn" by computer generating countless lines that scanned a screen more than 100,000 times per second. Shading, which gives depth to the picture, was part of the computer program.

Computer science. The plethora of remarkable devices attached to the computer should not disguise its awesome simplicity or the complexity of the intellectual developments in computer science. Computer science is the study of the phenomena arising from, and associated with, the computer. Research during recent months concentrated in

Chinese industrial computers on display in the Exhibition Hall at Shanghai demonstrate the growth of the computer industry in that country.

Courtesy, Image Analysing Computers, Inc.

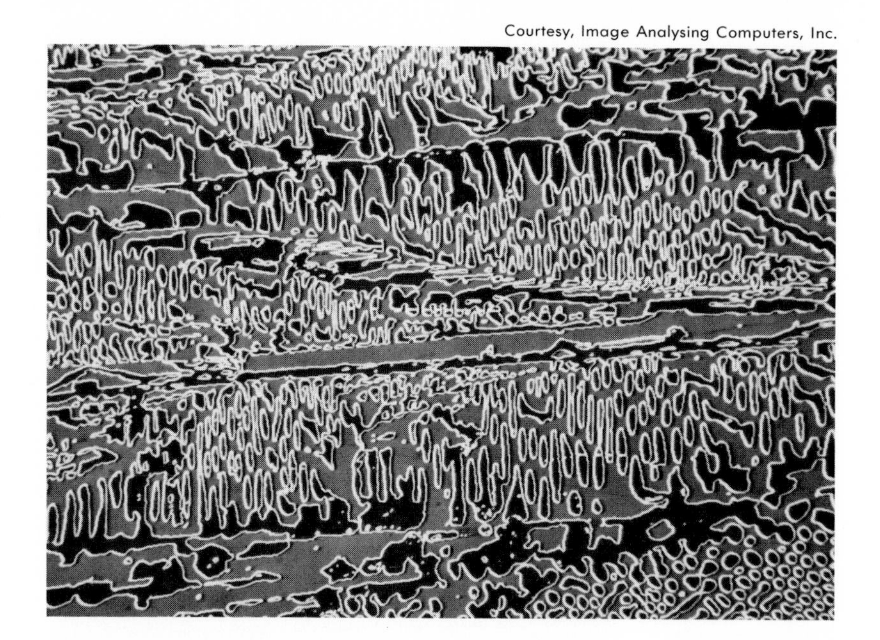

Fine structure in an aluminum-bismuth-copper alloy is displayed by an image analyzing computer. Such computers extract quantitative data from images of microscopic and macroscopic objects. They can, for example, instantaneously size every detected feature in an image.

the areas of artificial intelligence, complexity theory, programming languages, information retrieval, graphics, and numerical analysis.

A major goal of computer science for more than 15 years has been to create programs that will approximate, emulate, and ultimately rival aspects of man's mental abilities. Computer programs now play passable chess, but no one knows how much more complex the program must be or how long it will take, or even if it is possible, to create a program that will compete with grand masters. As each step of improvement is made, new problems arise requiring reprogramming and reorganization of the problem-solving process. Scientists realize, however, that a superb chess program, while it would be a tour de force, is not the issue and not the goal. Instead, they are working to master the art of constructing very complicated programs that extend the routine capabilities of computers.

Few tasks have provoked such a wealth of problems as the programming of vision. In the last few years enormous progress has been made in constructing programs that, in conjunction with scanning camera input and tactile output, gives the computer some ability in this area. Computers can now "see" limited fields of vision.

In conjunction with vision, some ability in processing languages is becoming widespread. Some researchers believed that semantics, the study of meanings, will be mastered by collecting sufficient knowledge expressed as a finite number of step-by-step procedures (algorithms). Meaning will not reside in tables (dictionaries) but in collections of cooperating procedures whose connections and contents extract meaning from syntax and context.

The belief that meaning resides in executable procedures led some computer scientists to attempt a restructuring of elementary education that places enormous emphasis on the inventing and writing of procedures by a child. Having these procedures control, through computer execution, pens for drawing and sound makers for creating music gives a child access to a creative power that he consciously generates. Some educators believed that such a facility would give a child more appreciation of his creative abilities.

Much current computer research was being done on complexity. A statement of one problem in this area was: "A computer cannot be programmed to do a certain task, *e.g.*, sorting N records, for all collections of N records in less than the order of N $\log_2$ N steps." In order to find these lower boundaries, generally quite difficult, new algorithms were being discovered that took less time to perform than those that had previously been assumed to represent optimal capabilities.

Some problems were shown to be essentially exponential in their time requirements, and these problems became barriers to the successful programming of important tasks. Thus, a fond goal of many computer scientists was the reduction of large classes of problems, such as program termination and automatic program writing, to a situation in which one general program proves a theorem. The best process known as of 1973 required exponential time on some and probably on most theorems of interest.

Special programming languages, such as FORTRAN, remained critical in computer science. They were important in easing access to the computer both for large groups of new users and for users confronting large groups of new problems.

Some of these new problems arose from the changing nature of the computer itself. As computers became more complicated, more complicated programs were required to manage them and new languages were being developed to make their construction easy.

—Alan J. Perlis

Earth sciences

Major developments during the year concerned the final phases of manned lunar exploration, completion of the Upper Mantle Project, advances in earthquake prediction, and a reappraisal of hydrology in light of unprecedented U.S. floods.

Geology and geochemistry

In April 1972 the crew of the Apollo 16 U.S. lunar mission explored the highlands north of the Descartes region and brought back to earth 213 lb of lunar rock and soil. Geochemical sensors indicated that the spacecraft had landed on a genuine highland site that is also characteristic of large expanses of highlands on the moon's far side. X-ray experiments detected high aluminum-silicon ratios, and a gamma-ray experiment showed little evidence of radioactive emissions over either the near-side or the far-side highlands. Visual observation and orbital photography confirmed the similarities of highlands on both sides.

The Apollo 16 samples were 75% breccias, with only a few samples of mostly plagioclase-rich rocks with components that confirmed a highland origin for anorthosites in powdery material at other lunar localities. The age of samples appeared to center around 4,100,000,000 years, bridging the gap between other highland samples and the postulated 4,500,000,000-year age for the lunar crust. (No rocks 4,500,000,000 years old had been collected anywhere on the moon.) The highland areas do not represent primitive lunar material.

In early December 1972 the Apollo 17 mission landed in the valley of Taurus-Littrow in the highlands. This was the final mission of the Apollo series, and no further manned landings on the moon were scheduled. The dark material covering the surface at the Apollo 17 site was first thought to be volcanic ash, but instead it proved to be breccia material, the same age as the highland breccias at the Apollo 16 site. These final moon missions indicated that the moon underwent a phase of intense cratering some 3,900,000,000 to 4,100,000,000 years ago, well after its initial formation. This was followed by voluminous lava flows that filled the maria, or seas, 3,600,000,000 years ago. Smaller meteorite cratering continued after that.

Mariner 9 photographs of Mars telemetered back to earth greatly modified earlier impressions of Martian geology. Southern Mars appeared old and rather stagnant like the moon. The northern hemisphere and polar regions had a livelier history, with evidence of shield volcanoes, enormous rifts, and chaotic terrain. There was evidence of fluids in channels, which might represent braided stream erosion or lava flowing just beneath the surface through breccia fields. Some workers even suggested that the terrain of Mars indicated feeble plate tectonics sometime in the past, although the absence of a significant magnetic field argued against any such activity at present.

The Soviet spacecraft Venera 8 landed on Venus in July 1972 and transmitted data for about an hour. Gamma-ray spectrometer measurements indicated that potassium, uranium, and thorium are present in amounts comparable to volcanic rocks on earth. This suggested that Venus, like the earth, Mars, and the moon, had undergone magmatic differentiation.

Marine geology. The research vessel "Glomar Challenger" continued its program of drilling in

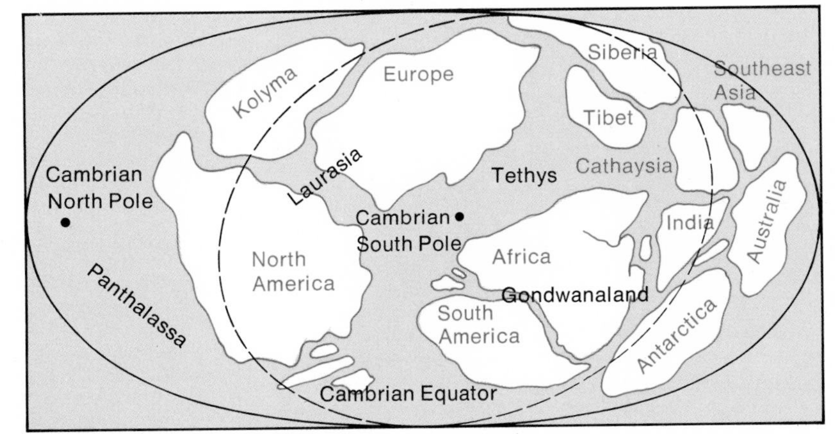

Adapted from Annals of New York Academy of Sciences (1972)

Drawing shows the world's continents and seas during the Cambrian Period, approximately 550 million years ago, as reconstructed by Rhodes Fairbridge of Columbia University.

the deep ocean floor, with the aim of penetrating the entire sedimentary accumulation at most of the drilling sites. Activity during 1972 was centered in the Indian Ocean. The oldest sediments encountered there were Early Cretaceous in age. There was evidence indicating that the Indian Ocean formed when the eastern part of the ancient continent of Gondwanaland broke up as a result of seafloor spreading in a general east-west direction. This spreading persisted throughout the Cretaceous. The Timor Trough first subsided during the late Pliocene.

After repair of a break in the drilling apparatus that occurred off Western Australia, the "Glomar Challenger" cruised to the region of the Ross Ice Shelf in Antarctica. Cores were taken at 11 sites along the way, bringing the total number of sites drilled to 274 since the program began in 1968. The most surprising results were those indicating that Antarctica had been covered with ice much longer than previously believed. It appeared that the continent had been glaciated for 20 million years, since the Miocene, rather than 5 million to 7 million years. Ice-transported pebbles were absent in Antarctic core samples older than 20 million years, a date that also marked the appearance of a cold-water microfauna.

There was a major glacial expansion about 5 million years ago, in the Pliocene, when the Ross Ice Shelf extended 200–300 mi farther out to sea than at present. Drilling results confirmed other evidence that Australia broke off from Antarctica about 50 million years ago and has been drifting northward on an average of two or three inches per year ever since.

Upper Mantle Project. This international effort officially ended in 1973 with the issuance of its final report. The study, coordinated by a committee headed by V. V. Beloussov of the U.S.S.R., began in 1964 with emphasis on three problem areas: (1) continental margins and island arcs; (2) the world rift system; and (3) rheology (deformation and flow) of the upper mantle. In addition, a 1964 suggestion by J. Tuzo Wilson of Toronto was adopted as a prime objective: "to prove whether or not continental drift occurred."

The final report affirmed that continental drift had indeed occurred, and the plate tectonics model seemed to be a reality. In this model, the interior of the earth consists of a central core extending to about half the earth's radius, a thick mantle, and a surface crust. The brittle crust is broken into many irregularly shaped plates that slowly move about with respect to each other on top of the solid yet viscous mantle.

Better understanding would be achieved when holes were drilled that actually penetrated to the mantle through thin parts of the oceanic crust. Meanwhile, however, combined geochemical and geophysical studies had led to an understanding of what the mantle's mineralogy must be like. Kimberlite pipes, such as the diamond-bearing igneous

For decades scientists have been puzzled by the orientation of the four thousand-year-old Great Pyramids of Giza, four minutes west of true north. A possible explanation is that the malalignment developed as the continents drifted apart in the general directions shown on the map, below.

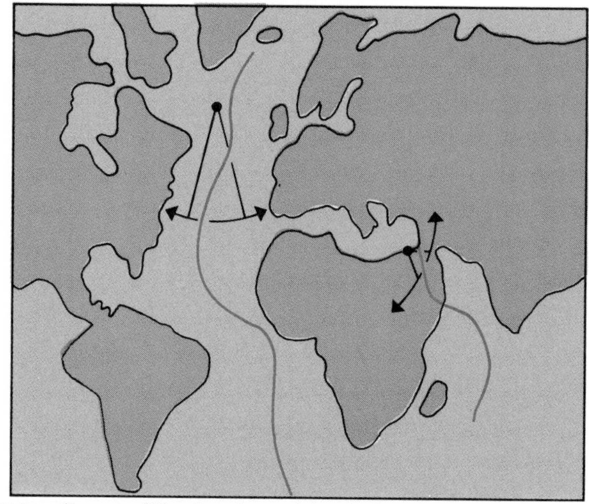

(Left) Courtesy, Museum of Fine Arts, Boston; (above) adapted from *Science* (1973)

intrusions of Africa and Arkansas, probably represent samples of mantle rock that have been injected toward the earth's surface along fractures. It seemed fairly certain that most compression, rifting, and metamorphism of rocks on the continents and sea floor resulted from major energy changes and shifts of rock masses within the mantle. Boundary discontinuities probably occur within the mantle at depths of 400 and 650 km, and uneven energy distribution along these discontinuities is ultimately responsible for most surface deformation.

Ancient continental drift. Not only has continental drift occurred, but the present oceans are young. Only the drifting continents contain remnants of the oldest rocks (Paleozoic and Precambrian), which in places have been badly deformed by various episodes of continental rifting, collision, and stress. The oldest rocks on the sea floor are Jurassic, so the present oceanic record of drifting spans only the last 4% of the earth's history. Within the continental plates, however, evidence of earlier episodes of plate tectonics has been preserved, and this evidence was the subject of vigorous investigation.

The Urals originated when two subcontinents collided 300 million years ago, closing a geosynclinal seaway. A proto-Atlantic Ocean opened when eastern North America rifted apart from Africa about 700 million years ago and began to close in the Ordovician. This was followed by a collision between North America and Africa about 700 million years ago, in the Devonian. The continental geography about 550 million years ago was reconstructed by Rhodes Fairbridge of Columbia University. (*See* figure, page 223.)

Geochronology. The boundary between the Miocene and Pliocene epochs is generally considered to be about 12 million years old. J. B. Gill and Ian McDougall of Australian National University in Canberra dated volcanic rocks from Fiji that are stratigraphically related to sediments containing marine fossils that define the Miocene-Pliocene boundary. According to their findings, this boundary is 4.9 ± 0.4 million years old, and from this evidence they suggested appropriate revision of the geologic time scale.

Xenon isotopes from deep gas wells in New Mexico are believed to have originated from iodine-129 and plutonium-244, both of which totally decayed and became extinct during the first 100 million years of the earth's history. Analysis of these isotopes resulted in a calculated age for the earth of at least 4.5 billion years, which was quite consistent with several other independent estimates.

—John M. Dennison

Geophysics

Earthquake prediction received much attention in the news media during the year as scientists made important advances toward this old and elusive goal. Among the advances were the confirmation of a Soviet finding of changes in seismic wave velocities prior to the occurrence of earthquakes, the explanation of this and other premonitory phenomena in terms of a common model of rock deformation preceding failure, and the first prediction of an earthquake by U.S. government scientists. This progress, coinciding as it did with increased U.S. government support for research related to the reduction of earthquake hazards, justified optimism that prediction of the place, size, and approximate time of many potentially destructive earthquakes might become a reality by the 1980s.

Seismic wave velocities. Earthquakes generate two types of seismic waves that are refracted down into the earth as they travel from the earthquake focus to an observer: compressional (P) and shear (S) waves. P waves travel faster than S waves and are the first waves recorded on a seismogram. For most rocks within the earth's crust, the velocity of P waves is about 1.75 times the velocity of S waves.

In 1969 Soviet seismologists reported variations in the ratio of P-wave to S-wave velocity prior to shallow-focus, moderate-sized earthquakes in the Garm region of the Tadzhik S.S.R. The variation was remarkably systematic from one earthquake to another; an initial decrease in the velocity ratio of as much as 10% was followed by a return to normal values shortly before the earthquake occurred. The time interval between the initial decrease and the occurrence of the shock increased with the size of the eventual earthquake; for quakes of magnitude 3.5 the interval was from two to three weeks, while for those of magnitude 5 it was about three months. No satisfactory physical explanation was offered.

After U.S. seismologists became fully acquainted with the Soviet findings, they set about to see if such data could be duplicated elsewhere. In a study of an earthquake swarm in the Adirondack Mountain region of northern New York, Yash P. Aggarwal, a Columbia University seismologist, and his colleagues observed decreases in the velocity ratio of P and S waves of as much as 13% before earthquakes in the magnitude range 1 to 3. In all respects, the premonitory variations for the Adirondack shocks were similar to those reported from Garm. The amount of the decrease in the velocity ratio was independent of earthquake magnitude, while the time interval between the initial decrease and the quake correlated with it, ranging from a few hours for magnitude 1 to about a week for magnitude 3.3.

Courtesy, Christopher H. Scholz, Lynn R. Sykes, and Yash P. Aggarwal, Lamont-Doherty Geological Observatory, Columbia University

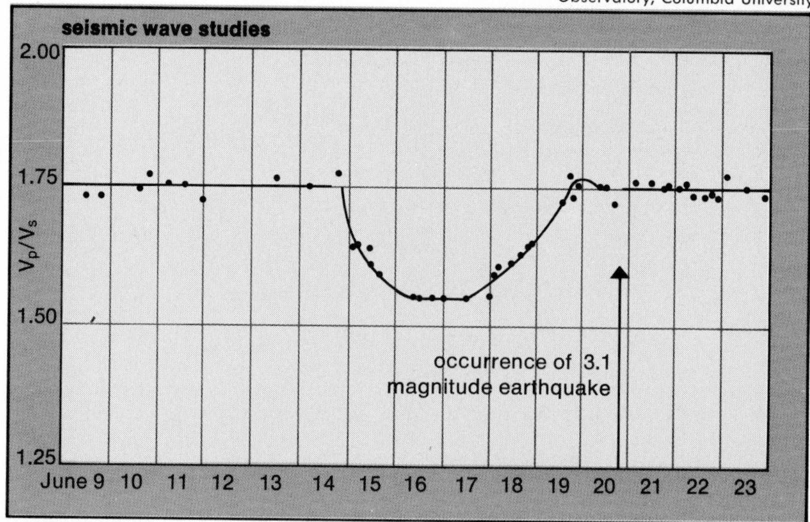

Changes in the normal ratio between the velocity of P (compressional) and S (shear) waves traveling through the earth have been detected before moderate-sized earthquakes in the Garm region of the U.S.S.R. and in the Adirondack Mountains. The graph shows the variation preceding an earthquake of magnitude 3.1 on the Richter scale. The finding appeared to open new possibilities for earthquake prediction.

A similar but less well documented change in velocities prior to the destructive San Fernando, Calif., earthquake of 1971 was reported by James H. Whitcomb and two co-workers at the California Institute of Technology. The magnitude of the earthquake was 6.6, and in this case the apparent decrease in the velocity ratio occurred about 3½ years prior to the shock. Most of the change was associated with a decrease in the P-wave velocity.

Dilatancy in fluid-saturated rocks. Geophysicists from Stanford University, Columbia University, and the California Institute of Technology developed a plausible physical model to explain the precursory changes in seismic velocities. The model was based on results of laboratory experiments on the behavior of rock subjected to large stresses. As the stresses applied to a rock specimen are increased, the volume of the specimen increases prior to ultimate failure. This phenomenon, termed dilatancy, results from cracks opening within the rock.

Laboratory experiments also indicated that the velocity ratio of P to S waves decreases with the expansion of cracks in dry rocks but increases in saturated rocks. Shallow earthquakes occur at depths where rocks are saturated with water. When cracks first form, they are dry, but they become wet as fluid flows into them. Deep within the earth's crust, fluid flow is likely to be slow compared with the rate at which rocks dilate. Hence the initial decrease in the velocity ratio can be explained by the formation of dry cracks, and the subsequent increase can be attributed to the slow migration of water into them. As the cracks fill with water, the fluid pressure within the cracks increases, reducing the effective stress that is resisting slippage on the fault. In effect, the water lubricates the fault so that an earthquake can occur. Finally, if the dilatant volume increases with the magnitude of the earthquake, then the duration of the velocity change would also increase since the migration of fluids would involve greater distances.

Other precursory phenomena were reported to occur on approximately the same time scale. In Japan the ground in the source region of a magnitude 7.5 earthquake suddenly rose five centimeters five years prior to the earthquake. In the Garm region the electrical resistance of the earth's crust measured over distances of several kilometers began to decrease a few months before moderate-sized earthquakes. In other instances premonitory tilting of the ground and variations in the rates of occurrence of small earthquakes were reported. These phenomena may also be explained in terms of dilatant behavior in water-saturated rocks, suggesting that dilatancy may be a widespread phenomenon that will provide a physical basis for earthquake prediction.

Seismic gaps. Most shallow-focus earthquakes occur in narrow seismic belts outlining the large mobile plates that comprise the earth's surface. These earthquakes result from the interaction of the plates as they collide, break apart, or slide past one another. Over a long period of time—a century or more—the entire boundary of a plate experiences earthquakes, but over a much shorter time span there are gaps in the shock pattern. Such gaps indicate where the driving forces acting on adjacent plates are momentarily insufficient to overcome the local resistance to slip.

The identification of a seismic gap provides a basis for predicting the location of a future shock. From the length of the gap, the magnitude of the shock can also be predicted. For example, in 1971 a 150–300-km-long gap in the seismic zone along the coast of southeastern Alaska and British Columbia was identified as a likely site for the next earthquake of magnitude 7 or greater in southern Alaska. In July 1972 a magnitude 7.25 earthquake occurred off Sitka.

The seismic gap concept was also the basis for the first precise earthquake prediction by U.S. government scientists. In January 1973 geophysicists with the U.S. Geological Survey predicted that a magnitude 4.5 earthquake would occur within several months on the San Andreas Fault in central California. The focus of the shock was given to the nearest kilometer. Potential damage from a shock of that size is minimal. Nonetheless, the prediction was significant because it was the first attempt to precisely predict the location and size of an earthquake from detailed seismic and geodetic observations made along the San Andreas Fault zone during the preceding several years.

Convective mantle plumes. The origin of the forces responsible for the movement of crustal plates over the earth's mantle is not known; however, convection currents within the mantle have long been considered likely. In this mechanism, the motion of the plates is the response to tractions exerted on their bottoms by the horizontal movements of convection currents in the upper mantle. Two types of convective motion within the mantle have been proposed: a pattern of convection cells with well-defined zones of upwelling and sinking; and mushroom-shaped plumes of upwelling with diffuse sinking over a broad area in the return flow.

A model incorporating plumes was advanced in 1971 by W. Jason Morgan of Princeton University. In this model, horizontal currents in the upper mantle flow radially outward from convective plumes, and the driving force on a plate is the combination of viscous forces from several plumes acting on its underside. The plumes transport heat and relatively primordial material upward through the mantle. The evidence presented to support the existence of plumes was largely circumstantial. The age of volcanos in many linear chains of volcanic islands, like the Hawaiian Islands, increases progressively from one end to the other. In 1965 it was suggested that this could be explained by the passage of oceanic crustal plates over stationary sources of molten rock in the mantle, the so-called hot spots. Morgan proposed that such hot spots are manifestations of convective plumes within the mantle. The plume model also predicts the observed differences between the type of rock found on oceanic volcanic islands originating over plumes and the type of volcanic rock dredged from mid-ocean ridges where the crustal plates are rifting apart and where upper mantle material is intruding into the rifts to form new crustal rocks.

Lava falls 150 ft high were part of a spectacular display of eruptive activity at Kilauea volcano in Hawaii during 1969–71. One model of the forces that move crustal plates over the earth's mantle posits the existence of convective plumes within the mantle that transport heat and primordial material upward. Volcanic activity occurs as crustal plates move over these "hot spots."

Courtesy, Dr. Donald A. Swanson, U.S. Geological Survey

convective mantle plume theory

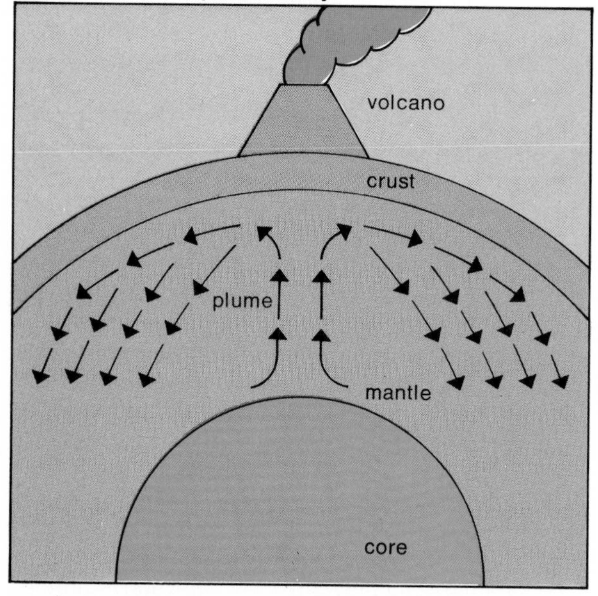

Schematic drawing of a convective plume indicates the movement of currents from the top of the plume along the underside of the crust. Plates would be moved by the combined force of several plumes.

To test the validity of this model, seismologists were studying the propagation of seismic waves that traverse the lower mantle to find evidence for the existence of plumes. Recently U.S. and Canadian seismologists reported regions of anomalous seismic velocities near the core-mantle boundary beneath Hawaii, Iceland, and the Galápagos Islands, all volcanic island chains. These findings were consistent with the existence of plumes. However, it must be recognized that significant lateral variations in seismic velocities have been found elsewhere in the mantle.

The geochemistry of surface volcanic rocks furnishes additional evidence. Where a plume lies beneath a rift zone along which two crustal plates are separating, the new rock formed in the rift at different distances from the center of the upwelling can be sampled. Such sampling along the Reykjanes ridge south of Iceland and in the Red Sea trough and Gulf of Aden adjacent to the Afar triangle in eastern Africa revealed variations in the relative abundance of rare-earth elements consistent with the upward convection of primordial mantle material by narrow plumes and the mixing of the primordial material with normal upper mantle material as it flows away from the plume.

The existence of mantle plumes could resolve the long-standing question of whether the earth as a whole has moved with respect to its rotational axis, or pole, in geologic time. Such movement is referred to as polar wandering. Past positions of the rotational pole can be determined by studying the magnetization of rocks and making use of the observation that the positions of the magnetic and rotational poles approximately coincide when averaged over geologically short intervals of time. A rock containing magnetic minerals acquires a permanent magnetization in the direction of, or in some instances opposite to, the local geomagnetic field existing at the time the rock is formed. If there were no movement between crustal plates, it would be possible to deduce ancient positions of the magnetic pole (and thus determine polar wandering) from the directions of magnetization in rocks of similar age and from diverse geographic localities. Plate motions render the problem insoluble unless the past movements of plates relative to some frame of reference in the earth other than the rotational pole can be determined.

Three geophysicists working at Princeton suggested that, if mantle plumes exist and are fixed relative to the mantle, passage of a plate over a plume would leave a permanent record in the form of igneous geologic features on the surface of the plate. Thus, the absolute motions of plates relative to the mantle could be determined from the history of igneous activity, and ancient magnetic pole positions could be reconstructed from rock samples obtained from different crustal plates. This method was applied to two igneous chains in Europe. The conclusion was that the pole has shifted about 23° with respect to the mantle during the last 50 million years.

Subsequently, an Australian geophysicist examined possible motion of the crust as a whole with respect to the earth's rotational axis and found no significant motion over the last 50 million years. At face value, the results from these two studies suggest that the crust has remained fixed with respect to the earth's rotational pole while the mantle has moved. This seems unlikely, however, and possibly indicates that the particular assumption of stationary plumes is incorrect.

—Robert A. Page

Hydrology

Perhaps the outstanding event of 1972–73, as far as hydrology was concerned, was the UN Conference on the Human Environment, which took place in Stockholm, June 5–16, 1972. There, representatives of 114 countries made it clear that from now on hydrologic problems—or those of other natural resources—could not be considered apart from the rest of the environment. Scientific programs in hydrology and proposals for the development and management of water resources would have to be related both to other physical processes and to the economic, political, and social pressures that bear on decisions regarding the use of water.

New interests in research. This increasing emphasis on environmental interrelationships affected the type of research that was being undertaken. The amount of water-related research was greater than ever before, but the emphasis had shifted from study of the traditional hydrologic elements to the interrelationships between the water cycle and the world in which it occurs. Recent research emphasized water quality and water resources planning, both multidisciplinary fields, while studies of the basic physical nature of water dwindled.

Concern for the environment was not limited to problems of the great outdoors. A large proportion of current interest centered on urban hydrology and the ways in which urbanization was affecting hydrologic parameters and their distribution. The difference between the hydrologic regimen of a sewer and a creek was greater than the simple difference in their basic functions.

Storm runoff from streets covered by concrete, asphalt, and brick, from square miles of roofs and from acres of parking lots, peaks faster and discharges more water than would have resulted from the same precipitation on the same terrain when it was covered with meadows, woods, brush, and grass. As an area becomes urbanized, the local climate changes and the chemical content of the water is modified, to say nothing of the changes in content and composition resulting from the addition of chemicals, pesticides, oil, and other pollutants. As more suburban areas are developed, the discharge of sediment rises enormously. In many instances the discharge of sediment from development tracts is thousands of times greater than from unbroken cover.

Rains and floods. Although much recent research had concentrated on probability analysis of various hydrologic events, such as storms, droughts, and snowfalls, the U.S. was hit by an unexpectedly vicious tropical storm in June 1972 and by damaging spring storms and floods in March and April of 1973.

Tropical Storm Agnes caused disastrous flooding in the mid-eastern states. In its wake it left 118 dead and damage that was estimated as high as $3 billion. The flooding was widespread because tremendous amounts of rain fell in basins where the underlying rocks could not have absorbed the water in any case and where earlier spring rains had left the ground already saturated. Much of the damage occurred on floodplains where extensive investments had been made in business and industrial construction. Unfinished construction, with bare earth exposed to rapid erosion, provided huge loads of sediment that clogged the streams, thus leading to further flooding.

Protective works had been built in the area to contain what had been the record floods up to that time, but Agnes broke all records. Hydrologically, it was probably impractical to build against any all-time maximum flood flow, but it should be possible

Photographs taken by ERTS 1 show the area around Las Vegas, Nev. (left), with Lake Mead in the upper right quadrant, and the lower Mississippi Valley (right). In its first year of operation, ERTS provided quantities of data of great value to the earth sciences.

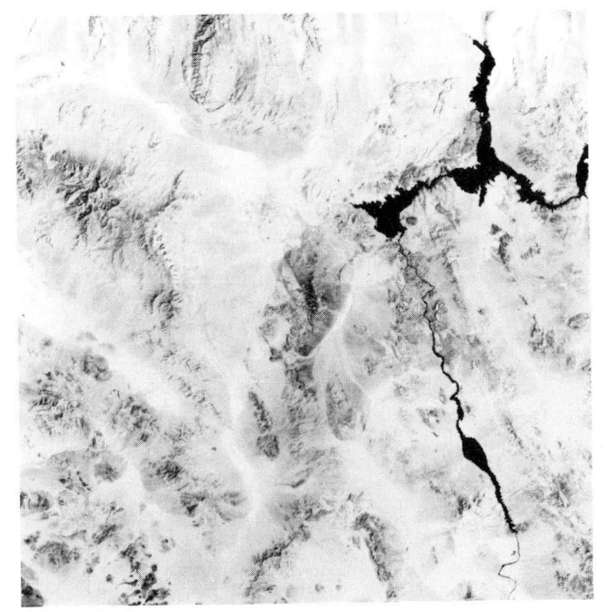

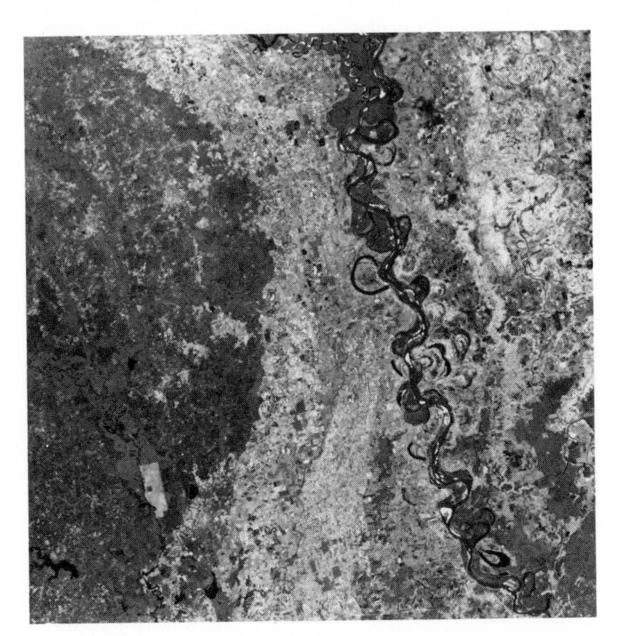

Courtesy, NASA

to minimize dollar losses by using the floodplains for less expensive and less susceptible structures. Floodplains could be used reasonably for parks, golf courses, hunting and fishing preserves, agriculture, and other appropriate and comparatively inexpensive purposes. It was no longer necessary to site towns and cities beside rivers in order to provide for water supplies, sewage dilution, and transportation.

In March and April 1973, torrential rains hit the Midwest and the valleys of the Mississippi River and some of its major tributaries. Locally as much as 30 cm (12 in.) fell in a single 48-hour period. Flood crests and flows set new records. Millions of acres of farmland were flooded, and estimates of damage to industrial, municipal, and agricultural areas ran into millions of dollars. Additional damage occurred along the shores of the lower Great Lakes, which were already unusually high.

Again, questions were raised as to whether the levee building, stream grading, and channel straightening undertaken since 1929 had helped or hindered the dissipation of major floods. Another aspect was the serious but as yet unevaluated damage to many small commercial fishing businesses in the Lake Pontchartrain, La., area due to diversion of silt-laden waters from the Mississippi River. The inflow of silt had completely or partially buried oyster beds, fish feeding grounds, and crab haunts, and the lowering of salinity and temperature had affected the fish-breeding cycle.

ERTS. In another sphere, large amounts of hydrologic data were being received from ERTS 1, the Earth Resources Technology Satellite, launched into a near-polar circular orbit on July 23, 1972. After six months of operation, the satellite had gathered a tremendous amount of high-quality multispectral scanner observations in several spectral intervals. Observations were normally presented in photographic imagery covering an area of about 185 sq km (115 sq mi) and had a resolution down to about 100 m (325 ft).

Barring cloud cover, the satellite was capable of providing observations of the same place on earth at least once every 18 days. Forty investigations dealing principally with potential water resources applications were using ERTS 1 data. ERTS 1 photographs or imagery were useful for monitoring and mapping, locating reservoirs and lakes (as small as a football field), identifying turbidity changes resulting from increased sediment loads in rivers, and observing the extent of sediment plumes where rivers empty into lakes and other bodies. In semiarid areas, the response of vegetation to water and streamflow and the amount and frequency of precipitation could be determined by comparing sequences of photographs.

While satellites were in the forefront of developmental research, older but still recent methods of hydrologic observation from airborne infrared cameras provided a better understanding of groundwater discharges in many places. Modern infrared imagery could indicate a difference in water temperature as small as 1°–2° C (2°–3° F), and the identification of leakage by infrared methods could help considerably in planning for the improved management of groundwater resources.

International developments. The joint U.S.-Canadian International Field Year for the Great Lakes completed its field data collection on March 31, 1973, and the data were being analyzed. This would provide the first comprehensive quantitative overview of the hydrologic cycle and the physical, chemical, and biochemical regimen of a large fresh-water lake and its basin. It should be of great value in the management not only of Lake Ontario, where the data were collected, but of the other Great Lakes as well. Many of the scientific conclusions might be applicable to large lakes throughout the world.

Over the preceding two decades, the World Meteorological Organization had increased its concern and responsibility in the field of hydrology. It accepted a global responsibility in operational hydrology and was considering the modification of its statutes so that eventually national hydrologic institutions could be represented in the organization on an equitable basis with national weather services.

—L. A. Heindl

Electronics

Important advances were reported in 1973 in methods of registering and displaying visual information by electronic means. Until recently, such displays relied mainly on increasingly sophisticated versions of the cathode-ray tube (CRT), first used for that purpose by K. F. Braun in 1897. During the three-quarters of a century since then, industrial production of CRTs rose steadily: the annual production of the most familiar type, the television picture tube, is in the tens of millions, and many more are manufactured for industrial and scientific uses ranging from radar to biomedical applications. Since the advent of electronic computers, CRTs have also found uses in a new field, the nearly instantaneous display of numerical and other data.

The present-day color-television picture tube shows vastly more information than the oscilloscope that Braun used to display simple waveforms. Yet the principle is the same: a beam of electrons is formed by heating an emitting cathode in the neck of a sealed glass bottle from which most of the air has been evacuated; the beam is focused onto a phosphor layer on the inside of the flat end of the bottle, where a luminous spot appears; and the spot is positioned by electric or magnetic deflection of the beam. (Many devices, from electron microscopes to electronic movie cameras, utilize variants of the CRT arrangement.)

The great advantage of this scheme is that the beam has almost no inertia: since the electric charge carried by each electron is very large compared to its mass, the beam responds immediately to the application of a relatively weak deflection field. For this reason quite rapid variations of the field can be faithfully reproduced. The limitations are in the materials: the cathode emits fewer electrons as time goes on, and so the picture becomes paler; the tube is bulky and fragile; the vacuum seal may fail; even the best phosphors are not too efficient in converting a fleeting electron beam into a light spot; and their efficiency deteriorates with time. Most radar and color-television displays are still best viewed in darkened rooms and cannot be readily projected onto a large screen.

In a growing number of applications the CRT's best feature, the instantaneous response to a continuously changing picture, is not needed. When a bank balance or an airline's departures and arrivals are to be shown, it matters little whether a new posting appears after a millisecond or a microsecond. Thus, during the year considerable efforts were under way to replace the CRT by different devices in which speed is traded off for smaller size, lower cost, greater reliability, and better viewing. Moreover, there was always the hope that one of these devices might eventually also come near to duplicating the CRT's fast-response characteristics and thus enable the inventor to tap the rich television market. As of 1973, no such device was available, but advances toward greater reliability, efficiency, and space savings were being reported from several laboratories intent on ending the long dominance by devices that depended on a thin electron beam moving in a vacuum.

Matrix displays. A prime candidate from solid-state electronics (the industry based on semiconductor rectifiers, transistors, and transistor circuits) is the light-emitting diode (LED). Junctions of certain semiconductor materials, such as gallium phosphide or gallium arsenide, produce a visible glow when a voltage is applied across them. Arrays of tiny LEDs can thus produce dots of light that can be individually controlled to form letters and numerals, in the manner of an advertising sign in which letters and words are spelled out by light bulbs. Since the well-developed technology of transistor circuits is directly applicable, LEDs

1947
1972

"In the shape of a small metal cylinder about a half-inch long, the transistor contains no vacuum, grid, plate or glass envelope.... Its action is instantaneous."

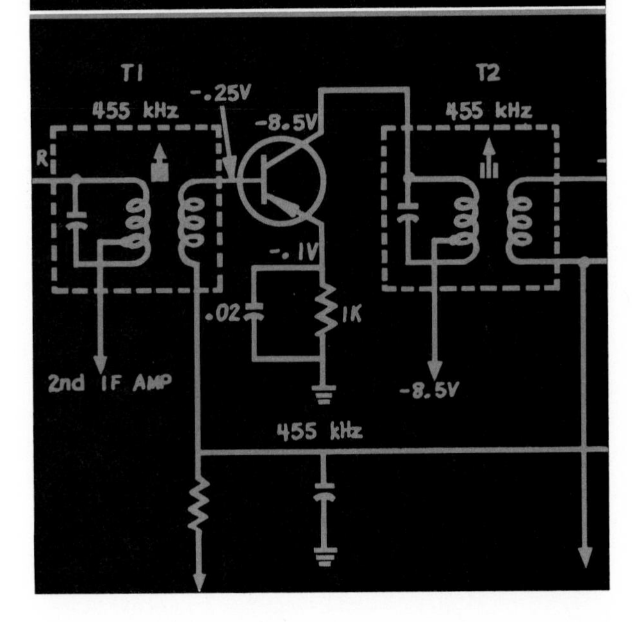

can be manufactured and assembled quite cheaply and have found a ready market in calculators, digital clocks and panel meters, and other consumer items.

It is already possible to perform certain simple programming operations in "on-board" circuits that are integral to an LED array. For example, a measured voltage or frequency can be translated into numbers in integrated circuits (tiny complexes of electronic components on semiconductor "chips") immediately adjacent to the array, and simple counting can also be done there. More complex graphic displays, such as television pictures, remained unattainable, not because of the failure of LEDs to respond to rapidly changing signals but because the requisite circuitry could not be provided at a cost competitive with the CRT. But the LED array was not the only display medium to challenge the conventional CRT. At least two others reached an advanced stage of development: the gas-discharge panel and the multiple-beam CRT.

In one form of the gas-discharge panel, the Burroughs Corp.'s Self-Scan, an array of tiny gas-filled cells is formed by an insulator sheet into which thousands of holes have been drilled at regular intervals. The array is then placed between two glass cover sheets, the three-layered arrangement then being sealed off at the edges and filled with neon gas. The rear sheet is grooved horizontally, and parallel conducting strips are inserted in the grooves; a second set of vertical strips, containing apertures aligned with the holes in the insulator sheet, is placed between that and the rear sheet. A set of thin horizontal wires (also aligned with the holes) on the front side of the insulator sheet completes the arrangement. Application of a small voltage between a pair of the horizontal and one of the vertical elements produces a tiny glow spot at the intersection, which is given a circular shape by the aperture configuration. A matrix of five by seven spots is sufficient to form any alphanumeric (letter or digit) character; displays containing up to 8 lines of 32 character positions each are available.

The other principal contender for the position of the conventional CRT is a variant of it, the multiple-beam CRT. Various schemes were proposed during the 1950s and 1960s to create a "flat" or thin CRT—one that could be hung on the wall like a picture. In such tubes, the source of the electron beam was installed in a corner of a flat glass box and the beam was started off along one edge. Unfortunately, the technical difficulties of bending the beam through two right-angle turns without producing a picture that was skewed or otherwise distorted were not overcome. But that problem was recently solved in the Digisplay vacuum tube being developed by the Northrop Corp. This device consists of a stack of parallel plates, with the entire rear plate acting as an electron emitter or "area cathode." Successive plates are apertured to form thousands of discrete beamlets, to modulate their intensities, to select those to be passed through, and to focus them onto the transparent front panel, on the inside of which a phosphor layer like that of a conventional CRT is deposited. The resulting device is thus not only suitable for alphanumeric displays—the largest completed to date can show 16 rows of 32 characters each, or 512 characters, with only 63 switching leads—but is capable of adaptation to such CRT characteristics as gray scale, sequential scanning, and color, as well as storage (memory). In 1973, a black-and-white TV display of good quality was demonstrated experimentally.

Courtesy, Owens-Illinois, Inc.

Halftone picture of a fish is projected on the back of the gas-discharge display panel of the Digivue display/memory unit. Because the panel is virtually transparent the picture can be superimposed optically on the unit's electronic alphanumeric display.

Solid-state imaging. Solid-state technology was also beginning to have a substantial impact on the art of transforming a scene into a train of electric pulses, as in television camera tubes. As mentioned previously, such tubes currently employ a variant of the CRT. In the most popular such device, the vidicon, the phosphor layer on the inside of the flat end of the bottle is replaced by a photoconductive layer deposited over a thin, transparent conducting layer, the signal plate. The scene to be transmitted is optically projected on the outside of the flat end, making each grain of the photoconductive layer more or less conductive in proportion to the amount of impinging light. An electron beam scans the layer from the inside, and a portion of it penetrates to the signal plate; how much gets through depends on the conductivity in each spot. The current that flows into the signal plate is thus a sequence of pulses that can be transmitted to another place and reassembled into the picture there.

In the solid-state version of this device, the photoconductive layer is replaced by an array of diodes fabricated from a large single-crystal slice of silicon. Conversion of light into electric signals takes place more efficiently than in the photoconductive vidicon, but resolution (the amount of detail that can be preserved) is not quite as good. The next step was expected to be the elimination of the electron beam altogether. An all-solid-state imaging device would also do away with the need for a vacuum envelope. In 1973 two approaches to this goal of "self-scanning" were in an advanced state of development, one employing photodiodes and the other, charge-coupled devices (CCDs).

The self-scanning photodiode image sensor was developed principally by Reticon Corp. of California. Self-scanning is accomplished by integrating the scanning circuits into the same silicon crystal that contains the photo sensors. The circuits serve to "interrogate" each diode in sequence to determine whether the light impinging on it is above or below a present level. Among the non-TV applications of the Reticon image sensor are industrial quality-control monitors that can automatically check defects, size, or position in production items as they pass by on a moving belt; readers that can scan along a printed page; and a miniature camera array small enough to be mounted on eyeglasses for use in mobility aids for the blind.

An even more promising self-scanning scheme employs CCD technology developed at Bell Telephone Laboratories. In a CCD element, a tiny charge is generated by the impinging light and is stored in the panel. The self-scanning feature depends on a property of CCDs that enables the charge from one element to transfer to its neighbor when a control "clock" voltage is applied, so that the charges along an entire row of elements can be read off sequentially at the end of the row. A modified version of the Bell Laboratories method was developed by Fairchild Camera and Instrument Corp. Both held promise for an all-solid-state imaging device to replace the beam-scanned models.

Projection display systems. As mentioned previously, the light output of a standard television tube is not sufficient to produce an acceptable projection display comparable with that attainable in almost any movie theater. One method of overcoming this is to use a larger nonstandard tube, one which operates at beam-accelerating voltages substantially higher than the usual 10–20 kv, so that the phosphor layer is illuminated more strongly and the resulting picture has enough intensity to be projected optically on a screen. For this procedure a fairly complex projection scheme is needed, usually involving a specifically corrected lens system of the Schmidt type used in astronomy, by which one can obtain undistorted images of a very wide field.

In all other TV projection schemes the functions of light generation and light control are separated: the CRT is used to control the light but is not called upon to produce it. An early example is the Eidophor projector, first developed in Switzerland during World War II, which remains the basis for a commercially available "light valve" capable of projecting a display changing at TV rates. A scanning electron beam produces ripples in a thin oil film; and a large display is produced by light from a separate source, refracted from the deformed surface. An analogous method, pioneered at the General Electric Co., utilizes diffraction rather than refraction: by creating optical gratings on a deformable medium, one can diffract and modulate the incident light, moreover, in a way that extends use of this method to color displays.

—Charles Süsskind

Environmental sciences

In the year that began with the UN Conference on the Human Environment, held in June 1972 in Stockholm, the hard choices posed by the environmental movement began to emerge more clearly. The conference itself had exposed a conflict of interest between rich nations concerned with the environmental effects of their high (if often wasteful) standards of living and poor countries eager to attain those standards. As the year wore on, even rich nations began to count the costs of a cleaner world, both in money and in terms of changes in life-style.

The Stockholm conference. Although the UN conference was largely advisory in nature, it did represent the first institutionalized attempt to deal with the problems of environmental pollution on a worldwide scale. The two-week gathering was attended by some 1,200 delegates from 110 countries. The Soviet Union was the only major nation that failed to send representatives, having withdrawn during the planning stages as the result of a dispute over the seating of East Germany.

The divergent interests of rich and poor countries were emphasized repeatedly by the speakers. Most notably, the less developed nations tended to define "pollution" in terms of low living standards, disease, hunger, and urban blight, while the more advanced countries concentrated their attention on problems of smog, solid waste disposal, air and water pollution, and other by-products of industrialized society. Nevertheless, agreement was reached on a number of specific points, including resolutions for continued worldwide discussion and cooperation through a new UN environmental agency, an Action Plan, and a Declaration on the Human Environment. Seven main recommendations submitted by the conference to the UN General Assembly invited member nations to:

1. Promote a convention to restrict dumping of harmful substances in the world's oceans.
2. Minimize the release of notably dangerous pollutants.
3. Organize worldwide monitoring of atmospheric, marine, terrestrial, and human health hazards.
4. Promote "genetic banks" to safeguard the existence of all species of animals and plants.
5. Halt all commercial whale catching for ten years.
6. Set up an international exchange service for information on the environment.
7. Increase emphasis on pollution control.

In addressing the delegates, Maurice F. Strong, secretary-general to the conference, pointed up the dangers that threatened rich and poor nations alike, particularly with regard to food supplies. By the year 2000 food supplies would need to be doubled in order to feed a predicted world population of some 7,000,000,000. Yet, in some cases, deserts were actually expanding, especially in less developed countries where population pressures led to farming on unsuitable land, with consequent erosion, interruption of the hydrologic cycle, and silting up of watercourses. At the same time, highly intensified farming in the developed countries brought its own problems of soil degradation and long-term pollution by persistent pesticides. Strong urged all governments to incorporate environmental considerations into their agricultural development programs and to conduct research, followed up by legislation, education, and the necessary financial measures.

"We should be thankful. What if oil and water did mix!"

Sidney Harris

International cooperation. During the year there were some signs that the impetus toward international cooperation was bearing fruit in specific measures. On Dec. 15, 1972, the UN General Assembly approved an international environmental program that was largely an outgrowth of the Stockholm conference. Strong was named as executive director of the program, which was to have its headquarters in Nairobi, Kenya. The political dispute that had marred the conference was settled when East and West Germany and the U.S.S.R. all became members of the 58-nation Governing Council, established to provide guidance in matters of general policy.

In other international actions, in November 1972, 57 nations signed the Convention on the Prevention of Marine Pollution by Dumping, prohibiting the dumping of oil, mercury, and cadmium compounds, some pesticides, and highly radioactive wastes into the oceans. Dumping of certain other dangerous substances, including scrap metal, could be done only with government permission, and the convention called for care in the disposal of a third group of substances.

In February 1973 the member countries of the Organization for Economic Cooperation and Development, acting under the first international agreement of its kind, undertook to control the widely used group of toxic chemicals known as polychlorinated biphenyls (PCBs). Under the terms of the agreement, use of PCBs was to be restricted to those specific applications, such as certain electrical equipment, where their nonflammable properties were of overriding value, and steps were to be taken to ensure safety in manufacture, transport, and disposal.

That the Soviet Union's failure to attend the UN conference was a matter of politics rather than lack of interest was further emphasized in September 1972, when the U.S.S.R. and the U.S. agreed to undertake 30 joint projects relating to the environment. The pact, an extension of an earlier agreement signed during Pres. Richard Nixon's trip to Moscow in May 1972, provided for the exchange of scientists and the dissemination of data to other countries. Among the programs were a joint air-pollution-control project using St. Louis, Mo., and Leningrad as models; a study of water pollution in lakes, with Lake Baikal in the U.S.S.R., Lake Tahoe in California, and one of the Great

Georg Gerster from Rapho Guillumette

Farmer near Rosario, Argentina, carefully avoiding a clump of trees in the middle of his field, shows an awareness of environmental values rarely matched in the so-called advanced nations.

Lakes as the subjects; a similar study of river basins; studies on combating oil pollution; and work on the protection of endangered animal species.

Early in 1973 the Soviet government announced the establishment of an environmental protection service that would monitor air and water pollution throughout the country. The agency, which was to be operated within the Hydrometeorological Service, was apparently designed to serve chiefly as a data-gathering organization, while pollution abatement was assigned to various government agencies charged with acting in their own fields. Thus, for example, the Ministry of Agriculture was to deal with problems arising from the use of agricultural pesticides.

A step forward in the international protection of endangered species of animals was taken in March 1973 when 80 nations signed a treaty forbidding commercial trade in 375 wild animals in danger of extinction and in all products, such as hides, that are derived from them. Among the species involved were alligators, tigers, jaguars, leopards, rhinoceroses, land turtles, and five species of whale. Earlier, in June 1972, the International Whaling Commission had rejected the UN conference recommendation of a ten-year moratorium on whaling, adopting instead an individual quota for each whale species. Previously whale quotas had been set in terms of the so-called blue-whale unit.

The automobile controversy. In August 1972 the U.S. Council on Environmental Quality issued a report stating that, although the nation's water was getting dirtier, the quality of its air was improving. Emission of carbon monoxide had dropped 4.5% during the year, according to the council, and particulates had fallen 7.4%. Nevertheless, the existing levels were still far above those set for 1975 by the Clean Air Act of 1970, and the controversy over automobile emissions, one of the major contributors to air pollution in urban areas, continued unabated.

Despite the protestations of the automobile industry, several studies concluded that the 1975 deadline of a 90% reduction in pollutant emissions could be met. A 1972 study conducted by the National Academy of Sciences said the industry could meet the requirements provided the public absorbed a $200 per car cost, along with increased operating expenses and lower engine efficiency. The Office of Science and Technology concluded that the cost of pollution control could be offset by sacrifices in nonessential conveniences and styling frills.

In February 1973 a committee of the National Academy of Sciences issued a new report in which it reaffirmed that the 1975 deadline was practicable and criticized the U.S. automobile industry for concentrating on the catalytic converter, which it described as "the most disadvantageous with

Courtesy, Dr. Joseph Shapiro, Limnological Research Center, University of Minnesota

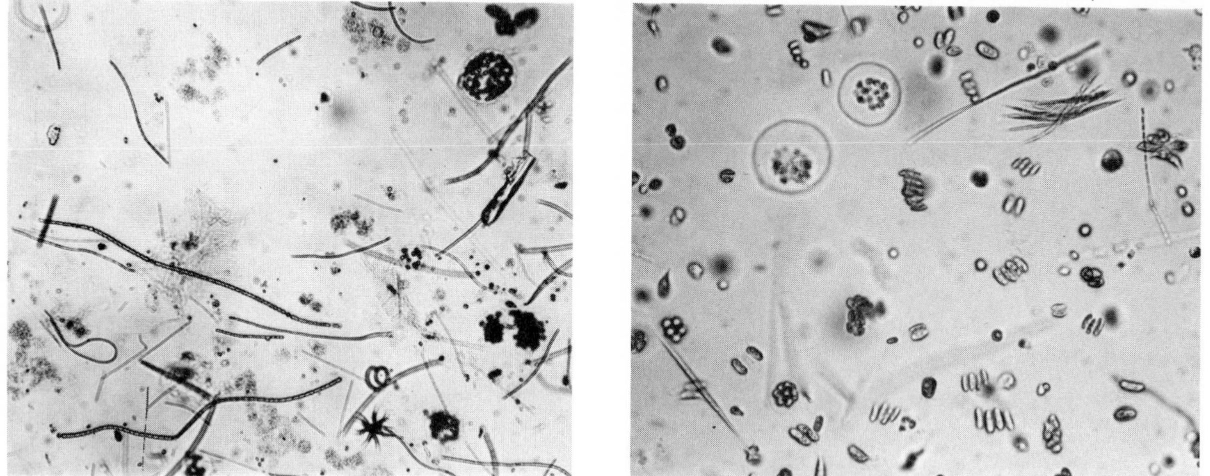

Blue-green algae, dominant in water from a Minnesota lake (left), were replaced by green algae (right) when carbon dioxide and nutrients were added. The experiment suggests that blue-green algae take over in polluted lakes because they are more efficient at utilizing CO_2 in low concentrations.

respect to first cost, fuel economy, maintainability and durability" among possible emission control systems. Embarrassingly, two Japanese-made cars, the Honda with a dual carburetor and the rotary-engine Mazda, already met the 1975 standards with considerable room to spare. (*See* Feature Article: ENGINES OF THE FUTURE.)

The U.S. industry, however, continued to insist that the 1975 deadline was not feasible. Lee A. Iacocca, president of Ford Motor Co., was quoted as saying that the industry had been "backed to the cliff edge of desperation" by the new requirements, and his company, together with General Motors Corp. and Chrysler Corp., took the matter to court in an effort of obtain a one-year extension. Ford had already been assessed a $7 million fine—one of the largest such judgments ever made—on charges that its employees had made unauthorized adjustments on 1973-model cars during testing in order to improve their emission-control performance. Because of the tampering, Ford—which had reported the employees' actions itself—had been forced to withdraw several thousand 1973 cars from the market.

As a result of the automobile companies' court action, William D. Ruckelshaus, then administrator of the U.S. Environmental Protection Agency (EPA), was ordered to reconsider his May 1972 decision denying the one-year extension. On April 11, 1973, he announced that the additional year would be granted. At the same time, he published interim standards for cars sold in 1975 and special interim standards for California, where the car-exhaust problem was most acute. Insisting that the effect on the campaign against air pollution would be "minimal," Ruckelshaus told a news conference:

> This is a terribly complex and important decision that involves the whole mix of our nation's struggle for a cleaner environment. Involved in this decision are billions of dollars, hundreds of thousands of jobs, probably the single most important segment of our economy, the largest aggregate manmade contributor to air pollution and the ambivalence of the American public's intense drive for healthy air and apparently insatiable appetite for fast, efficient, and convenient automobiles. . . . The ultimate effect of the decision will touch the lives of more than 200 million people.

The ambivalence to which Ruckelshaus referred was apparent in the public reaction to one of his earlier pronouncements. In the absence of a state plan for controlling air pollution in the Los Angeles basin, a federal district court, acting in accordance with the Clean Air Act, ordered the EPA to devise such a plan itself. On January 15, Ruckelshaus announced that the Clean Air Act standards could be met only through a stringent program, including gasoline rationing, that would reduce the use of automobiles by up to 82% during the summer months by 1977. Ruckelshaus himself admitted to having "grave doubts" that the plan was feasible in a region so heavily dependent on automotive transport and said the plan had been issued in order to stimulate further discussion and debate. A revised plan for Los Angeles was issued by the EPA in June, together with stringent transportation-control programs for 18 other urban areas. Among the proposed measures were limitations on gasoline sales in Los Angeles and northern New Jersey and a $4 a day offstreet-parking tax in downtown Boston.

No cleaner water. The Council on Environmental Quality report that had seen some grounds for optimism in the fight against air pollution had found little that was encouraging in the effort to clean up the nation's water supplies. Not only was water pollution from industrial sources and municipal sewerage systems as bad as ever, but the council found that pollution stemming from runoff from farms and construction sites had been underestimated. "Land runoff from farms and even urban land," the council wrote, "as opposed to discharges from cities and factories, has a much greater impact than we realized." Pointing up the problem was an EPA announcement declaring 27 drinking water systems ineligible as of April 17 to provide water to interstate carriers because their purification plants or operating procedures were considered questionable.

To some extent, federal programs aimed at combating water pollution became embroiled in the dispute between the legislative and executive branches over limitations on government spending and the administration's impounding of funds authorized by Congress. Late in 1972 the Water Pollution Control Act, providing new matching funds for municipal waste-treatment plants and prescribing strict standards for the regulation of industrial and municipal wastes, was vetoed by President Nixon on the ground that it would cost too much money. The bill authorized $24.7 billion over a three-year period, including $18 billion in grants to the states for sewage-treatment plants. Congress overrode the veto, but this was no guarantee that the administration would actually spend the money provided under this or similar programs. In April Congress failed to override a presidential veto of a bill directing the administration to restore a rural water-sewer grants program.

In addition to the sewage grants, the Water Pollution Control Act had provided a comprehensive program designed to eliminate all pollutant discharges by 1985 and make the waters safe for fish, wildlife, and recreation by 1983. One of the instrumentalities for achieving this involved permits for all dischargers of fluid waste certifying compliance with federal and state water-quality regulations. In March the *New York Times* reported that the program was proceeding more slowly than had been anticipated, largely because of the difficulties encountered in drawing up forms for the permit applications and in determining which and how many agricultural establishments should be included. As of late March the only applications on file with the EPA were from industries that had been required to file similar forms under an earlier reactivation of the Refuse Act of 1899.

The effort to clean up the nation's waterways by banning the use of phosphates in detergents received a setback during 1973, although the overall effect was not yet clear. Phosphates had been condemned by environmentalists on the ground that, when present in excessive quantities, they caused an explosive bloom of algae; the algae, in turn, used up the oxygen supply and brought about the "death" of the affected body of water. The process, known as eutrophication, occurred in nature, but in the absence of human intervention it took place over a very long period of time. The extensive use of phosphates in such substances as laundry detergents that were eventually discharged into lakes and streams threatened to speed up the process, and several jurisdictions had passed ordinances forbidding the sale of detergents containing phosphates within their boundaries. On March 6 a U.S. district court overruled Chicago's antiphosphate ordinance as an unconstitutional interference with interstate commerce. The court noted, however, that Chicago had failed to prove the ban was necessary and that similar ordinances in other jurisdictions might be upheld if it could be shown that they were justified by local conditions.

A pollution problem that had long been a cause of friction between the U.S. and Mexico concerned the salinity of the Colorado River. Under an international agreement concluded in 1914, Mexico received 1.5 million ac-ft of Colorado River water each year. However, the river was used extensively for irrigation by U.S. farmers. The water reaching Mexico was becoming increasingly saline, and the Mexican government claimed that cotton and other crops had been severely damaged as a result.

In June 1973 the *Wall Street Journal* reported that the Nixon administration was considering construction of the world's largest desalting plant in Arizona to purify the river waters before they reached Mexico. According to current plans, the plant would use the reverse osmosis process, whereby air pressure is used to force the water molecules in salt water through a permeable membrane while the salt ions are left behind. The water to be treated had an average salinity level of about 1,250 parts per million, compared with 35,000 ppm for seawater. In its present state of development, the reverse osmosis process was capable of purifying water with a saline content of up to 8,000 ppm, but U.S. Interior Department officials reportedly believed data obtained through operation of the new plant might make it possible to improve the process to the point where it could be used for seawater purification.

Pollution of the oceans themselves continued to be a cause for concern. In February 1973 the U.S. National Oceanic and Atmospheric Administration (NOAA) reported that 700,000 sq mi of ocean along the eastern coast of the U.S. and in the Caribbean were far more polluted than had been previously thought. Tests made by scientists on three survey ships during the summer of 1972 had revealed heavy concentrations of oil, tar, and bits of plastic, and more than half the samples of plankton taken in the survey had shown evidence of oil contamination.

In addition to signing the international convention to control dumping, the U.S. took a number of steps during the year designed to curb pollution in the oceans. In July 1972 President Nixon signed the Ports and Waterways Safety Act aimed at preventing oil spills at sea. In May 1973 the EPA banned the dumping of eight types of substances into the ocean and listed several dozen others that could be dumped only with "special care." The prohibited substances were high-level radioactive wastes, chemical or biological warfare materials, substances with unknown or persistent environmental effects, inert materials that would float or remain suspended, and materials with "more than trace amounts" of mercury and its compounds, cadmium and its compounds, oils, and organic substances containing a class of chemicals that included DDT. The U.S. also urged the Inter-Governmental Maritime Consultative Organization, a specialized agency of the UN, to set up a special committee to police the discharge of oil and other dangerous substances from ships at sea.

Elsewhere, pollution of the Mediterranean was reaching alarming proportions. In 1972 delegates from 11 nations, meeting in Paris, agreed to cooperate in a scheme to check oil pollution in the western part of the sea. A broader attack on the problem was considered at an international conference held in Beirut, Lebanon, in June 1973. The delegates called for immediate action, including the drafting of an international antipollution code and the possible creation of a Mediterranean environment fund. Speakers at the conference pointed out the dangers that had resulted from the 20-year European industrial boom. To the raw sewage routinely dumped into the Mediterranean had been added vast amounts of industrial waste that threatened to turn it into a dead sea by the year 2000.

The energy crisis. What some observers saw as an inherent conflict between the need to preserve the environment and the increasing demand for energy became more acute during the year. Symbolic of the dispute was the proposed trans-Alaska pipeline, which would bring oil from the North Slope fields to the ice-free port of Valdez. The project had been delayed for several years by conservationists who warned that it might severely damage or destroy the delicate Arctic ecological system. In 1972 the U.S. Department of the Interior had finally produced a thorough impact statement, as required by the National Environmental Protection Act, and then gave permission for construction of the pipeline on the ground that the country's need for fuel superseded the undoubted environmental dangers.

Construction was further delayed, however, when a federal appeals court ruled that Interior could not issue permits for the 146-ft right-of-way needed to build the pipeline, since existing legislation limited the right-of-way for such projects to 54 ft. In April 1973 the U.S. Supreme Court refused to review the decision. This appeared to put the matter back before Congress, where several members were known to favor a pipeline through western Canada to the Middle West. The dispute was

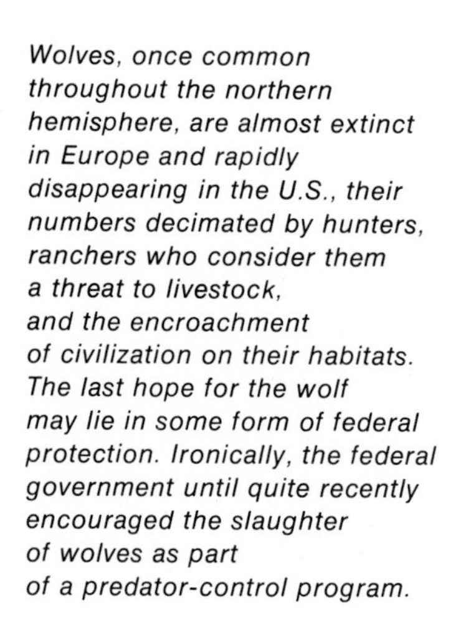

Wolves, once common throughout the northern hemisphere, are almost extinct in Europe and rapidly disappearing in the U.S., their numbers decimated by hunters, ranchers who consider them a threat to livestock, and the encroachment of civilization on their habitats. The last hope for the wolf may lie in some form of federal protection. Ironically, the federal government until quite recently encouraged the slaughter of wolves as part of a predator-control program.

Russ Kinne from Photo Researchers, Inc.

exacerbated by threats of a possible gasoline shortage in the U.S. during the summer. Proponents of the pipeline claimed that it would alleviate the shortage, while opponents charged that the shortage had been "manufactured" by the oil companies, partly to force construction. (*See* Year in Review: FUEL AND POWER.)

Whatever the causes of the gasoline shortage, it appeared certain that new sources of power would have to be found within the next decade. For some years the most feasible new source of clean and plentiful energy had seemed to be nuclear power, but during the year it too came under increasing attack. The dispute centered on the nightmare possibility of a "catastrophic accident," a failure within a plant that could release quantities of radioactive material and possibly wipe out the surrounding area. In February 1973 the EPA charged that the Atomic Energy Commission's environmental impact statement on emergency safety precautions in nuclear power plants was "inadequate." The AEC itself was engaged in a comprehensive review of its regulations governing emergency cooling of the nuclear core to prevent the release of radioactive materials in case there was an accidental loss of power plant cooling water. The power companies steadfastly insisted that their emergency precautions were adequate. Nevertheless, consumer advocate Ralph Nader and an environmental organization known as Friends of the Earth brought suit to close 20 power stations operating under AEC license, charging the AEC with a "gross breach" of its obligations to public health and safety. The 20 plants represented the bulk of the country's nuclear generating capacity.

Solid waste disposal. A report entitled "Cities and the Nation's Disposal Crisis," prepared by the National League of Cities and the United States Conference of Mayors and released in June 1973, reiterated the warning that the nation's cities were being inundated by garbage. The report estimated that by the late 1970s more than 46% of U.S. cities would have no place left to dispose of their trash.

The study estimated that each year cities were faced with the problem of disposing of 250 million tons of solid waste, including 28 billion bottles, 48 billion cans, 4 million tons of plastic, 30 million tons of paper, 100 million tires, and more than 3 million old automobiles. One of the principal culprits was packaging, some 90% of which was discarded. Between 16 and 24% of the nation's trash was reusable, including an estimated $5 billion worth of metals, but in actual practice only 1% was recycled.

The report put much of the blame on federal policies, such as depletion allowances and more favorable freight rates for virgin materials than for scrap, that discriminated against recycling. At least two earlier reports agreed with this finding. In February a committee of tax experts, reporting to the Council on Environmental Quality, urged the establishment of tax advantages for reuse. In March the Solid Waste Council of the U.S. paper industry found that the proportion of the industry's raw material that was recycled had actually fallen from 35% in 1945 to 21% in 1972, and added that, without tax incentives, it would probably decline to 17%. Approximately 90% of the cost of recycling paper went for sorting.

"Cities and the Nation's Disposal Crisis" pointed up other federal policies that it claimed were impeding solid waste disposal. Federal policies aimed at reducing air and water pollution, including prohibitions against incineration in major cities, had compounded the disposal problem. Further, although solid waste was a national problem, it had been left almost entirely in the hands of local authorities. Among other policies, the report called for some federal control of the packaging industry, for policies that would promote recycling and experimentation in disposal methods, and for support of cooperative disposal problems through the removal of interstate and intrastate barriers to the transportation of trash.

—H. L. Edlin; Richard Saltonstall, Jr.; David L. Smith

See also Feature Article: THE RISING DECIBEL LEVEL.

Foods and nutrition

Advances in food and nutritional science during the year ranged over a wide front, but the balance between food supplies and population remained precarious. A number of countries faced record high food prices and food shortages. Severe health problems resulting from lack of proper foods and, especially in less developed countries, outright hunger and even starvation were not uncommon.

There appeared to be no easy answers. Every day of the year, the world's 3,700,000,000 people consumed at least 4,000,000,000 lb of foods of plant and animal origin (dry basis; at least 8,000,000,000 lb on a fresh basis). Providing this food took up the greater part of the time and energy of the world's labor forces. Whether or not a country had ample food supplies to feed its population was a major dividing line between "have" and "have-not" nations.

The severity of the world's food problem was pointed up by Henry Labouisse, executive director of the United Nations Children's Fund (UNICEF), who stated that UNICEF's total budget in 1972 was

Central Soya Chemurgy Division

The foods shown at right were made from recipes that included Promosoy SL granular soy protein concentrate, a textured vegetable protein product high in nutrients, including all the essential amino acids. Textured vegetable proteins, approved for use in the federal school lunch program by the USDA, provided a low-cost method of enriching meat, seafood, and grain products and snack foods.

only $64 million for aid to 111 low-income countries "where some 780 million children under sixteen years of age live, and where poverty and its inevitable companions—hunger, sickness and ignorance—are all too prevalent." This amounted to less than ten cents per child per year.

In spite of severe difficulties, governmental agencies of most countries had developed ongoing programs aimed at providing food for the economically deprived, the handicapped, schoolchildren, and the elderly. In the U.S., for instance, in March 1973 about 25 million schoolchildren (out of a total of over 50 million) participated in federal school lunch programs. Of these about 8.7 million were receiving free or reduced-price lunches with the help of federal funds. Also in March 1973, 15.2 million persons—about 7.5% of the total U.S. population—were participating in family food assistance programs.

Food prices. Dominating the food scene in most countries were record high food prices, resulting from increased demand and smaller supplies. In the first quarter of 1973, for instance, food prices in the U.S. rose nearly 10%, and economists forecast an increase of 14 to 20% for the year as a whole. The price of soybean meal, used primarily for animal feed, rose from a low of around $96 per ton to over $250 in early 1973. This, plus large increases in the price of corn and other animal feeds, pushed up the retail price of red meat, poultry, and eggs.

As a result, the food bill for the 210 million people in the U.S. rose from about $125 billion in 1972 to an estimated $145 billion, or approximately $700 per person per year. This all-time record figure included soft drinks, candy, alcoholic beverages, and foods eaten away from home and purchased from vending machines.

Food labeling. The need to stretch the food dollar, as well as the prevalence in the marketplace of many new types of substitute, imitation, and "engineered" foods, led to a new emphasis on labeling. Many countries passed stronger regulations in this field. In the U.S. new regulations would provide the consumer with much more information about the nutritional composition of food products than had previously been available.

Even before the new regulations went into effect, many U.S. food companies were voluntarily following them. The new labels listed the number of servings per container, serving size, calories, protein and fat content, and provided information about the content of several vitamins and minerals. The term Minimum Daily Requirement (MDR) was to be eliminated from labels of food and vitamin products. Instead, nutrients were to be itemized under the U.S. Food and Drug Administration's new term, U.S. Recommended Daily Allowance (RDA).

The U.S. RDAs were based on the 1968 recommendations of the Food and Nutrition Board of the National Academy of Sciences. (The board's recommendations were to be revised in late 1973, and this would produce some complications.) The U.S. RDA listing had only three categories of age and sex groups, and included among the optional nutrients four newly added ones not on the 1968 RDA list: zinc, copper, biotin, and pantothenic acid.

The U.S. RDA values were not minimum requirements, but generous estimates intended to cover needs of individuals of different genetic backgrounds and different ages and levels of activity. Higher nutrient levels would be needed only in times of severe stress, illness, or convalescence. (The value of massive doses of nutrients, such as vitamin C, in the routine prevention of disease was considered speculative by most authorities.)

With the aid of nutrition educators, many food companies, schools, and public health departments were developing programs to help the consumer make the best possible use of the new labels. Although many new engineered foods would contain high levels of some RDA nutrients, the consumer would still have to depend on "natural" or traditional foods in order to obtain optimal amounts of all the essential nutrients—several of which were not on the RDA list.

Other new U.S. food labeling regulations included: (1) listing of the percentage of the major ingredient in a mixed food product, such as the level of beef in beef stew, fish in seafood cocktail, or orange juice in an orange drink; (2) listing of fatty acids, total fat, and cholesterol in certain foods; (3) a procedure for establishing new values for "protein quality"; (4) prohibitions of certain questionable nutritional claims; and (5) new regulations for listing all ingredients, flavors, and additives.

Products and handling. Food scientists and technologists throughout the world continued to search for ways to improve the quantity and quality of foods and, at least in more advanced countries, to increase food sales. These included the use of new plant strains, new methods of processing, new flavors, and better packaging and selling techniques.

One of the major technological advances of the year was the development of foods made with textured vegetable proteins. These low-cost products, usually derived from soybeans, were made by cooking and extruding, or forcing, the wet protein food through small openings. Added to other foods, they gave an acceptable texture to such products as protein-enriched macaroni, imitation meat products, and snack foods. Their great potential lay in the fact that they provided a cheap method for adding protein and minerals to meats and seafoods, common grains, bread and breakfast foods, snacks (such as french fried potatoes, chips, and cookies), soups, and many other items.

The availability of textured vegetable proteins, along with the rise in meat prices, led to the introduction of a number of substitute meat products. Hamburger with up to 30% soybean protein was widely used in U.S. school food service programs and was promoted in many retail stores. About 52 million lb (2,500 tons) of textured vegetable protein were used in the U.S. in 1972. This was less than 0.5% of the total red meat consumed, but the amount was increasing rapidly.

In the U.S. at least one milling company (Pillsbury) constructed a multimillion-dollar plant to produce high-protein wheat gluten products, similar to textured soybean products, for use in making engineered foods. Other companies were experimenting with protein from sunflowers, cottonseed, peanuts, corn, sesame, fish, bacteria, algae, animal wastes, green leaves, and other possible sources.

Hundreds of patents were issued on new food processes and new foods during the year. Those reaching the marketplace included new types of processed meat, compressed foods, egg substi-

Label meeting the FDA nutritional labeling standards lists the percentage of the U.S. Recommended Daily Allowance for ten nutrients contained in one cup of the product.

Courtesy, Del Monte Corporation

tutes, breakfast cereals, dairy products, infant foods, cereal and baked products, dietetic and health foods, whipping agents, frozen and dehydrated foods, fruit drinks, and convenience foods of all types. Among food scientists in universities, industry, and the government, recurring topics of interest were the effect of processing on the nutritive value of foods, physical and chemical properties of plant and animal foods (especially carbohydrates, fats, and proteins), protein and starch isolates, enzymes in food systems, food emulsifiers, microbiology of food products, biochemistry and physiology of fruits and vegetables, chemistry of flavors and food pigments, food waste management, food analysis, new methods of food processing and quality control, natural toxins in food, and flavor and sensory evaluation.

An important new international system of identifying individual items of food processors, known as the Universal Produce Code, was developed by the Grocery Manufacturers of America and six other trade associations. It was predicted that half of all food products would carry the code by mid-1974. The code, which used ten digits, would allow computerized handling of food packages from manufacturer through distribution channels, including supermarket registers with fully automatic scanners. It was expected to result in 65% faster checkout and inventory.

International organizations representing 92 UN member countries completed 67 recommended "Codex food standards." The first proposed standards, for edible oils, frozen peas, sweeteners, and canned sweet corn, were released in early 1973 for public comment. When adopted, the standards would apply to imports, exports, and domestic food products.

What was described as "the most advanced food system yet used for space flight" was developed, with the support of the U.S. National Aeronautics and Space Administration, for the orbital Skylab space missions. Designed to provide specific nutrient requirements under highly defined conditions, it included a wide variety of foods, served in open vessels and eaten with conventional knives and forks. Besides the usual dehydrated foods used on earlier space flights, frozen, heat-stabilized, and "intermediate moisture foods" (high in moisture but microbiologically safe) were used. It was hoped that this program would be of direct value in demonstrating how to provide nutrient needs to patients with special requirements.

Continuing studies were being made on the possible toxicity of chemicals in foods, whether naturally present or added artificially. It was estimated that safety testing of any proposed additive cost at least $250,000 and took from two to five years. Violet No. 1, a synthetic pigment used in many foods, was removed from the list of approved colors. The synthetic hormone diethylstilbesterol, used in fattening cattle, was also removed from approved lists during 1973. Derivatives of nitrates were being examined carefully for possible toxicity. Saccharin, the most widely used artificial sweetener, came under attack as the result of toxicity studies with rats, and many food scientists were concentrating on the development of new artificial sweeteners to replace it.

Advances in nutrition. Outstanding advances in the field of essential trace elements were made during the year. University of Wisconsin biochemists found that selenium is an essential constituent of the enzyme glutathione peroxidase, a vital cell component necessary for the prevention of oxidative damage to body cells. The discovery that fluorine, long known to be important in the prevention of tooth decay, was needed for growth of rats and reproduction of mice confirmed the essential nature of this element in the diet. It was hoped that these findings would allay the fears of those who objected to artificial fluoridation of public water supplies. Additional studies on the essential roles of silicon, vanadium, tin, and nickel in the metabolism of animal tissues were also made. Silicon was found to be involved in the formation of cartilage tissue and possibly in the aging of tissues. Nickel deficiency was found to affect hormone metabolism, body weight, cholesterol levels of blood, and red-cell levels in animals.

Other studies appeared to provide experimental proof that good nutrition during its mother's pregnancy and its first few months of life is vital to proper mental and physical development of a child.

Massachusetts Institute of Technology scientists showed that a marginal vitamin B_{12} intake during gestation in the rat had long-term effects on the offspring, providing an experimental model for studies of humans. Among other major developments were continued basic studies on the relationship of nutrition to heart and circulatory diseases, obesity, mental diseases, diabetes and alcoholism, infectious diseases and cancer, metabolism of bones and teeth, toxicity of drugs and oral contraceptives, and hormone metabolism. Special emphasis was given to new studies on vitamin D, calcium, magnesium, phosphorus, potassium, vitamin E, vitamin C, vitamin B_{12}, vitamin B_6, folic acid, zinc, iron, amino acids, fats, and carbohydrates.

There was a continued upsurge in nutrition education activities throughout the world. Meetings of international experts were held in Mexico, Peru, Hawaii, Washington (D.C.), Rome, and many other centers. The U.S. Department of Agriculture continued to support the largest U.S. program in community nutrition education, the "Expanded Federal Nutrition Education Program," with an annual budget of $50 million. At the same time, however, U.S. nutritionists were expressing concern over the announced ending of federal support of graduate training programs in nutrition science and the reduction of funds for support of research.

The dietetic and health food industry continued to grow at a rapid rate. Unfortunately, some popular health food magazines and books on weight control continued to spread misinformation along with accurate material. During the year many large supermarkets established "dietetic" or "natural" food sections offering a wide variety of foods usually associated with health food stores.

—George M. Briggs

Bill Gillette, courtesy, EPA Documerica

Strip mining for coal in Colorado leaves characteristic ridge-and-furrow pattern on the earth's surface. In 1971 production from such mines surpassed that from underground sources for the first time.

Fuel and power

Beginning in late 1972 and continuing into 1973 the American general public was awakened to the "energy crisis." And it was a rude awakening indeed for those in many parts of the country who suffered heating oil shortages, motorists who experienced local gasoline and diesel fuel outages, airports that ran out of jet fuels, industrial concerns whose normal natural gas supplies were interrupted or curtailed, and the many others affected by local fuel and power problems. The term "energy crisis" became a common household phrase during this period as the news media reported on fuel and energy supply difficulties with increasing frequency.

The failure of fuel supplies was for most Americans a new and uncommon experience, having occurred previously only during wartime emergencies. Throughout its history the United States enjoyed a large and diverse abundance of fuel and energy resources and the continuous availability of reliable fuel supplies at reasonable costs was largely taken for granted by most Americans. As a result, the U.S. greatly outpaced the world in energy consumption. As of 1973, the U.S. used about one-third of the world's commercial fuels and power, although it comprised only 6% of its population.

The energy crisis could be traced to the following: a huge and growing appetite for fuels and energy; an inadequate rate of development of new or improved fuels and power technologies; new environmental requirements and objectives that constrained the exploration, development, processing, transportation, distribution, and consumption of energy resources; imbalances in fuel demands caused by the regulation of natural gas prices, oil import controls, and other economic factors; overconfidence in the anticipated rate of development of nuclear energy; an inadequate research and development effort to develop clean synthetic fuels and solar, geothermal, and other nonconventional energy sources; and institutional deficiencies that precluded the most effective and coordinated energy policy and program planning, execution, and implementation. Singly or in combination, these factors inhibited an adequate rate of development of domestic fuel and power resources. As a result, the fuel and power situation rather abruptly changed from one that was primarily concerned with the management of an abundant oversupply of fuels to one of dealing with undersupply and scarcity.

The fuel and power problems about which Americans became so concerned in 1973 were not new to most of the other highly industrialized nations of the world. Western European countries, Japan, and others had coped with scarce supplies and a dependence on the less developed oil-exporting countries for many years. More than 90% of Japan's energy is supplied by petroleum imports, while Western Europe depends on such imports for about two-thirds of its energy. Together, these nations in 1973 were consuming only about three-fourths the energy consumed in the U.S., but their energy consumption growth rate was greater and their total consumption was expected to equal that of the U.S. in the early 1980s and exceed it thereafter. Their dependence on foreign oil sources was not likely to lessen within the next 10 to 15 years even though there were large and important oil and gas discoveries in the North Sea and other areas. The contribution of these new supply sources was expected to be offset by the expected increases in demands.

In the U.S., fuel and power problems continued to receive top priority attention by political and industrial leaders. U.S. Pres. Richard Nixon directed a number of his top advisers to develop a new series of comprehensive fuel and energy programs.

Among the more significant developments affecting fuel and power during the period under review were the following: (1) world petroleum and natural gas demands returned to normal growth rates and attained record new highs; (2) the balance of economic power in international petroleum trade shifted from the "buyers" to the "sellers," and the major oil-producing and exporting countries extended their control over oil concessionaires through "participation" and other agreements; (3) several countries instituted oil production and export limitations; (4) Canada announced a change in oil export policy that would limit future exports to the U.S.; (5) court-ordered delays continued to prevent construction of the trans-Alaskan oil pipeline and development of the North Slope oil and gas deposits; (6) U.S. oil and gas lease sales in the outer continental shelf were resumed following environmental delays; (7) the U.S. experienced further natural gas shortages, and curtailment orders were invoked in order to allocate available supplies to high-priority uses; (8) U.S. coal production rebounded during 1972 to increase by 7% over the strike-lowered level of 1971; (9) research and development efforts to produce clean synthetic fuels from coal were greatly expanded; (10) world electricity consumption continued to grow at very rapid rates; and (11) the first nuclear fast-breeder reactor demonstration plant in the U.S. moved a step closer to reality.

Petroleum and natural gas

World petroleum and natural gas demands during 1972 and early 1973 returned to near normal growth rates, rebounding from the relatively low rate of the previous year. The U.S. and other nations experienced supply shortages of certain petroleum products as the industry was hard pressed in meeting demands; it was subjected to greater political and economic controls by the large oil-producing countries with exportable surpluses and was faced with even greater environmental and other problems in its efforts to develop resources. In the U.S., curtailment orders of natural gas affected nearly one trillion cubic feet, or about 5% of all interstate pipeline natural gas sales, as demands greatly exceeded supply and environmental considerations continued to postpone development of Alaska's North Slope. In April President Nixon announced that mandatory quotas on U.S. oil imports would end on May 1.

Worldwide petroleum demand in 1972 averaged about 50.6 million bbl daily, or about 5.5% more than that of 1971. The U.S. continued in 1972 as the leading world consumer of petroleum products, increasing its consumption by 7.3% over that of 1971 to average 16.3 million bbl per day.

U.S. consumers in the upper Midwest and other localities experienced shortages in motor gasoline and middle distillate (home heating oils, jet fuels, diesel fuel, and kerosenes). This situation resulted from a number of contributing factors. The weather in those areas was colder than normal and arrived early, creating an abnormal demand. The limited availability of domestic crude oil supply and inadequate refining capacity to meet the surge in middle distillate demands created supply, transport, and marketing difficulties.

Demand for gasolines accelerated as environmental requirements for no-lead gasoline and the less efficient antipollution-equipped automobile engines increased consumption. Distillate fuel demand increased because it was used more to produce an acceptable substitute for high-sulfur coal and fuel oils through direct use and blending, as a replacement for natural gas and liquefied petroleum gases, and as a low-sulfur fuel for the growing number of turbines used for power generation. All of the distillates come from the same range of refinery products and one product cannot be increased without reducing the others. The government relaxed and in some cases removed petroleum product import quotas to relieve the supply problems; however, world distillate supplies were not as readily available as usual because of increased world demands for these same products and logistical problems in delivering them.

On the international oil scene, the balance of economic power shifted from the buyers to the sellers. Middle Eastern and other countries with large oil resources expanded their control over their concessionaires. Negotiating through the Organization of Petroleum Exporting Countries (OPEC), Saudi Arabia, Kuwait, Qatar, and the United Arab Emirates obtained participation agreements that provided them as of Jan. 1, 1973, a 25% ownership of the oil companies operating in their territory and an escalation to 51% participation by 1982. Iraq nationalized most of the properties of the Iraq Petroleum Company (IPC), the long-time oil concessionaire there. On Feb. 28, 1973, the government of Iraq announced that it had negotiated an agreement with IPC over its nationalization and reached a settlement of compensation claims. The accord reduced IPC operations to one small area in southeast Iraq and gave Iraq full control of the IPC areas that it had previously expropriated or nationalized.

Iran pressed demands for changes in its agreement with the operating consortium and announced its intention to refuse to renew the three five-year extensions of the pact when it expired in 1979. Bowing to this pressure, the consortium companies were reported to have reached a general understanding with the Iranian government in early March on a new agreement. The terms of the new accord would reduce the consortium's role to that of providing managerial services and would transfer essential control of the oil industry to the state-owned National Iranian Oil Co. In exchange, the consortium companies would be guaranteed a continuous supply of Iranian oil under preferential conditions for a period of 20 years.

Several oil-exporting countries instituted production controls and limited exports. Libyan production during 1972 dropped by 20% to 2.2 million bbl per day as a result of these controls. Venezuela established a production control and penalty system in the oil export tax-price schedule and set export levels for 1973 at about the same as 1972. Of particular significance to the U.S., the Canadian government announced in February that it was reassessing its oil export policy and would limit oil exports to the U.S. to levels that were clearly surplus to Canadian requirements. Until 1973 Canada had had excess producing capacity and its policy consequently was directed toward removing quota and other restraints on exports to the U.S., its sole export market. These exports had risen greatly in recent years to average almost one million barrels daily in 1972, an increase of more than 27% over 1971.

The hunt to discover and develop new oil sources continued throughout the world. The North Sea area continued as the most notable exploration area with approximately 20 new oil and gas discoveries reported during 1972. New finds were reported in Ecuador and it became an oil exporter for the first time in 1972. Other important discoveries and developments were made in Nigeria, Congo, and Peru.

The U.S. petroleum industry during 1972 made little progress in expanding domestic oil and gas resources, and in the face of rising demand the nation's dependence on foreign oil imports rose to nearly 29% of its supply. Domestic crude oil production in 1972 rose only fractionally to

"Soon shall thy arm, unconquer'd steam! afar
Drag the slow barge, or drive the rapid car."
Erasmus Darwin (1792)

Georg Gerster from Rapho Guillumette

Heliostats, mirrors moved by a clockwork mechanism so that sunlight is always steadily reflected in one direction, are arrayed at a solar furnace in Odeillo, France. The light rays are collected and concentrated onto the furnace itself to provide power for research activities there.

3,500,000,000 bbl and was at essentially maximum producing capacity. Development of the large known resources of the Alaskan North Slope was further postponed as an appeals court ruled that the U.S. Department of the Interior could not issue a pipeline right-of-way permit for a trans-Alaskan pipeline. The court held that the proposed right-of-way would be twice as wide as was allowable and that, therefore, the law precluded issuance of the permit. The U.S. Supreme Court refused to hear an appeal, and new Congressional legislation was planned by the government.

Environmental objections that had delayed lease sales in the federal outer continental shelf region were met, and two oil and gas lease sales were held late in 1972. These sales were in areas considered highly likely to contain oil and gas as evidenced by the cash bonuses in excess of $2,180,000,000 that were received from the successful bidders. Exploratory drilling operations were begun on the newly awarded leases as soon as permits were issued. By March 1973, two major oil discoveries on this new acreage had been reported, and more were expected from the numerous wells being drilled.

Although precise figures were not available, the marketed production and consumption of natural gas throughout the world was an estimated 42 trillion cu ft in 1972. Nearly 55% of this total was accounted for by the U.S. The Soviet Union was a distant second at about 18%, even though its proved natural gas reserves of 636 trillion cu ft were the world's largest and comprised more than one-third of the total. These large Soviet gas reserves are located mainly in the remote Arctic area of western Siberia and are essentially undeveloped. To develop them would require massive capital investments and advanced technologies not now available in the Soviet Union. Joint ventures with the U.S. and Japan to accomplish this development were under consideration.

During 1972, the U.S. experienced natural gas supply shortages that limited consumption to an estimated 22.6 trillion cu ft, an increase of only 2.1% over that of 1971 as compared with the average annual rate of growth of 5.8% attained until 1971. During late 1972 and early 1973, the U.S. Federal Power Commission (FPC) issued curtailment orders to interstate pipeline companies affecting nearly one trillion cubic feet of natural gas sales. The FPC orders curtailed deliveries to large industrial consumers in order to assure adequate supplies to residential and commercial customers. These curtailments affected about 5% of all gas sales and were expected to increase in the next few years. Such controls to assure allocations to priority uses during times of shortages were expected to continue.

The U.S. government and private industry acted to relieve the natural gas shortages by increasing traditional domestic supplies and by developing supplemental gaseous fuel sources. In addition to leasing highly prospective oil and gas lands in the outer continental shelf and accelerating coal gasification research projects, the government further relaxed and finally ended wellhead pricing regulations in order to improve incentives for industry exploration and development. The development of supplemental gaseous fuel sources was continued through expansion of projects to expand the importation of liquefied natural gas (LNG) and the manufacture of synthetic natural gas (SNG) from various petroleum feedstocks. The FPC authorized two long-term projects for the importation of about 1,000,000,000 cu ft per day of LNG from Algeria. Plans were announced for the construction of a total of about 40 SNG plants that would convert liquid hydrocarbons to gaseous fuels.

LNG and SNG projects are quite complex and require large capital investments in plants and transportation facilities. Therefore, the cost of these fuels is about two to three times greater than the current wholesale price of natural gas delivered to such large markets as New York City or Boston.

Coal

Despite changing fuel demands and environmental constraints on its production and use, coal continued as a leading primary fuel source. During 1972, world consumption of bituminous and subbituminous coal was estimated at nearly 3,200,000,000 short tons, equivalent to more than one-fourth of the world's gross energy consumption. Europe remained the principal coal producer on a continental basis, accounting for almost 58% of world output. North American and Asian countries each accounted for about 18%.

U.S. bituminous coal and lignite output rebounded to an estimated 590 million tons in 1972, an increase of 6.8% over that of 1971. This compared to average annual growth rates of only about 3% during the preceding ten years. The increase was largely the result of a rebuilding of stocks, which had been depleted during the work stoppage that closed most coal mines for six weeks late in 1971. Consumption of coal in the U.S. during 1972 rose 3.3% to 511 million tons. About 66% of this, or approximately 338 million tons, was consumed by utilities for the generation of electricity.

Air quality standards, sulfur regulations, and other environmental requirements continued to have direct and indirect impacts on coal. Many power stations were converted from coal to oil to comply with environmental regulations, which were expected to become more restrictive as state implementation plans of the Environmental Protection Agency become fully effective by 1975. Current coal consumption by power stations was being sustained largely because many existing plants were not yet subject to the state restrictions. Other factors directly affecting the coal industry during 1972 were a decline in overall coal producing capacity and an increasing use of low-sulfur coals.

Considerable progress in research and development of processes to produce clean, low-sulfur synthetic gaseous and liquid fuels from coal was reported. The U.S. Department of the Interior, through its Office of Coal Research and Bureau of Mines, implemented a greatly expanded joint industry-government coal research program in 1972. Four new pilot plants to develop processes for the production of high-quality, nonpolluting natural gas from coal were in operation or under construction, and a solvent-refined coal liquefaction testing facility was being built. These efforts were expected to lead to the development of technologies that would permit demonstration prototype plants to be built by 1976 and the first commercial plants to be operational in the early 1980s. These technologies would convert the vast coal resources of the U.S. into environmentally acceptable clean fuels.

Coal mining methods shifted steadily in recent years from underground to surface mining, the latter utilizing both strip and auger techniques. In 1971, the number of surface mines and their total production exceeded that of underground mines for the first time in the history of the industry. Surface mining is the most economical, safe, and productive method of coal recovery and the only presently feasible means of mining large deposits to produce the volumes needed for power generation or coal conversion plants.

Unfortunately, strip mining greatly disturbs the earth's surface and has devastated much land whenever proper reclamation and restoration practices have not been applied. This resulted in proposals that would ban strip mining altogether. Because of the continuing need for coal such action did not seem feasible. Instead, state and federal legislation and more strict rules to assure satisfactory restoration of mined lands and protection of environmental values were expected to be adopted.

Electric power

Despite generating-capacity growing pains and greater difficulties in meeting peak electricity demands, world consumption of electrical energy continued at exponential rates of growth to reach levels well in excess of five trillion kilowatt-hours in 1972. Electricity demands at these high growth rates require a doubling of generating capacity and output about every ten years. Electric power produced by utilities in the U.S. reached an estimated 1.7 trillion kw-hr in 1972, an increase of 7.1% over that of 1971.

The vast majority of electricity was generated in steam-electric plants using the fossil fuels (coal, oil, and natural gas) as the primary heat source. Nevertheless, most utilities looked to nuclear energy to supply the greatest share of thermal heat to meet future generating capacity. Nuclear power plants in operation in the U.S. at the end of 1972 had a capacity of about 14,700 Mw, only 3.8% of the nation's total generating capacity. Non-Communist nations outside the U.S. had an estimated nuclear generating capacity of about 14,100 Mw, less than 0.5% of their total capacity.

Despite the relatively insignificant current contribution of nuclear power, industry observers expected it to become the predominant source of electric power by the end of the century. In the U.S., plants with an aggregate capacity of 130,000 Mw were being built or planned, with orders already placed for reactors. By 1985, operable nuclear capacity in the U.S. was expected to reach 215,000 Mw and account for nearly 25% of the nation's total generating capacity.

The capacity of nuclear power plants in the European Economic Community was expected to reach 100,000 Mw by 1985 and account for one-third of all electricity output in that group of countries. An additional 160,000 Mw were expected to be in operation by 1985 in other non-Communist nations.

The first fast-breeder reactor demonstration plant in the U.S. moved a step closer to reality. The fast breeder is a technically advanced reactor that is expected to produce more fissionable fuel than it consumes. In November 1972, the U.S. Atomic Energy Commission (AEC) announced that the Westinghouse Electric Corp. had been awarded a contract for construction of the fast-breeder plant at a location near Oak Ridge, Tenn. The plant was estimated to cost about $700 million, of which $240 million would be paid for by the utility industry. The remainder would be financed by the AEC. The fast breeder was expected to become operational in 1980 after which it would serve as a prototype for commercial plants that hopefully would be in operation by 1986.

Even assuming that nuclear power can overcome the technological, economic, and environmental problems that have prevented it from achieving its expected level of development to date, the world will continue to rely for its power on fossil-fueled plants, and to a lesser extent on hydroelectric plants, to about the year 2000. By that time, nuclear fission will become dominant. Also, small but important contributions from nuclear fusion, geothermal, and solar energy sources are considered probable. Technological breakthroughs that would bring about economic development of these essentially inexhaustible energy sources would prove a boon to all mankind. Research efforts in all these areas are being expanded. (See *1973 Britannica Yearbook of Science and the Future* Feature Article: NUCLEAR FUSION: POWER SOURCE OF THE FUTURE?)

Future prospects

The energy crisis in the U.S. will continue as a major concern of the general public. Petroleum and natural gas shortages and electric power outages are expected to occur in local areas with in-

Collecting station of the Krestishchenskoye gas field in the Kharkov region of the Ukrainian S.S.R. helps furnish cities in the Ukraine with 11 million cu m of natural gas per day.

Novosti from Sovfoto

creasing frequency and severity. The new initiatives and accelerated programs to be instituted by President Nixon and the fuel and power industry are expected to help alleviate the immediate problems but will only be stopgap measures until domestic resource development, new technological solutions, and other long-term measures can become fully effective.

The following specific developments are considered among those most likely to occur in the coming years: (1) U.S. petroleum product shortages, especially for motor gasolines in the peak summer demand period, will continue; (2) demand for natural gas and supplemental gaseous fuels will increase in view of the growing requirement for clean, nonpolluting fuels, but the level of consumption will continue to be constrained by supply limitations; (3) relaxation of existing environmental standards and delays in the implementation of more strict antipollution regulations may be required in certain areas and during critical time periods to assure that energy shortages are averted or minimized; (4) new policies affecting strip mining, powerplant and other industrial plant siting, and other land-use planning and controls will become law; (5) world fuel supply problems will become more acute as fuel demands continue upward and confrontations between the exporting and the consuming nations become more frequent; and (6) the prices of fuels, especially petroleum, will rise sharply as competition for the available supplies becomes sharper.

—James A. West

Honors

The following major scientific honors were awarded during the period from July 1, 1972, through June 30, 1973.

Aeronautics and astronautics

Elliott Cresson Medal. The Franklin Institute of Philadelphia presents the annual Elliott Cresson Medal, established in 1848, for recognition of a notable discovery or original research adding to the sum of human knowledge, irrespective of commercial value. In 1972 a Cresson Gold Medal was given to William Powell Lear, chairman of the Lear Motors Corp. of Reno, Nev. An inventor, engineer, and industrialist, Lear was cited for his many significant contributions to radio and aviation, particularly for his development of the full-maneuvering automatic pilot (first marketed in 1949), and of the Lear jet, a popular business aircraft designed and developed in 1963.

Astronomy

Albert A. Michelson Medal. Established in 1968, the Franklin Institute each year presents the Albert A. Michelson Medal for outstanding scientific endeavor. Recipient of the 1972 Michelson Medal was Herbert Friedman, head of the atmosphere and astrophysics division of the U.S. Naval Research Laboratory. Recognized for outstanding and truly pioneering work in solar and X-ray astronomy, Friedman began his research in rocket astronomy in 1949 with his studies of the V-2 rocket. His subsequent work included the tracing of solar-cycle variations of X rays and ultraviolet radiations from the sun, obtaining the first X-ray and ultraviolet photographs of the sun, discovery of the hydrogen geocorona, and the measuring of ultraviolet fluxes of early-type stars.

Helen B. Warner Prize. The American Astronomical Society each year presents the Helen B. Warner Prize, established in 1952, in recognition of a significant contribution to astronomy made during the preceding five years by an astronomer under the age of 35. The 1973 award was presented to George R. Carruthers, a scientist at the U.S. Naval Research Laboratory, who was recognized for his work in ultraviolet astronomy, particularly the detection of absorption bands of interstellar molecular hydrogen. Carruthers was principal investigator for the Apollo 16's far-ultraviolet camera/spectrograph experiment covering the first observation of terrestrial upper atmosphere and geocorona from the lunar surface.

Vetlesen Prize. Administered by Columbia University, the $25,000 Vetlesen Prize was established in 1959 by the G. Unger Vetlesen Foundation for contributions toward understanding the origin and evolution of the earth and its place in the universe. The 1973 prize was given to William A. Fowler, physics professor at the California Institute of Technology, who was cited for his work in nuclear physics and its applications to astrophysics and geophysics. Fowler's laboratory research into the interactions of matter and energy made possible determinations of the processes by which stars evolve.

Biology

Louisa Gross Horwitz Prize. The College of Physicians and Surgeons of New York in 1967 established the Louisa Gross Horwitz Prize, to be administered by Columbia University and awarded annually, for outstanding basic research in biology or biochemistry. The 1972 award of $25,000 was presented to Stephen W. Kuffler, Robert Winthrop Professor at Harvard Medical School, who was cited for outstanding experiments that provided information of fundamental importance to the understanding of the nature of neuromuscular and synaptic transmission; the mechanisms responsible for inhibition in the nervous system; the functional organization of the retina and the visual system; and the role of neuroglia in the central nervous system.

Chemistry

AIC Gold Medal Award. The American Institute of Chemists presents annually its Gold Medal Award to recognize and stimulate activities of service to the science of chemistry or the profession of chemists or chemical engineers. The 1973 medal was given to Glenn T. Seaborg, associate director of the Lawrence Berkeley Laboratory of the University of California at Berkeley. Seaborg was cited for his key role in basic research as codiscoverer of several new elements of matter; in applied research in production of fissionable materials; in education as chancellor (1958–61) of the University of California, Berkeley; and in scientific administration and diplomacy as chairman (1961–71) of the U.S. Atomic Energy Commission.

Arthur C. Cope Award. Established by the American Chemical Society in 1972 under the terms of the will of Arthur C. Cope, renowned professor of Chemistry (1945–66) at the Massachusetts Institute of Technology, the prestigious new $40,000 Cope Award honors outstanding achievement in the field of organic chemistry, the significance of which has become apparent within five years preceding the award, and which represents new work not previously recognized through a major award. The first Arthur C. Cope Award was shared by Nobel laureate Robert B. Woodward of Harvard University and Roald Hoffman of Cornell University for their Woodward-Hoffman rules that enable organic chemists to predict correctly the feasibility and results of many experiments. Introduced in 1965, these rules of orbital symmetry, considered the most significant theoretical advance in organic chemistry in 30 years, had such impact that nearly 500 research papers published in 1971 alone made reference to this work of Woodward and Hoffman.

Beilby Medal. Representing the Royal Institute of Chemistry, the Society of Chemical Industry, and the Institute of Metals, the administrators of the Sir George Beilby Memorial Fund decided to make an award of a medal and £100 from the fund in 1972. The recipient of the 1972 Beilby Medal was Frank Pearson Lees, an Institution of Chemical Engineers industrial research fellow working on man-computer interaction in process control. Lees was given recognition for his work in chemical engineering with particular reference to studies of chemical process modeling, computer control, human operator control, and reliability engineering.

Davy Medal. The Royal Society of Great Britain awards the Davy Medal annually for the most important discovery in chemistry made in Europe or Anglo-America. The 1973 recipient of the medal was Arthur John Birch, professor of organic chemistry at the Australian National University, Canberra. Birch was honored for his distinguished biosynthetic studies of organic natural products and

Stanford Moore

William H. Stein

Courtesy, Rockefeller University

Courtesy, National Institutes of Health

Christian B. Anfinsen

for the development of new reagents for reduction processes. His pioneer work covered the use of metal-ammonia reductions in synthesis, and the production of total synthesis of the 19-nor steroid series to which belong oral contraceptives.

Franklin Medal. The Franklin Institute of Philadelphia presents its highest award, the Franklin Medal, annually for recognition of those workers in physical science or technology whose efforts have done most to advance a knowledge of physical science or its application. The winner of the 1972 award was George B. Kistiakowsky, emeritus Abbott and James Lawrence Professor at Harvard University, who was honored for a lifetime of pioneering research in organic chemistry. He was recognized especially for his precise measurements of important physical properties of unsaturated hydrocarbons, and for his development of techniques used in the release of nuclear energy.

Garvan Medal. The American Chemical Society each year presents the Garvan Medal, established in 1936 and consisting of a $2,000 honorarium and a gold medal, to a U.S. woman chemist selected for her distinguished service to chemistry. In 1973 the prize was presented to Mary L. Good, professor of chemistry at Louisiana State University in New Orleans, in recognition for her outstanding achievements as a chemist and educator. Good, author or coauthor of more than 50 published scientific papers, read approximately 35 other papers at symposiums both in the U.S. and abroad. She was instrumental in acquiring at least $250,000 in research grants and contracts while at Louisiana State.

Nobel Prize for Chemistry. The Royal Swedish Academy of Sciences chose three U.S. protein chemists for the 1972 Nobel Prize for Chemistry. Dividing the $100,000 award were Christian B. Anfinsen of the National Institutes of Health, and Stanford Moore and William H. Stein of Rockefeller University in New York City, who were cited for their fundamental contributions to enzyme chemistry. Working independently on the enzyme ribonuclease (RNAse), the three scientists developed the chemical structure and relationship between the structure and activity of the enzyme. In 1960 Stein and Moore completed their work on the analysis of the whole sequence of the 124 amino acid residues in RNAse, with Anfinsen contributing to the elucidation of the sequence and its relationship to the enzyme's biologically active conformation.

Perkin Medal. Presented annually by the American Section of the Society of Chemical Industry, the coveted Perkin Medal is awarded for outstanding work in applied chemistry in the U.S. Chosen for the 1973 award was Theodore L. Cairns, director of the DuPont central research department in Wilmington, Del., in recognition of his work on a series of organic compounds in which large numbers of cyano groups were attached to the carbon atoms. Cairns' pioneering efforts in synthesizing such chemicals led to the synthesis of tetracyanoethylene, the parent for a whole new structural class of chemicals he named cyanocarbons.

Priestley Medal. The American Chemical Society in 1922 established the annual Priestley Award, highest honor in U.S. chemistry, in recognition of distinguished services to chemistry. The 1973 recipient of the gold medal was Harold C. Urey, Nobel laureate, physical chemist and cosmochemist, and emeritus professor of chemistry at the University of California, San Diego. Urey was honored for his contributions of fundamental ideas and information to the present state of knowledge about the solar system and its history.

Roger Adams Award. Organic Syntheses, Inc., and Organic Reactions, Inc., in 1959 established the $10,000 Roger Adams Award, to be administered by the American Chemical Society and to be given every two years to recognize and encourage outstanding contributions to research in organic chemistry defined in the broadest sense. The 1973 recipient of the award was Georg Wittig of the Institute for Organic Chemistry, University of Heidelberg, W.Ger., who was cited for his exceptional contributions to man's knowledge of organic chemistry, particularly in the area of olefins, a broad class of compounds forming the basis of many commercial products such as fibers and detergents.

Earth sciences

George Davidson Medal. The American Geographical Society from time to time awards its George Davidson Medal for exceptional achievement in research or exploration in the Pacific Ocean or the lands bordering thereon. The 1973 award was received by F. Raymond Fosberg, botanist and special adviser on tropical biology at the U.S. National Museum of Natural History at the Smithsonian Institution. Fosberg, an authority on the ecology of coral atolls, was cited for valuable contributions in the field and especially to the literature, being the author of more than 400 papers ranging from a study of Polynesian grasses to a consideration of man as a dispersal agent.

Howard N. Potts Medal. The Franklin Institute awards annually the Howard N. Potts Medal for distinguished work in science or the arts. The 1972 recipient of the award was Jacques E. Piccard of Switzerland, oceanographic researcher and underwater vehicle designer, who was honored for his vision and skill in combining engineering principles for the development of oceanographic vehicles of greatly improved capability; for his meticulous attention to attaining safe, reliable operation; and for his personal accomplishments in operating those vehicles. Among his underwater vehicles was the submersible "Ben Franklin," designed in 1966, in which Piccard and his crew in 1969 studied the Gulf Stream drift, remaining submerged for 30 days while drifting 1,444 nautical miles at an average depth of 650 ft.

Massey Medal. The Royal Canadian Geographical Society annually presents the Massey Medal, established in 1959, for recognition of outstanding personal achievement in the exploration, development, or description of the geography of Canada. Winner of the 1973 award was Pierre Dansereau, professor at the University of Quebec at Montreal and director of the Montreal Centre for Ecological Research. Dansereau was cited for his major contributions in the fields of biogeography and ecology. One of his most recent studies covered the ecological aspects of the area of the new international airport for Montreal.

O. E. Meinzer Award. The Hydrogeology Division of the Geological Society of America presents annually the O. E. Meinzer Award, established in 1965, to an author or authors of an outstanding published paper of distinction advancing the science of hydrology or some related field. The 1972 award was made to Joseph F. Poland and George H. Davis for their paper "Land Subsidence Due to Withdrawal of Fluids." The paper was published by the Geological Society of America in *Reviews in Engineering Geology*, vol. II, pp. 187–269 (1969).

Medical sciences

Albert Lasker Medical Research Awards. The Albert and Mary Lasker Foundation each year presents a number of cash prizes in recognition of advances made in medical research. The 16 recipients of the 1972 awards of $2,000 were chosen solely for their research in cancer chemotherapy. Four were honored for the development of drugs for use in treating acute lymphatic leukemia: Emil Frei III of Harvard University (also cited for his work in Hodgkin's disease); Emil J. Freireich of the University of Texas in Houston (also cited for developing isolation techniques to protect cancer patients from infection); James F. Holland of Roswell Park Memorial Institute in Buffalo, N.Y.; and Donald

Rodney R. Porter

Courtesy, Oxford University

Courtesy, Rockefeller University

Gerald M. Edelman

Pinkel of St. Jude Children's Research Hospital, Memphis, Tenn.

For progress in chemotherapy for advanced Hodgkin's disease prizes were given to Paul Carbone and Vincent T. DeVita, Jr., both of the National Cancer Institute. Discovery of a chemical cure for gestational choriocarcinoma brought recognition to Min Chiu Li of Nassau Hospital in New York City and Roy Hertz of Rockefeller University, New York City.

Edmund Klein of Roswell Park Memorial Institute, and Eugene Van Scott of Temple University in Philadelphia were cited for work on certain types of skin cancer. Honored for contributions in the chemical treatment of Burkitt's lymphoma were four other scientists: Denis Burkitt of the Medical Research Council in London; Joseph H. Burchenal, Memorial Hospital for Cancer and Allied Diseases in New York City; John L. Ziegler of the National Cancer Institute; and V. Anomah Ngu of the Centre of Health Sciences in Yaoundé, Cameroon.

Isaac Djerassi, Mercy Catholic Medical Center at Darby, Pa., received an award for developing supportive treatment methods to counter side effects of chemotherapy. C. Gordon Zubrod, director of the division of cancer treatment at the National Cancer Institute, was awarded a special prize of $5,000 for leadership in creating "an effective national cancer chemotherapy program."

Kittay International Award. In 1973 the Kittay Scientific Foundation of New York established the $25,000 Kittay International Award, the world's largest prize in the field of mental health, designed to recognize an individual researcher in psychiatry whose work had practical clinical application. Recipient of the first annual Kittay award was Jean Piaget, Swiss child psychologist. Piaget, chosen for his analysis of the child's mental development, was also cited for pioneering work that opened new avenues for approaching normal and abnormal psychology in adults as well as children.

Nobel Prize for Physiology or Medicine. The Swedish Royal Caroline Medico-Chirurgical Institute awarded the 1972 Nobel Prize for Physiology or Medicine to Gerald M. Edelman of Rockefeller University in New York City and Rodney R. Porter of Oxford University. Sharing the $100,000 prize, the two scientists were honored for their independent achievements in the field of immunology. Their research on the chemical structure and function of antibodies revealed the structure and mode of action of these blood proteins, which defend the body against infection and disease. In 1959 Edelman demonstrated that antibody molecules consist of both light and heavy chemically separable, polypeptide chains; Porter showed that antibody molecules can be enzymatically cleaved into three fragments, two of which are identical. These discoveries paved the way for systematic and intensive antibody research throughout the world.

Stouffer Prize. Awarded by the Vernon Stouffer Foundation of Cleveland, O., the $50,000 Stouffer Prize is given in recognition of a scientist or scientists who made outstanding contributions to the understanding of hypertension and atherosclerosis. In 1972 the award was shared by four medical scientists: Vincent P. Dole of Rockefeller University, New York City; John W. Gofman of the University of California, Berkeley; Robert S. Gordon, Jr., of the National Institutes of Health, Bethesda, Md.; and John L. Oncley of the University of Mich-

Courtesy, Brown University

Leon Cooper

Wide World

John Bardeen

Courtesy, University of Pennsylvania

J. Robert Schrieffer

igan. The four recipients were cited for their research in determining how fat and cholesterol are transported in the blood—Gofman and Oncley for their pioneering studies of lipoproteins, Dole and Gordon for their discovery of the importance of the free fatty acids in the blood.

Physics

Enrico Fermi Award. The U.S. Atomic Energy Commission awards annually the $25,000 Enrico Fermi Award in recognition of outstanding scientific or technical achievement related to the development, use, or control of nuclear energy, and to stimulate creative work in the development and application of nuclear science. Recipient of the 1972 award was Manson Benedict, professor of nuclear engineering at the Massachusetts Institute of Technology, who was cited for pioneering leadership in the development of the nation's first gaseous diffusion plant, for imaginative contributions in the development of the nuclear reactor, and for the educating of nuclear engineers.

John Torrence Tate International Gold Medal. The American Institute of Physics presents periodically the John Torrence Tate International Gold Medal for distinguished service to physics on an international level. Selected for the 1972 gold medal was Gilberto Bernardini, director of the Scuola Normale Superiore in Pisa, Italy, for having contributed to collaboration and mutual understanding among physicists of many nations by his decisive and personal role in creation of the European Physical Society.

Nobel Prize for Physics. The Royal Swedish Academy of Sciences awarded the 1972 Nobel Prize for Physics jointly to John Bardeen of the University of Illinois, Leon N. Cooper of Brown University, Providence, R.I., and J. Robert Schrieffer of the University of Pennsylvania, in recognition of their basic contributions in the field of superconductivity. Between 1955 and 1957 the three U.S. scientists worked together in developing a microscopic theory of conductivity to explain why some materials become superconductive at supercold temperatures. This theory is described as the most important achievement since the quantum theory in advancing theoretical understanding of the universe.

Oliver E. Buckley Solid State Physics Prize. The American Physical Society each year awards the Oliver E. Buckley Solid State Physics Prize, established in 1952 by Bell Telephone Laboratories, for recognition and encouragement of outstanding theoretical and empirical contributions to solid-state physics. In 1973 the prize was given to Gen Shirane of the Brookhaven National Laboratory, Upton, N.Y., for his broad contributions to the understanding of structural phase transitions by means of inelastic neutron scattering.

Valdemar Poulsen Gold Medal. The Danish Academy of Technical Sciences from time to time presents the Valdemar Poulsen Gold Medal, established in 1939, for outstanding research in the area of radio techniques and related fields. In 1973 the tenth Poulsen Gold Medal was given to J. B. Gunn of the IBM Research Center in Yorktown Heights, N.Y. Gunn was recognized for his discovery of the Gunn effect, spontaneous oscillations of current that can occur in certain semiconductors when a steady voltage is applied to them, which provided the first solid-state source for microwave generation.

Science journalism

AIP-U.S. Steel Foundation Science Writing Award. The American Institute of Physics and the

United States Steel Foundation each year present jointly the AIP-U.S. Steel Foundation Science Writing Award to a journalist to stimulate distinguished writing and reporting in physics and astronomy. The 1973 award was made to Edward Edelson, science writer with the *New York Daily News,* who was cited for a series of articles that appeared in that paper April 10–14, 1972. The editors of the *Daily News* also received a certificate.

Bradford Washburn Award. Presented since 1964 by the trustees of Boston's Museum of Science, the Bradford Washburn Award is given for an outstanding contribution toward public understanding of science and appreciation of its fascination and the vital role it plays in our lives. Recipient of the 1972 gold medal and $5,000 honorarium was Walter S. Sullivan, science editor of the *New York Times,* who was cited as a journalist and author who "expands the mind, captures the imagination, and puts the often elusive facts of science firmly within our grasp."

Howard W. Blakeslee Award. The American Heart Association presents its annual Howard W. Blakeslee Award for the highest standards of reporting on diseases of the heart and circulatory system. Chosen for the 1972 award was Howard Sanders, senior associate editor of *Chemical and Engineering News,* for his articles "Artificial Organs—Total Artificial Hearts and Heart-Assist Devices," which were published in *Chemical and Engineering News* (April 5 and 12, 1971).

James T. Grady Award. The American Chemical Society each year presents the James T. Grady Award for Interpreting Chemistry for the Public to recognize, encourage, and stimulate outstanding reporting directly to the public, which materially increases the public's knowledge and understanding of chemistry, chemical engineering, and related fields. In 1973 the award was given to Orlando A. Battista, chemist, writer, and vice-president for science and technology at AVICON, Inc., Fort Worth, Tex. Battista was cited for his many magazine articles and books. His popular book for students, *The Challenge of Chemistry,* was published in 1959; a textbook, *Fundamentals of High Polymers* (1958) has been in continuous use in the Far East; his articles have appeared in *Reader's Digest* and other popular periodicals.

Miscellaneous

Nobel Prize for Economics. The Royal Swedish Academy of Sciences divided the $100,000 Nobel Prize for Economics in 1972 between Kenneth J. Arrow of Harvard University and Sir John R. Hicks of Oxford University for their pioneering contributions to the general economic equilibrium theory and welfare theory. The two economists' work applied to the problems of a higher standard of living and more balanced employment.

Westinghouse Science Talent Search. Science Service, through its Science Clubs of America, conducts the annual Westinghouse Science Talent Search in which high-school students compete for scholarships in the various fields of science. In 1973 a $10,000 four-year scholarship was won by Arvind Narain Srivastava, 16-year-old high-school senior from Fort Collins, Colo., for his project examining the possibility that the universe is finite. Second-place scholarships of $8,000 each went to Van Jay Wedeen of Midwood High School in Brooklyn, N.Y., and Joshua L. Rubin of the Bronx (N.Y.) High School of Science. Wedeen's work described the behavior of particles in a force field, and Rubin entered the contest in the field of mathematics.

Information science and technology

A prime goal of information science and technology is to investigate and implement ways of ensuring an efficient transfer of information from those that have developed new knowledge to those that have a need to know. In the past year a number of significant events occurred that should provide greater public accessibility to information.

Libraries as urban information centers. In order to provide better and more complete information services some library systems proposed to broaden their functions from simply loaning books and answering reference questions to becoming urban information centers that would provide residents with an information service about their community and where they might obtain help in their city, borough, district, and neighborhood. The library would inform people about what health services exist and would even make health clinic appointments when needed. Children would receive help with their homework from older student and adult volunteers. Senior citizens and handicapped and disabled persons would receive special attention.

A similar project was funded by the U.S. Office of Education. It was designed to create a consortium of five major public libraries in the cities of Atlanta, Cleveland, Detroit, Houston, and the borough of Queens in New York, with two neighborhood information centers in each area. This project might become a prototype for the future development of libraries into urban information centers.

New York Times information bank. One of the newest and most innovative information systems is the *New York Times* Information Bank. In the planning and design phases for six years, it consists of items selected from the millions of clippings from past issues of the *Times* and other newspapers and magazines. Specially trained personnel have been coding the index terms for each document into a magnetic storage system for computer processing. Simultaneously, the full texts of the documents were being filmed and stored on microfiche. When fully operational, the data bank will contain all news items and editorial matter from the *New York Times* plus selected material from more than 60 other newspapers and periodicals.

To retrieve this information, that is, to gather all news items on a given topic, the inquirer asks his questions by using a video terminal or a typewriter that can be connected to the *Times* computer as simply as dialing a telephone call. Once the terms have been verified, a computer search of the data bank will be initiated, and within a few seconds the abstracts of the documents in the system that meet the specifications will be displayed. After viewing the abstracts, the user will be able to request the full texts of the selected documents, and these will be sent to him on microfiche.

Science information systems. In 1972 the National Library of Medicine introduced MEDLINE. This information service enables the staff at medical schools, research institutes, and libraries to search for medical documents containing information on specific topics by means of a terminal connected to a large computer-based information storage and retrieval system. More recently, the board of regents of the National Library of Medicine approved recommendations to expand the network of biomedical communications and to develop it further in order to achieve an eventual goal of freeing the physician from total dependency on his own memory in clinical decision making and problem solving. This will be done by providing a computerized storage and retrieval system with more information and greater versatility.

The final achievement of this goal is far in the future, but meanwhile the information scientist is striving to improve medical practice by providing ever more efficient access to medical information. For example, the Stanford University Medical Center developed a computerized system to store, retrieve, and disseminate bone marrow examination reports. By means of this system, hematologists and pathologists in major medical centers can have quick access to pertinent patient information that is essential for the diagnosis, treatment, and research of blood disorders.

Another newly announced service is TOXICON. Sponsored by the National Library of Medicine, it is designed to respond effectively to the information needs of health professionals concerned with adverse drug reactions, poisoning cases, and

"Good morning, Frank. Hi Steve. How's it going, Mabel? You're looking good, Ed. . ."

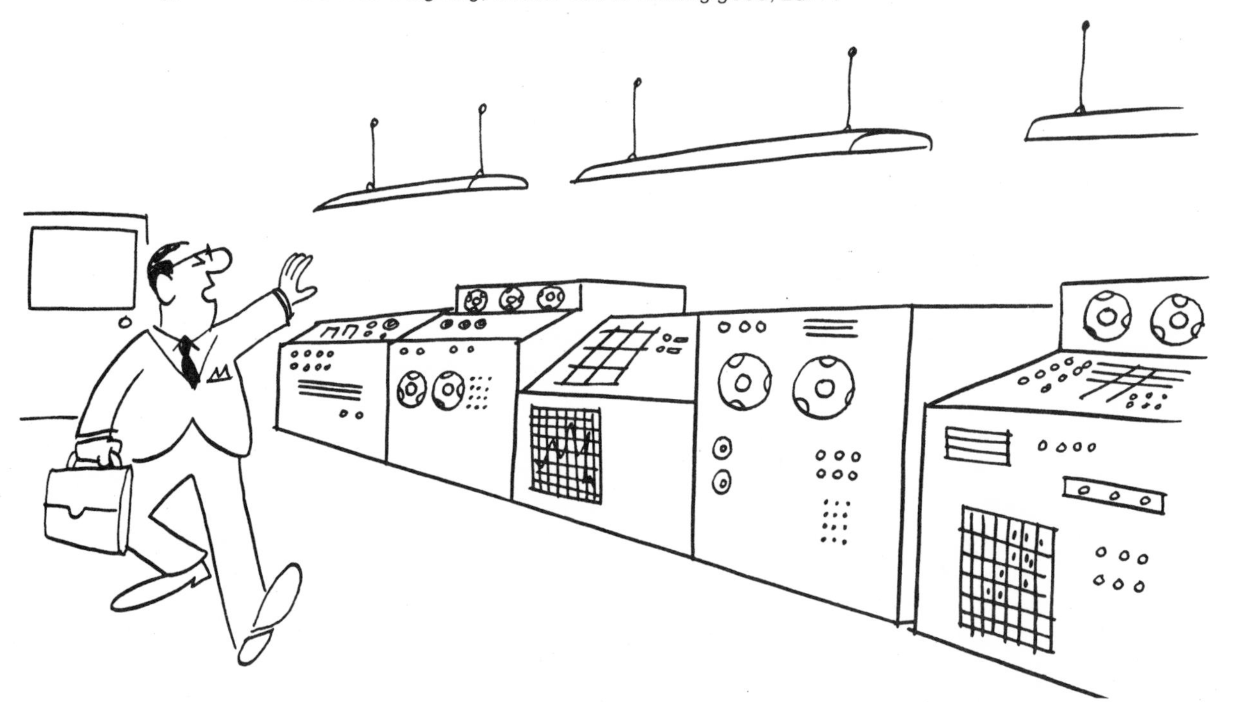

A. V. Johns, BULLETIN OF THE ATOMIC SCIENTISTS

effects of environmental chemicals and pollutants. The hub of the network is a computer that contains the TOXICON data base of more than 100,000 relevant citations and abstracts. This information can be accessed from a small typewriter-like terminal in various locations by means of telephone lines.

The two largest U.S. scientific abstracting and indexing services, the BioSciences Information Service of Biological Abstracts (BIOSIS) and the American Chemical Society's Chemical Abstracts Service (CAS), began cooperating in an extensive program to coordinate coverage and develop a degree of compatibility that will permit their services to be used effectively in combination. Coordination of indexing practices and the establishment of an effective means of interconnecting the indexes of the two services should eliminate much of the present duplication in the processing effort and thus assist both biologists and chemists in their search for information.

The American Institute of Physics also was developing new concepts in processing science information. The Current Physics Information Program aimed to improve the accessibility of physics literature by making it available in three different formats—printed journals, microfilm, and on computer-readable magnetic tape.

Psychologists also have an information problem, and in recent years they have had a computerized National Information System for Psychology to help them make more effective use of the information in their field. This sophisticated system consists of a comprehensive, well-indexed, and machine-accessible data base of primary journal publications and secondary abstract literature. When this data base is searched, the computer will print out relevant bibliographic citations and abstracts which are then sent to the person who requested them.

FBI National Crime Information Center. Scientists and businessmen are not the only ones who need fast access to current and accurate information. The FBI has a computerized system that links 104 state and local law enforcement agencies throughout the U.S. by means of a National Crime Information Center (NCIC). The center has been in operation since 1967, and by 1973 it stored more than three million records in its computerized files. These records are of wanted persons, stolen vehicles and license plates, and other identifiable stolen property. More than 75,000 file transactions are processed in an average day, and about 700 positive identifications result. This system is now being expanded to contain a Computerized Criminal History (CCH) File, which will allow for a more efficient coordination and exchange of criminal history information between state and federal agencies. The data in this file include fingerprint identifications, other information concerning the arrest of the offenders, and subsequent actions taken within the framework of the criminal justice system. The program includes safeguards to ensure the security and confidentiality of the system.

Ottawa, Ont., established a somewhat similar computerized Canadian Police Information Center with a network of terminals throughout Ontario and eventually to the other provinces. This network enables policemen instantaneously to receive information concerning stolen vehicles and wanted and missing persons. The system can be expanded to handle stolen property data as well.

International information systems. A particularly good example of the advantages accruing from information systems is a new service offered in the United Kingdom by the Scientific Documentation Centre. Its purpose is to help agricultural researchers and librarians keep abreast of the voluminous and specialized information in their field of interest. The service is called AG/PACK—*AG*ricultural *P*ersonal *A*lerting *C*ard *K*its. Each week a subscriber receives a set of index cards, each of which contains a reference to a recent article or report that has appeared in the international literature on agriculture or agricultural research. To prepare these cards, British university science graduates scan and index more than 7,000 items weekly from more than 3,000 periodicals and monographs published throughout the world. The subscriber prepares a profile of his specialized interests; that is, he describes his own interests by means of a set of standardized index terms. The cards he receives are those in which the index terms assigned to the article match the terms in his profile.

Some problems cannot be solved by one country alone. An interesting example of the utilization of modern information technology to solve complex, long-standing problems is seen in the work of the Worldwide Geometric Satellite Triangulation Program. The goal is to obtain precise measurements of the size and shape of the earth. Existing measurements are not accurate enough, and this is a particular problem for spacecraft navigation, surveying, and cartography. Even in 1973 there were many areas in the world that had never been adequately mapped or measured; the interior of Brazil is one example. The program called for the collection of measurement data by a worldwide network of observation stations and satellites and for analysis of the data by computers. The amount of data is enormous and the analysis complex. It was estimated by the U.S. National Aeronautics and Space Administration that about 1,350,000 pages of computer printout would be needed to record the results of the studies.

Future developments. A study was being conducted at the National Bureau of Standards aiming toward the development of computer-aided transcription of stenotype records. The normal court reporting system utilizes stenotypists to record the proceedings, after which these records are transcribed by typewriter. This procedure is both expensive and slow, but, most importantly, it is becoming increasingly difficult to find qualified court reporters, and a delay can slow down the judicial process. In the computerized system, the stenotype machine is modified so that a different electrical contact is closed when each key is pressed. These contact closures, which are codes, are recorded on an associated incremental magnetic tape recorder, and the tape is then processed by a computer programmed to translate the recorded stenotype symbology into typescript. For this purpose, the computer stores a symbol–meaning dictionary that gives the English-word equivalent to common stenotype symbols and a special glossary that lists the spelling of names and unusual words used in the recording session.

Unfortunately, although the concept is simple, the procedure is complex and many errors result. For example, there may be no match, as in the case where a symbol is not found in the dictionary or glossary. Some symbols have multiple meanings, and the automated program may not be able to select the correct word form. In other cases, a number of symbols may be needed to make up a word, and the computer program may be unable to recognize when to stop adding characters. And these are just a few of the more common problems. In their study, the bureau experimenters found that the error rate for automatic transcription was between 5 and 16%. The initial computer printout must, therefore, be manually edited and corrected, and this process is slow. However, the analysts believe that the system can be improved by training the stenotypists to use standard practices and by employing more advanced editing techniques.

—Harold Borko

See also Year in Review: COMPUTERS.

Marine sciences

The marine sciences scored many successes but also suffered numerous setbacks during the past year. The funding and priority situations in the United States were both altered so that some programs were accelerated and some were terminated. The very slow decision-making processes kept many scientists in limbo as decisions affecting their careers and programs were awaited. When these decisions were finally made, it was necessary to stop some programs abruptly.

One severe setback was the loss in a crash of the Convair 990 operated as a remote sensing platform by the U.S. National Aeronautics and Space Administration (NASA). This aircraft had obtained much valuable data for oceanography. In contrast, the success of U.S. astronauts in erecting an emergency radiation shield to cool the Skylab spacecraft and in deploying a stuck solar panel to obtain 3,000 additional watts for the craft made possible successful marine science experiments.

Pollution. New knowledge concerning the extent of marine pollution and the biology of deep-sea pollution was gained during the year. Further awareness of the extent of marine pollution in the Mediterranean, the North Atlantic, and the Antarctic oceans should accelerate efforts to stop various pollutants at their source.

La Commission Internationale pour l'Exploration Scientifique de la Mer Méditerranée (CIESM) held a meeting to study pollution in the Mediterranean and Black seas. The present status of pollution and the various steps being taken to reduce it, such as improved sewage plants and regulation of industrial discharges, were described. Concurrently, the Soviet Union began taking steps to clean up the Volga River, and, incidentally, restock it with sturgeon so that the supply of caviar could be increased. Success for all three bodies of water would substantially improve the quality of life in a large part of the world.

Similarly, the government of Brazil once believed that industrial expansion should not be hindered by regulations prohibiting pollution. It later modified this position, stating that "Development and preservation of the environment are not incompatible. The degradation of the environment is not necessarily the price of progress." This new policy appeared to be becoming effective in the nick of time, as unhealthy conditions in some parts of the country's rivers, estuaries, and coastal waters were reaching critical magnitudes.

The fight against pollution in the U.S. scored notable successes. In October 1972 Congress passed a multibillion-dollar law for aid in the construction of sewage treatment plants. U.S. Pres. Richard Nixon vetoed it, but Congress overrode the veto. Nixon then announced that only a part of the funds would be allocated. Those who were to have received the money took the matter to court and won a lawsuit that directed that the funds be made available.

These funds were expected to help many areas of the U.S. In particular, New York, New Jersey, and

Connecticut stood to benefit greatly. The waterways around New York City were being cleaned up. Henry L. Diamond, commissioner of environmental conservation of New York State, said that the additional funding should bring by 1975 the return of sturgeon and striped bass to many of the state's waterways, an edible shellfish once again in Jamaica Bay, and potable water in many areas where there is none now.

One of the very few new starts in the marine sciences was the Marine Ecosystem Analysis Program (MESA) of the U.S. National Oceanographic and Atmospheric Administration (NOAA). The New York bight was singled out for integrated physical, chemical, and biological oceanographic studies to define the total problem and indicate measures to solve it. Among other things the wave environment and the geology of the continental shelf were to be carefully studied.

The energy crisis involved marine scientists because it was creating the possibility that there might be drilling for oil on the continental shelf off the east coast of the U.S. Fearing the pollution that might ensue, people protested this. The facts, however, in many ways allay these fears, as offshore drilling was found to contribute only a very small percentage of the total oil pollution. M. P. Holdsworth estimated that 53% of the oil on the oceans was caused by tankers that flushed the residues in their oil tanks at sea by cleaning them with steam and hot water and pumping the mixture overboard. The older tankers were most responsible, with those of newer design accounting for only 8% of the 53%. The next most important source accounted for 32% and consisted of effluents from refineries and of sewage outfalls. The oil in the sewage outfalls came from the disposal of waste oil from automobiles. This one factor, for example, accounted for more than half of the oil in Long Is-

View of the deck of the "Glomar Challenger" shows the lower legs of the 142-ft drilling derrick and, beyond, the automatic pipe-racker where 24,000 ft of pipe can be stored. Vertical pipe above the derrick floor is the top end of a 15,000-ft drill string, then boring through the Mediterranean Sea floor near Sardinia. Samples from this drilling include a core of rock salt from 10,000 ft below sea level (below left); the vertical crack at the bottom is believed to indicate that the Mediterranean had once dried to that level. Further deep-sea core evidence of desiccation was provided by the convex-upward lamination of sedimentary rock (center), caused by shallow-water blue-green algae, and the light-colored nodules (right), anhydrites found in hot coastal deserts.

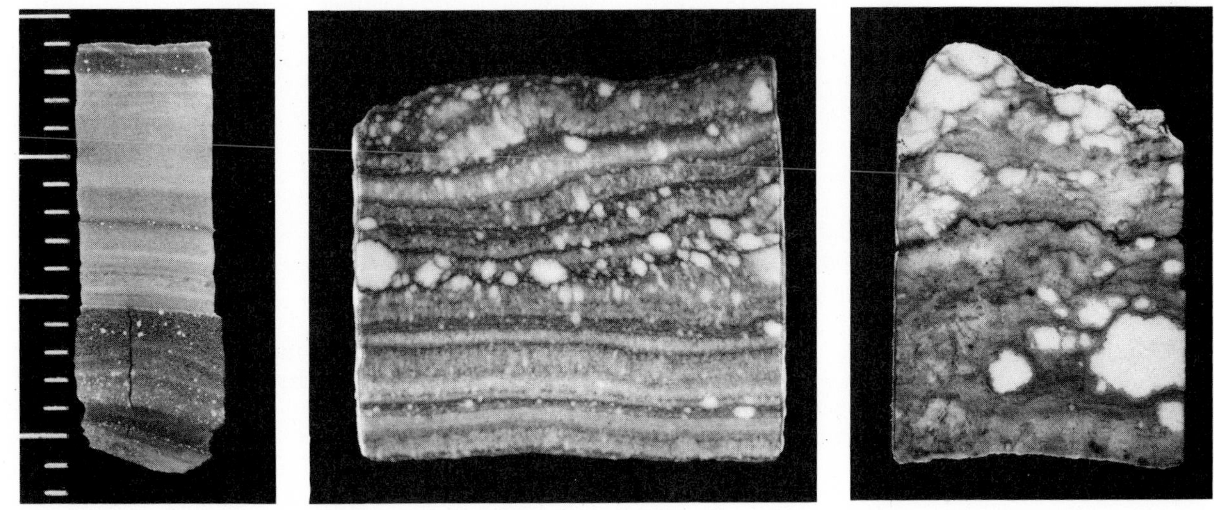

Courtesy, Dr. Kenneth J. Hsü, Swiss Federal Institute of Technology, Zurich land Sound. The third largest percentage, 11%, was accounted for by accidents to ships. Of a number of minor percentages (the above accounts for 96%), offshore oil-drilling operations provided a negligible amount and offshore drilling accidents totaled only 2%.

The large contribution of 53% from tank cleaning at sea was documented by a report issued by NOAA and prepared by Kenneth Sherman and three associates. Vast areas of the North Atlantic, north of Puerto Rico and Cuba and off the east coast of the U.S., totaling about 665,000 sq mi, were covered by floating clumps of a tarlike oil residue with concentrations ranging from light (0.1–2.9 g/sq m) through moderate (3.0–12.9 g/sq m) to heavy (more than 13 g/sq m). The total amount in all of the world's oceans was estimated as 86,000 metric tons.

Oil pollution in the ocean concentrates in those areas where currents converge or where there is heavy ship traffic. The main source of this pollution was being eliminated by improved design of oil tankers and by the development of better ways to flush fuel tanks and oil tanks. Many new devices were in the final development stages before full-scale implementation on ships. Thus, a steady improvement in the situation was expected to begin soon, and as the oil presently on the ocean is biologically and chemically degraded to less harmful forms, the oceans will cleanse themselves. A way to cleanse coastal waters would be for the coastal states to pass and enforce laws requiring that waste automobile oil be collected and either recycled or properly disposed of. The U.S. Coast Guard was developing systems for monitoring oil spills from ships in coastal waters, including new radars that detect the absence of capillary waves on a spill in time to identify the offending ship.

Other kinds of marine pollution were being carefully assessed by the International Decade of Ocean Exploration, headed by Feenan D. Jennings. Some of the goals of this program are: (1) to establish the concentration of selected important pollutants in living organisms, seawater, and sediments; (2) to understand the mechanisms and pathways by which pollutants move through the organisms, seawater, and sediment and the rate of this movement; (3) to determine the effects of pollutants on marine organisms; and (4) to predict the final effects of ocean pollutants. Important pollutants under study were the polychlorinated biphenyls, chlorinated hydrocarbons, and heavy metals.

The deep ocean was shown to be an unsafe dumping place for garbage and waste from the land. After the research submersible "Alvin" was recovered and brought up 5,000 ft from the sea floor, nearly a year after it sank, a meat sandwich was found aboard. At the high pressure and low temperature near the sea floor, biological decay had hardly proceeded at all. Bread, bologna, and mayonnaise seemed unspoiled. H. W. Jannasch and C. O. Wirsen placed a number of pure organic compounds inoculated with known bacteria at even greater depths. The rate of decay proved to be 100 to 600 times slower than comparable samples at the same temperature but at atmospheric pressure. Therefore, since natural recycling processes are nearly at a standstill in the deep ocean, these areas are not appropriate for dumping organic wastes.

"Glomar Challenger" and continental drift. The group of universities that jointly operates the "Glomar Challenger" continued to obtain important scientific results from the deep-sea drill cores obtained by the research ship. Recent operations near Antarctica confirmed the presence of oil and natural gas. Drill cores in the Mediterranean disclosed thick segments of solid sea salt, which led to a new understanding of continental drift. The sea salt could only have formed if the Mediterranean had been dammed at the Strait of Gibraltar and had dried up completely about six million years ago. The floor of the Mediterranean must then have been a huge equivalent of California's Death Valley. The Nile and other ancient rivers poured over the edges of the continental shelf, cutting deep canyons in the edge, a geological feature previously unexplained.

Finally the dam at the Strait of Gibraltar gave way and the Atlantic poured into the empty Mediterranean basin. In trying to visualize this event Samuel W. Matthews of *National Geographic* described a ten-mile-wide waterfall thousands of feet high. According to Kenneth J. Hsü, the water cascaded in at a rate of 10,000 cu mi per year for 100 years to fill up the Mediterranean.

The scientists of the Soviet Union followed the progress of the "Glomar Challenger" and contributed to the interpretation of some of the data. They agreed to join the JOIDES (Joint Oceanographic Institutions for Deep Earth Sampling) group and contribute financially and scientifically to further expeditions.

An interesting sidelight on the subject of continental drift and seafloor spreading was a study of the two largest pyramids at Giza, Egypt, by G. S. Pawley and N. Abrahamsen. The general alignment of the north–south sides of the pyramids is four minutes west of north, and the two researchers believe that the builders actually constructed them aligned to true north. They concluded that "continental drift is the most likely explanation, although somewhat implausible" because continental drift has been computed on geological time scales and the pyramids are relatively recent.

Skylab and ERTS. Remote sensing techniques from spacecraft were playing an increasingly important role in marine science. Data from the Earth Resources Technology Satellite (ERTS 1) provided useful information on coastal conditions. On Skylab 1 the Earth Resources Experiment Package of instruments (EREP) was an important part of the research. EREP included a combination scanning pencil-beam radar/radiometer, passive microwave radiometer, and altimeter, labeled S193. As the wind blows over the ocean it generates small capillary waves whose height increases with wind speed, making the sea rough. The rougher the sea the more radar energy is backscattered to the radiometer. When this backscattered energy is precisely measured, it in turn provides measurements of the wind speed over the ocean. Skylab 1 was designed to make thousands of measurements of this radar backscatter through clear air, thin stratus clouds, and thicker clouds. Simultaneous measurements of the passive microwave emission from the sea surface without clouds were expected to provide additional information on the wind speed. If clouds intervene, the passive signal will increase, and this measurement can be used to compute the effect of the clouds on the radar measurement and correct it so as to find the wind below the clouds.

Even before successfully deploying the stuck solar panel, the Skylab astronauts made several passes using S193. One was past Hurricane Ava in the eastern tropical Pacific on June 6. Concurrent aircraft flights through the hurricane provided extensive data that would allow scientists to check the accuracy of S193.

If this test is successful, an operational version could determine the winds at 40,000 points per day. For the Southern Hemisphere, which is 80% covered by oceans, these measurements would provide surface weather charts. Improved analysis over the oceans should provide improved weather forecasts for the entire earth.

Sea Grant. The Sea Grant program was cut back in funding. Nevertheless, the ongoing parts of the program made many contributions to marine science. Sea Grant developed a federal-state partnership to develop marine resources and use them wisely; it involved agencies, local organizations, and individuals in formulating needs and priorities, and it produced a nationwide network of marine-affairs units whose combined accomplishments far exceeded individual efforts.

Sea Grant programs often directly benefited many groups. Among them were improved mariculture methods, one being a new way to put air (oxygen) into water to stop fish kills. Ways to use the heated water from power-plant cooling to grow fish more quickly were described. Instead of returning it to a bay, lake, or estuary, the heated water can also be used to heat and irrigate land and increase the growing period. In some coastal waters it is beneficial to add heat; in others it is detrimental. Sea Grant studies have helped to discriminate between the two.

El Niño. Fishing is generally very productive off the coast of Peru because the usual current is a cold-water one flowing northward toward the equator. Moreover, the prevailing winds tend to cause an Ekman surface transport away from the coast, thus bringing cooler, nutrient-rich water to the surface and into the sunlight so that phytoplankton and zooplankton abound. These conditions result in abundant fish. Up to 1971 the annual catch of anchovies, used to make fish meal to feed cattle, was more than ten million tons.

However, the equatorial countercurrent, flowing from west to east, 600–700 mi north of the equator, can become strong enough near Panama and northern Peru to break out and flow southward off the coast of Peru. Warm, nutrient-poor water then replaces the productive waters, and the phytoplankton, zooplankton, and anchovies disappear.

This atypical current is called El Niño, and it occurs once every seven years on the average on the basis of past records. However, it has persisted since 1971 with the result that fish have been scarce and the fishing industry is about $200 million in debt. Thus, the vagaries of an ocean current have a tremendous economic impact and illustrate the need for prediction and better understanding of the ocean circulation.

—Willard J. Pierson

Mathematics

Problem solving versus axiomatization. Many mathematicians like to confine themselves to solving problems, preferably problems which are difficult, intuitively compelling, and at which many reputable mathematicians have already tried their hand, unsuccessfully. Other mathematicians attach more importance to method and to the elucidation of mathematical theories which, though already deeply advanced, are yet poorly understood. They have a predilection for a mathematics that restores order and simplicity in chaos, for finding a formalism or language with which theorems that seemed elusive and mysterious to their discoverers are rendered natural and intuitively clear. Such mathematics is sometimes called axiomatic or formalistic. The problem solver is concerned with establishing the truth of a theorem; the formalist wishes, in addition, to explain why it is true, to isolate simple underlying principles from which this theorem, and others to which the same principles apply, derive. Descartes, for example, was a formalist par excellence.

The progress of mathematics has always relied ultimately on the problem solvers. Often, however, after periods of major advance, a subject will enter uncharted and bewildering waters for which only a formalist can make it again seaworthy.

One of the great axiomatic periods in mathematics seems now to be drawing to a close. It began in the 1950s in the field of topology. This is a branch of geometry, systematically developed by Henri Poincaré about 1900, concerned with the properties of a geometric object that, roughly speaking, are unaltered by continuous deformations of that object. Topologists, largely in the United States and in France, axiomatized much of their subject during the 1950s. In so doing they created some sophisticated algebraic tools that subsequently were formed into an independent field of mathematics called homological algebra. Homological algebra, in turn, was rapidly axiomatized to accommodate its diverse manifestations, thus generating the field of category theory. Homological algebra and category theory constitute one of the great formalist constructs of the last two decades, comparable in importance, as a fundamental mathematical language, with linear algebra. Its applications have expanded from topology to innumerable other fields.

An especially important instance of this was the introduction of homological methods into algebraic geometry, initiated by the French mathematician Jean Pierre Serre in the mid-1950s and subsequently expanded to awesome proportions by another prodigious mathematician, Alexandre Gothendieck, also in France. Algebraic geometry studies geometric objects defined by algebraic equations, such as the circle defined by the equation $x^2 + y^2 = 1$. More precisely, it studies the sets of solutions of systems of algebraic equations in many variables. This study was deeply developed by the Italian school in the early 1900s, using very intuitive and geometric methods. More systematic algebraic foundations for the subject were later elaborated by Oscar Zariski and by André Weil, Weil being largely motivated by arithmetic applications. The subject was then completely refounded and permanently transformed by the work of Gothendieck, his students, and his disciples. Their work represents one of the great monuments of axiomatic mathematics, one whose far-reaching ramifications continue actively to be explored.

Some mathematicians hardly lamented the passing of this period of axiomatic zeal. They sometimes protested that an excessive attention to

Sidney Harris

"On the other hand, my responsibility to society makes me want to stop right here"

methodology for its own sake only concealed an evasion of the hard classical problems. They were pleased by the widespread return to such problems witnessed during the past few years. Nonetheless, many of these problems were solved only thanks to the refined tools which the axiomatists had wrought.

Algebraic K-theory. One of the progeny of these recent developments in topology and algebraic geometry is a young subject called algebraic K-theory. It grew originally out of certain algebraic questions (about matrix groups over rings) which were first raised by the British topologist, J. H. C. Whitehead, in 1939, in connection with his so-called simple homotopy theory. Algebraic methods adequate to treat these questions were only recently developed, inspired, ironically, by the ideas of a modern branch of topology called K-theory which has virtually nothing to do with simple homotopy theory. These methods also yielded significant information about a large variety of other natural algebraic questions which had previously been technically unapproachable. Important applications of them were made in group theory, the theory of quadratic forms, and in number theory.

Efforts to organize these methods into a coherent subject, and to answer the many questions naturally suggested by the methods themselves, have recently been gratifyingly successful. Furthermore, some fundamental connections between algebraic K-theory and number theory were revealed by the work of J. Milnor, J. Tate, S. Lichtenbaum, and J. Coates (all in the U.S.). This "coming of age" of algebraic K-theory was marked by a lively international conference on the subject at the Battelle-Seattle Research Center in September 1972.

Differential topology. The subject of differentiable manifolds, discussed at length in the *1973 Britannica Yearbook of Science and the Future,* remained a very active area of research on many fronts. One aspect of this research was concerned with the stability of dynamical systems. Such a system is said to be stable if a small perturbation of its initial state does not precipitate major qualitative changes in the manner in which it evolves. This is evidently a fundamental principle in many natural phenomena, much studied for example in the case of trajectories of moving bodies in classical mechanics. Considerable effort has been devoted to showing that stability is a "generic" property; that is, that it fails only in cases which, in a mathematically precise sense, are extremely rare. Pioneering work in this direction was done by S. Smale (U.S.) and D. Anasov (U.S.S.R.) and their students. However, Smale and others also exhibited examples in high dimensions where stability is not generic.

Important progress was also made in the study of singularities, points where continuity or differentiability (of a function, manifold, vector field) is violated. Among the aims of this study are, for example, to classify the topological type of a singularity, its stability properties under deformation, and the structure of the complete singular locus. Fundamental contributions to these important and difficult questions were recently made by F. Pham (France), R. Thom (France), J. Mather (U.S.), E. Brieskorn (West Germany), and many others.

Thom used mathematical ideas from differential topology as a basis for a proposed theory of morphogenesis, as exposed in his remarkable book *Stabilité-Structurelle et Morphogénèse.* In his model, singularities correspond to sudden changes of form in natural processes (metamorphosis) that are called "catastrophes." He offered numerous interpretations of this model in phenomena taken from such widely diverse fields as physics, biology, geology, and linguistics. Thom's theory is too mathematically demanding to have yet received the critical examination by empirical scientists that it merits, but it excited a great deal of mathematical interest. An important contributor to this field, C. Zeeman (U.K.), was among the first mathematicians to attempt serious application of topological ideas to qualitative questions in biology.

Government cutbacks. Along with the physical sciences, mathematics in the U.S. suffered substantial cutbacks in government support of research and of graduate students. While acknowledging that some reduction in the level of research activity and Ph.D. production was both inevitable and appropriate, many mathematicians believe that the actual cutbacks have been excessive in terms of the nation's basic scientific needs. They also criticize the government for failing to cushion adequately the effects of the cutbacks on advanced Ph.D. students, who have entered a saturated and dwindling academic job market.

—Hyman Bass

Medicine

Developments reported in the medical sciences in 1972–1973 more often than not dealt directly or indirectly with cancer or various infectious diseases. But at the same time, steady, if not always spectacular progress was made in a number of medical specialties. This review reflects these trends. An overview of recent developments throughout the medical fields is provided under the title *General Medicine,* which also reports significant achievements in cancer control and the fights being waged against a number of infectious diseases throughout the world. Three sections that follow, *Community Medicine, Immunology,* and *Venereal Diseases,* concentrate on specific aspects of these general areas. Developments in two other specialties are covered in *Dentistry* and *Ophthalmology.* A special section, *Room W-204,* shows some of the ways the technologies developed for the U.S. space program can be applied to one aspect of health care, that of the severely handicapped. Feature articles elsewhere in the *1974 Britannica Yearbook of Science and the Future* also provide detailed looks at various aspects of medicine. These are ACUPUNCTURE: MEDICINE OR MAGIC?; THE CLOCKS WITHIN US; CYCLIC AMP; DEATH AND DYING; and SCHIZOPHRENIA.

General medicine

Although modern medical science had all but eliminated many of the diseases that once decimated populations, illness remained man's constant companion. Its grip on humanity, however, was being loosened as medical researchers continued to probe for its causes and discover its cures. The past decade witnessed more medical progress than in the entire previous history of medicine. Recent months saw science unlock even more of the secrets of human health and increase man's knowledge of how to preserve and restore it.

Cancer control. Science had yet to identify the causes, let alone to discover the "cure" of cancer, a family of diseases characterized by rapid, uncontrolled cell growth. But it moved significant steps closer to both. Research over many years had identified several factors, such as cigarette smoking and exposure to asbestos, radiation, or certain chemicals, that predisposed toward cancer. Extensive efforts were under way in 1973 to identify as many of these environmental agents as possible. The International Agency for Research on Cancer, a World Health Organization (WHO) facility based in Lyons, France, was seeking the environmental factor underlying demonstrated regional differences in the incidence of cancer of the esophagus, which was particularly prevalent in limited areas, especially in the Middle East. In the U.S., the National Cancer Institute was to receive $75 million a year in federal funds for five years, triple the previous budget, for research on environmentally caused cancers. The aim of the program was to forestall any future reports like the ones made in late 1972 that linked exposure to airborne asbestos with the onset of cancers 30 years later. The discovery in 1973 that water supplies in Duluth, Minn., and 23 other Lake Superior communities contained high levels of asbestos, raised concern over whether waterborne asbestos also could cause cancer.

Cancer research. But many still suspected that the real villain in the cancer process was a virus. Although no true human cancer virus was isolated or identified, several studies in recent months tended to strengthen suspicion. For example, at Columbia University and the Bionetics Research Laboratories, Bethesda, Md., workers developed a new technique for identifying the genetic material of a virus in human milk. This material proved to be identical to the core of the virus known to cause breast cancer in mice. Researchers had established previously that the viruslike particles were common in the milk of high-cancer-risk groups of women.

Other researchers were able after many years of research, to establish firm links between some very common viruses, members of the herpes family, and several forms of cancer. One of the viruses under suspicion was herpes simplex, which some physicians estimated is carried by 99% of all people. In most people it remains dormant; in some it is activated, apparently by high fever, exposure to the sun, or emotional upset, to produce annoying but harmless blisters around the mouth and nose. The second suspected virus was a variant of the simplex known as the herpes Type II. Also quite common, it is responsible for painful infections of the genitals, thighs, and buttocks and is easily transmitted by direct contact. In the estimation of some medical scientists, genital herpes ranked second only to gonorrhea as the most common sexually transmitted disease (see *Venereal Diseases*, below).

Albert Sabin, a researcher at the National Institutes of Health (NIH), Bethesda, Md., who developed the live-virus polio vaccine, and Giulio Tarro of the University of Naples, Italy, implicated both herpes viruses in cancer after establishing "footprints," antibodies produced in response to the virus' presence or passage, in the sera of patients with cancers of the lip, mouth, nose and throat, kidney, bladder, prostate, cervix, and vulva. The response was found in 56 patients with such cancers but not in 81 patients with other malignancies or in 51 patients without cancer. Tarro and

Courtesy, Naval Research Laboratory

Cyclotron and beam transport equipment at the U.S. Naval Research Laboratory, Washington, D.C., are being modified for use in treating cancer. Researchers believe that the beams of high-energy neutrons generated by the cyclotron may prove effective in treating local tumors in cases where other radiations have failed.

Courtesy, Dr. Edward Reich, Rockefeller University

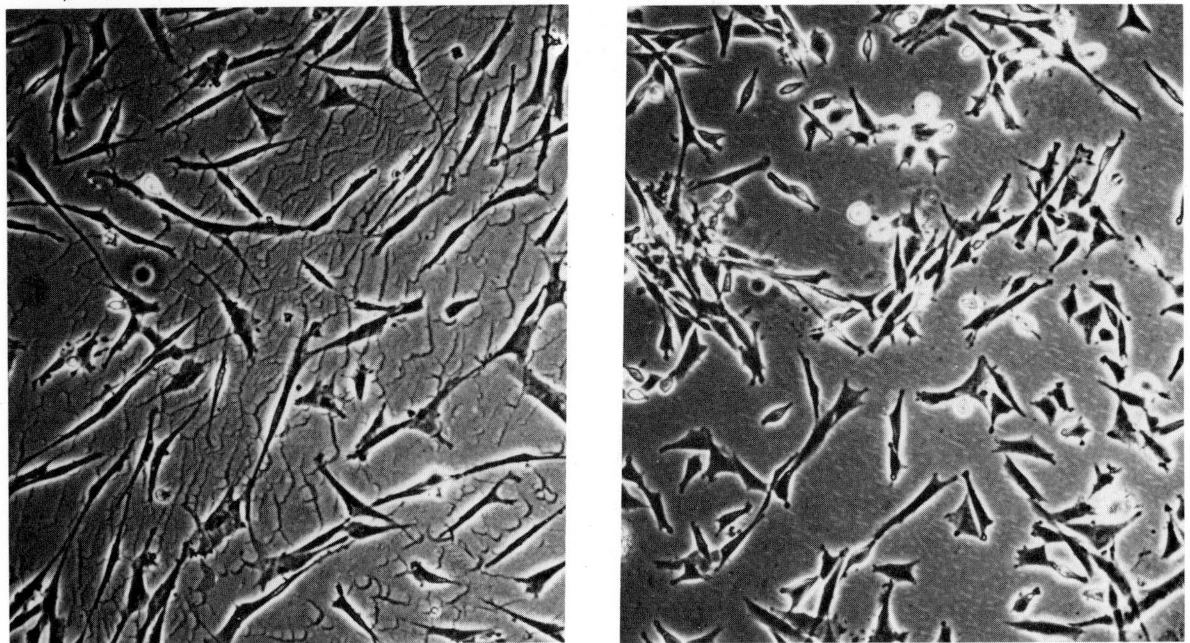

Fibrin, a whitish, elastic protein in blood serum, appears as wavy lines in a culture of normal cells (left) but is absent in a cancer-cell culture. Researchers at Rockefeller University were studying a protein that causes fibrin breakdown and is emitted only by cancerous cells.

Ariel C. Hollinshead of George Washington University, Washington, D.C., also achieved similar results in tests of cancerous tissues, rather than sera.

Despite these discoveries, the case against herpes simplex was anything but established. Millions developed cold sores each year, yet the rate of cancer was no greater among them than it was among those who did not. The case against the Type II virus was somewhat stronger. Some Type II viruses turned normal hamster cells into cancer cells in laboratory experiments. They also appeared more strongly implicated in cervical cancer, a common malignancy among women. Andre Nahmias of Emory University, Atlanta, Ga., studied 900 women known to have had genital herpes, compared them with 600 women who had not had the infection, and found the incidence of cervical cancer eight times higher in the first group than the second. In addition, at the University of Chicago, researchers detected core material of the Type II virus in human cervical cancer cells in such a manner that they concluded it to be integrated with the DNA of the human cells.

Much study was needed before the exact role of viruses in the cancer process could be determined. Nonetheless, research on ways of preventing normal cells from being transformed into cancer cells, either before or after they are infected by viruses, was producing results. At the University of California at San Francisco, pharmacologist Martin A. Apple and his colleagues established that two classes of drugs whose derivatives were used in the treatment of cancers were potent inhibitors of the action of the enzyme, reverse transcriptase. This enzyme was well established as the means by which the RNA genetic material of viruses known to cause cancer in animals can make DNA copies of itself in the cells of the animal it infects. This DNA might then be incorporated into the host's own DNA and, once activated, transform the cell into a cancerous one. Studies showed that the two drug groups, the anthracyclines and the cactinomycines, can prevent RNA viruses from reproducing, block their ability to transform cells they have infected, and protect animals that had been exposed to the viruses.

Work by another researcher provided a clearer understanding of how tumors grow once the cancer process has started. Judah Folkman of Harvard University Medical School found that solid tumors appear to require the presence of a substance called tumor angiogenesis factor (TAF) in order to grow beyond the size of a pinhead. In order to obtain nutrients and pass off waste products, tumors must develop their own blood supplies. They do this by producing TAF, which causes nearby capillaries (small blood vessels) to grow into the mass. When Folkman isolated tumors from nearby blood supplies in experiments, TAF's effects were can-

continued on page 268

Room W-204

In the spring of 1973, Room W-204 of the Huntsville Hospital in Huntsville, Ala., provided a prophetic glimpse into the future for the institutional care of the severely handicapped. The hospital had a year earlier joined the University of Alabama and the Marshall Space Flight Center at Huntsville in developing a unique facility for the care of such patients.

A radical departure in hospital care for amputees and paraplegics, the new room is furnished in such a way that even quadraplegic patients can now do for themselves things never before possible. The new facility is built largely upon technology deriving from the U.S. space program, especially its development and refinement of the technique of miniaturizing electronic controls.

"What we have done," says Wassie Griffin, Huntsville Hospital administrator, "is to adapt space technology to a real need in the hospital environment. To this end, we have applied the systems-engineering approach as well as the concepts of microminiaturized circuitry that the Marshall center's designers and engineers used in the Saturn V space carrier vehicle."

The severely handicapped patient controls his environment through a system of special switches that are activated by those functions that he still possesses. The signals from these switches are processed by a distributor and control unit and applied by it to various electrical appliances in the room.

"The switches used in the system are especially interesting," says Sener Sancar, assistant hospital administrator for medical engineering and project director for Room W-204. "At present we are using only six different types, but there are others that can be added because of the flexibility of the system."

Sight and breath switches. The most unusual of these devices is the sight switch. It was developed by Charles L. Nork, then of Hayes International Corp., Birmingham, Ala. The original idea behind the switch was to allow an astronaut to manipulate the controls of his spacecraft during periods of high acceleration when he could not use his hands, arms, and legs. However, it was never utilized for the purpose.

Perhaps a more accurate name for the device is eye switch, for it is operated by the movement of the eye rather than the sense of seeing. The unit consists of a pair of eyeglass frames on which are mounted two very small infrared lamps and sensors that detect a difference in reflectivity. One of each is mounted over the extreme outer edge of each eye. The lamps are adjusted to project a beam onto the sclera, or white, of the eye at a certain radial distance from the iris. Thus, most of the light is continuously reflected back into the sensors; while this level is maintained, nothing happens.

To operate the switch, a patient glances upward and outward. This moves the patient's darker iris into the beam. The difference in reflectivity between the sclera and iris is detected by the sensor, and the switch functions. Once adjusted to the patient, the switch is not activated by normal blinking or lateral eye movement.

Since even the most severely paralyzed patient can breathe, his breath can be used to operate another type of switch. This unit consists of two very small microswitches, the contacts of which need move only a millimeter or less in order to close. Small paddles on the ends of lever arms attached to the switches allow the faintest puff by the patient to operate the switches. The switches are mounted on the end of a flexible cable, like the neck of a study lamp, so that the unit can be positioned in front and slightly to the side of the mouth.

For patients who can move. For the less severely handicapped patient, other types of switches are used. With the head-motion switch, consisting of two microswitches mounted in a frame that rests beneath the head and neck, the patient can control appliances by moving his head very slightly to the left or right. The pneumatic switch consists of two small inflated pillows, pressure on which provides the force necessary to close and open the contacts. This switch is used by a patient who can move his shoulders, elbows, or knees and is placed beneath or in touch with these joints.

For a patient who can move his fingers and toes, a simple pushbutton switch, such as that which operates a doorbell, is used. Similar to the pushbutton operated by finger is a switch with a larger contact area for activation by a patient's toes or feet.

Each switch has two elements. One is used to start the sequencer of the distributor and control unit stepping through the channels available. The channel number is visually displayed for the patient. The same element stops the sequencer once the desired channel is reached. The other element starts and stops the appliance associated with the channel selected. These switches are connected through the distributor and control unit to the various electrical appliances in the room.

If the patient wants to turn on his television set, he activates the "system on, sequencer start" channel by any of the switches described above.

When the number 14 appears in the system's visual display, he operates the same element again, stopping the sequencer. He then operates the other element to turn on the set and start its channel selector switch. When the desired station is reached, he operates the same element again. With the program that he wants now on the screen, he repeats the process until channel 15 is reached, the channel that controls the volume.

Currently, Room W-204 has 12 appliances that the patient controls. These are a buzzer to attract attention; a call board with 12 "messages" (food, water, doctor, medicine, help!, etc.); nurse station intercom; telephone; curtains; bed adjustment; electric fan; page turner for books; radio; television; and two different lamps.

Designing the equipment. During the design of the electronic components, the engineers and technicians involved in the project kept patient safety uppermost in mind. Thus, the system operates on only five volts. Convenience was another important consideration. For example, since the situation could arise, particularly during the initial learning phase of operation, when the patient could become confused by the system, he can shut it off completely and swiftly. He need only hold down any of the switches for a period of three seconds.

Surprisingly few technical problems cropped up during the design and installation of the project. "There was no major problem in setting up this facility because all the hardware used was direct spin-off from NASA technology. Most of the equipment had been previously proven and 'debugged,'" said Wallace Frierson, consulting physician for Room W-204.

The facility is to a certain extent experimental. As such, it serves several important purposes. Obviously Room W-204 as it now exists permits hospital personnel to evaluate patient response and acceptance of it. In addition it gives a good idea of what a patient believes are the most important elements of his room environment and those which he desires to control. Equally important, the system provides data on how fast patients with varying degrees of handicap can learn to use it. Room W-204 also introduces and trains a variety of medical personnel to the system.

It must be stressed, however, that the room is not wholly an experiment. "This room is a functional, integrated part of our hospital," said Griffin. "It is available to any handicapped patient admitted to the Huntsville Hospital who can benefit from its specialized care features if it is available at the time of admission."

Future applications. Experience thus far with Room W-204 indicates that it is a valuable addition to the Huntsville Hospital. It embodies a concept that appears to have a great potential, especially for extended-care medical facilities. It seems ideally tailored to the needs of medical facilities that must minister to the needs of the severely handicapped over longer periods of time than the municipal hospital. Veterans' hospitals, rehabilitation centers, and other such specialized care facilities should benefit from its features.

"Any large hospital that sees acute injuries such as broken necks or other disease or injury resulting in paraplegia or quadraplegia would be greatly helped by this type of equipment. Institutions such as paraplegic centers would find great utility in this type of equipment. It makes the patient much more independent earlier in his illness and is a great morale booster for both acute and chronic paraplegics," said Frierson in commenting on the potential of Room W-204 for other applications.

—Mitchell R. Sharpe

Polio victim demonstrates device in use at Room W-204. Switches activated by her breathing operate the mechanism that moves her completely paralyzed arm.

Courtesy, NASA

continued from page 265

celled out and no capillaries grew into the mass. As a result, the tumors choked on their own wastes, withered, and died. Folkman continued to search for a way to prevent tumor growth by finding a substance capable of blocking TAF.

Treating cancers. Surgery remained the treatment of choice for a majority of malignancies, and for good reason. No other method permitted the removal of so many cancerous cells so quickly. But the trauma of surgery was being reduced, particularly for women suffering from breast cancer. Most surgeons still favored radical mastectomy, a disfiguring operation that included removal of the affected breast and surrounding tissue. About 70% of the women who underwent this procedure survived at least five years. But Vera Peters of Princess Margaret Rose Hospital in Toronto, Ont., reported equally encouraging results, at least in early-stage cancers, with lumpectomy, or removal of the cancerous lump alone.

Chemotherapy, or the use of drugs or hormones to control disease, was also proving effective against certain types of cancers. Most of the current anticancer drugs were cytotoxic, or cell-destroying, specifically at the moment of mitosis, or division. Since cancerous cells divide many times more frequently than normal cells, cytotoxic drugs were far more effective against malignant cells than healthy ones. In wide use against skin cancers and such systemic cancers as Hodgkin's disease and the leukemias, these agents had greatly increased the life expectancies of people suffering from these diseases. Immunotherapy, or use of the body's natural defense mechanisms, was also showing great potential as a modality of cancer treatment (see *Immunology,* below).

These approaches, coupled with techniques for earlier detection, had produced major reductions in cancer mortality, and the National Cancer Institute reported in late 1972 that survival rates from several types of cancer improved markedly. One of the most significant improvements occurred in leukemia, a major killer of children. Prior to 1949 only 4% of those with acute leukemia survived for three years; in 1972 combination chemotherapy permitted long-term remission in at least half of the cases.

The first phase of the implementation of the U.S.-U.S.S.R. agreement on health cooperation signed during U.S. Pres. Richard Nixon's visit to Moscow in May 1972 was an exchange of anticancer drugs developed in the two countries. The drugs would be subjected to clinical tests in both countries. The exchange was followed in November by the signing of an agreement to expand common efforts to fight cancer and by a second exchange—this time of viruses that specialists in both countries had found to be implicated in various cancers.

Contagious diseases. Infectious diseases continued to plague and puzzle physicians and their patients. Schistosomiasis, a disease caused by snail-borne liver flukes, or parasites, remained a major problem in many parts of Africa and South America. Cholera continued to be highly contagious and was endemic to India, Pakistan, and Bangladesh, where poor sanitation and the ravages of the Indo-Pakistani war made the disease a leading cause of death. The World Health Organization concentrated its major offensive against smallpox in the Indian subcontinent and in Ethiopia and the Sudan in Africa. Similar vaccination efforts in recent years had left South America and Indonesia virtually free of the disease. Increases in the number of known cases in 1971 and 1972 were attributed to better reporting and surveillance.

Modern sanitation practices and vaccination had made such highly contagious diseases historic footnotes in Western Europe and North America. Nonetheless, a serious scare developed in March 1972 when smallpox erupted in Yugoslavia. Health officials in many countries, especially in West Germany, which had a large Yugoslav labor force, quickly established successful immunization and quarantine programs. The possibility of a serious outbreak of poliomyelitis in the U.S. was forecast by various health organizations, which estimated that up to half of the children in most inner cities had not been vaccinated against the disease. The most serious outbreak of polio in seven years occurred in Connecticut in October 1972 when 11 children at a private Christian Science school suffered symptoms of the disease. Because of religious beliefs most of the student body had not been immunized.

Influenza viruses seemed capable of undergoing an infinite number of mutations. As a result, vaccines developed to fight one year's virus were usually ineffective against later models. In recent months, however, two groups reported major progress on the development of vaccines in anticipation of a major mutation and epidemic predicted for the late 1970s. NIH scientists combined Hong Kong viruses from the 1968 epidemic with structurally altered specimens from a 1965 strain to produce a hybrid strong enough to induce immunity to influenza but too weak to cause the disease itself. They also produced a virus that could be grown quickly in culture, and, just as important, be altered to match any mutations that the flu bug might undergo before its next assault.

An even more effective vaccine might be that developed by researchers from the Institut Pasteur in Paris. Aware that between its major muta-

Adapted from MEDICAL WORLD NEWS

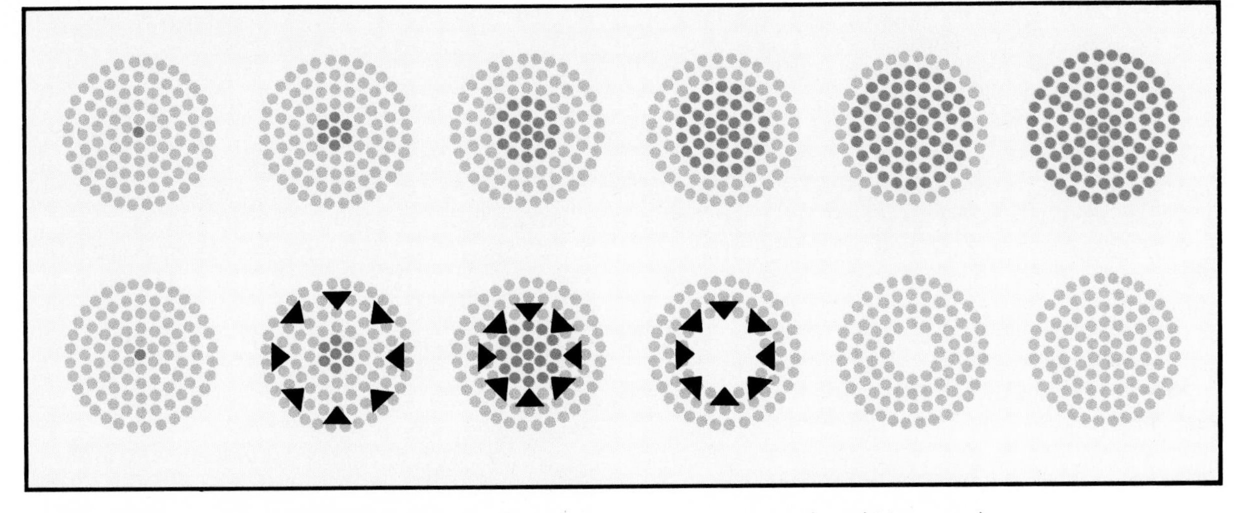

Herpes simplex, a viral cause of cold sores implicated in some cancers, was thought to spread over intercellular "bridges" before antibodies can attack (top row). Leukocytes, represented by triangles (bottom row), can surround the infection site, block the bridges, and kill invaded cells.

tions the flu virus undergoes subtle changes in its protein coating, a team led by Claude Hannoun tried accelerating this process of genetic drift by bombarding strains of the virus with different antibodies. They then produced a vaccine that was effective against the virus that had survived this process of unnatural selection, confident that they had anticipated all the minor changes the microbe was likely to undergo until about 1978. Samples of the vaccine prepared before the outbreak of the so-called London flu of the winter of 1972–73 proved 84% effective.

Also in recent months scientists at NIH established that the disease called intestinal flu is caused by a viral agent that in no way resembles the influenza virus. Properly termed acute infectious nonbacterial gastroenteritis, the disease, while not considered dangerous, usually proved to be very uncomfortable for 24–48 hours. The most common disease experience after colds and other respiratory infections, it is responsible for serious manpower and economic losses. Using immune electron microscopy that caused the virus particles to clump together to become visible, the NIH team, headed by Albert Z. Kapikian, found the agent to be very small and cube-shaped, similar to the rhinoviruses that cause the common cold.

In October 1972 the U.S. Public Health Service responded to a request of the Sierra Leone government to help fight an outbreak there of the mysterious and lethal Lassa fever, first discovered in Nigeria in 1969. No clue to the cause of the spread of the virus was available, and hence there was no way to forecast where else it might spread before work under way at the Center for Disease Control in Atlanta, Ga., produced a vaccine against it.

To aid efforts to curb a much more common disease, malaria, which was occurring in small but rising numbers among businessmen and tourists, WHO published the first comprehensive guide to areas and months of the year in which the disease's vector mosquito thrived. In the U.S. in the fall of 1972 public health departments in New England had to curb the fishing and sale of soft-shell clams affected by a "red tide" of algae. The organism caused a paralytic illness in humans that affected the ability to breathe.

In July 1972 it was reported that for 40 years the U.S. Public Health Service had been conducting a study to determine the full effects of syphilis on the human body by withholding all treatment for the disease. The study was begun in 1932 using more than 430 syphilitic black men from the Tuskegee, Ala., area, which at the time had the highest syphilis rate in the U.S. The study continued despite the discovery ten years later that penicillin could cure the disease and despite the wide availability of the drug by the late 1940s. Exact information on the number of deaths directly related to untreated syphilis among participants was unavailable.

Surgery. The field of surgery saw several significant developments during 1972–73. The widespread use of powerful microscopes enabled surgeons to work in parts of the body where operations were once believed impossible and led to increased use of such procedures as bile duct and auditory nerve repairs and the rejoining of severed fingers. Reconstructive heart surgery continued to expand in dozens of hospitals throughout the world, as more and more surgeons replaced diseased mitral valves with substitutes of Teflon and stainless steel, bypassed blocked coronary arteries, and repaired other defects. The nuclear-powered cardiac pacemaker, tested successfully in Europe since April 1970, began receiving its first U.S. trials in human subjects. The device was expected to perform for up to ten years instead of the 1½–2 years of conventionally powered pacemakers.

Organ transplantation was continuing, at least in a few medical centers, and was saving lives. A team at Stanford University headed by Norman E. Shumway was the only one performing heart transplants on a regular basis—roughly once a month on carefully selected patients. The major problem being encountered was the tendency of the recipient to develop something like atherosclerosis, the clogging of the arteries characteristic of heart disease, only at a much more rapid rate than in normal heart conditions. Careful following of the dietary and drug regimens prescribed for heart attack victims was proving to be effective treatment. A group led by Thomas E. Starzl at the University of Colorado reported experiencing progressively better success with kidney transplants, especially when relatives were donors. Their experience, however, was substantially better than that described by other centers. A University of Washington School of Medicine team led by E. Donnall Thomas described strikingly long survival rates among the few patients that had received bone marrow transplants either to correct aplastic anemia caused by failure of bone marrow to function or to arrest advanced leukemia, which attacks blood-forming tissues in marrow.

One procedure that more and more surgeons found themselves performing was vasectomy, or male sterilization. Some three million men in the U.S. had undergone the operation by the end of 1972; their ranks were expected to increase at the rate of about a million a year, as more men took advantage of this simple procedure, which could be performed under local anesthetic in a physician's office. But the procedure was not without its problems. For one thing, it had to be considered permanent. There was some success in rejoining severed vas deferens, but restored fertility occurred in only a few instances. Also, a study of the effects of the freezing of human sperm, a practice whose growth paralleled the use of vasectomy, indicated that freezing and thawing causes losses of up to 50–60% of the sperm's motility, or ability to move toward the egg. Finally, although the operation had no physiological effect on a man's potency, many men who felt that they had been pressured into having the operation reported that they became impotent.

Psychosurgery, the use of electricity, radiation, or ultrasound to destroy tissues deep in the brain,

Courtesy, Dr. Albert Z. Kapikian, National Institutes of Health

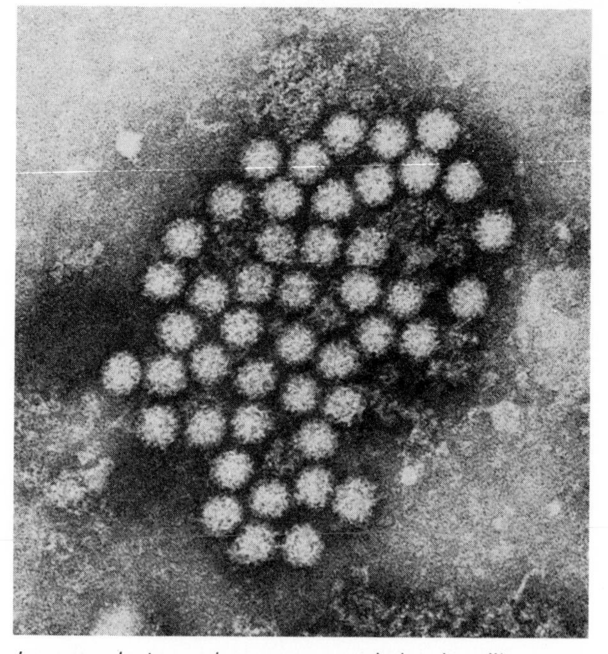

Immune electron microscopy revealed a virus-like agent of the brief, debilitating illness known as intestinal flu. The particle, called the Norwalk agent because it was isolated during an epidemic in Norwalk, O., in 1968, proved to be very small and cube-shaped.

was being used increasingly to treat uncontrollable epilepsy, schizophrenia, violent behavior, and even such conditions as severe depression and destructive hyperactivity. Better knowledge of the brain acquired in recent years permitted surgeons to isolate those areas thought to control behavior and to operate on them without destroying personality, as had occurred in the course of some of the lobotomies performed in the 1940s and 1950s. For example, a Harvard Medical School neurosurgeon, Vernon H. Mark, found that destroying tiny portions of the amygdala, an almond-shaped body in the midbrain, in patients who suffered from temporal lobe epilepsy or brain tumors that brought on seizures of violence, produced marked improvement in at least five cases and no apparent damage in seven others. At the University of Texas surgeons who used electric needles to destroy small bundles of cells in the cingulum area of the limbic lobe, or feeling brain, in order to treat schizophrenics and chronic alcoholics reported significant improvement in 50% of their patients.

Indications that these and other techniques provided effective control of violent or otherwise undesirable behavior increased public concern that psychosurgery could be abused and used indiscriminately to make permanent changes in the brain. A myriad of legal, ethical, and medical issues was raised in one case that came to trial in

Michigan, where a patient confined to a state institution for the criminally insane had agreed to undergo psychosurgery as part of a project studying new methods of psychotherapy. As a result of the publicity surrounding the case, the project was cancelled.

The question of whether human fetuses taken alive during abortion should be used for scientific research was another subject of considerable debate. Both the ethicality and scientific merit of keeping an aborted fetus alive for a few hours to conduct tests, which was being done in Europe, were subjects of controversy. In April 1973 the National Institutes of Health, which supported about 40% of all biomedical research done in the U.S., issued a statement saying it did not plan to support such projects. The issue was being studied as part of an overall probe into questions of human experimentation.

Other developments. Researchers at the Mayo Clinic of Rochester, Minn., found that chenodeoxycholic acid, a natural body chemical, dissolved gallstones. Such stones accounted for approximately 350,000 surgical operations a year in the U.S. alone. But the newly tried substance, the absence of which might be at least partially responsible for the formation of the stones in the first place, might eliminate the need for much of this surgery. In at least half of the patients treated in the Mayo Clinic study, the stones disappeared completely over a period of months.

New evidence that "speed" kills came from workers at the University of Southern California who had been puzzled by an increase in the numbers of young people suffering from strokes. They found signs of small vessel deterioration in the brains of 14 young people who mainlined amphetamines and detected similar problems in the brains of 100 other drug users. Tests of monkeys injected with methamphetamines showed irreversible damage in the areas surrounding the small blood vessels that was similar to the damage found in the brains of persons who died from strokes. Scientists at Johns Hopkins University, in what was described as a major advance in the understanding and treatment of drug addiction, demonstrated that narcotics produce their effects by attaching to specific sites in brain tissues.

Oral contraceptives, used by eight million women in the U.S. alone in 1972, were indicted as a possible contributing factor in strokes. The pill had long been linked with such minor complaints as weight gain, nausea, headaches, and breast tenderness, and held responsible for menstrual irregularity and temporary infertility in women who went off it. Studies in the U.S. and the U.K. had also shown that women who used the pill were more prone to clotting problems, leading to suspicion that the pill might also play a role in strokes. A recent study conducted by neurologists from a dozen major U.S. university hospitals reinforced that speculation, and concluded that women who used the pill were nine times more likely to suffer thrombotic or hemorrhagic strokes than women who relied on other means of contraception. This risk, however, might soon be reduced. Estrogen, a female hormone and a principal ingredient of oral contraceptives, was believed to aggravate hypertension (high blood pressure) and thereby lead to strokes. Drug manufacturers were, therefore, reducing the amounts of the hormone in their products.

The U.S. Food and Drug Administration (FDA) permitted a synthetic estrogen to be used as a "morning after" contraceptive in emergencies such as rape. The drug, diethylstilbestrol, or DES, was already approved for use in treating both cancerous and noncancerous conditions of the endometrium, the lining of the uterus. It had, however, been linked with the appearance of vaginal or cervical cancer in daughters of women who had used the drug during pregnancy and it had been barred from use as a growth hormone in cattle feed by 22 countries in 1972 when reports indicated it caused cancer in test animals.

As part of the FDA's program to establish the efficacy of nonprescription pharmaceuticals, a National Academy of Sciences-National Research Council committee reported that of 45 claims made for 27 cold remedies only four could be judged "effective" without reasonable doubt. The claims made for the products included relief of the nasal congestion, pains, fever, headache, and other characteristics of the common cold.

The need for care in the licensing of drugs was reinforced by the report of a Johns Hopkins University researcher who linked an epidemic of deaths among asthmatics in several countries with the sale, either by prescription or over the counter, of inhalators containing five times the amount of the drug isoproterenol that was approved for prescription sale in the U.S. and Canada, which experienced no epidemic. British health officials estimated that 3,500 asthmatics had died in six years, during which asthma became the fourth leading cause of death among children 10–14 years of age.

Another form of treatment to come under question was the use of adrenal hormones to treat hypoglycemia, or low blood sugar. In fact, the disease itself was the subject of considerable controversy. Some groups claimed that 10–20 million persons in the U.S. suffered in some way from the deficiency and that their symptoms, which

Courtesy, Dr. Wilhelm Z. Stern, Albert Einstein College of Medicine

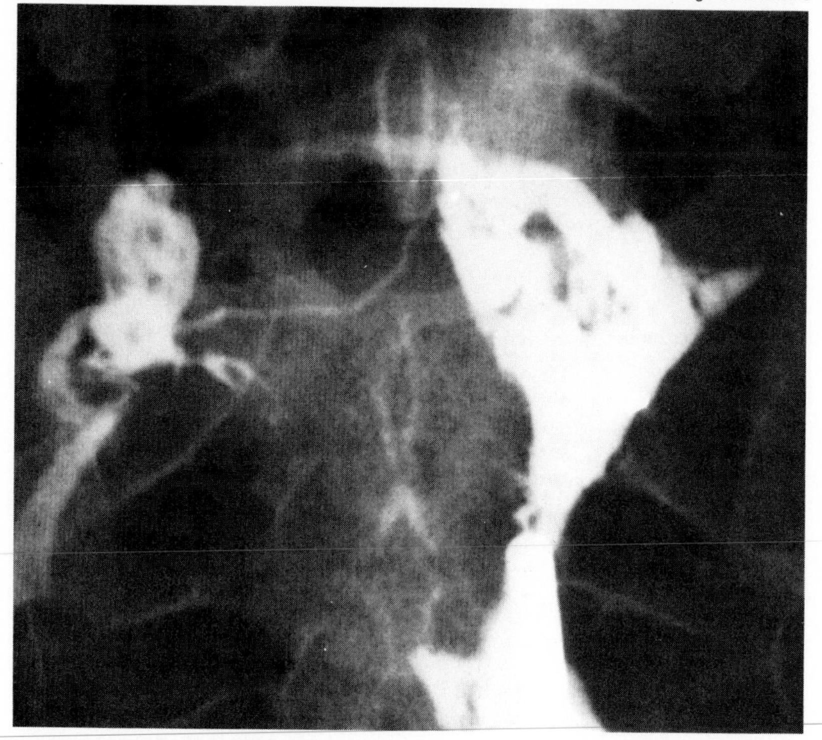

The filling effects typical of adhesions of the uterus stand out in a contrast radiograph. The tendency to develop such adhesions, known as Asherman syndrome, was high among women who had undergone two or more dilatation and curettages. The D & C was a common method of therapeutic abortion.

included depression and anxiety, perceptual problems, and fatigue and irritability, were responsible for such increasing social problems as alcoholism, divorce, juvenile delinquency, and drug addiction. In contrast, an editorial statement in the *Journal of the American Medical Association* in February 1973 claimed that few persons suffered from the disease. No objective, comprehensive study of these issues was available or underway.

Hypoglycemia could take either of two forms. Symptoms experienced before meals frequently indicated a tumor of the pancreas that leads to depletion of blood sugar by causing the pancreas to produce too much insulin. Symptoms appearing a few hours after eating could be the result of sporadic insulin output caused by various glandular disturbances, presumably including inadequate production of steroid hormones by the cortex of the adrenal gland.

Accepted therapies, given an accurate diagnosis of the type of hypoglycemia, included surgical removal of the tumor, or use of high-protein diets or drugs to stabilize sugar and insulin levels. What many physicians and researchers were questioning was the use of adrenal cortex extracts to bolster the patient's hormone level, especially when used to see if the patient would react, without first undertaking the expensive tests that would establish the deficiency.

Three new U.S. medical schools were scheduled to open in 1973, bringing to 26 the number of new schools since 1967. The school openings and the expansion of programs at existing schools had permitted enrollments to increase by about 1,000 students a year in recent years. Nonetheless, the shortage of medical doctors was estimated by the U.S. Department of Health, Education, and Welfare to be 50,000.

Hopes for lowering this figure were set back by the decreases in funding for medical education contained in the federal budget for fiscal 1974. The cuts consisted of a 25% decrease in what the medical schools had expected and a 40% loss of funding for biomedical research. The medical schools anticipated curtailment of innovative educational programs, of community health efforts in both inner-city and rural areas, and of medical school staffs, particularly of support personnel. In addition, mental health programs of all kinds would be affected by severe budget cuts.

—Peter Stoler

Community medicine

Community medicine is a term used to "identify the specialized knowledge and skills required in our emerging system of medical services, a system which is neither 'state medicine' nor 'socialized medicine' nor 'private medicine' but a combined public and private effort for comprehensive health care in every American community." It is concerned with the health status of individuals and groups; the incidence, prevalence, and natural history of disease; and the organization, financing, and delivery of health care.

Medicine has traditionally concerned itself with the human organism and its component systems in its attempts to understand, diagnose, and treat disease. Community medicine on the other hand has an interest that begins with the human organism and looks at it in terms of increasingly complex social systems. Although the overall concern of both fields is man, the perspectives differ, as do the approaches taken. The contrast in approaches is indicated by the differing scientific disciplines utilized in each field. Medicine has as its basic sciences biochemistry, microbiology, anatomy, pharmacology, physiology, and pathology. The sciences basic to community medicine include epidemiology, biostatistics, operations research, systems analysis, management sciences, sociology, political science, anthropology, and economics.

The vast scope of interests embraced by community medicine accounts for some of the confusion concerning what it comprises. No program in community medicine does or even can encompass the whole of these areas. The faculty and staff for such an ambitious undertaking would have to be enormous in terms of individuals and diversity and depth of interest. Accordingly, community medicine as practiced throughout the United States has limited its efforts to specific areas. Thus, while departments of medicine across the country are similar, departments of community medicine vary considerably, depending on the areas each has selected for emphasis in teaching, research, and service programs.

Community medicine evolved as a response to the limited scope and concern of public health and its virtually complete separation from the practice of medicine. Community medicine attempts to synthesize the two. It operates as a problem-solving process in the community, based on a philosophy of the health of individuals and families as seen from the consumer's perspective.

To some, community medicine has a much narrower meaning: the delivery of comprehensive health services to a defined population, such as in a neighborhood health center. While this definition is not contradicted by the concept articulated above, it is rather subsumed as the discussion focuses more broadly on health-care systems and all that they entail.

Community medicine as an established program is a fairly recent phenomenon. Departments of community medicine were established by schools of medicine in the United States during the 1960s. This move was largely in recognition of the fact that the traditional programs in medical education did not address the full range of existing problems. Notwithstanding its roots in public health and preventive medicine, community medicine must be viewed as a profession in its infancy.

Recent developments. Unfortunately, a discussion of the recent developments in community medicine does not lead to the specification of particular events or breakthroughs. There are seldom, if ever, scientific discoveries such as occur in molecular biology or clinical medicine, but rather the conduct of research that leads to program innovations and the launching of new health-care systems. In this section some of the program developments will be discussed.

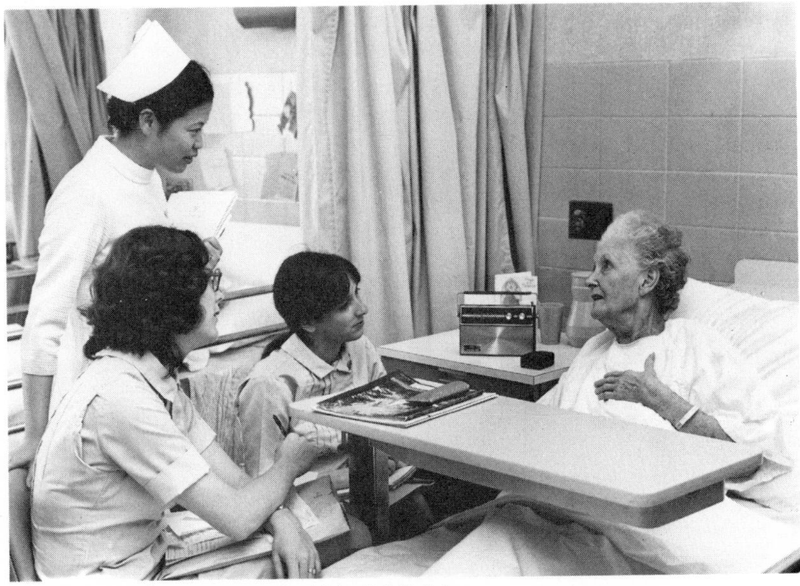

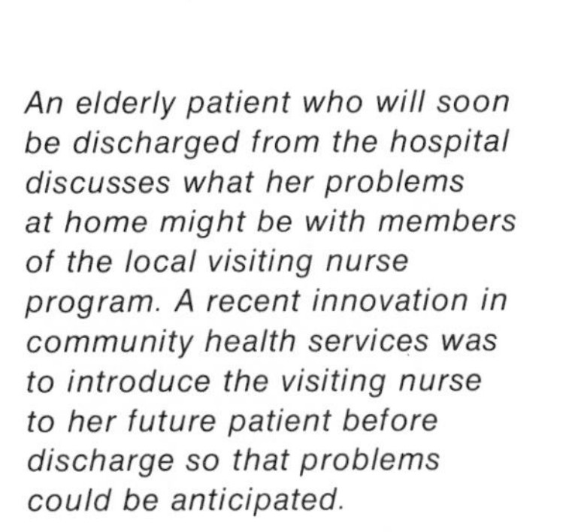

An elderly patient who will soon be discharged from the hospital discusses what her problems at home might be with members of the local visiting nurse program. A recent innovation in community health services was to introduce the visiting nurse to her future patient before discharge so that problems could be anticipated.

Camera Arts Studio, courtesy, Mt. Sinai School of Medicine

Much activity across the country has occurred in the planning, organization, and establishment of neighborhood health centers. These endeavors seek to provide comprehensive health care to a defined population, usually varying from 10,000–30,000. Teams of health professionals work in these centers along with new kinds of health workers such as physician's assistants, nurse practitioners, and community health aides. All of these centers began with outside grant support and as of 1973 were looking to ways of continuing their financing through Medicare, Medicaid, and other mechanisms of third-party reimbursement for services provided. These health centers often have brought much-improved service to the socioeconomically disadvantaged.

Washington, D.C., medical examiner James L. Luke views a homicide victim. Use of medical examiners, appointed forensic pathologists, instead of elected coroners was being recommended in the U.S. In addition to detecting murders, such specially trained physicians were able to detect various community health hazards.

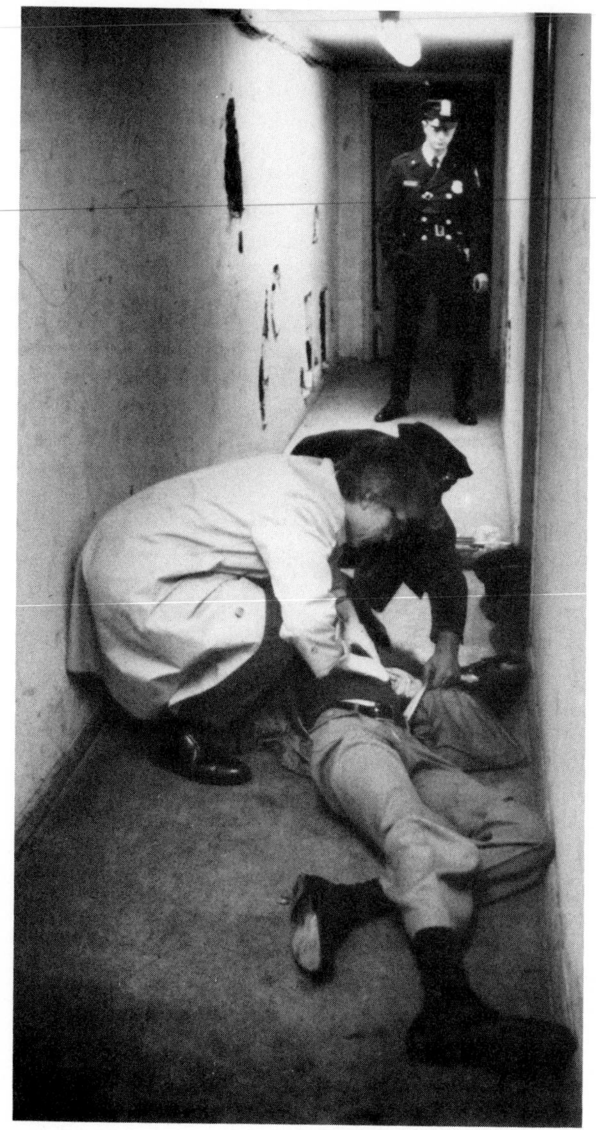

Joe Baker, MEDICAL WORLD NEWS

Extensive developments in health planning were occurring in many areas. In the U.S. these activities ranged from studies in the methodology of health planning to the training of health planners. Also, academic programs became increasingly involved with ongoing health-planning activities on an area-wide and regional basis. Approaches to health planning in other countries, particularly Sweden, the United Kingdom, and Yugoslavia, were being reviewed in order to assess their applicability to the U.S. Much adaptation would be required, as most nations with successful health planning had a structured health-care system controlled by the public sector. In the U.S., in contrast, health care was delivered on a primarily voluntary basis by thousands of private-enterprise practitioners.

Health-care administration is another area in which much activity was taking place. Management techniques developed in industry were being analyzed and applied to health-care systems. Previously, administrative considerations had always been de-emphasized in health and medical care. With their importance now being recognized, relevant research and educational programs were under way at about three dozen universities throughout the U.S.

Planning and operations of emergency medical systems had become a high-priority concern in the U.S. by 1973. Authorities had long recognized that this type of care could not be left to chance and required detailed formulation of strategies and operations. One of the major examples of this effort was the collaboration of several hospitals within individual communities to develop a comprehensive emergency medical system that would take advantage of the unique resources of each of the component institutions. In this area, particular note was being taken of the effective approach to emergency medical care in the Soviet Union.

Health-care financing remained a widespread concern as the debate in the U.S. intensified over national health insurance. Mechanisms for reimbursement that would yield the most effective and efficient health care were being sought. Relationships between third-party financing and payment by the government were being studied. Financing the full spectrum of health services to include extended care, home health service, dental care, and mental health, rather than just hospital care and some services by physicians, was being examined and tried.

Family medicine, the successor of general practice, was a concern of a number of departments of community medicine. Model practices were being established for service, education, and research interests. By and large, the focus was on delivery of comprehensive personal health services with a family orientation. Psychological, behavioral, and social aspects of illness were carefully considered both in diagnosis and treatment. A number of these practices incorporated physician's assistants and nurse practitioners in order to provide care on a more cost-effective basis.

Finally, health maintenance organizations (HMOs) were being planned and developed by a number of departments of community medicine. An HMO is generally either a prepaid group practice or a medical-care foundation. As is the case with the model family practices, these organizations were being developed as a setting for education and research as well as to provide services.

Future outlook. Recent trends in health affairs suggest rather strongly that the interests and approaches of community medicine will be increasingly critical in the years ahead. In all probability, the U.S. will realize national health insurance by the end of the 1970s or soon thereafter. At that time, the numerous disciplines involved in community medicine will almost certainly be called upon to assist in every kind of problem solving. Issues such as organization and relationships, mechanisms for resource allocation, and techniques for improving the effectiveness of health care will be of critical concern to the entire society.

Looked at in terms of the history of medical education, one can identify the decades to date in the 20th century as the era of scientific medicine. It is likely that the decades to come will be identified as the era of community medicine. In terms of a hierarchy of systems, the focus of priority is shifting from the molecular, tissue, and organ level of concern to clinical medicine and the levels of systems complexity represented by groups, institutions, and societies. Obviously all of these levels will be of continuing importance. The new emphasis, however, is mandated by the continuing evolution in patterns of illness and the needs for health care.

—William L. Kissick; Samuel P. Martin III

Dentistry

Children's dental care and the distribution of dental manpower continued to be high-priority items for the dental profession in the United States during 1972 and early 1973. Other developments included increased attention to acupuncture, a severe threat to the nation's dental schools because of proposed cutbacks in federal funding, and many promising findings in dental research.

Dental care problems. One of the key elements in the battle against tooth decay continued to be proper nutrition, and the U.S. dental profession in early 1973 expressed strong support for legislation that would establish a new program of nutrition education for children and regulate the in-school sale of sugar-rich foods and beverages that competed with the school lunch program. The profession's long-standing efforts to place the highest priority on dental care for children received new impetus through the introduction in Congress of a children's dental health bill. The measure would create a series of experimental dental care programs for indigent preschool children and those in the first five grades and would also provide one-time matching grants to communities and schools wishing to fluoridate their drinking water. Other governments recognized the same need. In Canada, for example, the government of Saskatchewan proposed a comprehensive "denticare" program for children.

The growing demand for dental care, spurred by a slowly increasing population, expansion of prepaid dental health plans, and greater public awareness of the importance of good oral health, continued to be a major concern of the profession. The latest estimates were that about 16 million persons were enrolled in dental prepayment plans in the U.S. and that by 1980 approximately 50 million would be covered.

In the U.S., as well as in other countries, the rising demand for dental treatment might lead to manpower distribution problems. One means being proposed to counter this situation was the expansion of dental auxiliary functions to relieve the dentist of most routine chores and enable him to see more patients. However, while the profession was studying this and other ways to increase the availability of dental care, the U.S. government proposed reductions in the funding of the health-professions' educational programs in fiscal year 1974 that could wipe out any hopes for improving the manpower situation. The American Dental Association declared such moves would hinder even the slightest expansion program of the dental schools, force drastic staff and faculty decreases, and deprive many students of the much needed financial assistance they were receiving under direct loan and scholarship programs.

Acupuncture, used in China for centuries, began to generate some interest in the dental research community. The ADA called for concentrated research on the possible use of acupuncture in dentistry but at the same time cautioned against its premature use pending further scientific evalu-

Courtesy, U.S. News and World Report

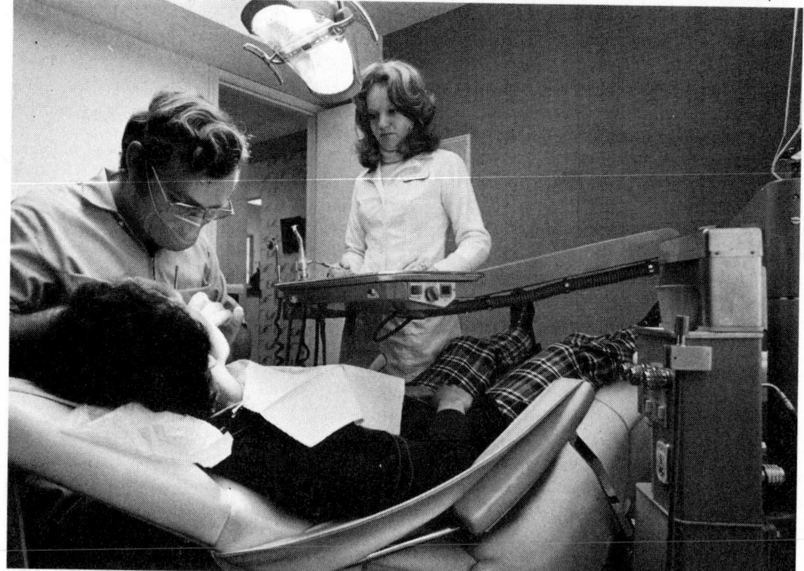

Form-fitting chairs were probably the most noticeable of recent advances that made a visit to the dentist a more comfortable experience. Other less visible improvements included high-speed drills, new painkillers, and new materials for preventing, repairing, or curing tooth and gum troubles.

ation of its effectiveness and safety. Acupuncture was used at least once, however, to demonstrate to a meeting of dentists its effectiveness as an anesthetic for tooth extraction.

Dental research findings. Experimental vaccination against tooth decay had generally proved to be disappointing, but in April 1973 dental scientists from the University of Connecticut and the National Institute of Dental Research reported encouraging although variable results in a new study. Three separate attempts to immunize rats against *Streptococcus mutans*, a bacterium that was strongly implicated as a prime decay-causing microorganism in human mouths, apparently achieved at least partial protection. While the scientists viewed their results as exciting and hopeful, they cautioned that in some cases immunization failed to occur. Other findings indicated that immunization against decay was a complicated task. For instance, although the salivary antibody apparently could inhibit the decay caused by *S. mutans*, it did not destroy these bacteria nor did it make the plaque (sticky masses of microorganism forming on teeth and considered to be a major factor in the development of decay and periodontal disease) disappear from the teeth. Furthermore, small sores developed in the animals' skin at the point of the immunizing injections.

At the University of California at San Francisco, dental scientist Jo Max Goodson provided a possible explanation of the bone loss occurring in periodontal (gum) disease and a possible way of controlling the destruction. He found that in periodontal disease gingival (gum) tissues contain ten times the normal levels of prostaglandins, hormonelike chemicals that can increase inflammation and cause bone to resorb or dissolve, and predicted that should drugs that inhibit prostaglandin production prove effective in controlling bone loss, natural teeth could be retained throughout life. Goodson tested vitamin E, which inhibits prostaglandin synthesis and is generally well tolerated, in 14 patients and found that after 21 days inflammation of the gums was reduced. He cautioned, however, that further studies must ascertain whether specific drugs could reduce prostaglandin levels in the gums and actually control bone loss.

Smokers have a higher incidence of periodontal disease than nonsmokers and the chief reason for this might be that they practice poorer oral hygiene, a University of Michigan public health expert, Richard C. Graves, found. A study of 388 men aged 35–59 years revealed that problems with receding gums were significantly higher and that personal oral hygiene was much poorer among smokers than nonsmokers. Previous studies suggested that smoking leads to the constriction of blood vessels in the gum tissues, thus reducing the circulation and speeding up the disease process. An interesting aspect of the study was the finding that those patients who had stopped smoking had lower levels of periodontal disease than either current tobacco users or nonsmokers. Graves theorized that these people had always been concerned about their health and had maintained good oral hygiene while smoking.

U.S. Army dentists suggested a new way to eliminate the frustrating habit of tooth grinding, or bruxism, by having tooth grinders practice grinding their teeth until they are uncomfortable. Prolonged tooth grinding can wear down the teeth and also cause painful misalignment of jaw bones. The technique could spare the dental and even psychiatric treatment sometimes required. By requiring a patient to practice consciously a habit that was largely unconscious, to the point of fatigue or slight discomfort, the patient might be induced to drop the habit.

A simple new technique utilizing a plastic sealant might be helpful in the treatment of fractured front teeth, according to Jorge Davila of the Eastman Dental Center, Rochester, N.Y. The new method called for the bonding of a plastic filling material to the broken tooth structure by means of an enamel coupling agent. The filling material is then shaped to resemble the original tooth and hardened by exposure to ultraviolet light for two to three minutes.

Although tooth implants were probably the most promising alternative to traditional bridgework and dentures, they were not always successful in patients because the materials, designs, and techniques were still imperfect. Lawrence Gettleman of the Harvard University School of Dental Medicine was experimenting with a new version of the polymer replica tooth implant that consisted of a polymethyl methacrylate resin, bone particles, and foaming agents to make the implant porous. The substance could be used either to make a replica of an extracted tooth for reimplantation or to coat metallic implant devices. Such implants had been tried successfully in an unsplinted, free-standing manner in baboons.

Researchers at the University of Southern California also reported encouraging results with implants using vitreous carbon, first used by the aerospace industry for various rocket parts. The substance appeared to have fewer of the drawbacks, such as infection or rejection, of previous implant substances. Vitreous carbon implants could be altered to fit differing situations in individual patients and inserted at an extraction site. Once the site had healed a crown or cap might be built on the implant to provide a tooth that will function well.

—Lou Joseph

Immunology

The great impetus behind immunological research in the previous decade was sustained in the early 1970s, but there was a distinct shift in the direction and emphasis of the work. Immunoglobulins, proteins that behave like antibodies in responding to foreign substances (antigens), had held the center of attention throughout the 1960s. The delineation of macromolecular structure and subunit composition of immunoglobulin G, the most dominant of these proteins, was followed by similar characterizations for immunoglobulins of other classes, which, in turn, set the stage for incisive studies of the function of immunoglobulins as antibody molecules. Moreover, investigations of the mechanisms of synthesis of the polypeptide chains and their assembly into antibody molecules became possible, and the way was opened to exciting research into the field of immunogenetics, genetic regulation of antibody production and distribution of antigenic determinants (allotypes) on the antibody molecules. (See *1971 Britannica Yearbook of Science and the Future:* MEDICINE, *Immunology.*)

These monumental achievements not only created a powerful and precise methodology, provided new insights, and generated fresh concepts, but, what was perhaps even more important, served to bolster the vigor of a whole scientific discipline. It was, therefore, most fitting that the 1972 Nobel Prize for Physiology or Medicine was bestowed on the two men who had determined the molecular structure of immunoglobulin, Rodney R. Porter and Gerald M. Edelman. (See HONORS: *Medical Sciences.*) Immunologic research in the early 1970s was focusing on the cells involved in the immunologic mechanism, and particularly on the specialization of cellular functions and the separation of cells with distinctive functions. This shift did not necessarily reflect a belief that little more could be learned about antibodies. Indeed, many immunologists were assiduously and successfully continuing investigations in that area.

Humoral and cellular immunity. The two branches of the body's immune machinery are the humoral and the cellular. Immunoglobulins, or more precisely, antibodies in the circulatory system, are the principal agents of humoral immunity; the lymphocyte, a type of white blood cell, is a pivotal component of both types of immunity. Humoral immunity encompasses the processes of antibody production and numerous antibody effects. Some of these effects are favorable, such as neutralization of bacterial toxins, prevention of attachment of viruses to susceptible cells, and sensitization of bacteria to destruction (lysis) by complement, a complex defense mechanism in sera. Other effects are adverse; for example, the formation of antigen-antibody complexes can, in combination with complement, damage tissues.

Cell-mediated immunity, on the other hand, is manifested through such effects as phagocytosis (by which leukocyte blood cells engulf and consume foreign bodies) and antigen clearance, cellular infiltrations of lymphocytes, monocytes (phagocytic cells), and other cell types, and positive, delayed skin reactions to a variety of sensitizing

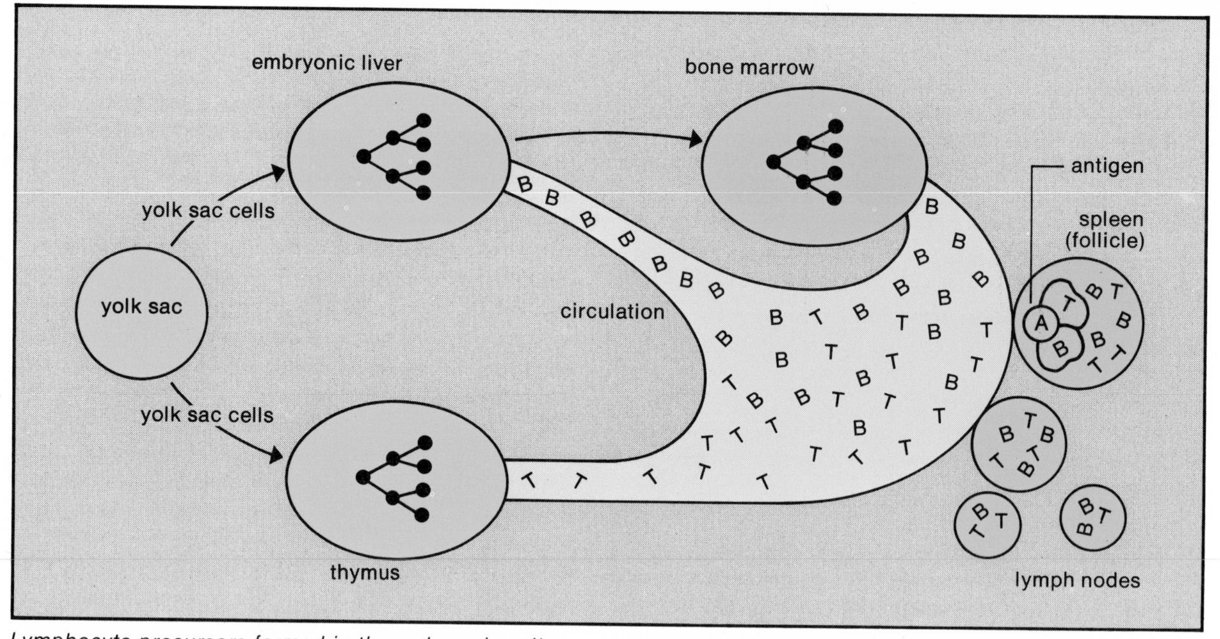

Lymphocyte precursors formed in the embryonic yolk sac are sent to the thymus, where T cells are formed, and to the liver, and later the bone marrow, where B cells are formed. B and T cells taken up from the circulation by the spleen and lymph nodes work together to recognize antigens and synthesize antibodies.

antigens. Cellular immunity is considered vital and, sometimes, more important than humoral immunity in the body's resistance to certain bacteria (tubercle bacilli), mycotic (fungal) agents, and certain viruses.

Much of the current interest in cell-mediated immunity, however, was based on its role in the body's resistance to cancer and in its rejection of organ homografts. In addition, technological innovations had increased the capability to isolate cells, to recognize fine details in the architecture and topography of membranes of lymphocytes and macrophages, to detect minute amounts of immunologically reactive substances, and to measure cellular actions and reactions with great precision. As a result, the lymphocytes were proven to be the primary elements of specific recognition of antigens and were demonstrated to exist in distinctive classes and, possibly, other hierarchic subpopulations with specialized functions, some relevant to the humoral immunity (antibody production), others to cellular immunity, and still others to forging a link between the two branches of the immune mechanism or to imposing regulatory controls on them.

B and T cells. Lymphocytes originate from stem cells that migrate during fetal life from the yolk sac to the thymus, liver, and bone marrow. In the chick they also migrate to a specialized organ in the intestinal tract, the bursa of Fabricius. Although final maturation of lymphocytes occurs in peripheral lymphoid organs such as the spleen, initial differentiation coupled to proliferation takes place in these primary sites. In addition, bone marrow remains the source of stem cell renewal.

Cells that have undergone initial differentiation in the thymus are designated T cells, while those that bypass the thymus are called B cells. In the case of the chicken, B stands for bursa-derived; in the case of a mammal, for bone-marrow-derived. According to recent evidence there are within these classes cells that perform different types of functions or respond to different stimuli. It was not known whether these different kinds of cells are derivatives of parallel lines of differentiation or whether they represent stages within a single pathway of differentiation.

Among the end products of maturation of B cells are the plasma cells, whose main function is the production and secretion of antibody molecules, a property that clearly sets apart B cells from T cells. The T cells of the mouse possess on their surfaces a unique antigen, the Θ antigen, that is absent from the surfaces of mouse B cells. B cells, on the other hand, carry relatively large amounts of antibody-like receptor sites on their surfaces, whereas T cells seem to be lacking or much more deficient in this respect.

Although T cells do not produce or secrete antibody molecules, they collaborate with B cells in this function but in a way that remained obscure: not all antigens require this participation of T cells.

Antigens that can stimulate B cells toward antibody production are termed thymus independent, a distinction that is sometimes a matter of physical state of the antigen. It was demonstrated, for example, that antigens in monomeric form require the participation of T cells but are thymus independent in polymeric form. Earlier studies had made use of haptens (small molecules that by themselves are not immunogenic) conjugated to proteins to construct immunogens (prepared antigens) with two major antigenic determinants (one, the hapten; the other, the protein carrier) on a single molecule. These investigations clearly established the phenomenon of the hapten-carrier effect in which T cells react with the carrier portion of the antigen and B cells with the haptenic determinant. The two cells were presumed to be linked by the composite antigen. Such cooperation required that both cells be in close proximity.

More recent experiments suggested an alternative mechanism of cooperation in which T cells stimulated by antigen shed some of their surface receptors, referred to by Australian scientists as immunoglobulin T. The receptors become complexed with the soluble antigen and become bound to auxiliary cells, such as macrophages (large phagocytes), which then present the antigen in a polymerized form to the B cells.

These types of interaction were considered antigenically specific. There were other studies of recent vintage indicating that T cells can release soluble factors upon activation by antigens different from the protein carrier. These antigens include alloantigens, histocompatibility antigens on cell membranes, also called transplantation antigens, that subserve the function of discrimination between "self" and "non-self." These factors can amplify the B cell response in what appeared to be a nonspecific manner. One such factor was partially characterized as a protein with a molecular weight of 75,000, which is smaller than an antibody molecule.

New information on T cells. The versatility of T cells comes into full prominence when their role in cell-mediated immunity is examined. There they recognize antigens and respond to them either by direct or indirect action. New information from experiments with cell cultures was providing correlates of cell-mediated immunity in living systems. Such studies indicated that lymphocytes from immunized animals (sensitized or "educated" T cells) show greater effects in response to certain antigens than lymphocytes from animals that have not been immunized. "Uneducated" cells can be immunized or sensitized by exposure to antigens. One of the important effects exerted by activated T cells is the killing of target cells bearing antigenic determinants to which the cells have been sensitized. This is a specific type of cytotoxicity.

In the activation of T cells the lymphocytes, which are in a relatively small resting stage, become enlarged (lymphoblast transformation) and enter the proliferative stage that is preceded by increased synthesis of RNA, protein, and DNA. This blast transformation reaction (BTR) can be assayed by measuring the increased uptake of radioactive precursors by the activated cells. Because this reaction can be induced by cells carrying membrane antigens different from those present on the responding T cells, it can be used to detect transplantation antigens. If little or no BTR occurs in a culture of lymphocytes from the prospective donor and recipient of a transplant organ, the two individuals are considered to be histocompatible and the transplant is given a good chance of success. The same principle has been employed for the detection of new antigens, in-

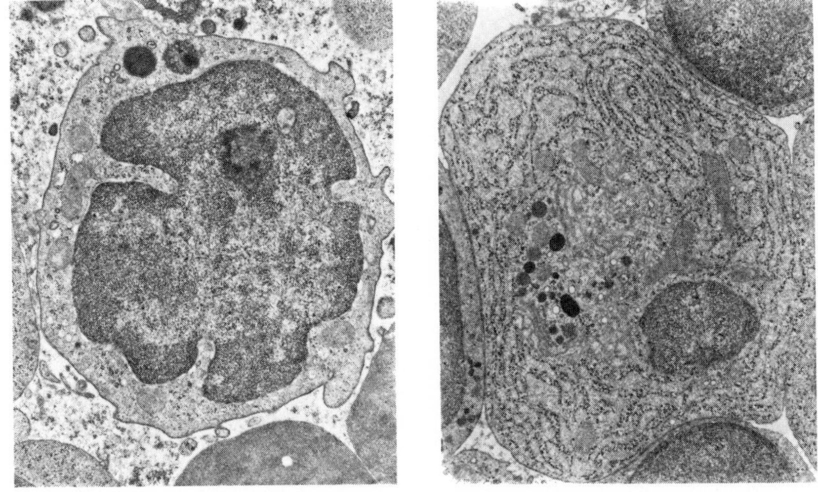

Courtesy, Dr. Joseph D. Feldman, Scripps Clinic and Research Foundation

A lymphocyte (left) is changed into a blood plasma cell (right) during an immune response. The cell increases in size, with a relative decrease in the size of the nucleus (dark masses), and ribosomes (small black dots) and other intercellular bodies involved in antibody production begin to appear.

cluding tumor antigens characteristic of cancer cells.

BTR can also be induced by nonspecific stimuli, exemplified by unique proteins (lectins derived from plants) with affinity to bind certain sugars. Examples of such proteins are phytohemagglutinin (PHA) and concanavalin A (Con A). BTRs induced this way serve as convenient means for distinguishing T cells from B cells. B cells apparently do not respond to these two agents but do undergo BTR when exposed to bacterial endotoxin, which does not cause BTR in T cells.

Another distinction between T cells and B cells was demonstrated recently with regard to their role in the cellular activation associated with killer action. Laboratory studies on the enhancement of tumor growth by complexes of antibody and tumor antigen demonstrated that killer effect on the part of T cells can be blocked by an antibody specific for the target cell or by antigen-antibody complexes. In contrast, antigen-antibody complexes are required for the induction of the killer effect on the part of B cells.

The activation of T cells also results in the production and release of soluble mediators of cellular immunity called lymphokines, an array of substances with beneficial or adverse effects on the operation of both branches of the immune mechanism. These include factors that induce blastogenic transformation, which, according to recent findings, may be formed by cells that are not undergoing cell division; substances that activate macrophages or inhibit their migration, thereby leading to their concentration in areas of immunologic combat; interferon, a basic protein produced on exposure to a virus that restrains virus replication; lymphotoxin, which, at least under experimental conditions, can damage tissue cells; and the factor that apparently causes delayed skin reactions. Some of these factors undoubtedly contribute to the destruction of target cells, but it should be kept in mind that the killer action of T cells requires close contact with the targets, indicating a different mechanism of action.

Negative interactions. Cooperation of lymphocytes in the evocation of the immune response is a highly important interaction, and its discovery was a milestone in modern immunology. Laboratory resolution of these problems was providing insight into the role of cell-mediated immunity in the medical problems of tissue graft rejection and resistance against cancer. Blast-transformation and cytotoxic reactions had also become valuable tools in several domains of research.

Equally important was a new awareness that there may be a negative type of interaction in which one type of lymphocyte suppresses the action of another type. This may provide regulatory control of the two arms of the immune mechanism. Very recent studies indicated that T cells are responsible for the induction of tolerance, or nonresponsiveness, to antigenic stimulation, that they can suppress the production of antibodies by B cells, and that lymphocytes can inhibit the BTR induced in T cells by PHA. In these experiments it was not ascertained whether the suppressors were T cells or B cells, but it was clearly demonstrated elsewhere that B cells derived from a chicken's bursa could inhibit the blastogenic transformation of T cells. It was quite likely, therefore, that both classes of lymphocytes participate in the modulation and regulation of immune processes.

The large spectrum of reactions performed by T cells suggested that there may be T cell subpopulations with specialized functions. This supposition rested on the demonstration of subsets of T cells with such differential characteristics as the propensity to localize in certain organs, the capability to recirculate, and the capacity to respond differentially to PHA and Con A.

Receptor sites. B cells possess on their surfaces a variety of receptors, the foremost being those that recognize and react with antigens. These are composed of immunoglobulin molecules similar to the ones that will be produced in large quantities and secreted when the B cells become differentiated to plasma cells. B cells also have receptors for a portion of the immunoglobulin and for at least one component of complement. B cells are also capable of binding to antigen-antibody complexes, which is not the case with T cells. This distinction made possible the recent separation of T cells from B cells in experiments where lymphocytes were passed through columns charged with antigen-antibody complexes.

The separated populations could be studied in relation to their specialized functions. It was demonstrated, for example, that both T cells and B cells could bind antigen but that T cells were less efficient; they required prolonged periods of time, apparently because of either a lower concentration of binding sites on their surfaces or a lower binding affinity per site. While it was relatively easy to demonstrate immunoglobulin-like receptors on B cells, it was rather difficult to demonstrate them on the surfaces of T cells. Australian investigators, however, succeeded in demonstrating, by means of highly sensitive techniques, the presence of immunoglobulin similar to monomeric immunoglobulin M on membranes of T cells.

—M. Michael Sigel

Ophthalmology

Perhaps the one advance in ocular therapeutics with which the public at large was most familiar was the introduction of the so-called "soft" contact lens. Originally designed as a bandage for diseased corneas and only secondarily given optical consideration, these new contact lenses received a prominent place in lay literature, even including the *Wall Street Journal.*

Hard contact lenses, in one form or another, had been used for decades. For over a quarter of a century, methyl methacrylate had been the standard plastic from which corneal contact lenses were fashioned. After its introduction, fitting techniques as well as lens design and manufacture underwent tremendous changes. Modern hard lenses were about three millimeters smaller and one-third as thick as those made in 1948. Adaptation to contact lenses had become much easier, although problems still existed. Nevertheless, corneal swelling, warping, and (rarely) scarring still occurred at an alarming rate. There was a definite need for contact lenses that could minimize or eliminate these difficulties, and it had been hoped that soft lenses might provide the answer.

Certainly adaptation time was reduced to almost nothing by these pliable, water-absorbing lenses, and corneal warpage and swelling were greatly decreased if not eliminated. But this measure of success had not been achieved without some sacrifice. The problems inherent in soft lenses in their present form could have been anticipated from the nature of the materials used to make them. Soft materials defy all efforts at compounding good optical surfaces; vision, therefore, is somewhat less sharp than with the hard lenses, especially if the wearer has any degree of astigmatism. They are porous and absorb water, but they absorb chemical and possibly microbial contaminants with the water and must be hydrated and sterilized daily in a ritual 20-minute boiling saline bath.

The idea was a good one, however, and the search for the ideal soft contact lens continued. The U.S. Food and Drug Administration was investigating a lens made of silicone rubber. This lens was somewhere between hard and soft lenses in rigidity, and so could have a fairly good optical surface; it was not water-absorbing, so it did not need boiling; and it seemed to allow oxygen to pass through it so that the layer of living cells on the corneal surface would not undergo any oxygen deprivation. Whether this lens would be the ultimate solution was something that only time and experience could answer.

Ultrasound. Among the recent techniques that had captured the imagination of ophthalmologists was the use of ultrasound in "echo-ophthalmography." Although work with ultrasound dated back to the mid-1950s, its use in ophthalmology did not gain widespread acceptance until about 1970 because of the lack of sufficient experience and standardization. Since then, ultrasonics has become part of the daily routine in many centers, and each year an increasing number of facilities were adding it to their equipment inventories.

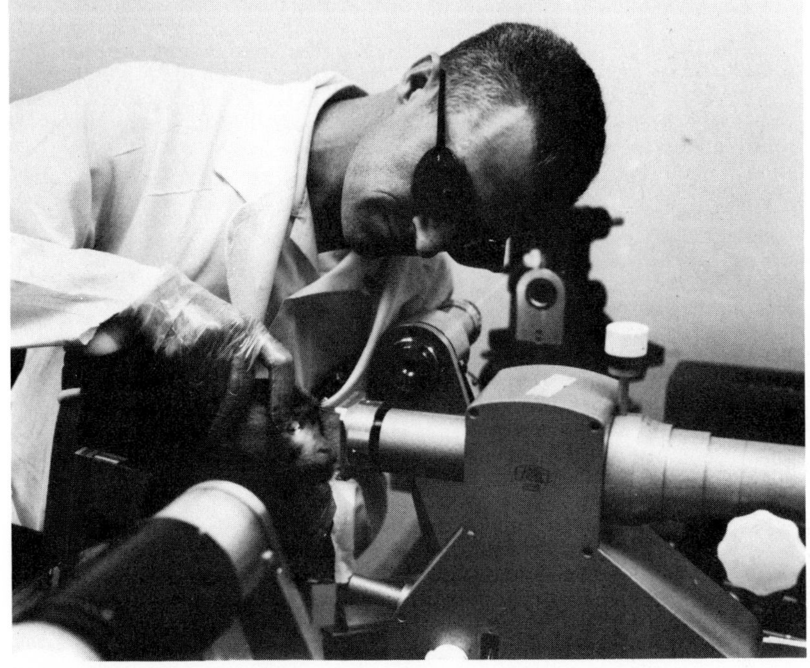

Courtesy, Stanford Research Institute

Uses for lasers in the treatment of glaucoma are tested on a monkey by Arthur Vassiliades at the Stanford Research Institute. Lasers formed the basic element in an experimental holography technique that measured exact distortions in the eyeball caused by glaucoma's characteristic build up of fluid pressure. They could also be used to puncture the iris to relieve the pressure.

Ultrasound shares many characteristics with sonar, used in the detection of underwater objects. Simply stated, a crystal (transducer) converts short bursts of electrical energy into high-frequency sound waves. When these waves strike objects of varying acoustic impedance or density, some of the sound energy is beamed backward into the crystal, which also serves as a receiver, and produces "spikes" on the sound beam baseline that are recorded on a cathode-ray oscilloscope. This tracing can be photographed to provide a permanent record for later analysis. This is time amplitude ultrasonography ("A" mode), and is the most frequently used technique. It is unidimensional, showing vertical spikes on a horizontal baseline; it is simple to learn and use; and it is relatively inexpensive. Scanned intensity modulated ultrasonography ("B" mode) is a bidimensional process that provides a silhouette of the eyeball and associated structures. It is a valuable procedure, but it is much more difficult and expensive.

The principal value of ultrasound in medicine is to demonstrate graphically the presence and nature of structures not normally visualized by standard procedures such as direct observation and X ray. When opacities in the normally clear media of the eye—the cornea, lens, and vitreous—make it impossible to see lesions within the eye, ultrasound can demonstrate their presence and location. Even when certain growths and space-occupying lesions of the interior of the eye can be seen, their exact nature cannot always be determined. An echo-gram profile of the questionable area can often permit differentiation of benign and malignant lesions, a differentiation that is of vital importance since biopsy is virtually impossible.

Soft tissue structures behind the globe are seldom demonstrable by X ray. Cysts and tumors in the bony orbit can cause an eye to bulge forward and can compress blood vessels and nerves. Such conditions can lead to loss of vision or can limit eye motion. Abnormalities on an echo-gram can be extremely helpful in determining both the probable cause and the reasonable mode of therapy in such cases.

Ultrasound was also being used in ocular therapeutics, most recently in the removal of cataracts. Normally, all eyes have a clear lens, but under certain conditions, such as aging or trauma, the lens may become opaque and obscure vision. This is called a cataract, and it must be removed if good vision is to be restored.

Since the diameter of a lens is about ten millimeters, an opening in the eyeball at least that large must be created to permit easy extraction. And since the incision must be placed at the junction of the sclera (white) and the cornea (clear), it must be curved, giving a linear dimension of about 19 mm. Viewing the cornea as a clock, this opening would extend from 9 to 3 o'clock, or 180°. This substantial wound must be made secure and must heal. This takes time, and patient rehabilitation must be cautious and slow. Complications, though infrequent, do occur. It is reasonable to assume, therefore, that if the same result could be achieved with a smaller incision, more rapid ambulation and fewer complications would result.

It is possible to manipulate a small ultrasonic probe into the eye through a very small (two–three millimeter) incision after cutting away the capsule (envelope) on the front of the lens, and to use it to emulsify (pulverize) the lens. This process is called phako-emulsification, and it leaves a milky suspension of lens particles in the eye that can be aspirated

The patient with a diseased cornea (left) received an experimental prosthesis that, six years later, permitted nearly normal vision and appearance (right). The artificial device consisted of a "bolt" inserted through a hole in the cornea and held from behind by a "nut," and use of a cosmetic contact lens.

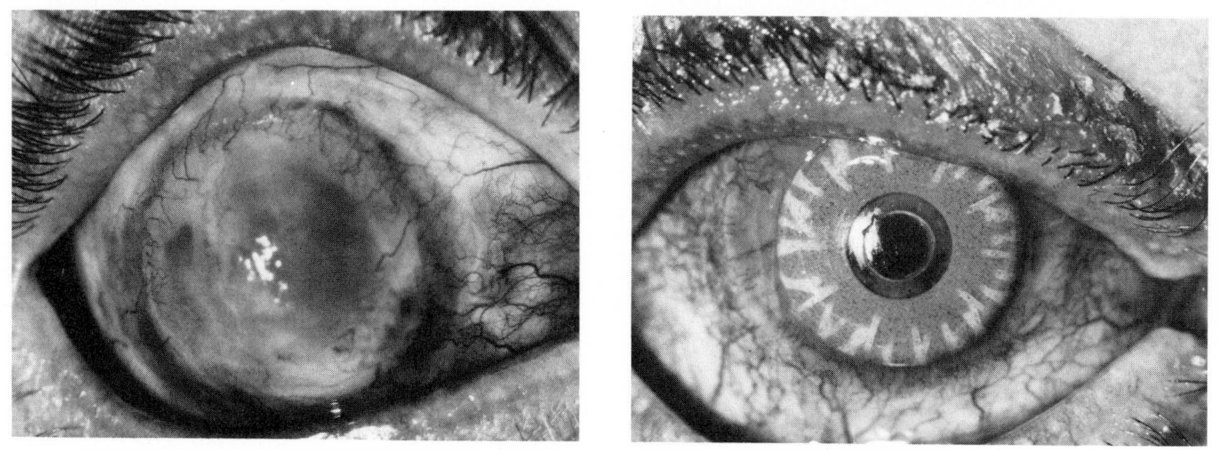

Courtesy, Dr. Hernando Cardona, Columbia-Presbyterian's Edward S. Harkness Eye Institute

and irrigated out through the small wound. The back capsule of the lens can then be teased out through the wound, and the wound can be closed with a single suture. This technique showed some promise, although it was far from ready for general use. Long-term follow-up on patients would tell whether the theoretical advantages of this procedure justified abandonment of old, proven techniques.

Fluorescein. In 1959 sodium fluorescein was used for the first time in ophthalmology to demonstrate normal and abnormal circulatory states in the retina. Shortly thereafter, photographic techniques were evolved to record the observed phenomena. The basic principle underlying all procedures utilizing fluorescein is that it glows or fluoresces under blue light of wavelength 465 mm. Since its molecules are relatively small, it escapes uniformly and rapidly from capillaries into the extravascular tissues and is reabsorbed equally rapidly. Where damaged or abnormal blood vessels exist, there is excessive leakage and accumulation of dye and a characteristic fluorescence of the tissue. This property permits the precise location of leaking blood vessels in a number of ocular diseases and allows differentiation between benign choroidal nevi (pigmented moles) and malignant melanomas.

Fluorescein angiography, as this process came to be called, became a standard procedure in virtually all ophthalmologic centers for the diagnosis of retinal pathology and for the teaching of normal and abnormal retinal physiology. It was not until about 1970, however, that fluorescein came into use in the study of the circulation of the iris or colored portion of the eye. With this technique, something of the nature of the iris circulation could be demonstrated for the first time in the living, and many disease processes peculiar to the iris could be studied in closer detail or at least from a fresh point of view. There was little reason to doubt that certain systemic disease processes may be associated with some variation in the iris circulation, and fluorescein angiography could become a simple screening test for many diseases that affect small blood vessels throughout the body.

Laser technology. Space age technology brought the laser (light amplification by stimulated emission of radiation), and since light and optical devices have always been the special interest of ophthalmology, it was only natural that the laser was soon being applied to the treatment of ocular pathology.

Lasers have many theoretical advantages in the treatment of ocular disease. Laser light is monochromatic, *i.e.,* the emitted light is all of a specific wavelength, depending on the properties of the generating substance. This may be a solid, such as a ruby, or a gas, such as argon. Many pigments absorb only specific wavelengths, and if the wavelength preference of a particular pigment is known, it is easy to select the appropriate laser. Laser light has no chromatic aberrations, permitting a very discrete, small area of impact, and it delivers a very high intensity of energy per unit time, so that small "burns" can be made in an instant. All this makes it possible to treat target sites while minimizing the possibility of damage to surrounding tissues.

There are two principal pigments in the eye, melanin (in the iris, retinal pigment epithelium, and choroid membrane) and hemoglobin (in the red blood cells). Melanin and hemoglobin are present in differing concentrations in different tissues, and the color of the tissue depends on how it absorbs or reflects various wavelengths of light. Keeping this in mind, it is possible to understand the applications of laser technology in ophthalmology.

In the treatment of localized retinal detachments, heat generated in retinal pigment epithelium cells leads to scarring that can be used to bind potentially separate sensory (visual) retina to the pigment epithelium and thus wall off gradually enlarging retinal separations.

By selecting light of the specific wavelength that is absorbed by hemoglobin, blood in abnormal vessels can be heated to the point where protein coagulation occurs, with a resultant occlusion or blocking of the vessel by solid coagulum. When the coagulum heals, a permanent scar-like occlusion remains. This is of great importance in sealing off the blood source to leaking vessels, thus reducing hemorrhage and/or swelling in the retina. This particular application is most promising in the treatment of the blood vessel disease that sometimes accompanies diabetes mellitus. Prior to the advent of laser photocoagulation, very little could be done to stop this major cause of blindness.

Malignant pigmented tumors within the eye have almost always indicated removal of the eye. Using a laser, it is often—though not always—possible to destroy these pigmented cancers, thereby salvaging at least some visual function. The value of this for a one-eyed patient who develops a malignancy in his remaining eye need not be underscored.

—Barton Lyle Hodes

Venereal diseases

An epidemic increase in the incidence of several sexually transmitted diseases continued in 1972 and 1973 in much of the world. There were 718,401 new cases of gonorrhea reported in the United States during the year ending June 30, 1972, one reported case (and perhaps three unreported ones)

Drs. Ivan L. Roth, H. Farzadegan and W. J. Callan, University of Georgia

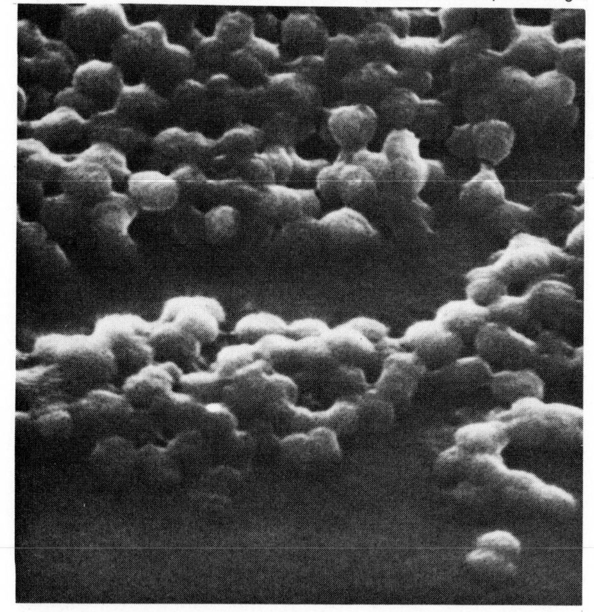

A colony of Neisseria gonorrhoeae, *the bacterium that causes gonorrhea, is seen in a scanning electron micrograph. The typical arrangement of the bacterium is two spherical cells attached to one another.*

for every 39 men aged 20–24. In 1970 a significant increase in the number of reported cases of syphilis in the U.S. had been noted for the first time since 1965, and further increases continued through 1972. Male homosexuals comprised up to two-thirds of all cases of syphilis reported in some large cities.

The growing problem of venereal diseases in the U.S. could be linked to the introduction of oral contraceptives in 1960, to the availability of intrauterine devices in 1964, and to a corresponding increase in sexual promiscuity that had been documented among young adults, particularly among women. Despite these profound changes and the worldwide "pandemic" of venereal diseases, basic research activities concerning both sexual behavior and the diseases themselves remained meager. However, important contributions to the understanding of certain venereal diseases were made during the past year.

Gonorrhea. While it had been known that asymptomatic female carriers of *Neisseria gonorrhoeae* were a major reservoir of gonorrhea, the epidemiology of gonorrhea was further clarified by studies in Norfolk, Va., and Memphis, Tenn., which showed that 10–20% of newly infected males also become asymptomatic carriers for periods up to one year, and by the demonstration of asymptomatic gonorrhea in about 1% of U.S. servicemen returning from Southeast Asia. The scope of the gonorrhea problem was underscored in a study sponsored by the Center for Disease Control (CDC) in Atlanta, Ga., that showed that salpingitis (inflammation of the Fallopian tubes) occurred as a complication in approximately 15% of female patients with gonorrhea. Bloodstream invasion by the gonococcus, usually resulting in arthritis, was found in 1% of a large series of patients with gonorrhea in Seattle, Wash., and in 1–3% of patients in Scandinavian countries. New perspectives of the spectrum of the disease were the reported occurrence of gonococcal throat infection in 20% of females and homosexual males with gonorrhea who practiced fellatio; and the description of a new syndrome in infants born to mothers with gonorrhea called the gonococcal amniotic infection syndrome.

Progressive gradual increases in resistance of the gonococcus to the penicillin and tetracycline antibiotics made treatment of gonorrhea more difficult. Studies established that resistance of gonococci to multiple antibiotics (cross-resistance) could be lost simultaneously and then reacquired by single-step genetic mutations. The loss was not associated with a detectable loss of the extrachromosomal genetic material in gonococci, and the mechanism remained unexplained. In hopes of curbing an increasing number of gonorrhea treatment failures, the U.S. Public Health Service in February 1972 revised upward the doses of penicillin recommended for the treatment of gonorrhea. A new antibiotic, spectinomycin hydrochloride, had been approved by the Food and Drug Administration in 1971, and an antibiotic combination containing a sulfonamide and trimethoprim was undergoing therapeutic trials in the U.S. during 1973.

Several investigators were involved in studies of immunity to gonorrhea. Field trials of an unpurified gonococcal vaccine were actually underway in Canada, under the supervision of Louis Greenberg of the Canadian Communicable Disease Center, who reported a detectable antibody response in 90% of those vaccinated. However, the protective benefit of such antibodies remained uncertain. In fact, it was not yet certain that any protective immunity to gonorrhea resulted even from natural infection. Most investigators, therefore, focused attention upon studies of the immune response during naturally acquired active infection, or upon isolation and purification of specific protein or polysaccharide components of the gonococcus that might evoke a protective antibody response when employed in a vaccine. A polysaccharide component recently isolated from a very closely related organism, *Neisseria meningitidis* (one cause of spinal meningitis), by Emil Gottschlich and others at the Walter Reed Army Institute of Research, was being employed in an effective vaccine to prevent meningitis in military recruits.

Although it appeared unlikely that a similar component would be found in *Neisseria gonorrhoeae,* one important step in characterizing other antigenic components of the gonococcus was the observance of minute hairlike projections from the surface of gonococcal cells examined with the electron microscope. These projections, called pili, were found only on the surface of virulent strains of gonococci, and might be related to infectivity of the organism. Use of purified pili antigen to detect a serum antibody rise during active gonococcal infection might provide the basis for a diagnostic serologic blood test for gonorrhea, which had eluded investigators for years. Another type of diagnostic blood test, employing the indirect fluorescent antibody (IFA) technique used for the diagnosis of syphilis, was also found to be promising for the diagnosis of gonorrhea. The technique was employed by Richard J. O'Reilly at the CDC to demonstrate the appearance in vaginal secretions of local secretory immunoglobulin A antibody to the gonococcus during the course of uncomplicated gonorrhea.

Since no experimental animal was known to be susceptible to gonorrhea, studies of gonorrhea had to be carried out in human volunteers. Recently, however, workers at the CDC reported that symptomatic genital gonorrhea was produced in chimpanzees with gonococci obtained from infected humans and that small laboratory animals could be infected by injection of the bacteria into hollow chambers implanted beneath the skin. There were other reasons to anticipate new advances in gonococcal research during the next one or two years. Several measures to prevent the transmission of venereal disease were being evaluated and the genetic variations responsible for acquisition of antibiotic resistance, or for loss of virulence and infectivity in strains of gonococci that are repeatedly grown in the laboratory, were being studied. Work was also in progress on typing systems by which different strains of gonococci could be distinguished. Attempts to isolate highly purified antigenic components of the gonococcus might lead to the discovery of strain-specific antigens that could be used in such a typing system, in new diagnostic serum tests, or even in a gonococcal vaccine.

Syphilis. Few diseases of man have caused as much suffering, or achieved as much historical, social, and cultural significance as syphilis. Yet knowledge concerning the basic biology and chemical and antigenic structure of the causative organism, *Treponema pallidum,* remained very limited. One of the most provocative observations made during the past decade was the recognition of the persistence of *T. pallidum* in the lymph nodes of patients, and in rabbits, long after presumably adequate penicillin treatment for syphilis. More recently, persistence of treponemes in the aqueous humor of the anterior chamber of the eye and in the cerebrospinal fluid were reported following high doses of penicillin. This apparent indifference of

Courtesy, Dr. John Swanson, University of Utah College of Medicine

Electron microscope techniques detect the minute projections called pili adhering to surfaces of gonococci, the bacteria that cause gonorrhea. Because pili were found only on virulent strains of gonococci, they were suspected of being related to the organism's infectivity.

T. pallidum to penicillin, particularly in the late or latent stages of syphilis, did not necessarily imply resistance of the organism to the drug; in 1971 the CDC indicated that penicillin G remained as effective as when it was first used for syphilis nearly 30 years before. Whether treponemal persistence was harmful was not clear, but the treatment of late stages of syphilis now required further study.

A few laboratories were engaged in the search for a syphilis vaccine. Suggestive evidence of immunity in rabbits was observed following a very prolonged series of vaccinations with large numbers of *T. pallidum* rendered nonvirulent by exposure to X-rays or by other methods. It also became possible to infect chimpanzees with pinta, a disease of humans that is closely related to syphilis, and researchers were interested to see whether such an infection would modify the chimp's susceptibility to experimental infection by *T. pallidum.* Although immediate prospects for a syphilis vaccine seemed poor because of the inability to cultivate and purify *T. pallidum* in an artificial environment, the factors involved in the immune response to syphilis would certainly soon be better defined.

Current studies suggested that the immune response may actually enhance certain harmful manifestations of syphilis. The kidney inflammation that sometimes occurs in the secondary stages of syphilis was shown to be associated with the deposition of antibody and complement (an inflammation-promoting substance) in the glomerular tufts of the kidney. New methods for the study of cell-mediated immunity were being evaluated. These included the macrophage-inhibition test, lymphocyte transformation, and passive transfer of cellular immunity in syngeneic (genetically identical, inbred) animals.

Other sexually transmitted infections. The five diseases traditionally called "venereal" are gonorrhea, syphilis, chancroid, lymphogranuloma venereum, and granuloma inguinale. Only the first two appeared to be major problems in the U.S. However, the list of other microorganisms shown to be sexually transmitted in humans was growing. Five diseases that are usually transmitted by sexual intercourse and that appeared to be increasing in epidemic proportions were genital herpes, trichomonas vaginitis, pubic lice, genital warts, and nongonococcal urethritis. Other infections caused by microorganisms that are sometimes transmitted sexually included yeast infection of the genitalia, scabies, viral hepatitis, molluscum contagiosum,

Pubic lice, or "crabs," was one of the infections usually sexually transmitted that was, apparently, increasing in epidemic proportions in 1973. The causative organism, Phthirus pubic, *resembles a tiny crab with clawed legs that cling to hair. Although its bite causes severe itching, the crab louse, unlike the other lice that infect man, transmits no known disease.*

Courtesy, King K. Holmes, University of Washington

and infection due to cytomegalovirus (closely related to herpesvirus). Significant events occurred in recent months in studies of genital herpes, nongonococcal urethritis, viral hepatitis, and cytomegalovirus.

The two sites most commonly infected by *Herpesvirus hominis* are the mouth and the genitalia. Recent studies showed that the type that causes oral infection, Type I, could be differentiated from the strain that causes genital infection, Type II. Some studies suggested that infection of the cervix by herpesvirus, particularly during adolescence, might be associated with an increased risk of cervical cancer years later. It was clear that many women could harbor *H. hominis* in the cervix for prolonged periods following acute infection, and that this virus might infect the newborn during passage through the birth canal, causing essentially untreatable illness.

An interesting method reported for the treatment of the less serious but painful local herpes ulcerative lesions on the genitalia, called photoinactivation treatment, involved the application of a photoactive dye, such as Neutral Red, to a lesion, permitting the dye to form a complex with the viral nucleic acid (DNA). The treated lesion was then exposed to fluorescent or incandescent light, allowing the dye to absorb photons of light energy and, presumably, leading to a reaction that causes single-strand breaks in the viral nucleic acid.

Nongonococcal urethritis (burning on urination, discharge from the penis but without gonococcal infection) was probably the most common sexually transmitted disease of men in both the U.S. and the U.K. Occasionally, NGU is complicated by a form of arthritis known as Reiter's syndrome. In recent studies in London and Seattle, a group of small intracellular microorganisms known as *Chlamydiae* was linked to many cases of NGU. The agent is closely related to the pathogens causing a chronic eye infection known as trachoma and also to the agent responsible for lymphogranuloma venereum. It is inhibited by tetracycline antibiotics, which are known to be effective in the treatment of NGU. Implication of the *Chlamydiae* as important genital pathogens was greatly facilitated by the development of markedly superior tissue cell culture techniques for primary isolation of these organisms. At the present time, it seemed likely that NGU might result from infection by several different organisms, including *Chlamydiae* and possibly "T"-strain mycoplasmas ("T" refers to the tiny colonies formed by these microorganisms on artificial culture medium), and perhaps even other unidentified organisms.

It was long recognized that hepatitis A virus is shed in feces, and that male homosexuals were at high risk of developing its disease. Recently it was also observed that the occurrence of hepatitis B (serum hepatitis) is also much higher among male homosexuals than among heterosexuals of either sex. Hepatitis B virus was also demonstrated in menstrual blood, and a group of investigators at Baylor University suggested that transmission of the disease, which may persist in the blood for long periods, may result from sexual exposure.

Cytomegalovirus (CMV), which produces an illness akin to infectious mononucleosis, was recovered in high proportion from semen as well as from the genital (cervical) secretions of young women. Although most women who carried CMV in the cervix had no symptoms, studies at the University of Alabama indicated the virus was transmitted to about one-third of the newborn children of women with genital infection. CMV infection had become one of the most common congenital infections, was thought to be a leading cause of deafness, and might have other long-term effects upon the nervous system that were currently under study.

Research goals. Despite the obvious seriousness of venereal disease problems, basic research in the field lagged behind that expended in many other areas of biological science. Under the stimulus of increased federal spending, however, it seemed inevitable that the bringing together of vigorous biochemical, immunologic, and genetic methodology would produce an explosion of new information concerning gonorrhea and syphilis during the next few years. Whether the ultimate goals of cultivation of *T. pallidum,* and the production of immunity to gonorrhea and syphilis by vaccination, were achievable with existing technology remained to be seen.

—King K. Holmes

Microbiology

An increasing awareness of the applied aspects of microbiology appeared to be developing in the scientific community by mid-1973. One heard and read more and more about the use of microorganisms to help solve environmental, industrial, nutritional, and medical problems. Thus, while this article covers highlights of the more significant advances in microbiology during recent months, it also attempts to predict future developments.

Environmental and applied microbiology. Some substances regarded as atmospheric pollutants result from natural origins rather than from manmade technology. A case in point was the occurrence of atmospheric sulfur compounds, principally from industrial sources but also in the form of

Courtesy, Kenneth W. Adolph and Robert Haselkorn, The University of Chicago

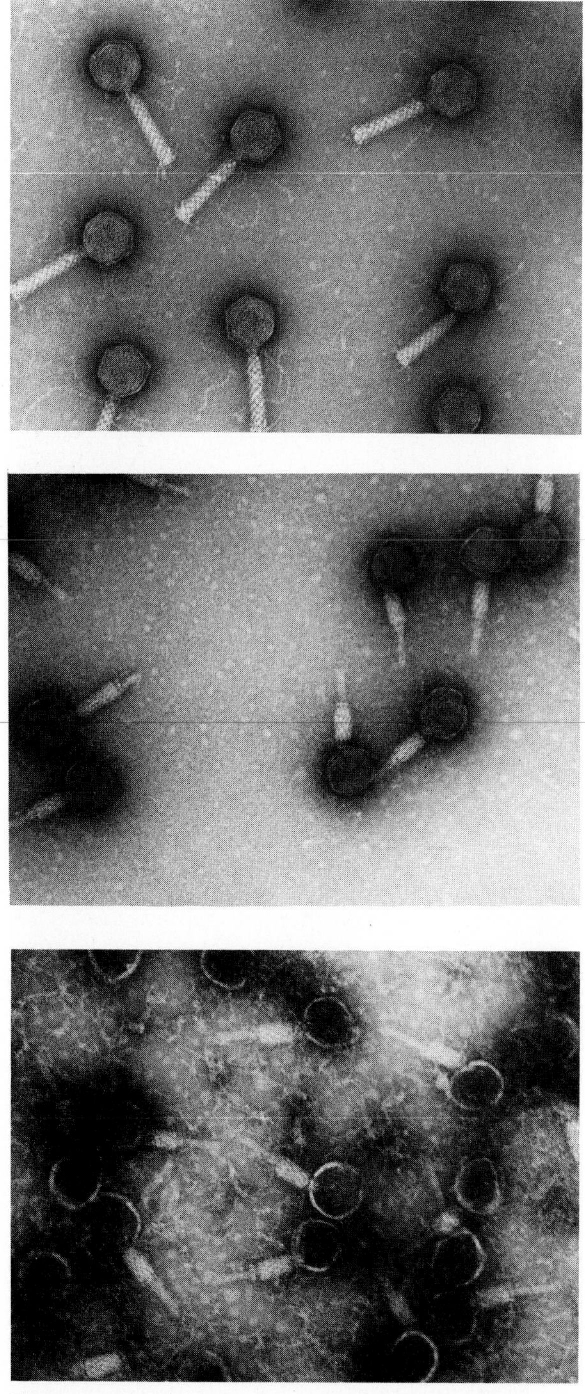

The means by which a newly discovered virus, N-1, infects the blue-green alga Nostoc muscorum *was shown in chemicals that mimicked collision of the virus with the algal cell surface. Prior to infection the viral "head" contains a single DNA molecule and is attached to a "tail," a sheath, surrounding a thin tube (top). Upon contact (center), the sheath contracts to expose the tube, which in nature would be forced into the host cell. The DNA is then ejected from the viral head (bottom). Magnification: approximately 100,000 times.*

bacteriogenic sulfur released from muds by anaerobic bacteria, those living in the absence of free oxygen. On a seasonal basis, the bacteriogenic source of atmospheric sulfur compounds may rival the industrial source in importance, and, during peak periods of microbial activity, may even be the dominant source in an area. Recent studies showed, for example, that bacteriogenic sulfur was 10% that of the smelters on an annual basis in Salt Lake City, Utah.

Scientists at Woods Hole (Mass.) Oceanographic Institution verified the implications of earlier deep-sea experiments of the dangers of dumping waste materials in the ocean because of their slow rate of decomposition. The work also indicated the unsuitability of shallower coastal waters for the dumping of untreated sewage and waste materials. Decomposition is less efficient and slower than on land. Moreover, the containment of the matter dumped in ocean waters is uncontrollable.

Microorganisms had long been known to be capable of concentrating substantial quantities of inorganic substances from their surroundings, such as metals of the rare earths class. Recently, scientists showed that bacteria can take up and concentrate lead and cadmium. Cadmium, especially, is a dangerous pollutant that can adversely affect health in a number of ways. (See *1973 Britannica Yearbook of Science and the Future* Feature Article: THE NOTORIOUS TRIO: MERCURY, LEAD, AND CADMIUM.) The uptake of cadmium by microorganisms might have significance in relation to the concentration of this toxic metal in food chains. These observations also suggested that bacteria could be used to remove metals such as these from certain environmental sources. Methods might also be devised to use bacteria to biologically "mine" metals of commercial importance; they had, in fact, been used recently to recover uranium compounds from low-grade ores and slag from mines. Similarly, Soviet engineers were attempting to use bacteria that utilize methane as a source of nutrition in deep mines to remove methane from the air.

Research had been in progress for a number of years to use radiation to preserve foods and medical supplies. The predictable seemed now to be happening: mutant forms of bacteria had been selected from the irradiation-susceptible population with many times the normal bacterial resistance to radiation. The bacteria appeared not to become resistant to the radiation as such but rather to develop more efficient cellular mechanisms to repair the radiation damage.

The bacterium *Bacillus thuringiensis* causes disease in at least 150 insect species. It was being tested over about three million acres in the U.S. via broad-scale aerial dispersion to determine its effects in controlling the gypsy moth, spruce budworm, and elm spanworm.

Experiments were under way in several laboratories to evaluate the usefulness of recycled animal wastes as animal feed ingredients. Animal wastes contain protein, minerals, products that yield metabolizable energy, and roughage materials. Recycled poultry litter, for example, was used experimentally to supplement feeds for laying hens and for cattle. Similarly, cattle manure was added to silage for cattle feeds. The U.S. Food and Drug Administration did not sanction the use of animal wastes as food for animals; it regarded them as adulterated because they could be expected to contain drugs and drug metabolites and toxic heavy metals, and feared that they might contain parasites, disease organisms, and microbial and fungal toxins.

Medical microbiology. One of the most interesting stories in microbiology in recent months was the elucidation of the effects of the toxin (poison) produced by *Vibrio cholera,* the bacterium that causes Asiatic cholera. This was made possible by recent experimental efforts in areas unrelated to cholera. The epidemic of cholera currently rampant on the Indian subcontinent and in parts of Africa stimulated research on the biochemical, immunological, and clinical aspects of this disease, characterized by a profuse diarrhea brought on by a toxin that is elaborated by *V. cholera* in the intestine. Researchers discovered that the toxin stimulates the enzyme adenylyl cyclase, which catalyzes the formation of an excessive amount of the cytoplasmic molecule cyclic AMP, which in turn, induces the hypersecretion of water and salts. The best known function of cyclic AMP is its mediation of the effects of a variety of hormones and other biologically active agents. (*See* Feature Article: CYCLIC AMP.)

Cholera toxin was further shown to mimic the action of prostaglandins, which also cause the production of cyclic AMP by stimulating adenylyl cyclase. Prostaglandins are a family of lipid (fatty) acids whose presence in many, if not all, mammalian tissue and ability to modify the actions of other naturally occurring physiological mediators had been studied in detail only in the last decade. Research was under way to prepare a vaccine that would induce immunity against the effects of the toxin. The present vaccine, composed of the killed bacterium, was not entirely adequate because the immunity it induced lasted only a short time and its protection was incomplete.

"....no more pleasant sight has ever come before my eyes than these many thousands of living creatures all alive in a little drop of water."
Anton Van Leeuwenhoek

One area of potential importance in microbial studies might be based on the identification of infectious agents that are smaller than viruses. Called viroids, these bits of low-molecular-weight

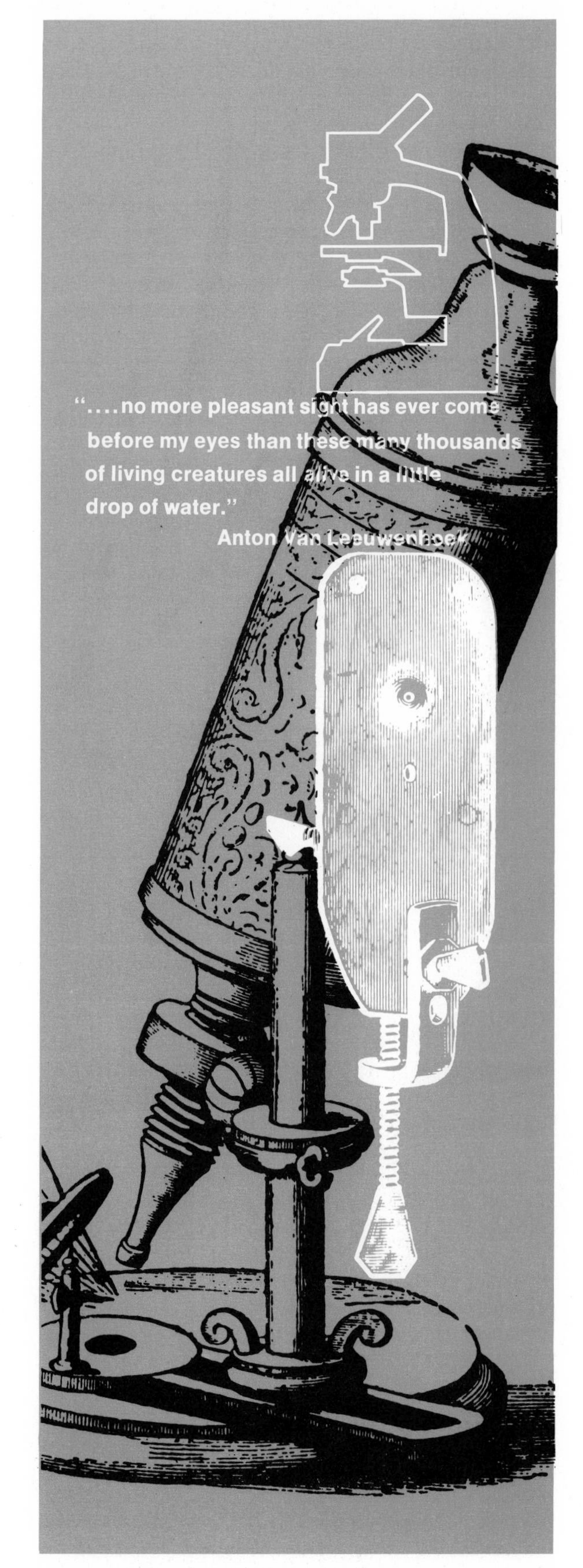

RNA had been established as causative agents of at least three plant diseases since 1971. These findings suggested that viroids might be involved in other plant, animal, and human diseases for which no infectious agent had been identified. (*See* Year in Review: MOLECULAR BIOLOGY, *Genetics.*)

Microbiology of the future. Microorganisms have been used to enhance the quality of life since the dawn of civilization. Bacteria, yeasts, and molds preserve foods, enhance their taste, and convert natural foods to more desirable products. Many additional applications of the microbe to the benefit of mankind took place during the last 30 years. For example, over 50 antibiotics produced by microorganisms became available to combat infections of humans, animals, and plants. Efficient industrial processes for the microbial production of vitamins, plant growth factors, enzymes, amino acids, flavor substances, and complex sugars were also developed during that period. The microbial conversion of the steroid molecule to yield new drugs introduced a new phase of biotechnology in which bioconversions were used as adjuncts to chemical steps in the manufacture of products.

Recently the oil and other industries undertook the conversion of petroleum products and fermentable organic wastes into edible microbial cell material. However, the actual application of this "single-cell" protein production to the world's food needs remained to be worked out. Biological control of insects also seemed possible. More than 300 different viruses were known insect pathogens, and hundreds of different bacteria, fungi, nematodes, and protozoa attack insects. Such microorganisms were host specific and, thus, considered to be safe to other forms of life; they spread rapidly, killed quickly, and exhibited predictable control. The replacement of chemical herbicides was also distinctly possible. Research on bacteria, viruses, and fungi that attack specific plants was part of the U.S. Army's biological warfare program. Further efforts could study and produce microbial herbicides for the control of undesirable plants. The safety of microbial pesticides, their control, and

Grant Heilman

Aerial spraying with chemicals remained the principal method of protecting valuable crops, such as this apple orchard, from insects and plant pests. However, host-specific microorganisms—safer, quicker agents of control—might soon be widely used once their full environmental impact could be thoroughly tested.

their impact on the environment would have to be tested thoroughly before they could be used extensively. Nevertheless, it seemed safe to predict that the control of insect and plant pests by biological agents was near.

Technical discoveries that would further the applications of the microbe to the benefit of mankind, particularly in medical areas, might include the increased use of enzymes of microbial origin in analytical chemistry and in medicine, dentistry, pharmacology, and industry; economical microbial production of essential amino acids; better and safer antibiotics, especially for gram-negative bacteria; nontoxic agents to treat diseases caused by fungi; microbial agents active against infections caused by protozoa; antiviral and antitumor agents of microbial origin; and microbial products for use in treatment of diseases not caused by microorganisms.

One aspect of microbiology that appeared not to be in the future was the use of microorganisms to solve the energy crisis, at least as far as methane production was concerned. While several bacteria produced methane as a major product from the oxidation of simple organic compounds coupled with carbon dioxide reduction, some of the organic compounds that served as nutrients for these bacteria, such as methyl or ethyl alcohol, were excellent fuels themselves. Other nutrients were too expensive or would have to be first produced by other microorganisms. The conversion process would produce enormous quantities of acetic acid that would have to be disposed of. There would also be staggering engineering problems involved not only in handling the microbial cell mass of hundreds of tons, but also in supplying the nutrients and maintaining sterile conditions. Methane might also be produced by fermenting waste materials, such as garbage, municipal sewage, and industrial wastes. But, in addition to posing the same problems mentioned above, hydrogen sulfide would be produced in the fermentation of most waste materials. Since fuels rich in sulfides were already major contributors to atmospheric pollution, this would not be permitted.

—Robert G. Eagon

Molecular biology

Use of the instruments, techniques, and breakthroughs that had been noteworthy in past years permitted steady progress to be made between mid-1972 and mid-1973 in all disciplines studying life at the molecular level. Several aspects of cellular life could now be understood in considerable detail.

Courtesy, Drs. H. T. Bonnett, Jr. and E. H. Newcomb, University of Wisconsin

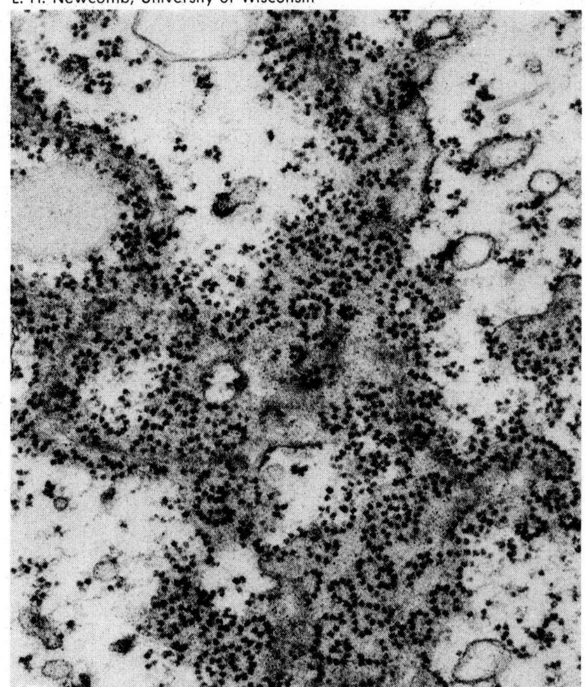

A longitudinal section from a surface cell of a radish root shows numerous ribosomes, sites of protein synthesis, forming whorl- and rosette-shaped patterns on the endoplasmic reticulum, the network of membranes in cytoplasm. (Enlarged: 37,000 times.)

Biochemistry

Advances in many different areas of biochemistry during the past year furthered understanding of the relationships between the structure and function of the chemical constituents in living things. The diversity of biochemical research was evident from several particularly noteworthy studies that were underway, including those on transfer-ribonucleic acid, a special type of nucleic acid that plays a vital role in the biosynthesis of proteins, on biological membranes and some of the protein constituents of the membranes from specific types of cells, and on vitamin C.

Transfer-RNA studies. Many investigations over the past 10–20 years helped to establish the overall picture of protein biosynthesis. It was generally agreed that the genetic information in DNA is first transferred to three different types of ribonucleic acids (RNAs), designated as transfer RNA (t-RNA), messenger RNA (m-RNA), and ribosomal RNA (r-RNA). These nucleic acids in conjunction with many proteins, including a variety of enzymes, act in concert to allow the stepwise synthesis of specific polypeptide chains, which in turn become proteins with unique structures and biological functions. In protein synthesis, the amino acids, which are the fundamental chemical units of all proteins, must first combine with t-RNA to form amino acid–t-RNA complexes.

Much information accumulated in recent years about the structure and function of t-RNA. In the mid-1960s R. Holley and co-workers at Cornell University first established the sequence of the individual nucleotides in a t-RNA from yeast, and since that time about 40 t-RNA molecules of different kinds and from several sources have been sequenced. All have a molecular weight of about 25,000 and contain approximately 80 individual nucleotides. Each chain of all t-RNA molecules can be depicted by a two-dimensional cloverleaf pattern, in which specific segments of the chain are paired to form a stem and other segments form loops.

Although the cloverleaf pattern was useful in suggesting how portions of t-RNA molecules may be combined so as to provide specific structures that endow the molecule with several functions, in itself the pattern did not indicate how the molecule is arranged in space. A group of investigators at the Massachusetts Institute of Technology (MIT), led by A. Rich and S. H. Kim, reported in December 1972 the three-dimensional structure of the t-RNA from yeast that combines with the amino acid phenylalanine. These workers deduced the structure from X-ray diffraction analysis of t-RNA crystals, and, although similar crystallographic studies had given considerable insight into the three-dimensional structures of several proteins in the last few years, this was the first ribonucleic acid for which the three-dimensional structure was established.

It was found that phenylalanine t-RNA has a structure consistent with the predicted cloverleaf pattern but different from that proposed in any previous model. The t-RNA molecule contains two segments of double helix, each about one turn in length and formed by the pairing of purine and pyrimidine bases much in the same manner as first found in DNA. The two double helical segments are oriented to each other at approximately right angles so as to form an L shape. Four loop regions are clearly formed, two of which are near the corner of the L and another, the so-called anticodon loop, which must interact with m-RNA, at one end of the L. The other end of the L is the acceptor end, which combines with the amino acid phenylalanine. Researchers found it particularly noteworthy that the structure could account very well for the various properties of t-RNA in solution, including the way it is cleaved by enzymes, how it is specifically modified by certain reagents, and how it binds with nucleotides.

Biological membranes. The membranous structures in a living cell are involved in the organization of the chemical constituents in the cell and in the regulation of metabolic processes. The surface of a cell is formed by the plasma membrane that not only serves as a barrier between the internal and external environments of the cell but also permits the selective transfer of substances in and out of the cell. The plasma membrane is connected to the

The configuration of transfer-RNA nucleotides was thought to resemble a cloverleaf (left) until X-ray crystallography revealed the three-dimensional structure (center) of one t-RNA to have two twisted, double-stranded regions that formed an L-shaped pattern. A new diagram (right), transforming the cloverleaf to show the physical connections in the molecule, was, therefore, proposed.

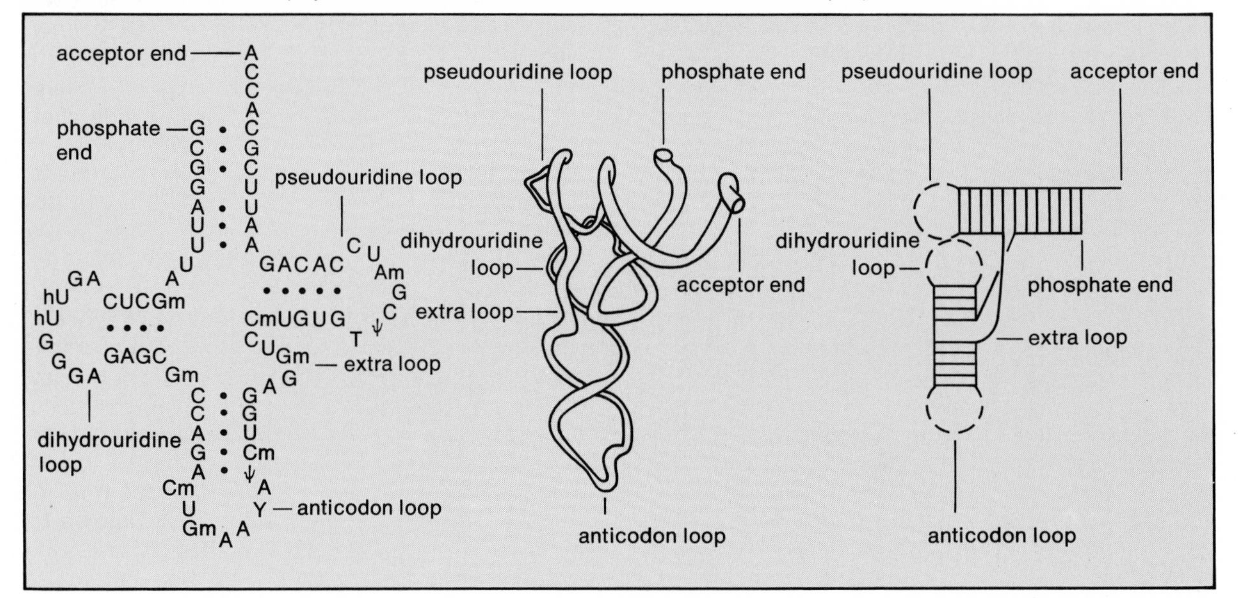

interior of the cell by another membranous network, called the endoplasmic reticulum, where many special functions, such as protein synthesis and assembly, occur. Other specialized membranes are associated with subcellular structures, including the mitochondria, the nucleus, and lysosomes. It was generally agreed that cell membranes, irrespective of their location and function, are complex structures resulting from association of specific proteins with a variety of lipids (fatty substances). There is very little carbohydrate in membranes and when present it is usually bound to a specific type of protein, the glycoproteins.

Although there continued to be much discussion as to the actual detailed structure of cell membranes (see *Biophysics*, below), it was recognized that membranes from specific cells might often contain a major protein component uniquely associated with that membrane. For example, J. Reynolds and H. Green at Duke University in Durham, N.C., examined myelin membranes from porcine (pig) brain, and found several proteins unique to this special material. One of the proteins, however, the so-called basic protein, which can induce an allergic disease resembling multiple sclerosis in experimental animals, accounts for 25–30% of the total protein in myelin, the fatty cover of some nerve cells.

Another example was provided by the protein component of retinal rods, light receptors in the eye. H. Heitzman at Yale University showed that 80% of the protein present in membranes from bovine (cattle) rod outer segments was a single polypeptide species identical to the polypeptide chain of the visual receptor protein, rhodopsin. Similar results were obtained for frog rod outer segments by W. E. Robinson and co-workers at the University of Wisconsin. The researchers concluded that the predominance of specific proteins in specialized membranes undoubtedly endows that membrane with specific properties.

Other studies provided insight into the organization of specific proteins in the biological membrane. Unlike many proteins, which were thought to act in a cell much as they would in solution in the laboratory, membrane proteins might exert their specific effects by virtue of their environment and location within a membrane. Studies by P. Strittmatter and co-workers at the University of Connecticut revealed the structural organization of cytochrome b_5 in the membranes of the endoplasmic reticulum of mammalian cells. Cytochrome b_5 is a protein functionally associated with an enzyme, cytochrome-b_5 reductase, and is a vital component for oxidation-reduction reactions in the cell. Strittmatter's group showed that cytochrome b_5 contains two domains, one of which lies wholly outside the membrane and is in contact with an aqueous environment. The other domain anchors the protein to the membrane and binds with the lipid components of the membrane. The two domains are joined by a small length of polypeptide chain.

A similar type of structural organization was suggested for the major glycoprotein component of erythrocyte (red-corpuscle) membranes by V. Marchesi and co-workers at the National Institutes of Health (NIH), Bethesda, Md. This protein was shown to contain about 65% carbohydrate by weight, which is attached to a protein backbone as discrete polysaccharide chains attached to one-half the length of the protein chain. It was believed that half of the protein containing the carbohydrate is located largely outside the membrane proper. Since the carbohydrates have blood-group-specific determinants, the protein itself must be located on the external or outer surface of the cell. That portion of the chain that does not contain carbohydrate was found to contain a stretch of 23 amino acids that are very hydrophobic (lacking in affinity for water) and would provide a structure capable of combining with lipids. This suggested that this region anchors the protein to the membrane. Thus, two domains appeared to be present in this glycoprotein, just as in cytochrome b_5, one extending away from the membrane into the aqueous environment outside the cell and the other penetrating the lipid-rich membrane.

Considerable attention was also given to the metabolic reactions associated with the synthesis of glycoproteins, such as those found in biological membranes. Of particular interest was further understanding of the mechanisms for addition of the carbohydrate groups to the glycoproteins. It was believed that the polypeptide chains of glycoproteins are synthesized in a manner identical to that of other proteins and that the carbohydrate groups are added only after the intact polypeptide chain has been synthesized. For some glycoproteins, it appeared that the carbohydrate groups are transferred to the protein from nucleotide-sugars (the sugar part of a nucleic acid unit) by the following reaction: nucleotide-sugar + protein → sugar-protein + nucleotide. By repeating this reaction, several individual sugars from different types of nucleotide-sugars could be added to the protein to produce the glycoprotein with its appropriate carbohydrate side chain.

On the basis of recent studies it appeared that synthesis of some glycoproteins might be a more complex process. Of particular interest was the possible role of vitamin A. It had been known for many years that vitamin-A deficiency markedly affects virtually every organ. But, except for its

Courtesy, Dr. A. V. Crewe, University of Chicago

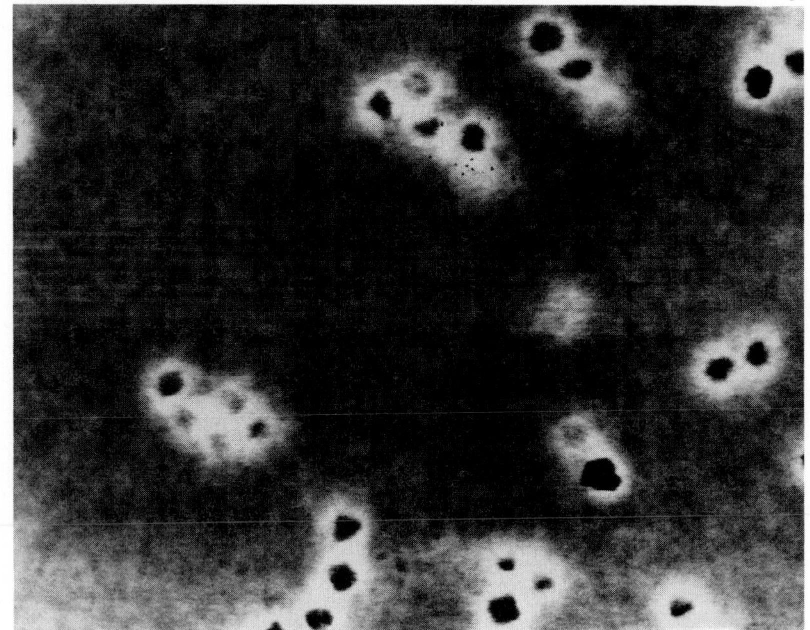

Scanning electron microscopy shows the dark cores of ferrous hydroxide within the protein coats of bovine ferritin. A molecule about 140 Å in diameter, ferritin stores the body's excess iron.

involvement in vision, vitamin A's metabolic function in many tissues remained obscure. On the basis of studies in the laboratories of G. Wolf at MIT and L. DeLuca at NIH, it appeared that carbohydrate derivatives of vitamin A are synthesized in liver. These workers showed that an enzyme system in rat liver catalyzes the formation of retinylphosphomannose from retinol (vitamin A) and guanosine diphosphomannose, a nucleotide-sugar. Evidence was also obtained that nucleotide-sugars containing glucose and glucuronic acid may also react with retinol to give similar retinylphosphosugars. Similarly, T. Helting and P. A. Peterson at the University of Uppsala, Sweden, reported the synthesis of a retinylphospho-galactose derivative in mouse mast-cell tumors. It was interesting that the vitamin A-sugar derivatives resemble chemically the carbohydrate derivatives of polyisoprenols, which were also shown a few years before to play a role in the synthesis of oligosaccharides in bacterial cell walls.

Vitamin C. Vitamin C was identified in 1932 and established to be an essential nutrient in the diet of humans; however, during the past few years some question was raised about the exact amounts of vitamin C that humans require to maintain good health and development. Linus Pauling at Stanford University was the first to popularize the question in 1970 in his book *Vitamin C and the Common Cold*. He suggested, on the basis of several arguments, which were often indirect, that human needs for the vitamin may well have been underestimated by a factor of ten or more, and that perhaps a much higher dietary intake could alleviate the susceptibility of humans to the common cold. Clearly, evidence against Pauling's suggestions could be cited, but it was equally clear that the questions he posed could be answered with considerable certainty by further research on the biological action of vitamin C, including its dietary requirements.

Recently Man-Li S. Yew at the University of Texas reported the results of studies designed to reevaluate dietary levels of vitamin C. Groups of guinea pigs (the only animal other than primates that must ingest vitamin C) were fed a diet complete in all known essential nutrients supplemented at different and wide levels (up to 1,000-fold) with vitamin C. The animals were examined in several ways, including growth before and after surgical stress, recovery times after anesthesia, scab formation, wound healing, and production of amino acids in skin during wound healing. Based on the results obtained, Yew estimated that guinea pigs need about 5 mg vitamin C per 100 g body weight per day, a value much higher than that required to prevent scurvy, the disease produced after prolonged intake of diets deficient in the vitamin. Under stress, the guinea pigs appeared to have an even higher requirement.

The implications of these studies for human dietary requirements could not be judged unequivocally, but Yew believed that the results suggested that the required intake of the vitamin may be 10–20 times greater than the usual recommended daily allowance. It was believed, however, that daily requirements may vary greatly among individuals and that young persons may require a much higher intake than supposed. Yew's studies did not indicate exactly what the recommended daily allowance for humans should be. In addition, enough doubt was raised from them so that a more exact answer would have to be obtained by further research.

—Robert L. Hill

Biophysics

The Fourth International Biophysics Congress held in Moscow Aug. 7–14, 1972, gave scientists from the East and West a chance to compare notes in such areas as the structure and functions of proteins, nucleic acids, membranes, and free radicals in biological systems. Neuronal organization, biophysics of motility and of reception, and medical applications of biophysics were other areas of interest.

New techniques. Many varieties of light and electron microscopes were commonplace in biophysical laboratories. Recently, a scanning X-ray microscope was constructed using the soft (weakly penetrating) X-ray portion of the synchrotron radiation from an electron accelerator as the source. By using a quartz ellipsoidal condensing mirror, hard X rays and high-energy bremsstrahlung (literally "braking radiation," or radiation produced by the rapid deceleration of high-energy electrons) were not reflected, but low-energy X rays, from about 100 eV (electron volts) to 3,500 eV (3.5 KeV), were focused into an intense, narrow beam. The X-ray beam was collimated (made parallel) by passing it through a pinhole approximately 2 microns (1 micron, $\mu = 10^{-6}$ m) in diameter in a gold foil. The emergent beam scanned one point at a time of a specimen that moved in a zigzag pattern perpendicular to the axis of the X-ray beam. The intensity of the characteristic fluorescent X rays emitted from each point of the specimen was detected with proportional counters. Because the counters could be set to select only X-ray energies corresponding to radiations of a given element, a picture could be formed of the distribution of that element in the specimen.

The scanning X-ray microscope appeared to be well suited for use with biological, even live, specimens because it operated in an atmospheric environment and required little or no special sample preparation. There were still many problems to be resolved, such as resolution (about 2 μ), discrimination between adjacent elements, and the range of elements that could be examined by fluorescence (those lighter than potassium). But, when perfected, the scanning X-ray microscope promised to be a welcome addition to the biophysicist's tools for determining biological structure.

Fourier transform nuclear magnetic resonance. The introduction in the past year of Fourier transform (FT) methods began to revolutionize nuclear magnetic resonance (NMR) spectroscopy, making possible studies that previously were too time-consuming to undertake. In conventional continuous-wave NMR spectroscopy the entire spectrum is examined for resonances by slowly varying a "monochromatic" frequency from a continuously operating source; in FT-NMR a short intense pulse of radio-frequency energy is applied to the sample. Because this pulsing procedure gives a broad-

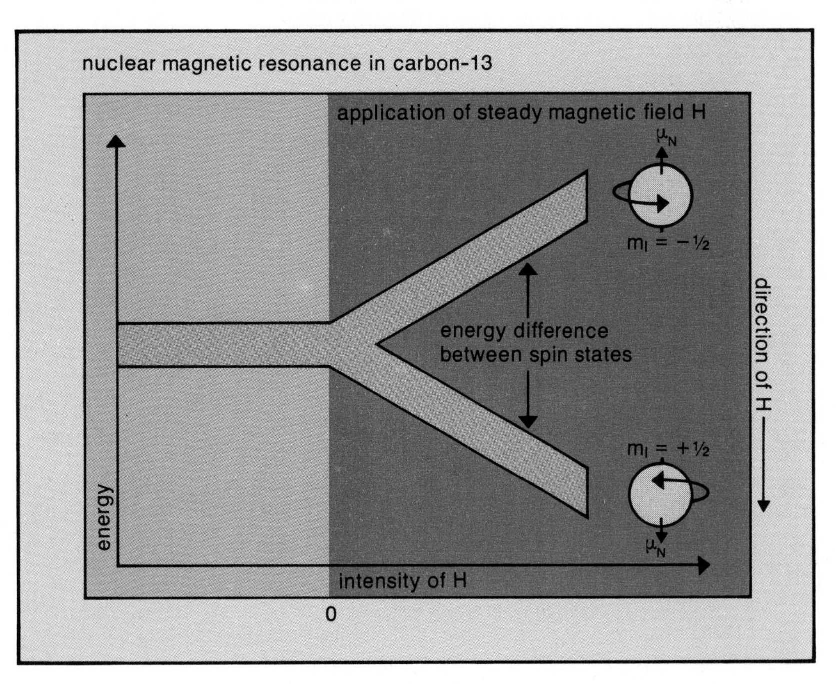

A nucleus can spin at certain distinct quantized energy states, designated by m_I numbers; for carbon-13, $m_I = -1/2$ and $+1/2$. The combination of spin and charge causes a nucleus to have a magnetic moment (μ_N), which interacts with an applied steady magnetic field so that it is either parallel ($+1/2$) or antiparallel ($-1/2$). When an oscillating electromagnetic field of proper vector and frequency is applied, nuclei in the lower energy state absorb this energy and "flip" to the higher state. Nuclear magnetic resonance spectroscopy depends on observing these transitions.

band, rather than a monochromatic, source of irradiation, each pulse is equivalent to a slow sweep through the entire spectrum in conventional continuous-wave NMR.

An experiment that would require about 5,000 sec to obtain a complete spectrum with a resolution of 1 Hz (one hertz = one cycle per sec) by conventional carbon-13 (^{13}C) NMR spectroscopy could be done in 1 sec using FT methods. The rapid acquisition of data using FT methods also made it possible to improve the sensitivity of NMR spectroscopy by computer "time-averaging" the signals, that is, by coherently adding the weak NMR signals to increase their intensity relative to the noise signals. If data from a thousand repetitive scans were necessary for the desired signal-to-noise ratio and resolution in the final ^{13}C NMR spectrum, the procedure would take about 15 min using FT but about 60 days by conventional means.

With the advent of FT methods, ^{13}C NMR blossomed. Although ^{13}C was in short supply and was considerably less sensitive than protons (^{1}H), there were several reasons why ^{13}C FT-NMR spectroscopy could be more useful than ^{1}H NMR for studying the structure and conformation of large organic molecules, especially those of biological interest. The most important of these was that the carbon atoms make up the backbone of organic molecules and are usually more sensitive than protons to the intramolecular environment and less sensitive to the intermolecular environment.

Researchers found that the amino acid residues are virtually independent of their positions in the long peptide chain in denatured proteins, leading to well-resolved spectra but little information on amino acid sequence. By 1973 only two proteins, ribonuclease A and lysozyme, had been investigated in native form by ^{13}C NMR. In lysozyme

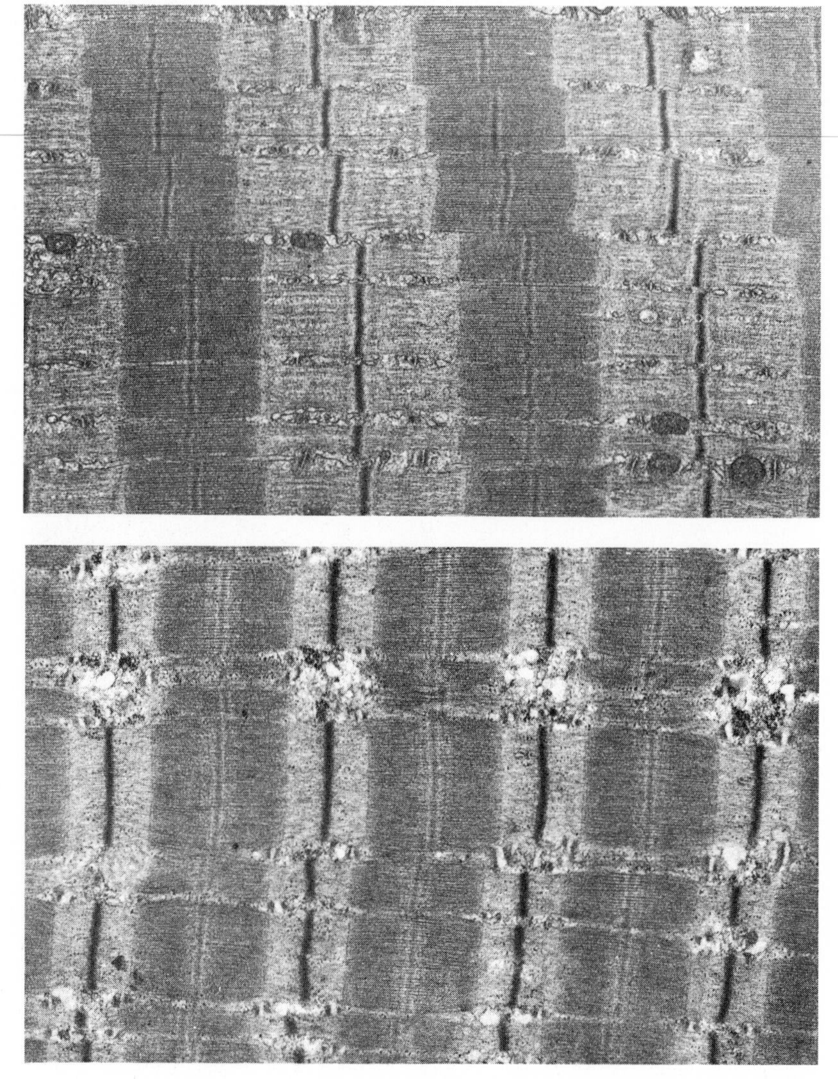

The mechanism by which muscles contract was becoming understood at the molecular level. Electron micrographs show voluntary (striated) muscle from adult rat legs in a relaxed state (top) and partially contracted (bottom). The dimensions of the dark wide A bands do not change during contraction. The widths of the lighter I bands and the distance between the narrow black Z bands is reduced. (Magnification 12,000 times.)

Courtesy, Dr. Donald A. Fischman, University of Chicago

it was shown that the same kind of amino acid, when located in different parts of the polypeptide chain, could give rise to different chemical shifts because of differences in local environments.

Membrane study techniques. Studies of membrane structure and function continued to occupy a prominent niche in the biophysical milieu. In addition to the well-established techniques of X-ray diffraction and electron microscopy, two new techniques, ^{13}C NMR and spin labeling, were being added. Earlier studies indicated that membranes consist of a bimolecular phospholipid core (two molecules of fatty complexes), large portions of which are covered with proteins and other molecules. Some ^{13}C NMR work using natural membranes showed several broad resonances. But most of the ^{13}C NMR studies used synthetic membranes made of lecithin, one of the phospholipids. Well-resolved NMR spectra were obtained only when experiments were carried out under conditions that kept the interior of the membrane in a liquid rather than "crystalline" form. Spin-lattice relaxation times varied with the position of the carbon in the molecule and also depended on whether the lecithin was in a bilayer, in a micelle (aggregation), or in solution. The carbon chemical shifts, however, were nearly independent of the state and nature of aggregation of the lecithin.

Electron spin resonance (ESR) spectroscopy using nitroxide free radicals was gaining wide acceptance as a probe of the physical state of artificial and biological membranes. Fatty acids labeled with nitroxide at various positions along the hydrocarbon chain were the most widely used type of probe (commonly called spin labels). These were added either *in vivo* or *in vitro* to membranes, and they readily associated with the membrane lipids. Spin-label studies using both biological membranes and pure phospholipid bilayer membranes produced essentially identical ESR spectra under a number of different conditions. From these studies it was concluded that many biological membranes contain phospholipid bilayers, as X-ray diffraction and differential thermal analysis studies had suggested.

Other recent spin-label experiments indicated that the incorporation of proteins into synthetic phospholipid bilayers results in an ordering of the phospholipid chains, that is, an increase in their rigidity. It was suggested that this is caused by the penetration of the proteins into the hydrocarbon-chain region of the membrane and by the binding of a layer of phospholipid onto the surface of the protein molecule.

Both NMR and spin-label studies indicated a transition in the fluidity of membranes as the temperature was lowered. It was thought that a connection between these lipid dynamics and biological activity might be the reason why fatty-acyl chains of bacteria, fungi, plants, insects, and marine organisms become richer in unsaturates as their growth temperature is lowered. Recent spin-label studies indicated that indeed the membrane from bacterial cells grown at 55° C was more rigid and viscous than the membrane from identical cells grown at 37° C. These results were consistent with the idea that the organism adjusts the lipid composition so as to keep its lipid fluidity in a range favorable for the functioning of membrane-bound enzymes.

Muscle studies. Work continued on the mechanism of muscle contraction. It was well known that the voluntary (striated) muscle contains filaments of thin protein molecules (actin) alternating with thick protein molecules (myosin). During muscle contraction the myosin molecules bridge over to actin filaments. The "tails" of the myosin molecules compose the backbones of the myosin filaments, and the "heads" of the molecules do the bridging. The myosin "heads" were thought to attach perpendicularly onto the actin filaments and then to tilt about 45°, drawing the actin filaments along. By attaching, tilting, and reattaching, and always tilting in the same direction, the myosin filaments on both ends of a muscle segment (sarcomere) move the actin filaments to the center of the sarcomere and cause it to contract. The energy for this contraction is supplied by the hydrolysis (breakdown) of the energy-rich molecules of adenosine triphosphate (ATP).

Another protein in muscle that was of interest was tropomyosin, which was thought to keep muscle fibers from contracting during the resting state. X-ray diffraction studies showed that tropomyosin is in one position on the actin filament during rest and moves to another position during contraction. It was suggested from these results that tropomyosin keeps muscle fibers from contracting by blocking the attachment of myosin to actin when the muscle is at rest. Thus, the resolution of the mechanism of muscle contraction at the molecular level was well on the way, but the resolution at the atomic level still lay ahead.

Regeneration findings. It has been known for some time that in animals capable of the regeneration of tails, limbs, or even heart tissue there was a larger proportion of nerve tissue and more electrical activity in the area of regeneration than in similar areas of animals without regenerative capacity. This suggested that regeneration was brought about by electric currents in the tissue. If this were true, then would it be possible to cause regeneration in animals normally unable to regenerate by applying electric currents?

In a recent experiment parts of the right foreleg were amputated from rats. The rats were then divided into four groups. One group had no further treatment, but the three other groups had current-generating devices giving low, medium, and high currents, respectively, implanted in the amputation sites. Those rats with little or no current applied showed no regeneration while those receiving high currents showed bone destruction. However, those rats receiving medium currents of 3–6 nanoamperes (1 nanoampere = 10^{-9} amperes) showed clear evidence of regeneration. Broken bones were also found to heal faster when stimulated by direct current. In other recent work cartilage regeneration in the joints of mammals was enhanced by this technique, suggesting a possible application to the treatment of arthritis.

—Jeremy E. Baptist

Genetics

At the molecular level, genetics and virology had merged by 1973 to the point where it became a distortion to separate studies that may have been initiated from very different points of view and that might have implications in either field. This could be illustrated by the continuing perplexities about scrapie, long recognized as a contagious affliction of sheep that progresses very slowly, with obvious involvement of the brain, and is eventually fatal. It attracted attention because of its resemblance to several human diseases, such as multiple sclerosis and kuru. Despite assiduous efforts, the scrapie virus had not been isolated. In fact, it resisted rigorous treatments that would be expected to destroy most agents containing either RNA or DNA and was extraordinarily resistant to ultraviolet light, presenting an effective cross section equivalent to a molecular weight of only 150,000—barely enough for a single gene, if it indeed consisted of nucleic acid.

These confusing facts provoked considerable speculation about whether scrapie was caused by a virus at all—if by virus is meant a particle containing informational nucleic acid and protected by a protein coat. Perhaps scrapie might be an enzyme normally present in tissue cells but bound to the membrane; it could then be postulated that once freed, this enzyme would be able to release similar particles from normal cells. Until self-propagating mutational variants of scrapie could be found, such a hypothesis remained tenable—as it had with respect to other viruses before genetic techniques were developed for them.

It might be premature, however, to abandon the idea that scrapie is a remarkably small virus (called a viroid), growing very slowly, and, perhaps, using host-cell materials for its coat to minimize the informational requirements of its minute core. H. K. Narang, B. Shenton, P. P. Giorgi, and E. J. Field, of the Medical Research Council's Demyelinating Diseases Unit at Newcastle (Eng.) General Hospital, published high-power electron micrograph figures showing tiny sausage-shaped particles with an estimated molecular weight of about two million and a dense core whose size, they said, "accords well with that of the target nucleic acid estimate" determined in previous studies. These particles were found only in the neurones of scrapie-infected rats. Obviously much more work was needed to verify whether they indeed corresponded to a mini-virus and what the role of similar agents might be in other diseases and in genetic transmission.

Another lead to possible clarification of viruslike agents came from studies of two plant diseases, potato spindle tuber disease and citrus exocortis, which may be the same, as judged by their effect on tomato plants. Different workers obtained disparate estimates of molecular weight of the infectious agent, ranging from 25,000 to 120,000—the latter was already very small compared to other known viruses. They agreed, however, that the agent is a free ribonucleic acid, similar in many ways to transfer RNA (t-RNA), which activates specific amino acids for transfer to a growing protein chain. Unlike scrapie, the RNA of the plant viroid would break down.

Artificial viroids. A number of experiments had begun on the evolution of artificial viroids in simple, cell-free systems, using the known enzymology of nucleic acid replication. The assumption that any nucleic acid sequence would act as a passive template for information-duplication catalyzed by a polymerase, was not quite correct. The RNA polymerase associated with the bacteriophage Qβ proved to be moderately specific in terms of the templates it would accept. (A bacteriophage is a virus that destroys bacteria.) However, Sol Spiegelman and his associates at the Institute of Cancer Research, Columbia University, isolated a fragment only 218 nucleotides long that was accepted even more efficiently than the virus molecule (MDV-1) from which it was selected.

Using a somewhat different system, namely the enzyme transcriptase (also known as DNA-dependent RNA-polymerase) of the bacterium *Escherichia coli* (*E. coli*), C. K. Biebricher and L. E. Orgel demonstrated the selection of a specific sequence of an RNA out of a random mixture of copolymers. This sequence functioned much more effectively than natural RNA as a template for the enzyme, which could be "fooled" into accepting RNA rather than DNA by replacement of magnesium ion salts by manganese ions in the incubation mixture.

For rather complicated reasons, the sequence analysis of RNA proceeded more rapidly than that of DNA. Thus, the analysis of a chain of 218 units in length, though still a remarkable tour de force, was clearly "within the state of the art" in 1973. New methods for handling DNA were, however, being developed rapidly: K. Murray (Edinburgh) gave partial sequences for distances of over 100 nucleotide units in the lambda bacteriophage.

At a slightly more complex level, the possibility of physical mapping of the *E. coli* chromosome was advanced by taking advantage of the special properties of the bacteriophage Mu-1. This "mutator" virus evidently consists of a linear DNA molecule about 37,000 bases long that, remarkably, can insert itself at any of a great many locations of the total *E. coli* DNA (a hundred times larger). The total DNA remained difficult to manage and M. T. Hsu and N. Davidson (California Institute of Technology, Pasadena) concentrated on a fragment, so-called F-Lac, which is about 146,000 units long. An F-Lac with an inserted Mu-1 should, therefore, measure about 180,000, and was observed to do so under the electron microscope. By melting the DNA of F-Lac together with the F-Lac/Mu-1, heteroduplex complexes were formed in which the F-Lac portion was paired but the Mu-1 was not and seemed in the micrographs to appear as a side-branch filament. The point of insertion could also be recognized genetically by the apparent "mutation" of the adjacent genes.

Cyclic AMP and genes. Recent years had seen much attention to the remarkable range of functions of cyclic AMP, both in eukaryotic (nucleated-cell) and bacterial systems, where it is associated primarily with "catabolite repression." (*See* Feature Article: CYCLIC AMP.) Every week still new roles for cyclic AMP were being found at levels very close to primary gene expression and integration. For example, cyclic AMP deficiency in various mutants of *E. coli* either promote lysis (autonomous growth of virus DNA) or lysogeny (its integration), depending on which genes are first activated or inhibited.

In other studies, E. M. Wise, Susan P. Alexander, and Marilyn Powers (at Tufts University, Medford, Mass.) showed that the transformability of cells of *Hemophilus influenzae* by DNA can be enhanced 10,000-fold by the addition of large doses of cyclic AMP to the medium. If this procedure could be generalized, it might make many other bacterial species amenable to genetic study by DNA transformation.

Hybrid studies. The fusion of somatic cells matured into a well-established technique for studying chromosome differences between genetically marked tissue cell lines, including differences between mouse and human cells. An increasing number of genetic markers concerned with clear-cut metabolic differences were being associated with specific human chromosomes.

The most reliable method of securing hybrid cells was to coat them with ultraviolet-light-killed parti-

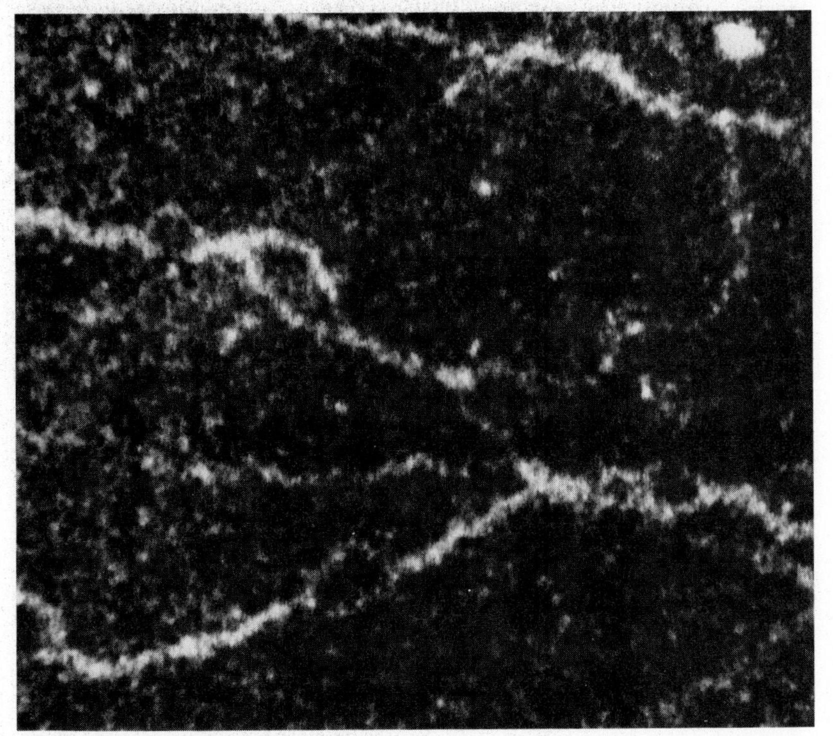

Courtesy, Dr. A. V. Crewe, University of Chicago

When A. V. Crewe of the University of Chicago displayed photographs made by the scanning electron microscope he had developed in 1970 to be able to "see" atoms, he proposed that the equipment might also study the structure of DNA. Crewe's own micrograph of unstained DNA from the T4 bacteriophage magnified about 1.6 million times captures single- and double-stranded regions of the nucleotide chain.

Courtesy, M.T. Hsu and N. Davidson, California Institute of Technology

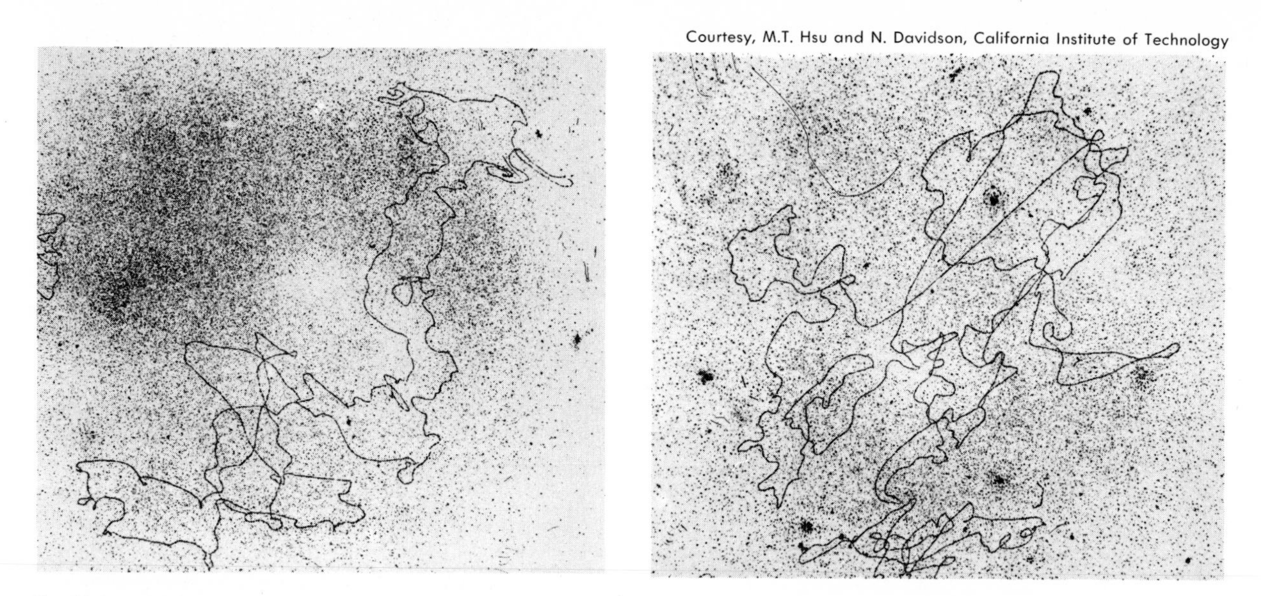

The F-Lac, a circular fragment of 146,000 base–pair units of DNA of the bacterium E. coli *(left), shows the effects of insertion of the 37,000-unit DNA of the bacteriophage Mu-1. Electron microscopy proved that insertion took place by revealing a molecule of about 180,000 units (right).*

cles of Sendai virus, whose coats contain cell-surface-dissolving enzymes that encourage fusion. Elaine G. Diacumakos and E. L. Tatum showed that specific pairs of cells can be made to fuse by gentle microsurgery. This procedure made it possible to ensure that a particular pair of cells in a controlled phase of development or of cell division had been fused, and avoided hypothetical complications that might be connected with the use of the killed virus.

Similar techniques were being applied to investigating the genetic basis of differentiation and of cancer. A group of workers at the National Institutes of Health led by Marshall Nirenberg hybridized nervelike cells of a tumor, neuroblastoma, with cells derived from liver. The hybrid cells were allowed to segregate chromosomes, resulting in a wide variety of tissue cell lines that retained different neuronal functions. These observations reinforced the accumulating evidence that tissue differentiation is propagated as a change in chromosomes, although many questions remained about the initial causes and the reversibility of these changes, not to mention their ultimate chemical basis.

Furthermore, although a chromosome might carry essential information about tissue specification, which can be thought of as a differential mask that inhibits some gene functions and elicits others, this was not to say that the site of inhibition was at the gene, where messenger RNA (m-RNA) is transcribed or DNA amplified, or at later stages of transport of the m-RNA from the nucleus to the cytoplasm, perhaps via the nucleolus. Gene amplification had been demonstrated for the DNA controlling the r-RNA (or ribosomal structural RNA). However, the silk gland of the silk moth, a tissue that is remarkably specialized to produce huge amounts of a specific protein, does not show differential amplification of the corresponding DNA.

While limited aspects of differentiation could be selected for study by these techniques in animal cells, the fusion of plant cells opened the door to synthesizing artificial hybrids that could then regenerate whole plants. This approach was pioneered by E. C. Cocking of Nottingham University, and carried in 1972 to the point of actual synthesis of a tobacco hybrid by Peter S. Carlson of the Brookhaven National Laboratory, Upton, N.Y. (*See* Year in Review: BOTANY.) In principle, even distant species of plants could be hybridized by cell fusion. Besides their obvious implications for improvement of food crops, these techniques might be equally important in allowing special plants to be developed for medicinals and for other precious compounds. Not only chromosomes but genetically autonomous chloroplasts could be reshuffled in this way, which would be the most important and dramatic application of "genetic engineering."

Carlson also reported that tobacco protoplasts would support the growth of tobacco mosaic virus, which promised to be a useful laboratory tool for the study of plant viruses. More startling, these protoplasts also took up the DNA of a bacterial virus (T3) and exhibited some of the enzymes coded by that DNA. Conversely, Cocking showed that yeast protoplasts would support the multiplications of tobacco mosaic virus. The absorption of the usual host range of such viruses is an important tool for studying these specifications—and should caution researchers about the possibility of inventing new diseases by the indiscriminate disposal of insect viruses for pest control.

Genetic recombination. Genetic recombination, by natural processes that generate evolutionary variety, remained the most efficient way to produce different gene complexes. Besides sexual recombination, which had been exploited by representation of all forms of life, bacteria are also amenable to transduction (carriage of genes by virus particles) and transformation (by purified DNA). However, these forms for recombination are restricted to closely related varieties or species. Thus, *E. coli* could be crossed with *Salmonella typhimurium*, but there was no useful way of integrating mouse or human DNA fragments into a bacterial cell for closer study of its functions.

Within the year, major leaps were made that promised an early conquest over this hurdle. At Stanford University, V. Sgaramella showed that DNA molecules of the *Salmonella* virus P22 can be made to join end-to-end under the influence of a DNA ligase obtained from another virus, T4. The terminal sequences of these molecules were quite variable, but the only requirement for this joining was meticulously cut flush ends. Synthetic polymers were also joined to P22 by similar techniques by A. Nussbaum, Hoffmann-LaRoche, Inc., Nutley, N.J. At Stanford, A. Jackson, R. H. Symons, and P. Berg spliced the DNA of the monkey tumor virus, SV40, into the sequence of a bacteriophage DNA, lambda. This procedure took advantage of certain nucleases that would unite DNA at specific sequences that occur in both viruses. These specific united ends could then recognize one another's complements and form DNA duplexes stable enough to be sealed by the single-strand ligase of *E. coli.* These examples of molecular translocation were not yet verified to result in well-defined, self-replicating clones of modified viruses, but could be expected to be at any moment.

A new future for engineered genes. Some concern was being expressed about the possible hazards of enabling a tumor virus to multiply freely in bacterial hosts. In fact, all work involving mammalian viruses, especially, had to be conducted with meticulous attention to the possibility of escape of new mutants, or indeed of familiar plagues. Nothing, however, would be worse for public health than to forbid such research. If we did not study viruses in the laboratory, we would be helpless against outbreaks of new diseases, such as the Marburg virus or Lassa fever of recent years. Molecular translocation might also be used for the genetic engineering of beneficial viruses, including safe, nonpathogenic viruses for use as vaccines.

Engineered viruses were also prepared as agents for gene therapy by many researchers. The basic idea was to produce a harmless live vaccine virus augmented with a few normal "human genes." This vaccine would elicit not antibodies but the production of a missing enzyme, thereby functionally augmenting the missing or faulty DNA in some tissue cells of a genetically diseased child; it would not be expected to be a repair genetically transmitted to offspring cells.

With luck, such a feat might be managed technically and, if no other resource were possible, might have to be stressed as a major approach to coping with tragic diseases. Each step would be a formidable challenge, but of a kind that might attract vigorous effort and could yield valuable by-product knowledge. The isolation of specific genes might be facilitated by previous purification of m-RNA from tissue cells and differential binding of DNA using the RNA from both normal and diseased cell lines. The vaccine virus would have been purified and authenticated in cell cultures. For therapeutic use, however, the vaccine must be demonstrated to be both safe and effective. In this regard researchers faced theoretical doubts that could only be answered empirically by clinical investigations, which in turn pose serious pragmatic and ethical difficulties. For some of the more prevalent genetic diseases such as sickle-cell anemia or cystic fibrosis, virus-gene therapies might be competing with chemical (drug) treatment even as scientists learn more about the underlying biochemistry of either.

More reliable alternatives, however, were on the horizon: cell transplants might augment the missing functions and, more surely, there would be the continued development of prenatal diagnosis and selective abortion. These steps would not guarantee the birth of healthy children, merely the preemption of serious disease.

—Joshua Lederberg

Obituaries

The following persons, all of whom died between July 1, 1972, and June 30, 1973, were noted for distinguished accomplishments in one or more scientific endeavors. Biographies of those whose names are preceded by an asterisk (*) appear in *Encyclopædia Britannica.*

Artsimovich, Lev A. (1909—March 1, 1973). A leading Soviet nuclear scientist, Artsimovich was a pioneer in the atomic energy program of the U.S.S.R. He graduated from the Belorussian University in Minsk in 1928 before becoming a staff member at the Physico-Technical Institute in Leningrad, a center for early nuclear research, where he worked with Igor V. Kurchatov, head of the atomic program. In 1951 Artsimovich, as chief of the Thermonuclear Laboratory at the Kurchatov Institute of Atomic Energy in Moscow, began work on controlled nuclear fusion and with his group developed the first tokamak device as a step toward achieving fusion power. He was a member of the Soviet Academy of Sciences from 1953 and secretary of the academy's department of physico-mathematical sciences from 1959. He received the Stalin Prize in 1953 and the Lenin Prize in 1958, and was made a Hero of Socialist Labor (Soviet highest civilian award) in 1969.

Ashby, William Ross (Sept. 6, 1903—Nov. 15, 1972), British cyberneticist and pioneer in studies of the organization and control of complex systems, Ashby was the discoverer of the Law of Requisite Variety. He analyzed the mechanisms of biological homeostasis and supplied mathematical formulations that explained some baffling features of brainlike activity. Ashby was director of research at Barnwood House Hospital, Gloucester, from 1947 until 1959, during which time he also wrote *Design for a Brain* (1952) and *An Introduction to Cybernetics* (1956). In 1960–61 he was director of the Burden Neurological Institute in Bristol, and from 1961 until 1970 he served as professor of cybernetics in the electrical engineering department of the University of Illinois, Urbana. Ashby became a fellow of the Royal College of Psychiatrists in 1971.

Belozersky, Andrei (Aug. 24, 1905—Dec. 31, 1972). A biochemist and vice-president of the Soviet Academy of Sciences, Belozersky served as senior biochemist at Tashkent State University and at Moscow State University, where he became head of the plant-biochemistry department in 1960 and chief of the laboratory of organic biochemistry in 1965. Belozersky was a three-time winner of the Order of Lenin; he also held the Red Banner of Labor and was a Hero of Socialist Labor.

Bowen, Ira Sprague (Dec. 21, 1898—Feb. 6, 1973). A noted U.S. astronomer, Bowen was director of joint operations of the Palomar and Mount Wilson Observatories from 1948 until his retirement in 1964. While an instructor on the staff of the California Institute of Technology, he became involved in planning the Palomar project and directed the final testing and completion of the 200-in. Hale telescope. Bowen was also a consultant in designing other large telescopes such as the 120-in. lens at Lick Observatory, and the 84-in. optical telescope at Kitt Peak National Observatory near Tucson, Ariz.

Hartley, Sir Harold (Sept. 3, 1878—Sept. 9, 1972). A British scientist whose contributions to the fields of crystallography and electrolytes gained for him many governmental appointments, Hartley also served as president of the British Association for the Advancement of Science (1950), president of the Institute of Chemical Engineers (1951–52, 1954–55), chairman of the Energy Commission of the O.E.E.C. (1955–56), and president of the Society of Instrument Technology (1957–61). In 1966 he was the recipient of the Kelvin Medal in recognition for his great services in furthering the union of science and technology.

Hutchinson, John (April 7, 1884—Sept. 2, 1972). British botanist, Hutchinson began his career as a gardener in the Royal Botanic Gardens at Kew in 1904, then served as keeper of the museums at Kew from 1936 until 1948. His studies included the classification and evolution of flowering plant families, and were reported in a number of books, including *Common Wild Flowers* (1945), *The Genera of Flowering Plants* (1964), and *Evolution and Phylogeny of Flowering Plants* (1969). Hutchinson illustrated his many books with his own artwork. His honors included the Linnaean Gold Medal and the Victoria Medal of Honour of the Royal Horticultural Society.

Katz, Louis Nelson (Aug. 25, 1897—April 2, 1973). Polish-born U.S. medical researcher, executive, and widely known specialist in heart diseases, Katz was head of the Cardiovascular Institute at Michael Reese Hospital in Chicago, and from 1967 emeritus professorial lecturer at the University of Chicago. He pioneered in researching the cause of hardening of the arteries and developed clinical electrocardiography. He served as president of the American Physiological Society (1956–57) and of a number of other medical associations. Katz was the author of several books, including *Nutrition and Atherosclerosis* (1958), and many articles on medical research.

Leakey, Louis Seymour Bazett (Aug. 7, 1903—Oct. 1, 1972). British anthropologist, Leakey was also an expert in prehistoric archaeology and paleontology. Born to a missionary couple working with the Kikuyu tribe in Kenya, Leakey graduated from St. John's College, Cambridge, Eng., with firsts (highest honors) in modern languages, archaeology, and anthropology. During four expeditions to East Africa between 1926 and 1935 he traced the main sequence of prehistoric cultures in Kenya, and from 1937 until 1940 he conducted a study of the Kikuyus. In 1945 Leakey became curator of the Coryndon Memorial Museum in Nairobi, and later founded the National Museum's Centre of Pre-History and Palaeontology adjoining the museum. His discoveries included the Kanam jaw and Kanjera skulls, the remains of Miocene apes (on Rusinga Island), and an almost complete skull of *Proconsul africanus*, the earliest ape skull known to date. In 1959 at Olduvai gorge in Tanganyika (later Tanzania), Leakey's wife Mary discovered the skull of *Australopithecus (Zinjanthropus) boisei*, and the following year their son Jonathan found the first remains of another hominid even closer to the human line, called *Homo habilis*. In 1961 Leakey unearthed the upper jaw of *Kenyapithecus wickeri*, 14 million years old and considered to be one of the earliest hominids. His writings include *Stone Age Cultures of Kenya Colony* (1931), *Stone Age Races of Kenya* (1935), *Adam's Ancestors* (1935), *Stone Age Africa* (1936), *Olduvai Gorge* (1951), and *Olduvai Gorge 1951–61* (1965).

Mackenzie, Melville Douglas (June 29, 1889—Dec. 1, 1972). A British epidemiologist, Mackenzie served as chief U.K. delegate to the first six World Health Assemblies, and chairman of the World Health Organization executive board during 1953–54. He was senior medical officer to the Nansen Russian Famine Relief Administration in the Volga Valley in 1922 and a member of the ministry of health (1926) and of the League of Nation's Health Organization (1928), serving as acting director of the League's Epidemiological Bureau in Singapore (1936). Mackenzie was the author of a standard work, *Medical Relief in Europe* (1942).

Millionshchikov, Mikhail Dmitrievich (Jan. 16, 1913—May 27, 1973). Internationally known Soviet physicist and spokesman for scientific contact with the West, Millionshchikov was an authority in the field of turbulent flow of fluids and gases, on the theory of filtration and applied gas dynamics, and in the area of atomic engineering. He became deputy director of the Kurchatov Institute of Atomic Energy in Moscow in 1960, and served as vice-president of the Soviet Academy of Sciences from 1962. He was active in the Pugwash movement and was chairman of the Soviet Pugwash Committee at the conference in Venice in 1965. His honors included the Stalin and Lenin Prizes, four Orders of Lenin, and the Order of the October Revolution.

Pecora, William Thomas (Feb. 1, 1913—July 19, 1972). U.S. geologist and public official, Pecora graduated with honors in geology from Princeton University in 1933. He joined the U.S. Geological Survey in 1939 and served as its director from 1965. In May 1971 the Senate confirmed Pres. Richard M. Nixon's appointment of Pecora as undersecretary of the U.S. Department of the Interior. He was the author of more than 50 scientific publications, mostly based on his worldwide field trips. In 1965 Pecora was elected to the National Academy of Science and to the American Academy of Arts and Sciences. He received a Rockefeller Public Service Award in 1969.

Rabinowitch, Eugene (April 26, 1901—May 15, 1973). U.S. physical chemist, atomic scientist, and writer, Rabinowitch was a founder (1945) and editor of the *Bulletin of the Atomic Scientists*. From September 1972, having taken leave from his post as professor of chemistry and director of the Center for Science and the Future at the State University of New York at Albany to accept a Woodrow Wilson Fellowship at the Smithsonian Institution, he was occupied with a project on scientific revolution and its social implication. Rabinowitch was senior chemist (1944–46) on the Manhattan Project, which produced the first controlled nuclear reaction leading to the development of the atomic bomb. He was professor of botany and biophysics at the University of Illinois, Urbana,

Louis S. B. Leakey

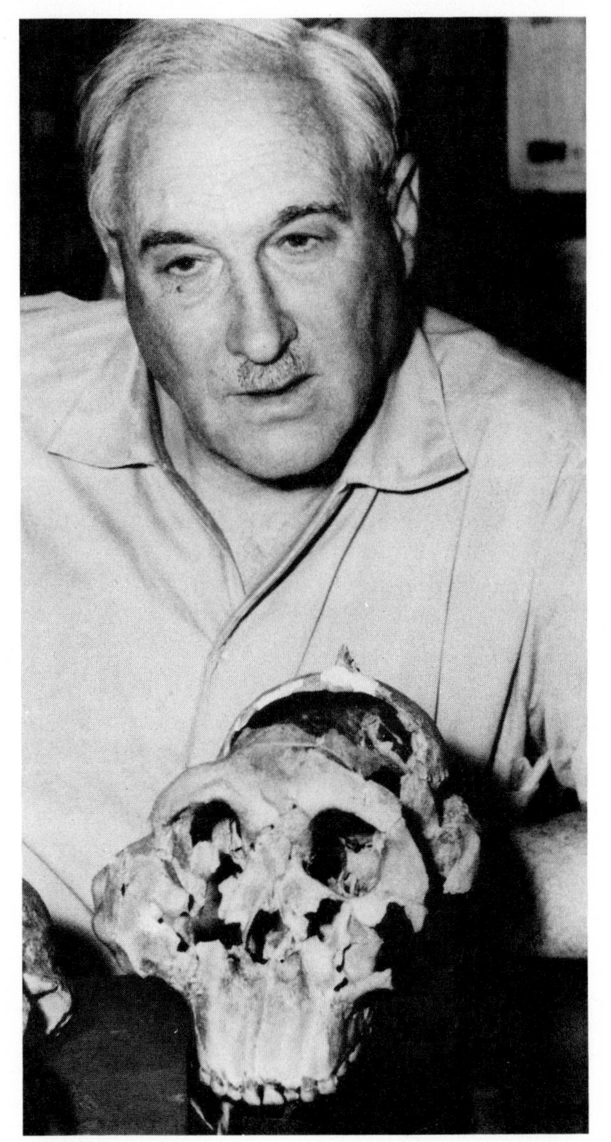

UPI Compix

Courtesy, Sikorsky Aircraft, United Aircraft Corporation

Igor Sikorsky

from 1947 until 1968, when he went to Albany. Included in his writings were a monograph, *Photosynthesis and Related Processes* (3 vol., 1945-56), and the books *Minutes to Midnight* (1950) and *The Dawn of a New Age* (1963).

Richards, Dickinson W. (Oct. 30, 1895—Feb. 23, 1973). U.S. physiologist who shared a Nobel Prize for Physiology or Medicine, Richards in 1945 became director of the First Medical Division at Bellevue Hospital in New York City, and in 1947 was appointed Lambert Professor of Medicine at Columbia University. He retired from both posts in 1961. His work in cardiac physiology paved the way for open-heart surgery. For his research in that field Richards was awarded the Nobel Prize for Physiology or Medicine in 1956, along with André F. Cournand, also of Columbia, and Werner Forssmann of West Germany. Richards published two books, *Circulation of the Blood: Men and Ideas,* with A. P. Fishman (1964), and *Medical Priesthoods and Other Essays* (1970).

***Shapley, Harlow** (Nov. 2, 1885—Oct. 20, 1972). Called the dean of U.S. astronomers, Shapley was long connected with Harvard University, serving as director of the observatory from 1921 until 1952. In the latter year he was appointed Paine Professor of Astronomy, becoming emeritus professor in 1956. Granted his Ph.D. at Princeton University in 1913, Shapley became a staff member at the Mount Wilson Observatory in California, remaining there until 1921 and doing extensive work on the Cepheid variables. He was the author of numerous popular books including *Flights from Chaos* (1930), *The Inner Metagalaxy* (1957), *Of Stars and Men* (1958), and *The View from a Distant Star* (1963).

***Sikorsky, Igor Ivan** (May 25, 1889—Oct. 26, 1972). U.S. aviation pioneer and prime developer of the helicopter, Sikorsky also built multiple-engine planes and created the flying boat. He built his first helicopter in Russia (his birthplace) in 1909, but it failed to fly. A later U.S.-built model, his VS-300, got off the ground in 1939. Sikorsky had designed the two-engine and four-engine planes, used as bombers in World War I, before going to the United States in 1919. By 1923 he had founded the Sikorsky Aero Engineering Corp., which in 1924 produced its first successful twin-engine, 14-passenger plane. This was followed in 1928 by the S-38, a ten-seat amphibian used by the U.S. Navy and by Pan American World Airways. In 1929 the Sikorsky company joined United Aircraft Corp. and in 1931 brought out the S-40, a four-engine, single-wing flying boat. The S-40, largest U.S.-built plane at that time, was flown by Pan American on its Latin American runs. The VS-300 helicopter, first displayed to the public in 1940 and later much modified, was to become the all-purpose workhorse of military flying.

Stevens, S. Smith (Nov. 4, 1906—Jan. 19, 1973). U.S. physicist and authority on the physics of sensory perception, Stevens was associated with Harvard University from the time he was granted his Ph.D. there in 1933. He became a full professor in 1946, in 1949 director of the Psychological Laboratories, and in 1962 was appointed professor of psychophysics and director of the Laboratory of Psychophysics. He formulated Stevens' Law, which states that the magnitude of a sensation produced by a stimulus grows as a function of some power of the intensity of the stimulus. Stevens, editor of *Handbook of Experimental Psychology* (1951) and co-author of several books on hearing, was the recipient of many awards, including the 1972 Rayleigh Gold Medal of the British Acoustical Society.

*Theiler, Max** (Jan. 30, 1899—Aug. 11, 1972). U.S.-South African immunologist and Nobel Prize winner, Theiler was associated with the Rockefeller Foundation from 1930 until 1964, when he became professor of epidemiology and microbiology at Yale University. After receiving his early education in South Africa, Theiler studied at the London School of Tropical Medicine, taught at Harvard University (1922–30), and then transferred to the virus laboratory of the Rockefeller Institute for Medical Research (which later became Rockefeller University). He was named director of the laboratory there in 1951. At the Institute he furthered his research into the cause of yellow fever and helped prove that a filterable virus was the causative agent. By 1936 Theiler and his co-workers had produced a vaccine, known as 17-D virus, which they tested on themselves to prove its effectiveness. Additional successful testing led to mass production of the vaccine in the mid-1940s. For his work in developing the yellow fever vaccine, Theiler was awarded the 1951 Nobel Prize for Physiology or Medicine.

*Tupolev, Andrei Nikolaevich** (Nov. 10, 1888—Dec. 23, 1972). Soviet aircraft designer Tupolev planned and built a wide variety of civil and military aircraft, culminating in the Tu-144 supersonic transport (SST), which first flew on Dec. 31, 1968, two months ahead of the Anglo-French Concorde SST. Educated at the Moscow Higher Technical School, Tupolev in 1918 was appointed assistant director of the Central Aerohydrodynamic Institute. In 1922 he designed his first aircraft; a later model made a historic Moscow–New York flight via Siberia in 1926. After the advent of jet propulsion his most significant designs included the twin-jet Tu-104, and the Tu-114, powered by four turboprops. When introduced in 1961 the Tu-114 became the largest and heaviest aircraft in use. Tupolev was a member of the Soviet Academy of Sciences and an honorary fellow of the Royal Aeronautical Society. He received numerous state prizes, and in 1959 was the recipient of the Gold Medal of the Fédération Aéronautique Internationale.

Photography

A new compact Polaroid camera was introduced during the year. The camera was a folding single-lens reflex, with dimensions of 1 × 4 × 7 in. (2.5 × 10 × 18 cm) when closed. In addition to eliminating the bulk of the older Polaroid Land cameras, it had the advantage of permitting development to take place without timing and without any waste products.

After exposure, the print was expelled from the camera by a motor drive and the image began to appear almost instantly. The sensitive material consisted of an integral structure of some 17 layers, including a mordant or fixer. A wide margin on the print contained a version of the Polaroid processing pod which held, among other reagents, the viscous processing solution. The major ingredients of the pod were alkali and titanium dioxide (titania). The alkali, which permeates the dye layers, becomes oxidized, and the oxidation products immobilize the dye on the image. The unoxidized dye pigment passes through the titania layer to the mordant, where it is held and appears as a positive image against the white background. New kinds of indicator dyes were synthesized which absorb light at low pH values but become colorless in conjunction with alkali. These dyes were incorporated into the viscous reagent so that, in effect, development takes place under a dark curtain that is bleached away during the process.

Films and accessories. Since the negative size of the popular Kodak Pocket Instamatic camera was 0.5 × 0.7 in. (13 × 17 mm), an improved emulsion was necessary to produce color prints of good quality in the 3.5 × 4.5 in. (9 × 11.5 cm) size. The Kodak monochrome film, Verichrome pan, was altered to give finer grain and a sharper image, and Kodacolor II was introduced with corresponding improvements.

Two special-purpose transparency color films were introduced by Eastman Kodak. SO-456 was a slow, high-contrast material with a green-sensitive layer at the top of the monopack in order to achieve high sharpness. Ektachrome Duplicating Film 6120 replaced the firmer 6119 and was from one to two stops faster than the earlier material.

The Fujica ST 701 and 801 incorporated a method of light measurement employing silicon photo cells, heavily filtered to give correct color response and using a field effect transistor circuit to restore sensitivity. The advantage of the system was that the silicon cell does not display the "memory" of the conventional cell. Meter response in dim light was also appreciably faster.

A fundamental advance in electronic flash units was introduced in the Rollei E36RE. The previous system depended on a light sensor, which, when activated by reflected flash light from the subject, actuated a shorting device in parallel with the capacitor storing the electrical energy. This meant that when a flash was interrupted before the capacitor was completely discharged, the energy not converted into light was dissipated as heat. The new system had a switch-off device in series with the capacitor, so that unexpended charge remained in the capacitor.

Lenses. More firms adopted multicoating of lens elements to achieve reduction in flare and avoid internal reflections. Carl Zeiss of Oberkochen introduced 23 new lenses at the Photokina '72 in Cologne, W.Ger. The Sonnar Superachromat 250-mm *f*/5.6 lens virtually eliminated the secondary spectrum, a feature that was of particular value with high-sensitivity panchromatic film, infrared film, or false color film. No refocusing was needed when the sensitivity of the film extended beyond the visible spectrum. The 15-mm *f*/8 three-element Hologon, covering the amazingly wide angle of 110° with minimum distortion, was introduced in a new focusing version of the Leica Model M. In 1966 most authorities had considered such a modification to be impossible.

Another lens of extreme wide angle was the Zeiss 15-mm *f*/3.5 Distagon with 14 elements. A filter turret built into the lens permitted the use of four filters, and the lens could be used in single-lens reflex cameras. A general trend was the inclusion of floating elements that do not move with the rest of the lens during focusing, thus retaining lens corrections through an extended focusing range. The Asahi Optical Co. Ltd. of Tokyo exhibited a 135–600-mm *f*/6.7 zoom with 15 elements and a 1,000-mm mirror lens for the 6 × 7 format. Canon made extensive use of floating element construction in many of its new lenses and employed fluorite elements in a 300-mm FD *f*/2.8.

Underwater photography. For several years steady progress had been made in underwater photography. Equipment became available permitting men and women to make lengthy excursions into the water world or even to live for extended periods on the ocean floor, making scientific

Surface of a tungsten crystal magnified 1,375 times was photographed through a microscope as part of a research project to find better ways of making tungsten filaments for light bulbs.

Courtesy, General Electric Research and Development Center

Flip Schulke

A 16-mm Kodak K-100 movie camera fits inside French Underwater Systems housing, which protects the lens with a dome port. Angle of coverage of the camera is 94°. Viewfinder is at the top.

observations, searching for minerals, and laying the groundwork for farming of the sea.

Three optical problems confront the underwater photographer: the blue color of water acts like a blue filter, altering the true colors of underwater subjects; particles suspended in water cut down clarity and, thus, the sharpness of the photographic image; and water, coming in contact with the flat viewing port of an underwater camera housing, creates optical aberrations.

The effect of the blue in the water may be partly solved by allowing as little water as possible between the camera and the subject and by utilizing artificial light to counterbalance the "blue-filter" effect. Particles in suspension are a somewhat more complex problem, since water can assume the coloration of the impurities, and light intensity can be reduced by scattering or redirection of light rays. Again, the practical solutions for the underwater photographer are to get close to the subject, using a wide-angle lens, and to utilize artificial light. If possible, the artificial light should be kept at a 60° to 90° angle from the camera to minimize backscattering, caused when the light source reflects back into the camera lens from the suspended particles.

Underwater camera lenses are usually protected by a flat glass or Plexiglas "window," called a flat port. Objects photographed through a flat port will appear to be one-third larger and one-fourth closer than they actually are, and the field of view will be narrowed by one-third. If the photographer backs away from the subject in order to restore the full field of view, more water comes between the camera and the subject. Further, the flat port causes undesirable color aberrations, degrades the sharpness of the photographic image, and distorts its shape.

Many solutions to these problems have been tried. One of the most elementary and least costly is to use a hemisphere, with concentric inner and outer surfaces. The majority of these hemispheres, commonly referred to as "domes," are made of Plexiglas, which has good optical properties.

The dome port, used in conjunction with a conventional "in-air" lens, sees an "apparent" image that is much closer to the camera's film plane than the actual image. Most "in-air" lenses cannot be mechanically focused close enough to place the apparent image in sharp focus, and this must be corrected by the addition of a supplementary positive (converging) lens placed between the "in-air" lens and the dome port.

The Ivanoff corrector, invented and originally manufactured in France, was designed specifically to correct the aberrations of conventional lenses when used in underwater photography. It is basically a reversed Galilean telescope, in which the front concave lens is used for the watertight window and the second element is a positive lens positioned behind it. A separate Ivanoff corrector must be constructed and matched to each "in-air" lens with which it is to be used. The ultimate in correction devices, however, is the "in-water" lens

Flip Schulke

Underwater photograph of a test pattern shows so-called "pincushion" distortion of the outer edges. This picture was taken through a flat port, a flat piece of Plexiglas that protects the camera lens from the water. By using a dome port, a hemispheric-shaped Plexiglas window, the distortion can be corrected.

designed specifically for underwater use. The first of these was Nikon's 28-mm underwater lens, designed for the Nikonos (formerly known as the Calypso) camera. A series of completely corrected "in-water" lenses, called Elcan, was being manufactured by Ernst Leitz Canada Ltd. The latest "in-water" lens was Nikon's 15-mm *f*/2.8, designed for use with the Nikonos 35-mm underwater amphibious camera.

Nearly any "in-air" camera can be used in an underwater housing, although housings have to be designed to fit the shapes of specific cameras. The only "in-water" still camera was the Nikonos, which can be immersed in water without a protective housing. Many "in-air" lenses can be utilized with the Nikonos in conjunction with a dome port adapter. With single-lens or twin-lens reflex cameras, viewing underwater is accomplished by looking directly through the reflex-viewing prism. With cameras that lack reflex viewing, such as the Nikonos and many movie cameras in housings, optical underwater viewfinders are used. Focusing can also be done through the prism of a reflex camera, which is why reflex cameras are so popular in underwater photography. With nonreflex type cameras, the distance must be guessed, placing a good deal of reliance on the depth of field of the lens.

(*See also* Feature Article: MAN'S NEW UNDERWATER FRONTIER.)

—N. F. Maude; Flip Schulke

Physics

Major developments in physics during the past year included new discoveries in the areas of super conductivity and surface physics that were expected to help achieve a cleaner environment and relieve the energy crisis. Measurements of the shapes of nuclei supported theoretical insights concerning the motions of individual protons and neutrons.

High-energy physics

Unified theories of weak and electromagnetic interactions. The past year of work in high-energy physics included important steps toward a unification of two seemingly unrelated sets of phenomena known as electromagnetic interactions and weak interactions. As a result of this work, physicists may soon have a complete theory of these two sets of processes as well as an understanding of several previously puzzling relations among the properties of subatomic particles.

Electromagnetic interactions are responsible for many processes involving electrically charged particles such as electrons. These processes all originate in the fundamental electromagnetic interaction in which a charged particle emits or absorbs a photon, one of the particles of which a light beam is composed. For example, the electrical force between two charges is produced by the emission of a photon by one charge and the absorption of the photon by the other. All of these processes are described by a theory known as quantum electrodynamics. The predictions of this theory, about such things as the frequencies of light emitted by atoms, have been confirmed to great accuracy by various experiments.

The weak interactions are involved in certain transformations and decays of subatomic particles into each other, including one step in the hydrogen fusion reaction that generates energy in stars. They are called weak interactions because of the typical rate at which they occur, which on the time scale of most subatomic processes—about one-trillion-trillionth of a second (10^{-24} sec)—is very slow. Since the strength of an interaction is a measure of the probability that the interaction will occur under standard conditions in a given time interval, it is clear that the longer the time needed for an interaction to take place, the weaker the interaction. By this measure, the weak interactions are about one-hundred-thousandth (10^{-5}) the strength of typical nuclear interactions.

Physicists had gradually developed an approximately accurate description of the weak interactions among particles of low energy. However, there was no theory available that could be used to make predictions of very high accuracy, as quantum electrodynamics could do for electromagnetic interactions. Nor was there an adequate theory of the weak interactions among particles of high energy. For some time physicists had hoped that some of the ideas used in quantum electrodynamics could be applied to create a theory of weak interactions which would not be limited in these ways, and, eventually, to establish a unified theory of the two sets of phenomena.

There are several important differences between weak and electromagnetic interactions, which were stumbling blocks in past efforts to unify them. One such difference is the much longer time scale of the weak interactions. Another is that electromagnetic interactions can take place between particles that are great distances apart, whereas weak interactions occur only between particles separated by distances less than the size of an atomic nucleus. Finally, as discovered in 1956, the weak interactions are not symmetric under mirror reflection, while the electromagnetic interactions do obey this symmetry.

On the other hand, it is known that at least one similar property exists for the two sets of interactions. The electric charges of various subatomic particles always have values that are simple integer multiples of the same unit charge (numerical value of the charge of the electron), even though the particles may be quite different in other respects. Similarly, certain numbers called weak coupling constants, which measure the relative strengths of weak interactions of various particles, were found to be simply related (in a mathematical sense), even when the particles differed widely in other properties. No explanation was formerly available for either of these facts, but they are a key element of the newly developed unified theories.

One step toward unification came with a suggestion made some years ago that weak interactions occur in two steps, the emission and reabsorption of a still-undiscovered particle of large mass called the W boson. This hypothesis has the advantage of allowing an explanation of several of the cited differences between weak and electromagnetic interactions. For example, the much smaller range of weak interactions would follow from the fact that the distance an exchanged particle can travel decreases as its mass increases. Since photons have zero mass, they can be exchanged between objects arbitrarily far apart, whereas the massive W bosons can only be exchanged between objects very near each other.

However, until recently the W boson hypothesis remained somewhat indefinite in that the exact mass and several other properties of the W boson, including the number of different types, were uncertain. In the new unified theories of weak and electromagnetic interactions, for which the original stimulus was work done by Steven Weinberg of the Massachusetts Institute of Technology, these deficiencies were mostly removed. This work revealed several types of W boson, the masses of which are arranged so that the strength of the weak interactions among particles of high relative energy is equal to that of the electromagnetic interactions among similar particles. In order for this to be the case, the W bosons must have masses of at least 37 GeV (billion electron volts), or sufficiently high that they would not have been detected in previous experiments.

Another advantage of Weinberg's type of theory is that any combination of weak and electromagnetic effects, for high or low energy, can be calculated as accurately as desired. Therefore, the theory can be confronted more directly with experimentation. The unified theories also provide an explanation of the simple relations that exist among the electric charges of different elementary particles and among the weak coupling constants of different particles. These relations are consequences of the fact that these theories must satisfy certain symmetry principles in order to be mathematically consistent, and the symmetry principles are only satisfied when the electric charges and weak coupling constants of different particles are simply related.

Several theoretical problems remain with the

unified theories. One is to harmonize them with theories of the strong interaction, which have their own symmetries, in such a way that no contradictions with known experimental results arise. Theorists are actively working on this and other questions, while experimentalists are planning searches for some of the new particles and new phenomena that the theories predict.

Proton-proton scattering at the ISR. One of the main reasons for building the intersecting storage rings (ISR) at the European Organization for Nuclear Research (CERN) laboratory in Geneva was to make measurements of proton-proton collisions at higher energy than had been possible. As a consequence of relativity theory, a head-on collision between two protons moving in opposite directions in the two beams of the ISR, each with an energy of 26 GeV, has the same effect as a collision between a proton at rest and a proton of 1,500 GeV, an energy level beyond the range of any present accelerator.

In the past year, experiments were carried out at the ISR to measure the total cross section in collisions between two protons, each of which had a variety of energies between 12 and 26 GeV. This total cross section may be thought of as the area presented by a single proton to another proton moving toward it. For conventional experiments, in which one proton is at rest, a measurement of the total cross section can be done by measuring the number of protons scattered out of the incident beam and dividing by the total number in the incident beam and by the number of target protons per unit area. However, in a colliding beam experiment this technique cannot be used. Instead, one measures the number of scattered protons and divides by the luminosity, which is a measure of the number of protons in the volume where the two beams overlap.

The problem in measuring the proton-proton cross sections at the ISR arises from the difficulty in determining this luminosity. Two different experimental groups attempted the measurement, each using different methods to estimate the beam luminosity as well as different techniques to measure the number of scattered protons. However, their final values for the cross section of various energies agree well with each other, increasing the confidence in the accuracy of the result.

The results that were obtained by each group show an increase in the proton-proton cross section of about 10% as the energy of each proton

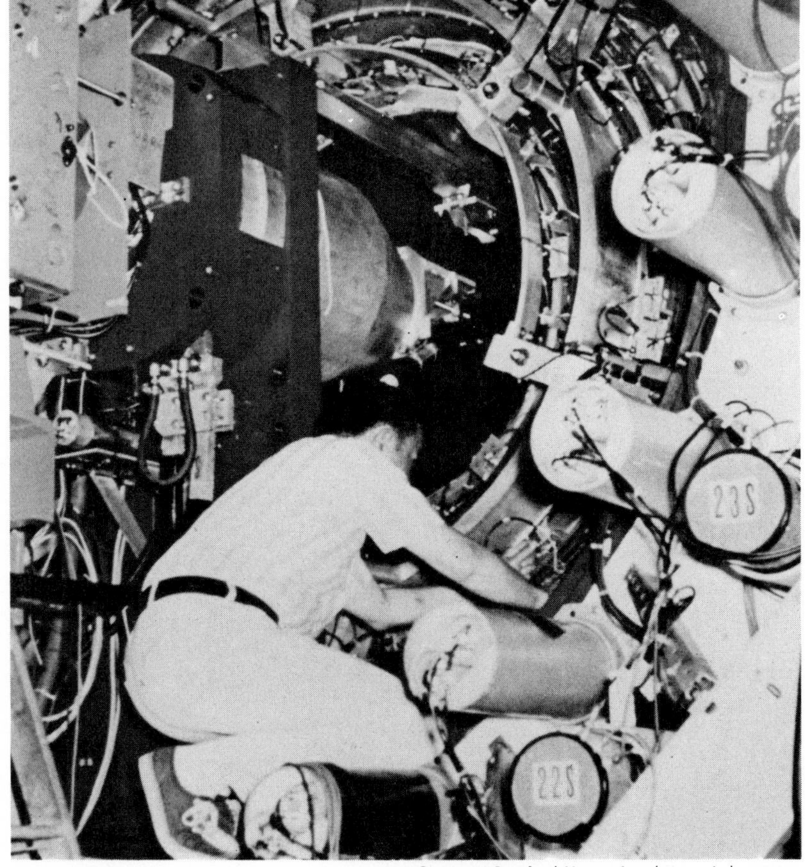

Engineer adjusts part of a 200-ton particle collector built for the first colliding-beam experiment of the SPEAR program at the Stanford Linear Accelerator Center. The experiment was designed to study electron-positron annihilation and the resulting hadron states.

Courtesy, Stanford Linear Accelerator Laboratory

Courtesy, Lawrence Berkeley Laboratory, University of California

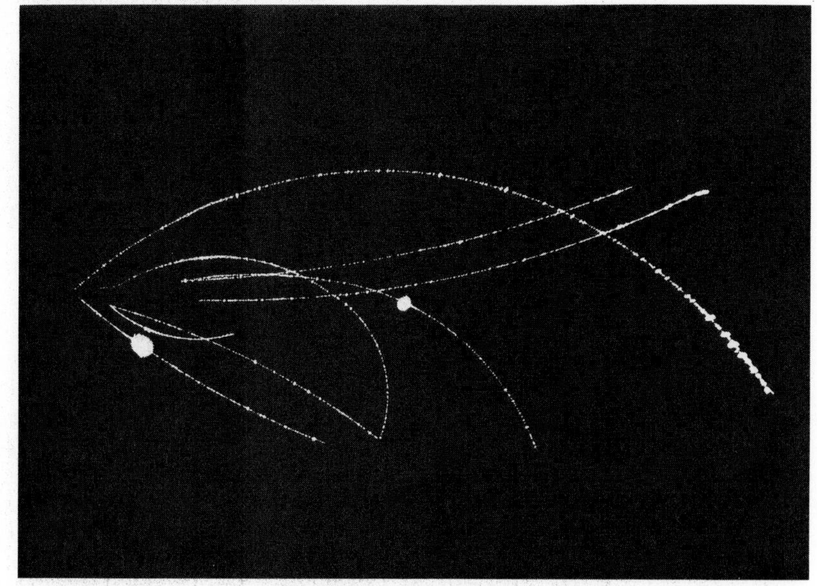

Collisions of subatomic particles in a streamer chamber ionize inert gas between conductive plates that act as charged electrodes. Such reactions give rise to streams of sparks. In the photograph a K^- meson, a particle of antimatter, is colliding with a proton.

rises from about 12 to 26 GeV. This must be compared to the result at lower energies, which showed a cross section that is approximately constant for energies per proton between 3 and 12 GeV. Based on these results, many physicists expected that the cross section would remain constant at higher energy.

The ISR results on the increasing size of proton-proton cross sections raise a number of obvious questions: (1) Does this increase continue at still higher energies? (2) What is the precise mathematical relation between the cross section and the proton's energy? (3) Do the cross sections for scattering of other particles exhibit behavior similar to that of the proton?

It is likely that answers to the first two of these questions will not be forthcoming immediately, because higher-energy protons are not available at the ISR and no experiments with present single-beam accelerators can attain equivalent energies. Possibly, the solution will come from building intersecting storage rings containing higher-energy protons. This might be done at the U.S. National Accelerator Laboratory (NAL) in Batavia, Ill., where 500-GeV protons are available.

The third question would be even harder to answer if the relevant energy for the increase in cross section is the same for different particles. However, some researchers believe that reactions involving different particles should be compared at different energies, proportional to the particle's mass. In that case, an increase in cross section could be expected in pion-proton collisions at much lower energy than in proton-proton collisions, because the pion mass is only one-seventh of the proton mass. This energy range is accessible at present at the NAL, and it will be interesting to see if the phenomenon occurs there. In any case, the ISR results already show that very-high-energy scattering will contain more of interest than might have been expected just a short time ago.

—Gerald Feinberg

Nuclear physics

Until the mid-1950s, lacking any better information, physicists assumed that nuclei were spherical. As better measurements became available, it soon became evident that most nuclei were shaped like cigars and a few even resembled doorknobs. During 1972 and 1973, new and much more precise measurements showed that even more peculiar shapes are the rule. Some nuclei have pronounced polar and equatorial bulges, while others have bulges at middle latitudes. What was particularly gratifying about these new results was that they could be understood on the basis of very recent theoretical insights into the motions of the individual neutrons and protons in the nucleus acting under the influence of the strong nuclear force. In some nuclei these peculiar shapes were predicted—and found—to be stable. In others the shape oscillates systematically through a variety of forms. And in all nuclei different shapes appear as differing amounts of energy are added to the nuclear system.

Valence protons and neutrons. This discovery had important consequences in achieving a better understanding of how neutrons and protons move in nuclei. About 1970 physicists began to find that many nuclei could be understood as comprising a few valence neutrons or protons moving in orbits outside of closed cores or shells. Surprising simplicities appeared during the past year. The nuclei oxygen-16 and lead-208 are among the best closed cores in nature. Oxygen-17 and fluorine-17 correspond, respectively, to a one-valence neutron and one-valence proton in orbits outside of the oxygen-16 core; lead-209 and bismuth-209 are similarly related to the lead-208 core. Fluorine-18 and bismuth-210 both have a valence neutron and a valence proton outside of their respective cores, but their behavior is strikingly different. In fluorine-18 the two valence particles interact strongly, while in bismuth-210 they interact hardly at all. It has become clear on the basis of very recent studies that this is simply because the lead core is much bigger than the oxygen one—in effect, the bismuth valence particles can not find one another whereas in fluorine they can not get away from one another.

Much of our knowledge of such structure comes from reactions in which a single neutron or proton can be systematically added to or removed from a target nucleus. The simplest of these is the so-called deuteron stripping reaction, in which an incident deuteron (the nucleus of heavy hydrogen, containing one proton and one neutron) breaks up on impact with a target nucleus, with the neutron being captured and the proton emerging to provide surprisingly detailed information concerning the orbit into which the neutron had disappeared. In this simple picture the target nucleus provides the core about which the valence particle orbits.

But, as explained above, if energy is added to the core, it can assume shapes quite distinct from that of its normal ground state. Studies in the past few years demonstrated clearly that many nuclear states can be understood as valence particles orbiting about these excited cores rather than about the normal ones. Nuclear physicists have long wondered as to whether in a simple stripping reaction, of the kind just mentioned, it was justified to ignore so-called higher-ordered processes that would involve these excited core situations. In a normal stripping reaction the neutron is simply captured by the target core and the proton emerges. But

Courtesy, Lawrence Berkeley Laboratory, University of California

Super-HILAC (heavy-ion linear accelerator) at the Lawrence Berkeley Laboratory, Berkeley, Calif., is redesigned to make it the world's first machine potentially capable of creating virtually all the transuranium elements proposed in theory. It will accelerate all elements to energies high enough to overcome the electromagnetic repulsion in similarly charged nuclei in the targets being bombarded. Physicists hope that the collisions of such heavy particles will produce larger nuclei than previously observed.

suppose that on its way into the nucleus the deuteron exchanged energy with the target, raising it to one of the excited states, and *then* transferred the neutron; or suppose that the neutron was captured in the simplest fashion but the emerging proton—on its way out—exchanged energy with the core and raised it to an excited state without greatly affecting the captured neutron in its valence orbit.

Physicists recognized that all three processes could happen simultaneously, and, as in any other coherent process, it would be possible to detect the interference between the different amplitudes. Until this past year no compelling evidence for such effects was found, but during 1972 dramatic interference phenomena were finally detected in a wide variety of nuclear reactions.

The fact that even in nuclear physics neither the neutron nor the proton can be considered as elementary entities turned up quite unexpectedly in another area of nuclear physics during the year. One of the best established facts in nuclear physics is that the density of nuclear matter is constant. This implies that the volume of a nucleus is directly proportional to the number of neutrons and protons present (the mass number A); consequently, the radius must be proportional to the cube root of A. One of the most reliable methods of measuring a nuclear radius is through the scattering of very-high-energy electrons from the nucleus in question. Such measurements were carried out systematically at Stanford University and elsewhere. It was with great surprise that physicists found that systematic measurements on the calcium isotopes from calcium-40 to calcium-48 (differing only in the addition of neutrons—from 20 to 28 in number) showed an apparent decrease in radius rather than the expected increase.

This remained an outstanding puzzle for several years. It was resolved early in 1973 by a group at Massachusetts Institute of Technology which correctly noted that the neutron is not really an unstructured entity; it rather has an inner shell of positive charge surrounded by a more diffuse negative shell (clouds of pions) such that overall it is electrically neutral. In the high-energy electron measurements this structure was found to be masking the normal radius behavior.

Fission. From many points of view, both scientific and practical, one of the most important nuclear phenomena is that of fission. And, strangely enough, until very recently it has been one of the least understood. At first it was believed that upon absorption of a neutron a uranium nucleus, for example, became unstable against a deformation of its normal football shape so that a central neck developed; this was followed by a rapid separation of the original nucleus into two fission fragments, with the release of much energy. Central to this picture was the idea that, once initiated, fission occurred almost instantaneously. It was thus with great surprise that in the mid-1960s a research group at Dubna in the U.S.S.R. discovered many fission cases where this did not occur. The fissioning nucleus appeared to pause for periods as long as a second (an enormously long time if measured in appropriate nuclear time units) at some intermediate point in its progress from a football shape to two fragments.

With this clue, an international group including Soviets, Danes, Chinese, Swedes, and Americans quickly evolved a theoretical picture that could explain such behavior within the framework of existing understanding of the motions of the component neutrons and protons in heavy nuclei. In a recent experiment by a research group in Munich, W.Ger., important new supporting evidence for the theory was found.

Under appropriate conditions any deformed nucleus can be forced to rotate, but in contrast to more familiar larger objects it can have only certain isolated angular velocities, or spin rates, allowed to it. When it jumps from one to another of these spin rates, it gains or loses energy, which appears

Stable compressed column of plasma in the Scylla IV theta-pinch confinement device is photographed by a laser-interferometric technique. The plasma is stably confined in the radial direction, escaping only by diffusion out the ends of the tube.

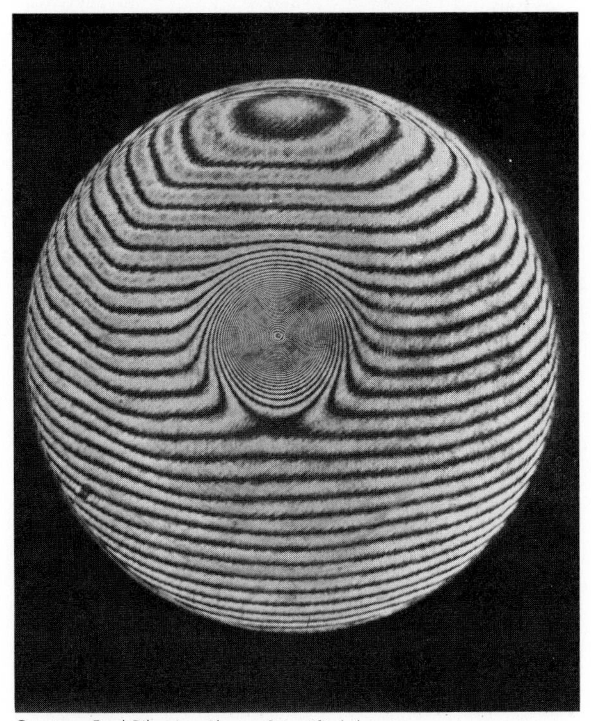

Courtesy, Fred Ribe, Los Alamos Scientific Laboratory

typically as a gamma ray. In a very difficult experiment the Munich group was able to detect the transitions between these different spin rates not only of the original football nucleus but also of the nucleus during its pause on the way to fission. This provided strong evidence for the correctness of the new understanding of the fission mechanism. Quite apart from its importance because of this alone, the new evidence gave nuclear physicists greater confidence in their understanding of many other complex nuclear phenomena.

Heavy-ion interactions. The fission process is characterized by the separation of a complex nucleus spontaneously, or after absorption of a neutron, into two relatively large nuclear fragments. Once the fission process is initiated, there is no way to influence its course. If the situation is reversed and the nuclear fragments are brought together under controlled conditions, substantial new information can be gained. Heavy-ion interactions can achieve such a reversal, and for this reason they are growing rapidly in popularity in nuclear physics.

Rotating shield for a prototype fast-breeder nuclear reactor in Dounreay, Scot., is lowered into position in the reactor roof above the core. The shield contains the control and shutdown rod mechanisms for the reactor.

Courtesy, United Kingdom Atomic Energy Authority

During 1974, as the super-HILAC (heavy-ion linear accelerator) at the Lawrence Radiation Laboratory at Berkeley, Calif., becomes operational, it will be possible to initiate the long-awaited experimental search for the so-called supertransuranic elements. No known natural process could have produced these elements, but the best present calculations predict them as being possibly stable once they are made. The production involves the fusion (inverse fission) of heavy fragments such as calcium and plutonium or palladium and erbium; if successful, it will open up entirely new fields in both physics and chemistry.

Heavy-ion interactions have other attractive features as physical probes. The spin or angular momentum is one of the most fundamental properties that can be used to label any quantum state. In nuclear physics it has also been one of the most directly measurable labels, by means of study of the angular relationships between projectiles inducing nuclear reactions and the subsequently emitted reaction products. With the recent availability of high-quality beams of heavy ions it has become possible to study nuclear states with angular velocities that are a factor of ten higher than any previously accessible. In effect, these higher spin rates are achieved when there are glancing collisions between the target nuclei and these heavier ion projectiles.

Increasing the angular velocity by so large a factor makes it possible to place stress on the material involved in entirely new ways. It has long been known, for example, that as the nucleus spins faster and faster it stretches under centrifugal forces and the moment of inertia (a convenient parameter which measures the distribution of the matter involved with respect to the spin axis) changes in measurable fashion. In the past, in going from lower to higher angular velocities, this change in moment of inertia has been systematic and quite small, reflecting the fact that the forces which bind the nucleus together are extremely strong. Within the past year, however, measurements in many laboratories showed that at very high angular velocities a dramatic change takes place in the moment of inertia; in rare-earth nuclei such as dysprosium the moment of inertia can change suddenly by as much as 50%.

If the nucleus were behaving like a more familiar solid body, this would imply that it had suddenly undergone a marked increase in its centrifugal stretch—had "yielded" in the usual material-science sense. Other evidence, however, suggests that the situation in the nucleus may be quite different. Precision measurements of rotating nuclei have always shown a moment of inertia substantially smaller than would have been expected if all the nuclear matter had been in motion about the spin axis simultaneously, that is, if the nucleus were behaving like a rigid body. What seems to be happening in these cases is that only part of the nuclear matter participates in the rotation. This is not an unfamiliar phenomenon in the physics of fluids, for example, where so-called irrotational flow is frequently observed. In the case of a liquid drop, for example, it is possible to establish surface waves that run symmetrically around the droplet. Viewed from outside it is impossible to tell whether one is dealing with a pair of these waves at opposite ends of a droplet diameter or whether the whole droplet has been deformed into a cigar shape and is spinning end over end as a whole. The moments of inertia, however, are strikingly different in these two cases, with that for the pair of waves being very much smaller.

The dramatic change in the moment of inertia observed in dysprosium may be interpreted as reflecting an abrupt change from one to the other of these two kinds of motion. At low spin the neutrons and protons in the center spherical core region of the nucleus apparently are not participating in the rotation, but at some critical angular velocity they become coupled to the surface rotation and the nucleus begins to rotate as a solid body. This poses a fundamental and central problem for nuclear structure theory inasmuch as it should be possible, from microscopic considerations, to predict at what angular velocity this abrupt change occurs. Resolution of this puzzle can lead to important new insight into the coupling between collective and individual particle aspects of many-body systems throughout physics and to the detailed internal structure of nuclei in particular.

—D. Allan Bromley

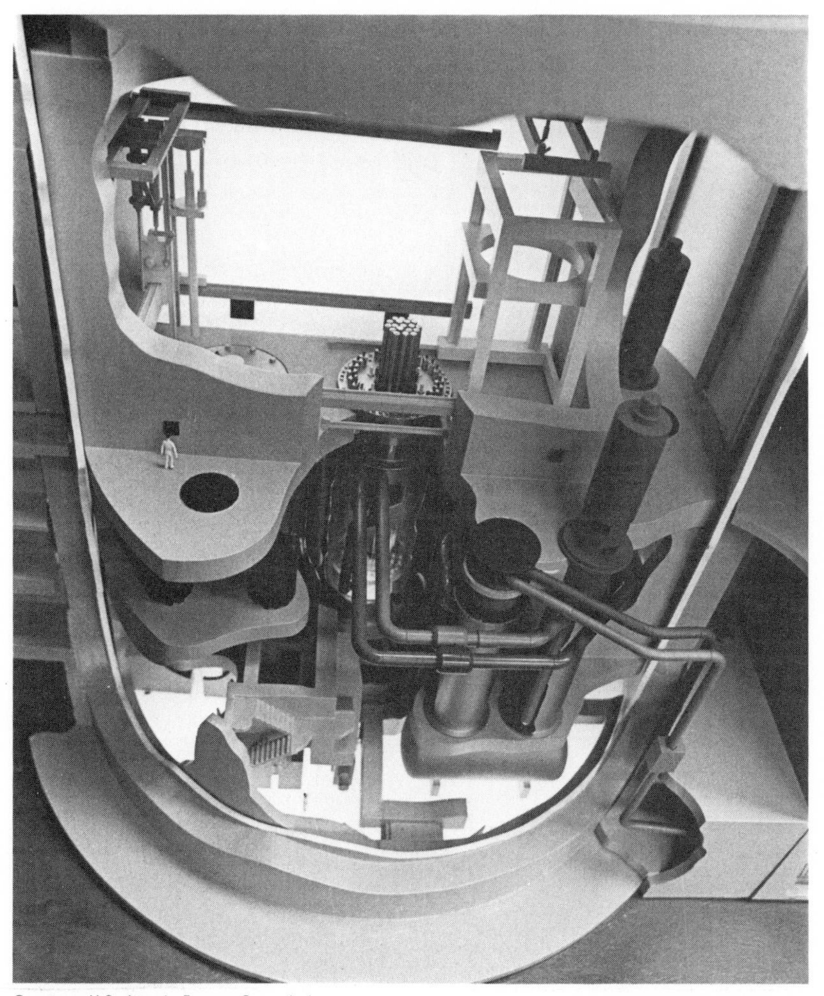

Courtesy, U.S. Atomic Energy Commission

Model shows the type of power plant of a fast-breeder nuclear reactor that private industry and the U.S. government planned to build jointly for operation by 1980. A breeder reactor is designed to produce more fissionable material than was originally supplied to it.

Solid-state physics

Because solid-state physics deals directly with the building blocks of society—solids—it often is strongly affected by matters of public policy. In 1973 two closely related sets of societal problems were having a noticeable effect on the direction of research in the field, protection of the environment and the energy crisis. This report will concentrate on certain areas of work related to and being stimulated by possible contributions to those problems.

One such area is superconductivity. As will be discussed, superconducting materials give the possibility of increasing the efficiency in generation and transmission of electrical power and in contributing to vastly improved transportation systems. Another area is that of surface physics. Improved understanding of it could lead to better control of harmful emission from automobile exhausts and to the development and utilization of new sources of fuel.

Superconductivity. In 1911 Heike Kamerlingh Onnes was amazed to find that when mercury was cooled below 4.2° K it lost all resistance to the flow of electric current. Since that time this phenomenon, named superconductivity, has fascinated physicists for two reasons. First, they wanted to understand the mechanism that gave rise to this phenomenon. Second, they were fascinated by the possibilities of its practical applications. However, from the beginning, it was apparent that applications would be practical only if the transition temperature—below which superconductivity occurs—could be raised to a relatively high level. This was necessary to keep the cooling process from being too expensive.

Presently, superconducting alloys suitable for practical applications have transition temperatures no higher than about 10° K. Greatly improved large motors and electric generators are now being made with these materials, and work is beginning on electrical transmission lines and levitated trains. However, cooling to such low temperatures is both expensive and clumsy. As a result, there is a strong desire to find superconductors with much higher transition temperatures. Consequently, a considerable sensation took place at the American Physical Society meeting in early 1973 when a group from the University of Pennsylvania, led by Alan J. Heeger, reported sudden and large changes in conductivity as the temperature of a sample was lowered below about 60° K. The materials in which this was seen were organic, of a class called charge-transfer salts. The particular materials were based on tetracyano-*p*-quinodimethan (TCNQ), the phenomenon of interest occurring in a modified form of its tetrathiofulvalene (TTF) salt.

The Pennsylvania group suggested that the sharp increase in conductivity, starting at about 80° K, was due to the beginning of a transformation into the superconducting state. If nothing else happened, conductivity would have become infinite at about 58° K. However, the group's argument goes, imperfections in the crystals prevented this and forced the conductivity to drop precipitously at about 50° K. If these arguments prove correct, it follows that more perfect crystals may maintain superconductivity below 58° K.

As can be expected, the possibility of superconductivity in this temperature range excited sustained interest. The fact that the experimental data was very suggestive but not definitive made the subject tantalizing. (TTF) (TCNQ) differs from ordinary metallic superconductors in that it grows in narrow, needle-like crystals and the conductivity appears to be basically along one dimension rather than three as in conventional superconductors. As a result, it appears that the fundamental mechanism of superconductivity may be significantly different from that in conventional crystals. Theoretical

Cavity is formed in water when a spinning steel ball is dropped into it from a height of ten feet. Spiral striations on the cavity were studied to determine the rotational speed of a U.S. Navy missile.

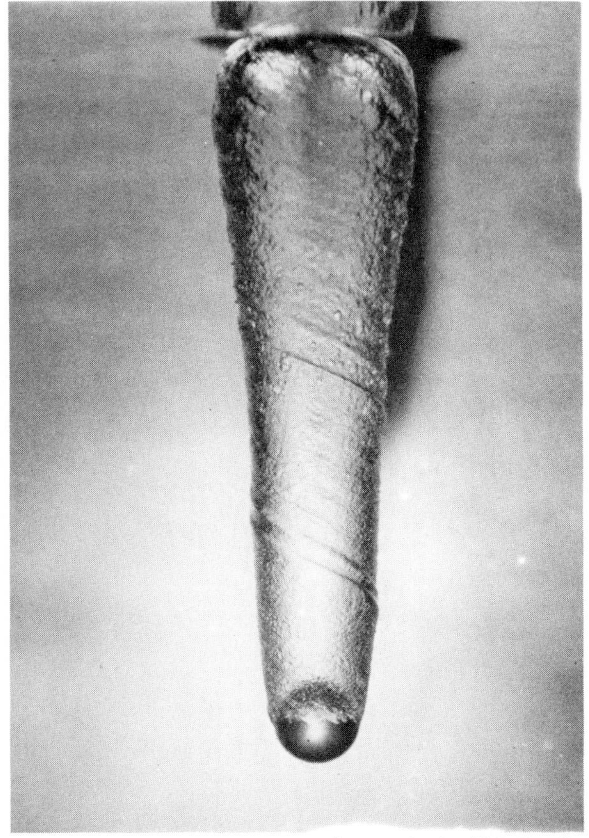

Courtesy, Dr. Gerald G. Mosteller

Courtesy, Stanford Linear Accelerator Center

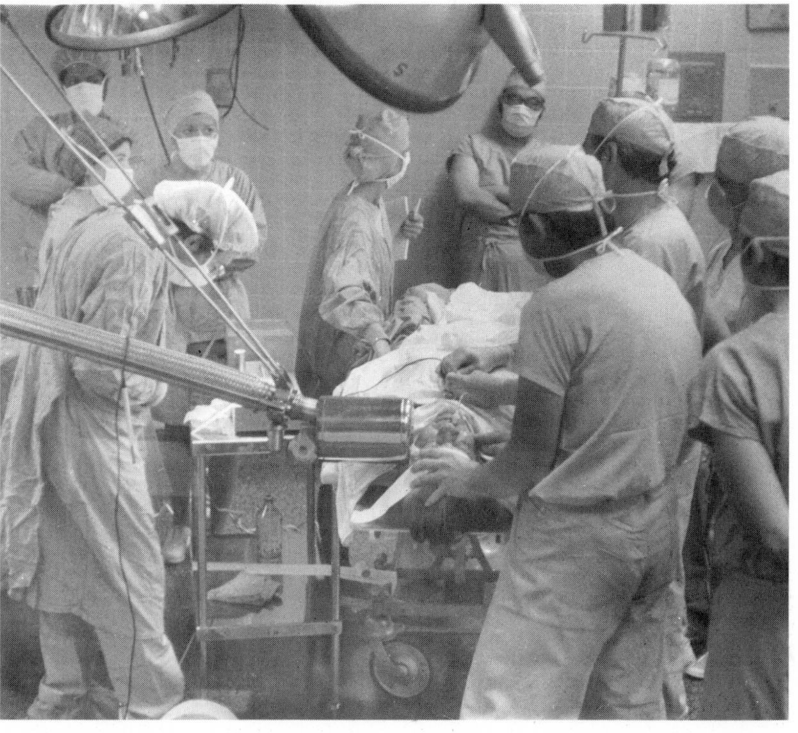

Superconducting magnet cooled to near absolute zero by liquid helium is used at UCLA hospital to treat a patient with a tumor of the tongue. A silicone-iron compound, injected into the patient, is held in place by the magnet until the compound hardens in blood vessels near the tumor. This destroys the tumor by cutting off its blood supply.

models appeared quickly to account for the observed behavior, and much new experimental work was under way. It seemed apparent that a burst of activity in this field and a rapid development and testing of concepts would take place in late 1973 and in 1974. The optimists hoped for the first really practical high-temperature superconductor, whereas the skeptics hoped that some new knowledge would be gained.

Surface physics. What happens to electronic states at the surface of a solid when a crystal that occurs there is abruptly terminated? Is there a new and different electronic structure that characterizes the surface? What happens when foreign atoms or molecules come up to the surface and are adsorbed on it?

Since the development of modern solid-state physics in the 1930s, physicists have been aware of these problems. However, it seemed appropriate to first concentrate on understanding the bulk of the materials before the even more difficult problem of the surface was tackled. However, increasing emphasis recently has been placed on understanding surfaces. There are several reasons for this. As more understanding was gained of bulk properties, the understanding of the surface became an increasingly challenging problem. Furthermore, much of the theoretical and experimental knowledge of the bulk could be used as a foundation for studying the surface. Another factor was the development of a set of experimental tools uniquely suited for the study of the surface. These gave the possibility of providing decisive experimental data. Finally, it has become apparent that if a greater knowledge of surfaces and surface reactions can be obtained, major contributions may be made to the solution of important problems of society. For example, such knowledge might lead to the development of improved catalytic materials, which, in turn, would effectively control the automotive exhaust gases that pollute the atmosphere.

The attack on surface physics can be divided into two parts. First is the experimental and theoretical work aimed at understanding the clean surface. The second comprises the understanding of what happens to the surface and the adsorbed species when foreign atoms or molecules attach themselves to surfaces. It is this adsorption process that is critical for most practical applications; however, it is necessary to understand the clean surfaces before the more complicated situations of adsorbed species can be completely understood.

The experimental tools are themselves of considerable interest. One set of these is designed to determine the crystal structure of the last layer of atoms on the surface. The structure of this outer layer may be different from that of the bulk material. To study the crystal structure at the surface, a low-energy monoenergetic electron beam is directed at the surface and a picture is made of the reflected electrons. Because of the low energy of the electrons used this experimental method is identified by the acronym LEED (low-energy electron diffraction).

One can also use LEED to study the changes in structure as foreign atoms are adsorbed on the surface. Figure 1 (left) shows the characteristic pattern from one crystal face of tungsten, while figure 1 (right) is from the same crystal face of tungsten after it had adsorbed enough oxygen to fill half of the available surface sites. The more complicated LEED pattern after the oxygen adsorption is caused by the more complicated symmetry of the surface atoms after the oxygen adsorption. By analysis of the LEED patterns, it is possible to determine in detail the position of adsorbed atoms on the tungsten surface.

Researchers also developed a second set of experimental tools involved with the electronic structure at the surface. With these tools, energy is fed into the surface in such a way that electrons escape from the solid into a vacuum where their kinetic energy can be measured. By knowing the detailed nature of the excitation and the energy of the excited electron, one can deduce the energy of the electron before excitation.

For a clean surface, such information is essential in determining the changes that take place in the electronic states as one approaches the surface. This experimental data can be compared to theoretical calculations of the surface electronic structure. As foreign atoms are adsorbed on the surface, one can also follow the changes in the electronic structure. Conversely, researchers can also use such studies to determine when the surface is clean and when it has been contaminated.

Three principal methods are used to liberate electrons from the surface: electromagnetic radiation, electron beams, and high electric fields. For the electromagnetic radiation, monochromatic radiation is used and the wavelength is varied from the visible to the X-ray region so that both the valence and core electrons can be studied. This field of study is called photoelectron spectroscopy (PES). It can be used to study both the surface and bulk electronic states. Figure 2 shows a spectra taken with ultraviolet radiation from silicon. In that figure the number of electrons escaping with a given energy is plotted upward and the energy of the electrons horizontally. The cross-hatched peak lying at the highest energy results from electrons residing in states induced at the surface of the silicon. The rest of the structure is determined by bulk states. When the sample is exposed to small amounts of oxygen, the surface states disappear quickly but the bulk states are practically unaffected.

Exciting the solid by an electron beam has been used most successfully as a method for detecting small amounts of foreign atoms on otherwise clean surfaces and in identifying the impurity atoms by means of their specific electronic levels. However, the tool is being refined and extended to other types of analysis. Another method of testing for foreign

Figure 1. Patterns of one surface crystal face of tungsten before (left) and after (right) adsorption of oxygen atoms are revealed by low-energy electron diffraction (LEED) technique. The more complex pattern at right results from the more complicated symmetry of the surface atoms after oxygen adsorption.

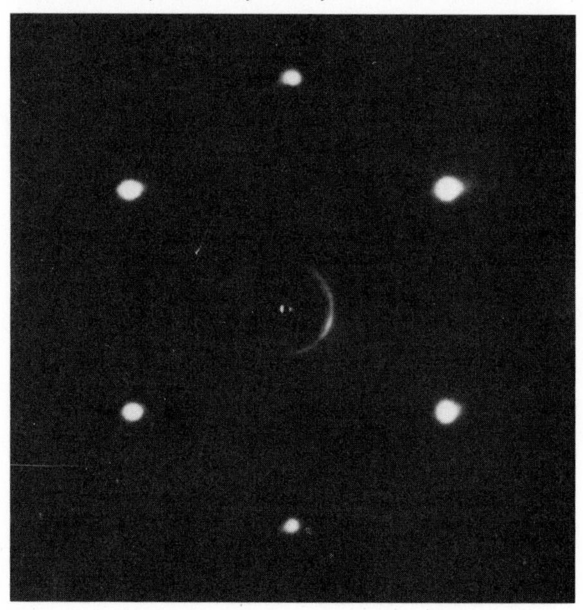

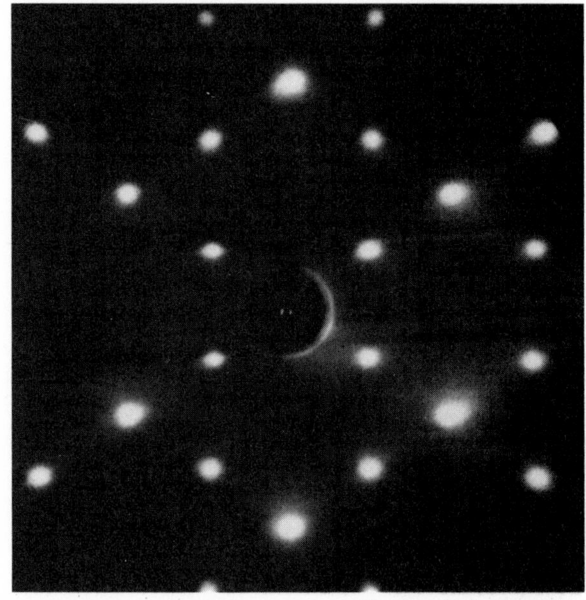

Courtesy, Varian Associates

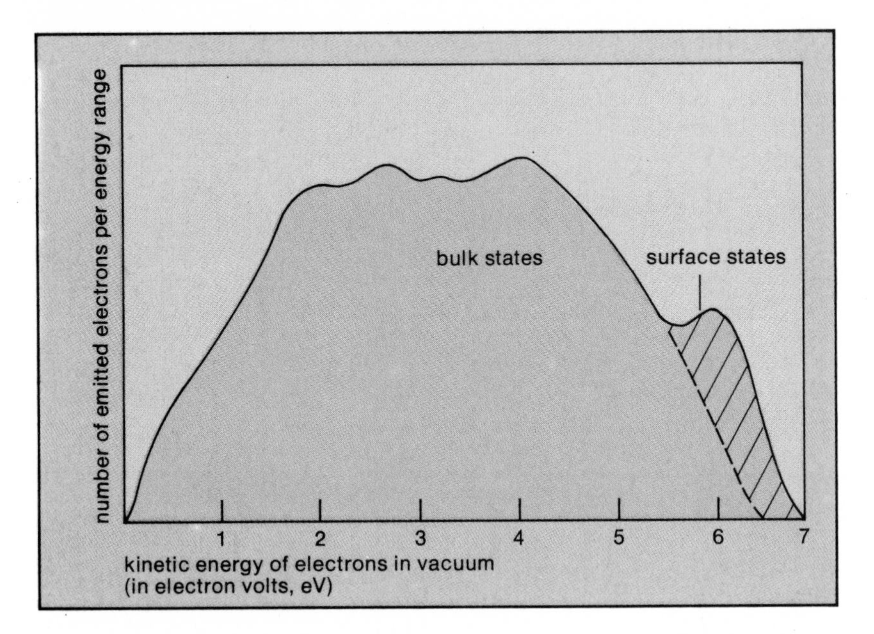

Figure 2. Graph shows the effect of beaming ultraviolet radiation at a sample of silicon. The number of electrons escaping with a given energy is plotted against the energy of the electrons. The cross-hatched area at the highest energy indicates the electrons residing in states induced at the surface of the silicon.

atoms is to use a beam of atoms to knock other atoms off the surface. The atoms knocked off are analyzed and identified with a mass spectrometer.

Another electronic technique is that of ion neutralization spectroscopy. In this technique, a low-energy beam of ions is applied to the surface. This results in the excitation of electrons from the surface. By measuring the energy of these electrons researchers can obtain information on the energy states of the clean and contaminated surfaces.

In field emission spectroscopy (FES), a very high electric field is applied to the surface of a solid, allowing electrons to escape into the vacuum. Again, the energy levels of the pure and contaminated surfaces can be determined by this method.

Through the use of all of these methods, the experimental knowledge of surfaces is growing swiftly. For example, during 1972 strong filled bands of surface states were completely resolved for the first time using ultraviolet photoelectron spectroscopy in both semiconductors (silicon, germanium, and gallium arsenate) and a metal (tungsten). The results from silicon are shown in figure 2 where it has been possible to distinguish clearly the surface electrons from the bulk electrons. The results from the other semiconductors and the metal were quite similar. In similar manners, the electronic states induced by various foreign species adsorbed on surfaces have been detected. It appears that work in surface physics will expand in the coming years and that there will be a strong commitment to make contact with practical problems such as that of catalysis. Only time can tell how successful these efforts will be, but the beginnings are encouraging.

—William E. Spicer

Science, General

The world of science was probably less startled than the general public when the Royal Greenwich Observatory in Sussex, Eng., announced on June 19, 1972, that the earth was running down at the rate of three-thousandths of a second per day. Greenwich Mean Time is based on the diurnal rotation of the earth; atomic clocks, on which many scientific experiments depend, are far more precise. Therefore, it became necessary for the Greenwich Observatory to insert a "leap second" between June 30 and July 1, 1972, and another between Dec. 31, 1972, and Jan. 1, 1973. That the scientific community was not unduly alarmed by the earth's rotational fatigue could be traced not only to its acceptance of Newtonian physics but also to the fact that its own world seemed to be running down too.

Although the output of the U.S. scientific community was as impressive as ever, there was considerable evidence that the federal government had withdrawn from its former position of close concern and that the public at large was becoming disenchanted with scientific research and its technological products. The first indication of a significant change in attitude at the highest levels of government was the suddenly visible intention on the part of the second Nixon administration to encourage the departure of some of the top scientific officers in the government—including the director of the National Institutes of Health and the assistant secretary of the Department of Health, Education, and Welfare for health and science. By mid-January 1973 U.S. scientists could see few trusted faces attached to the body politic. Although science still had friends in government, it was becoming increasingly difficult to locate individuals at the policy level with emotional ties to academic research.

OST and PSAC: death by reorganization. These early forebodings were dramatically borne out on January 26, when Pres. Richard Nixon issued his first message on governmental reorganization. Swept away in a general housecleaning of the White House policy-making staff was the Office of Science and Technology, an organization that had been the linchpin of the federal science-policy apparatus since its establishment in 1962. Although the demise of OST had been expected by many political observers in Washington for at least six months, its sudden disappearance sent shudders of apprehension throughout the scientific community.

The 1972–73 issue of the *U.S. Government Organization Manual* listed the following functions of OST:

1. Evaluation of major policies, plans, and programs of science and technology of the various agencies of the Federal Government, giving appropriate emphasis to the relationship of science and technology to national security and foreign policy, and measures for furthering science and technology in the Nation;
2. Assessment of selected scientific and technical developments and programs in relation to their impact on national policies;
3. Review, integration, and coordination of major Federal activities in science and technology giving due consideration to the effects of such activities as non-Federal resources and institutions;
4. Assuring that good and close relations exist with the Nation's scientific and engineering communities so as to further in every appropriate way their participation in strengthening science and technology in the United States and the free world; and
5. Such other matters consonant with law as may be assigned by the President to the Office.

Even the most loyal friend of the Office of Science and Technology would admit that it was never able to carry out all its assigned functions. But there was a general acceptance—within the academic world, at least—that it had been the principal means by which scientific judgments were brought to bear on major national policy issues and through which the reasonable needs of the scientific community could be brought to the attention of the White House.

We've been a transplanted heart since 1957, and there have always been subconscious forces of rejection. We've managed to stay here and pump blood into the system, but we've never been compatible with the cells around us. Finally, the rejection mechanism overcame the ability of the heart to pump the blood.—*Senior staff officer, Office of Science and Technology, February 1973.*

The disappearance of the scientific leadership in the executive branch and the removal of any formal scientific input at the White House level raised the question of whether science, as such, had been downgraded by the Nixon administration. To the press, to some members of Congress, and to many scientific institutions, the answer seemed to be affirmative. Rep. John W. Davis (Dem., Ga.), chairman of the House Subcommittee on Science, Research, and Development, declared that, insofar as it affected science, the reorganization plan was "disastrous for the nation's economy and for our country as a whole." He said that it represented a "downgrading of science and engineering in national policy" and predicted that its effects would soon be felt in several areas of national life, includ-

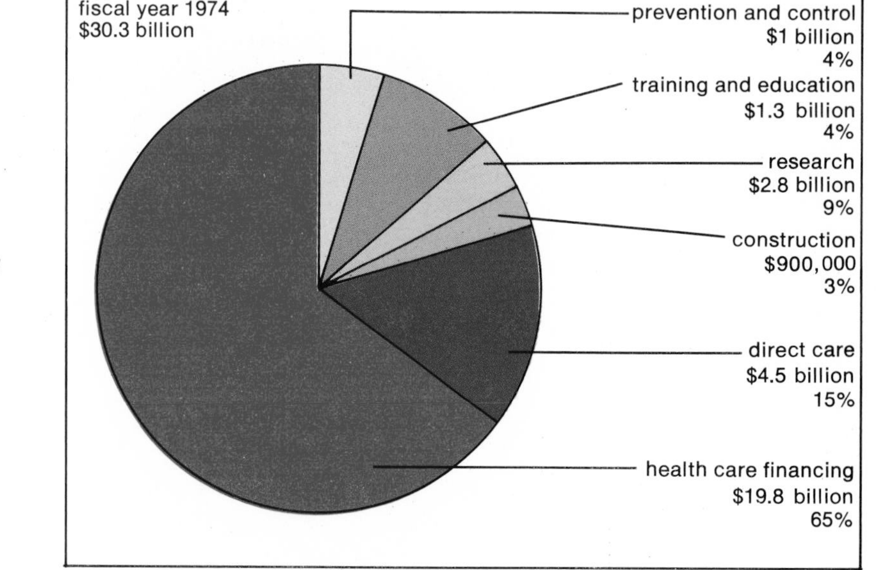

Proposed U.S. health care budget for fiscal 1974 provoked criticism for reducing federal aid to such programs as assistance for treatment of alcoholism and the construction and modernization of hospitals and hospital facilities. Its supporters claimed that it placed primary emphasis on those programs that directly support individual economic well-being and access to services.

"If we ever intend to take over the world, one thing we'll have *to do is synchronize our biological clocks."*

Sidney Harris

ing the country's balance of trade and standard of living.

Unmentioned in the president's reorganization message was the President's Science Advisory Committee. PSAC (pronounced "pea-sack"), once a minor agency in the Defense Department, had been reconstituted by Pres. Dwight Eisenhower in 1957 in response to Sputnik. By launching the first artificial earth satellite, the Soviet Union had challenged U.S. technological leadership and, by implication, its military supremacy. The membership, drawn principally from the academic community, was called upon to offer technical judgments to the White House. During the Eisenhower administration, PSAC was both visible and effective in dealing with many of the national concerns involved in the rapid strengthening of U.S. science. Under the leadership of Jerome B. Wiesner, it continued to grow in prestige throughout the Kennedy years, and was regarded as one of the principal instrumentalities that made the 1963 limited nuclear test-ban treaty possible.

PSAC seemed to slip from view in the Johnson era, however, and during the first Nixon administration there were long delays in the appointment of new members. By the end of 1972 it was quite clear that the resignations of all PSAC members, submitted by custom at the conclusion of a presidential term, would be accepted and no new appointments would be made. Without so much as a formal announcement of its decommissioning, the flagship of academic science in the government had been placed in mothballs.

Many explanations were offered for the demise of PSAC and OST. There was general agreement that the president's moves were simply a de jure recognition of a de facto situation. PSAC had been close to moribund, and even the senior staff members of OST were willing to acknowledge that the agency's participation in the policy and planning functions of the White House was usually pro forma. Most often, it occurred at the instigation of OST itself rather than the Office of Management and Budget or the National Security Council. The most cynical asserted that OST had been struck down because it represented to the White House those sectors of the academic community that opposed most administration policies, particularly in Vietnam. Others argued that policy-making—insofar as the White House staff was concerned—was primarily a management function. There was no more reason to give science a special place in that apparatus than any other specialized segment of society. Spokesmen for labor and agriculture, for example, operated at Cabinet level—why not science and technology as well?

New roles for R and D. For whatever reason, many of OST's functions were assigned to the director of the National Science Foundation (NSF). Although the fiscal 1974 budget contained no provision for giving the foundation additional funds or manpower to carry out these new assignments,

Courtesy, Scientists and Engineers for Social and Political Action

Antiwar pamphlets protesting the bombing of Indochina were circulated at the annual convention of the American Association for the Advancement of Science (AAAS) in December 1972 at Washington, D.C. Such protests in recent years have become common occurrences at AAAS meetings.

the director, H. Guyford Stever, was officially named "science adviser" and asked to report in the future to George Shultz, the secretary of the treasury and assistant to the president for economic affairs.

The new arrangement created new opportunities and new problems. By linking the science adviser to the highest ranking federal official specifically concerned with economic policy, the president seemed to be implying that a fundamental policy decision had been made with regard to federal support of science and technology—namely, that its central function was to promote the economic strength of the country. Recent moves within the NSF had gone far to increase that portion of the foundation's budget devoted to applied research, assessment of R and D (research and development), and experimental programs to advance technological innovation. The new alignment seemed to reinforce that trend.

On the other hand, there were many government agencies that supported applied research and technological development related to their central missions, among them the Departments of Defense, Transportation, and Health, Education, and Welfare. The NSF was the only federal agency whose central mission was the support of fundamental research—the search for new knowledge for its own sake. There was concern that the new responsibilities given to the foundation's director might weaken federal support of academic research. That concern was deepened when Sen. Edward M. Kennedy (Dem., Mass.) introduced a bill to create, within the foundation, a subagency devoted entirely to nonmilitary applied research and provided with a massive budget equivalent to Department of Defense support for military research. (*See* below.)

Ever since the middle of the Johnson administration, the federal government had increasingly favored socially relevant research and the short-term payoff rather than the long-term potential of basic research. This trend became even more pronounced in the federal budget for fiscal 1974. As John D. Holmfeld, science policy consultant to the House Committee on Science and Astronautics, noted in the newsletter *Public Science:*

Within the civil R & D sector, sharp changes are taking place with long-range implications for federal science policy. The stress on payoff and relevance as criteria for project selection, espoused last year by former Presidential Science Adviser Edward E. David, Jr. as the "strategic approach," is continuing. But its implementation and elaboration is leading in new directions. One such direction is that projects with short-term payoff are favored over those with long-term application. This criterion is being applied independent of judgments about the long-term needs for the technology that is expected to emerge, and in some cases simply amounts to a deferment to a later point in time. For example, research on air-traffic-control systems needed for the late seventies and eighties and on space nuclear systems is being phased out. Secondly, R & D projects which can be identified as having completed the R & D phase are actively being sought out, and further development on these projects to increase performance or to carry out full-scale demonstration is being halted.... Thirdly, the policy of getting American industry to carry an increasing share of total national R & D costs is being pursued across a broad front.... Finally, in a small number of cases, public demand and public acceptance are probably important factors in leading to a substantial concentration of resources. This appears to be the case with the big funding increases for research on cancer, heart disease, and energy production and with the continuation of substantial efforts in space exploration and military R & D.

Within the NSF itself, the 1974 budget showed a conscious effort to recognize the agency's unique role in supporting individual research in the basic sciences. The funds devoted to such support rose from $261 million to $275 million—the largest dollar increase of any segment of the NSF budget. However, the Research Applied to National Needs (RANN) program was authorized to request an additional $9 million, bringing its total to $79 million for a larger percentage increase on a smaller base.

The bulk of the increase to be requested for RANN was earmarked for energy research and earthquake engineering. Other reflections of current interests were a 20% increase in requested funding for the Experimental R & D Incentives Program ($15 million–$18 million) and a tripling of requested funds for the Program of Institutional Grants for Research Management Improvement ($1 million–$3 million).

In other agencies, military R and D continued to rise; the space budget underwent a 12% reduction; the Atomic Energy Commission began to concentrate its smaller budget, notably in the liquid metal fast-breeder reactor and the National Accelerator Laboratory at Batavia, Ill.; and increases in the budget of the National Institutes of Health were limited to the Cancer and the Heart and Lung institutes. The government's attitude toward R and D was summed up in its *Special Analysis* of the fiscal 1974 budget:

The funds in the 1974 budget for R & D recognize the need for continuing the application of scientific and technological resources to meet pressing civilian problems where R & D can make a contribution. At the same time, the 1974 budget recognizes that how we spend our resources is just as important as how much we spend. All programs are not equally urgent.

The start of a dry season. The scaling down of federal support for the national science effort was, of course, the continuation of a trend that had become apparent during the Johnson administration. From 1953 to 1964, the annual growth rate of federal support for R and D was an impressive 14.8%. Between 1964 and 1972 it averaged only 2.4%—considerably less than the rate of inflation. Over the same period, the proportion of the gross national product devoted to federal R and D support grew from 1.43% in 1953 to 3.04% in 1964 and then dropped to 2.41% in 1972. To those scientists who had entered their profession in the early or middle years of the 1950s, the situation was just getting back to normal. But to those who entered their fields in the late '50s and early '60s, the overall impression was undoubtedly one of sudden and frightening austerity.

The effect of the budgetary cuts was considerably amplified in late 1972 and early 1973 by decisions within the Office of Management and Budget to withhold, as an anti-inflation measure, some of the funds already appropriated by Congress to scientific agencies. The NSF, for example, was prevented from spending approximately 10% of its fiscal 1973 budget. Hardest hit by this device were anticipated expenditures for the improvement of science education.

By and large, the training of scientists received short shrift from the Nixon administration. In the face of complaints that available scientific manpower was underutilized, it could see no need to subsidize the production of additional manpower. The scientific community retorted that the apparent surplus of scientists was a product of administration cutbacks in research support, but the Office of Management and Budget was not impressed. One of the hardest-hit areas was that of training grants from the National Institutes of Health. For many years these grants had been a mainstay of the system for producing biomedical scientists and—to a certain extent—of support for training hospitals.

Some of the nation's leading scientists gathered in Washington in the spring of 1973 to protest the cutbacks, among them James D. Watson, internationally renowned molecular biologist, and Michael DeBakey, the noted Houston, Tex., heart surgeon. Testifying before a Senate committee, Watson protested the shift in emphasis from training grants to targeted research: "This way of proceeding represents a puerile understanding of how good science is done and how its discoveries have been directed toward human application." He pointed out, in addition, that "almost every important new discovery comes from someone under thirty-five and who at the moment of his breakthrough is essentially unknown to the outside world."

The scaling down of federal support also resulted in a mobilization of protest activity within the traditional scientific societies. The president of the American Astronomical Society and the headquarters offices of the Federation of American Societies for Experimental Biology—among others—strongly urged their members to develop close contacts with their political representatives on Capitol Hill. Others tried to reach the public directly. A resolution, passed unanimously at a meeting of the American Society for Neurochemistry in March 1973, declared in part:

> Because of the appreciable time lag between fundamental discoveries and their practical applications, and because of the long training period for a scientist, 7 to 12 years after entering college, the effects of the present policies are not yet apparent to the general public. The scientific community, however, is well aware of and concerned by these effects. Promising research projects are being cut back and abandoned. The knowledge and skills of highly trained and highly dedicated persons are lost as competent researchers are forced to abandon their chosen fields or leave science altogether. Their replacements are not being trained, and many with the rare combination of necessary talents are compelled to enter other fields.
>
> The community of scientists which has made the United States the recognized international leader in biomedical research is a unique and invaluable national resource that is being put in jeopardy. Over a quarter of a century was required to bring it to its present eminence. Once lost, it may well be impossible to restore within our generation. Therefore we urge the present policies be changed while there is time.

There was, however, a pervasive concern within the leadership of the scientific community that the government's desire to economize reflected a general turning away from science on the part of the public. In these circumstances, specific pleas for public support could indeed fall on deaf ears.

Public opinion. What disturbed the scientific community was the generally antiscience view emanating from several segments of the intellectual community outside of science. Many of the more vocal members of these groups, when they looked at the world about them, saw much that they did not like—and, in their view, what they did not like was traceable directly or indirectly to science-based technology. In addition, the scientific apparatus was coming under increasingly suspicious scrutiny from many members of the press who had previously held more favorable attitudes toward it. Finally, there was evidence of a gradual loss of interest in science on the part of the people who dominated the communications media. One tabulation indicated that in all of 1972 only 18 hours of prime-time television were devoted to science, and 12 of those were on public television.

For the past several decades, a small but increasing minority of scientists had deliberately set out to cultivate public understanding of science. Within the NSF, the Program in Public Understanding of Science received modest but useful sums—between $500,000 and $1 million annually. Most of this was spent on the assumption that the more the public knew about the scientific endeavor the more they would appreciate it and contribute to its support. In 1972 and 1973, however, it became clear that to know science was not necessarily to love it.

In 1972 the NSF responded by diverting some of its funds from the making of such films as *The Birth and Death of a Star* to support a Workshop on Goals and Methods of Assessing the Public's Understanding of Science. According to its final report, the "rough consensus of the workshop [was] that a good, comprehensive measurement of the public understanding of science is the most urgently needed research right now."

There were some reassuring indications that the current concern over public disenchantment with science and technology had been overstated. A small but well-structured survey of three communities in eastern Massachusetts revealed, in October 1972, that 83% of the sample questioned thought technology had done more good than harm overall. But half of those questioned felt that "a lot of people are critical of technology." The authors concluded:

> In general, our survey does not reveal the kind of rampant antitechnology, antiexpert sentiments that we hear so much about from social critics and the mass media. The majority of the sample see technology as more beneficial than harmful and regard such things as machines, television, and computers as bringing comfort, awareness, efficiency, and a high standard of living. While they favor a greater allocation of national resources to social than to technological programs, they do not want to stop technological progress. They are, however, less positively inclined toward such more remote technological developments as the SST, the space program, and the ABM than they are toward computers and automation.

It might be that at least some members of the public were more interested in science than the scientists themselves. In the January-February 1973 issue of *American Scientist*, a distinguished scientist noted that, on a recent evening at Harvard, a scholarly discussion of astronomical research had outdrawn an address by consumerist Ralph Nader. A few months later, however, an address by Nader drew the largest audience at a four-day meeting of the American Physical Society in Washington, D.C.

The Science Policy and Priorities Act. The questioning of "remote" technologies revealed by the Massachusetts survey was shared by a number of influential thinkers. J. Herbert Hollomon, director of the Center for Policy Alternatives at Massachusetts Institute of Technology, visited the NSF in late 1972 to urge its officers to pay more attention to the potential role of government leadership in advancing U.S. manufacturing technology. He pointed out that, as of that time, 75% of the components of all black and white television sets, 50% of the components of all color television sets, nearly 100% of all tape decks and recorders, 75–80% of all camera parts and sophisticated lenses, and 97% of all motorcycles sold in the U.S. were made in other countries. Except for a few areas of "massive high technology," he declared, the technology of Europe and Japan and their ability to use it was equal to and perhaps superior to that of the U.S.

Hollomon and others traced the loss of U.S. leadership in civilian technology to overemphasis on such highly specialized fields as aerospace technology and military weapons systems. Although proponents of these technologies emphasized that many of the devices they had developed had found commercial uses (*see* Year in Review: MEDICINE: *Room W-204*), it was becoming more and more difficult to sell that justification to Congress.

It was in partial response to this argument that, on Aug. 9, 1972, the Senate Labor and Public Welfare Committee unanimously reported out Senator Kennedy's National Science Policy and Priorities Act of 1972 (S. 32). Despite vigorous opposition from the Nixon administration, the bill sailed through the Senate 70–8. It was not acted upon by the House before Congress adjourned, but was reintroduced in the Senate on the first day of the 1973 session.

The bill was, potentially, the most important piece of science legislation to be introduced in any recent Congress. Title I would give explicit authority to the NSF to develop national policies for applying science to national problems and would broaden composition of the National Science Board to include more technical and industrial representation. Title II would establish a Civil Science Systems Administration (CSSA) within the National Science Foundation to do research, design, testing, evaluation, and demonstration of civil science systems capable of providing improved public services in such areas as health care, public safety, public sanitation, pollution control, housing transportation, public utilities, communications, and education.

Title III would authorize the NSF to plan and assist in the transfer of scientific and technical manpower from research and engineering programs that had been terminated or significantly reduced to other, civilian-oriented research and engineering activities. Title IV declared as national policy that scientists and engineers should be protected as much as possible against forfeiture of pension rights or benefits as a consequence of job transfers or loss of employment resulting from changes in federal policy. Finally, the bill set as a goal growth of civilian R and D proportional to the gross national product and at parity with military R and D. The bill would authorize, over a three-year period, the expenditure of $50 million under Title I,

"Pleasant working conditions, the latest equipment, lots of money—you won't even realize you're working on biological warfare."

Sidney Harris, BULLETIN OF THE ATOMIC SCIENTISTS

$1.2 billion under Title II, and $560 million under Title III.

In reintroducing the bill in the 93rd Congress, Senator Kennedy declared:

> If there is one overriding lesson I have learned in my four years as chairman of the National Science Subcommittee, it is that the potential of science is nowhere being matched by its performance.
>
> We can cruise on the moon's surface—yet we cannot commute from suburb to city without traffic jams, air pollution, and no parking at our destination. We can build beautiful, enclosed shopping malls in the suburbs—yet we cannot begin to cope with the housing crisis in the cities. We can design high-speed computers to process billions of bits of data instantly—yet we cannot teach all our children to read effectively.
>
> Why is it that cities have to be put on pollution alert, so that mothers have to be concerned about their children playing out of doors? Why is it that household appliances continually break down and that their repair is not only costly but often unreliable?
>
> Why is it that thousands of American children burn to death each year because they wear highly flammable fabrics? Why is it that decent housing is increasingly out of reach of more and more of our citizens? Why is it that millions of Americans are undernourished in an age of affluence? Why is it that there are only kidney dialysis facilities for two thousand of our citizens when fifty thousand need such care? . . .
>
> The list is endless, but the lesson is clear: We have the scientific knowledge, but we have not made the concerted effort necessary to put it to use for the benefit of all our people. . . .
>
> We are all aware that productivity in American industry has been lagging in recent years, especially in comparison to Japan and Western Europe. Yet how many realize that part of the answer for this lies in American underinvestment in civilian research and technology? A Department of Commerce study shows that Western Europe, when its Gross National Product is only one third of the U.S. GNP, has a third more technical personnel engaged in civilian research. And Japan, with only half the population of the United States and one-seventh of our GNP, had seventy per cent as many scientists and technical personnel employed in civilian research and development.
>
> If we make the necessary national commitment to tackle these problems, if we provide the Nation's scientists and engineers with the wherewithal to do the job, I am confident they can solve many of these problems and make a giant step forward on earth toward making this the kind of society we want for our children, and their children to come.

The opposition. Supporters of Senator Kennedy's bill were confident that it would pass the Senate again, but no one expected it to have an easy time in the House. Ominously, it had not even been reported out of committee in the previous session. Furthermore, the House was the locus for the opposition.

That opposition took several forms. When the bill was first introduced, the Nixon administration took the position that its managerial innovations were either unnecessary or counterproductive. Edward E. David, Jr., then director of OST, rejected any implication that "there is no planning or direction at the top for civilian technology." In September 1972, NSF Director Stever told the House Committee on Science and Astronautics:

> I would note that many of the problem areas identified in S. 32 are already the responsibility of a mission agency of the Government. These agencies are closer than NSF to the problems and to the environment in which specific solutions must be applied. . . . In short, the Foundation is [already] concerning itself with those civil science areas for which NSF has a unique capability.

Other concerns focused on the effect that vastly increased funding in the civil technology area might have on the role of the NSF in supporting fundamental research across the board. "The problem with CSSA," Stever observed, "is that with its sizable outlays, the tail will wag the dog."

"Let 'em eat irradiated cake!"

Editorial cartoon by Pat Oliphant © "The Denver Post," reprinted with permission Los Angeles Times Syndicate

Philip Handler, president of the National Academy of Sciences, worried that placing CSSA within the NSF

> could seriously divert the attention of the top people of NSF from their concern with basic research. The realities of a bureaucrat's life are that you expend your time and effort in those areas where you are most vulnerable. The Congress does not expect immediate results from fundamental research. But the kind of research supported by CSSA will be of the short-range variety and visible results will be expected in the near-term. The problems that will arise from the need to generate apparent payoffs rather quickly in the civilian technology programs—plus the host of new problems that Guy Stever will face as Science Adviser—would not leave many hours of the day to worry about the state of basic research.

Similar anxieties were voiced by Rep. Charles A. Mosher (O.), ranking Republican on the House Committee on Science and Astronautics. The president's reorganization plan, he said,

> will wrench NSF away from its traditional role as the promoter and guardian of basic science in the federal government. . . . [Passage of S. 32] would greatly add to the danger for basic research because the applied research and development projects undertaken by the CSSA could swallow up NSF's basic research mission.

Even if these concerns were allayed—for example, by adopting the proposal of Rep. Alphonzo Bell (Rep., Calif.) that NSF be divided into two separate but roughly equal administrations with at least 40% of appropriated funds going to basic research and education—there would remain the problem of where the new money was to come from. Rep. John W. Davis (Dem., Ga.), whose Subcommittee on Science, Research, and Development had jurisdiction over the House version of the bill, observed that "the mood of the country and that of Congress render almost nil the chance for legislation that would require over a billion dollars during the next two or three years."

Senator Kennedy indicated he was willing to compromise somewhat on the dollar authorizations, but there were few in the House leadership who shared his optimism that the bill would become law in anything like its present form. However, Senator Kennedy did achieve a measure of success in a related field. He was appointed the first chairman of the Technology Assessment Board, which would direct the activities of the newly established (though still unfunded) Congressional Office of Technology Assessment. (See *1971 Britannica Yearbook of Science and the Future*, Feature Article: PRIORITIES FOR THE FUTURE: GUIDELINES FOR TECHNOLOGY ASSESSMENT; *1973 Britannica Yearbook of Science and the Future*, Year in Review: SCIENCE, GENERAL.)

A worldwide problem. If the need to stimulate the system whereby applied research fed industrial growth was apparent in the United States, it was even more apparent in the less developed countries. The basic problem was that most of these nations lacked the financial and human resources —and sometimes the resolve—to maintain a large research effort. Thus it was that the United Nations Advisory Committee on the Application of Science and Technology to Development proposed to the UN Economic and Social Council—and, throughout 1972 and 1973, to governmental and nongovernmental organizations around the world—a "World Plan of Action." In the words of the chairman of the Advisory Committee, the plan

> envisages in particular the attainment of three targets by the year 1980: (1) that the developing countries should by the end of the decade, aim at spending 1 percent of their GNP on research and development and scientific and technical services; (2) that the developed countries should devote 0.05% of their GNP to international aid in the field of science and technology; and (3) that 5 per cent of the non-military research and expenditures of the developed countries should be allocated to R & D on problems of concern to the developing countries.

The committee selected a few areas for priority treatment: (1) development of high-yield seed varieties for specific regions, an objective already achieved with some cereals; (2) applied research in marine fisheries; (3) less-hazardous pesticides, including biological controls; (4) development of forest resources and of tropical fibers; (5) arid-land research; (6) atmospheric research, making use of functioning weather satellites; (7) materials research for shelter and road building; (8) research in such fields as metallurgy, ceramics, and glass manufacturing, chemical and textile manufacturing, the design of equipment and industrial plants for small-scale manufacturing and handicrafts, and repair and maintenance services; (9) control of schistosomiasis, a debilitating and often fatal disease spread by water-borne snails; and (10) research on human fertility and contraception.

The scientific community retained its confidence that, insofar as science and its applications could help to solve the most pressing problems of the world, the answers could be provided by research. The anxiety that pervaded the scientific community stemmed rather from the fear that their confidence was not shared by the policy-makers and the public. The result was a great deal of soul-searching.

Why had the world suddenly become such a bleak place for many young scientists? One set of answers was offered in the British publication *New Scientist* by Ludwik M. Celnikier, a French astrophysicist:

> Perhaps science became too expensive, too separated from reality or was itself changing reality in an uncomfortable way; it may be that we had overproduced and oversold it. Surely we had provoked consumer resistance. . . .

For the first time in their careers, scientists are having to assess objectively the significance of their work, their future, and their personal security. Were we doing science, or simply going through a daily ritual? What was it in anti-proton physics, or protein chemistry, or Moonrocks that made us think to study them? In what way does our own work satisfy—is it fun, intellectually stimulating, or useful? . . .

Today, the capacity for research and the demand for certain kinds of manpower are determined to a great extent by murky political machinations. . . . It takes several years to make a skilled aeronautical engineer, while one year is sufficient to take away the need for his peculiar abilities. . . .

The future, I believe, lies with the wandering scientist. He will be flexible, adaptable, so as to better profit from whatever overlap there might be between his talents and the opportunities available: chemists may work as petrologists or ecologists, thermodynamicists will dabble in laser applications or systems programming. . . . Individuals may find that they are better teachers than research workers, more creative as interpreters than as originators, more content as participants in a society their skill helped to shape than as mediocre hunters of a fundamental chimera they can never hope to find. They must above all learn to think. . . .

Splash some water on a hot electric stove. The water breaks up into separate little globules dancing about precariously and making a fine sizzling noise. They survive for a remarkably long time, because the liquid evaporates to form a protective insulating layer between themselves and the hot plate. Watch how the drops rush hither and thither, bounding from each other, shrinking.

This is the year of the evaporated scientist.

—Howard J. Lewis

Transportation

The rapid recovery of the U.S. economy during the year brought renewed interest in technological innovations in transport, some of which had been slowed because of the tightening of funds for research. With both domestic and international traffic showing sharp gains, carriers were in a more optimistic mood and eager to try new and better ways to move both freight and people.

Serious challenges to transport technology were posed by the ever tightening regulations on safety, pollution, and noise, as well as by the worsening fuel supply. This was particularly acute in the U.S., where more than 50% of all petroleum products consumed went to some form of transportation. (*See also* Year in Review: FUEL AND POWER.)

Aviation. Most airlines dropped their options to purchase the Anglo-French Concorde supersonic transport, the cost of which had increased to an estimated $55 million–$60 million per plane. The reasons given were primarily economic, although noise and pollution factors obviously were involved. The U.S. Federal Aviation Administration issued a regulation prohibiting civil supersonic flights over the U.S. and its territorial waters. The National Aeronautics and Space Administration (NASA) expressed serious doubts as to whether the Concorde or the Soviet Tu-144 SST would be economically successful, primarily because of the restrictions on payload and range needed to meet standards of acceptable noise during takeoff and landing. Nevertheless, NASA believed these problems could be overcome through further research and testing. The Concorde's backers still expressed optimism and predicted sales beyond the expected, though limited, British and French airline markets. The Tu-144 was expected to begin scheduled passenger service on Aeroflot routes by early 1975. Production models were to have a longer fuselage than the prototype, bringing the total number of passengers to 140.

The world fleet of huge four-engine B-747 jumbo jets neared the 200 mark, and another 26 were on order, but the even newer fleet of somewhat smaller, wide-body, tri-jet L-1011s and DC-10s was growing faster. More than 80 were in service, and another 210 were on order. While virtually every one of these aircraft was flown in a passenger configuration, their belly cargo capacities were so huge that each one added about 67% of the capacity of a conventional jet freighter. For U.S. airlines alone, the combined fleet of these aircraft had more than doubled the total cargo-lift capacity in only three years.

The only B-747F freighter was operated by Lufthansa German Airlines, although Seaboard World Airlines, Inc., an air cargo carrier, ordered three for delivery starting in August 1974. The Lufthansa plane carried payloads of about 95 tons in 28 specially designed ten-foot-long containers in the main deck and smaller containers and pallets in the lower deck. World Airways, Inc., a large U.S. supplemental air carrier, was planning to put three B-747Cs into service, thus becoming the first airline anywhere to operate the convertible version. The B-747C could be converted quickly to all freight from mixed passenger-freight.

The A-300B European airbus program moved ahead swiftly, with prototypes completed and test flights begun. The twin-engine aircraft, specifically designed to capture the large short-to-medium-distance travel market, would be able to carry from 200 to 345 passengers. The shorter-range version, expected to be priced at about $15 million per plane, was scheduled to enter service in 1974, with a longer-range version scheduled for early 1975. Backed primarily by the French and West German governments (43% each), the project was also supported by the Netherlands and Spain, plus a British firm. As of early 1973, orders totaled 16, plus 22 options. The competitive potential for this type of aircraft was so great that all three U.S. aircraft manufacturers were seriously considering development of twin-engine airbuses, possibly through a modified DC-10 or through a completely new design.

Two big boosts were given to short takeoff and landing (STOL) aircraft. The Canadian government, as a "national program," authorized about $80 million for research and development of such aircraft. It foresaw a 300-plane market, mainly overseas and for U.S. commuter airlines, and its goal was to start deliveries of an economically feasible craft by late 1975. Initial emphasis would be on production of the four-engine, 48-passenger, turboprop DHC-7, which it hoped to sell for $2.3 million each.

The other boost for STOL aircraft came from the U.S. Air Force, which awarded two approximately equal contracts, totaling $181 million, to Boeing Co. and McDonnell Douglas Corp. for development of four jet STOL prototypes for later production at a unit cost not to exceed $5 million. In civilian configurations, the planes would be able to take off and land on runways 2,000 ft in length and would carry about 150 passengers on trips up to 500 mi, although trips between 200 and 300 mi would probably be the most economical.

Highway transport. After extensive in-service tests, Greyhound Lines Inc. announced that the turbine engine had proved successful in intercity buses and that it planned to have ten Turbocruisers in revenue service by mid-1973. Greyhound also announced a long-range program to convert all of its 5,000 buses to turbine power. The 300-hp engines, built by the Allison Division of General Motors Corp., had four-speed automatic transmissions and were said to be virtually free of pollution and vibration. They also offered other advantages that more than offset the higher initial purchase price and greater fuel consumption. The engines' life between major overhauls was expected to be about double that of diesels; they could use a wide range of fuels; they did not need water system and oil changes; they weighed half as much as comparable diesels; they presented no muffler or cold-weather starting problems; and they were much quieter.

Despite this enthusiastic endorsement, General Motors announced that it would delay for one year large-scale production of turbine engines for marine and large truck use in order to build a new block capable of providing more horsepower. Trucking industry spokesmen were cautiously optimistic about the turbine. However, its potential value in twin-trailer operations over long continuous-driving routes, plus its expected ability to meet future emission standards, could well result

Soviet Tu-144 supersonic airliner undergoes flight testing. On June 3, 1973, one day after this photograph was taken, the plane crashed into a small village while flying at the Paris Air Show, killing the crew of six and eight residents of the town.

UPI Compix

Courtesy, Association of American Railroads

ACI (automatic car identification) label is welded onto a freight car. When this car passes an ACI scanner, a beam of white light emitted by the scanner is reflected back to it as colored light by material on the label. The scanner then processes this optical information and sends it in digital form to a central office of the railroad. Thus, railroads can determine the locations of their cars.

in a rapid switchover, given the relatively short operating life of truck tractors.

Another major development involved the sudden success of the Wankel rotary engine in the Japanese-made Mazda and the growing commitment by General Motors to use this type of engine in its future cars. GM had already announced plans to put Wankels in 100,000 of its 1975-model Chevrolet Vegas, as a $400 option. If this was successful, output would be doubled the following year and the engine would be used in other models, with future possible doubling of output the year after that. Some analysts predicted that by 1980 rotary-engine models could account for as much as 75% of GM's auto output. Other automakers disagreed, although the engine's proponents pointed to several advantages. These included the much simpler design, fewer parts and lighter weight, quieter and vibration-free performance, and considerably more power than a comparable piston engine. These features, they claimed, gave the rotary engine a big edge in the competition to meet pollution and noise standards. (*See* Feature Article: ENGINES OF THE FUTURE.)

Trans-Alaska pipeline. The long delay in starting construction of the 789-mi, 48-in. trans-Alaska crude-oil pipeline from the North Slope fields to the ice-free port of Valdez had some major effects. One was to triple the original $1 billion cost, largely because of adjustments and technological improvements made in an effort to meet environmental objections to the heated pipeline, as well as the adverse climatic conditions. The U.S. Supreme Court, in refusing to review a court of appeals decision that current law did not allow the Department of the Interior to approve the required width of the right-of-way, put another roadblock before the project. Congressional action would be needed before construction actually got under way.

If and when it was finally built, the line would provide for optimum response to any change in conditions, and although it would be almost entirely automated, it could be manually controlled if necessary. The oil flow through the 12-pump system would be monitored continuously and controlled from a center at Valdez. While the builders believed the possibility of a major oil spill was remote, they took steps to ensure fast and accurate detection of any leaks or spills and to establish a plan of corrective action that included special shutdown procedures, immediate dispatching of maintenance crews, prompt recovery of any spills, and provision for rehabilitation and restoration of all affected areas. Further studies were being made to refine these already elaborate procedures.

According to current plans, part of the line would be buried in the conventional mode of pipeline construction, part would be buried with special systems to reduce or prevent heat transfer to the ground, and part would be placed aboveground on gravel berms or piles. The pipeline would be built to withstand stresses from internal pressure and from thermal, bending, and seismic forces. To prevent flotation, it would be either concrete-coated or anchored at river crossings and in certain flood areas, and the buried portions would be electrolytically protected from corrosion. Aboveground portions would be thermally insulated to minimize the drop in oil temperature that would otherwise occur during a shutdown.

The line would cross three mountain ranges, many rivers and streams, and a known active seismic fault. Extensive research by the owner companies resulted in refinements designed to permit the pipeline to withstand direct earthquake effects, including actual shaking by seismic waves and movement of the earth surrounding it. Equal attention was given to facilities at the southern terminal, the harbor area at Valdez, where the two million barrels of crude oil that would move through the line each day would be stored before being transferred to tankers for shipment to U.S. mainland refineries. The storage tanks and connecting pipelines would be designed so as to be virtually immune to damage from earthquakes, tidal waves, and fires.

Railroads. The efficiency and economy of rail freight service were being improved through the increased use of unit trains, which haul single commodities such as coal or grain in a fast shuttle-type service, thus avoiding classification yards and the costly and time-consuming coupling and recoupling of cars. Despite the empty back hauls, this type of service proved so popular with shippers and so efficient that the total number of unit trains in the U.S. was estimated to have reached several thousand. Over 90% of rail coal shipments moved in this manner, and the list of commodities hauled was expanding rapidly to include all types of grain and chemicals, ores, lumber, petroleum products, sand, gravel, and new autos and auto parts. Industry observers believed that an entirely new market for such service might be opening up in the form of removal of urban wastes to outlying dumping or rehabilitation areas.

Closely related was the use of fast, run-through trains handling mixed freight in two-way traffic over two or more railroads, also bypassing yards and minimizing locomotive and freight-car switching and handling. In just two years the number of such trains had increased to at least 80 per day. Both the unit and run-through trains offered excellent opportunities for railroads to increase pro-

"Truck assemblies" constitute the undercarriages of the cars used in San Francisco's Bay Area Rapid Transit (BART) system. Each assembly includes wheels, axles, and the suspension system. Similar units were being designed for rapid transit use in New York City and Washington, D.C.

Courtesy, North American Rockwell

ductivity through shorter transit times, faster turnaround, and greater utilization of equipment.

The U.S. automobile industry took another step toward its goal of moving all new autos in completely enclosed transport equipment. Chrysler Corp. announced that the Santa Fe Railway had developed a new, enlarged, fully enclosed trilevel rail car designed to carry 12 to 15 automobiles. A sizable number of trilevel cars were already in use, but most of them were open and subject to vandalism and pilferage. Already tested and in regular use were a rail car that hauled compact autos in a vertical "bumpers-up" position, and one that hauled luxury cars in single, stackable containers.

Amtrak, the U.S. government-run rail passenger service, reported modest gains in intercity passenger travel by train, but its estimated losses in fiscal 1973 amounted to about $128 million. While it was planning some cutbacks in service and in unprofitable routes, it was also planning extensive improvements in equipment and some major service expansion. Late in 1973 it expected to be operating four new, modern, high-speed, turbine-powered passenger trains along the Chicago–Milwaukee, Wis., and Chicago–St. Louis, Mo., corridors. Two had been built for the Canadian National Railways by United Aircraft Corp. but became surplus in a restructuring of service. Two other United Aircraft Turbotrains operating on lease between New York and Boston were to be sold to Amtrak for $2.3 million each and continued in that service.

The other two turbine-powered trains to be used in the Midwest would be leased from ANF-Frangeco, a French firm, with an option to buy at a price of $2.3 million each. These second-generation trains were expected to outperform the ones that had been operating for over two years between Paris and Cherbourg, a distance of 232 mi, with 99% reliability. The new five-car trains would seat 304 persons each, be capable of speeds of 125 mph, and yet able to stop from 100 mph in 3,200 ft. Independent springing of the wheel trucks would permit the trains to maintain high speeds through curves and still give passengers a comfortable ride.

Water transport. Three of the new Seabee barge-carrying ships were put into service by Lykes Bros.

Courtesy, Lykes Bros. Steamship Co., Inc.

Seabee barge-carrying ship "Doctor Lykes" of the Lykes Bros. Steamship Co. discharges its cargo at Bremerhaven, W.Ger. Two of the barges at the rear of the ship are lowered by elevator for removal to the dock. The giant Seabees, 875 ft. long, were the largest U.S.-flag cargo vessels in 1973.

Courtesy, Boeing

Interior configuration of Boeing 747C allows plane to be used as an all-passenger, all-cargo, or part-passenger and part-cargo transport. When carrying cargo only, the aircraft has a maximum payload of 233,700 lb.

Steamship Co. in the Gulf-European trade. These huge ships, the largest U.S.-flag cargo vessels ever built, were 875 ft long, and each could transport as many as 38 large barges loaded with up to 24,000 tons of cargo or as many as 1,800 20-ft containers. They could cruise at 20 knots and could load or discharge cargo in as little as 13 hours.

Another type of barge-carrying ship, the LASH (lighter aboard ship), introduced into commercial service two years earlier, proved so popular that 22 were expected to be in operation by the end of 1974. Pacific Far East Line, Inc., was receiving deliveries on an order for 200 fiberglass barges to be used in its LASH ships. They were 40% lighter than comparable steel barges and were expected to cut maintenance costs as much as 80% because they were easier to repair and had no corrosion problems.

With containership capacity already excessive and barge-carrying fleets nearly completed, the next major emphasis by shipyards would be on jumbo oil tankers and LNG (liquid natural gas) ships. It was expected that large numbers of these ships would be required by Western nations to meet the growing demand for imported petroleum and LNG. While estimates varied considerably, the need for additional oil tankers by U.S. petroleum companies could reach the equivalent of 350 ships of 250,000 deadweight tons each. Another 30 smaller tankers would be needed to haul Alaska oil if the trans-Alaska pipeline was built. The larger tankers would have drafts too great for U.S. ports on the East and Gulf coasts, and two or three offshore ports were being planned to handle them (*see* Year in Review: ARCHITECTURE AND BUILDING ENGINEERING, *Berths for Giants*).

The LNG ships were large, complex, and expensive. Designed to transport LNG at temperatures of −260° F (−127° C), the ships had to have tanks strong enough to prevent leakage of a cargo compressed on a ratio of 600 cu ft of gas into one cubic foot of liquid. Fifteen such ships were in operation, about half of them built by the French. Thirty, each about the size of a 120,000-ton oil tanker and costing around $100 million, were planned for construction in U.S. yards alone.

There was renewed interest in large passenger-carrying hydrofoil craft during the year, and orders for a number of them were either made or pending. The Boeing Co., which for many years had worked on hydrofoils with the U.S. Navy, announced development of a new design that promised to open up a large waterborne commuter and tourist market. Boeing expected to sell 11 of its new craft, provided that applications for construction loan and mortgage insurance were approved by the U.S. Maritime Administration.

The new Boeing hydrofoil, designated the Jetfoil 929, would come in two models, one designed to carry 250 passengers in commuter service, the other to carry 190 tourists at speeds of 45 knots for distances up to 450 mi. The craft would feature fully submerged foils, an automatic control system, and water-jet propulsion powered by turbine engines. They would be able to maintain top speeds in the open seas in conditions with waves up to 12 ft high,

Sidney Harris

"The supporting beams and evidence of tracks lead to the conclusion that the caveman used this tunnel for the express line of his subway system."

and would be able to come to a full stop in less than 500 ft with no passenger discomfort. Emissions would meet U.S. government standards, and waste fluids would be stored in holding tanks for discharge into sewers at dockside. Authorities expected that the cost would range from an estimated $3.5 million to $5.1 million, depending on the facilities desired.

Urban mass transit. The first 28-mi segment of the 75-mi, $1.4 billion San Francisco Bay Area Rapid Transit system, known as BART, opened in 1972, and the entire system was scheduled for completion by the end of 1973. This was the first U.S. rail transit system in almost 70 years that had been built as a completely new project. The federal government provided about $200 million toward the total cost, emphasizing research and testing of new concepts and methods for possible application in other urban areas.

BART featured many technological advancements. Some difficulties were encountered in the initial operations, but the builders expected that they would be soon overcome. The trains had attendants in the cab, but they were not needed for operations, which would be completely automatic when the system was completed. Speeds would reach 80 mph and would average about 45 mph, including stops at the 34 passenger stations in San Francisco, Oakland, and 30 smaller cities and suburban communities. In many cases, commuter travel times for the riders were expected to be cut in half.

The size of trains could vary from two to ten vehicles, with each car individually powered by four 150-hp electric motors. A ten-car train could carry 720 seated passengers and twice that number including standees. Boarding would be facilitated by self-service ticket purchasing. Multiple-trip tickets, another time-saver, could be inserted into electronic computers that would automatically deduct the cost of each trip.

The system was about evenly apportioned among surface, elevated, and underground or underwater mileage. Innovations introduced in the construction of the underground section included new soft-ground tunneling machines that were expected to be of use in subway construction elsewhere. The trans-Bay tube linking downtown Oakland with downtown San Francisco was 3.6 mi long and consisted of fabricated sections, each as long as a city block, that were sunk into place. The tube was designed to withstand seismic action—a necessary requirement because it extended between two earthquake zones—through use of a unique joint at its terminal structure that permitted movement in all directions.

In 1967 the National Academy of Engineering completed a study of urban buses and made recommendations for design requirements that would ensure passenger comfort, good service, and compliance with the noise and pollution standards then being developed. The Urban Mass Transportation Administration then contracted with Booz-Allen Applied Research Inc. (BAAR) to manage a $24 million program to design an advanced prototype of a 40-ft urban bus in line with these recommendations. BAAR in turn subcontracted with American Motors Corp., General Motors' Truck and Coach Division, and Rohr Industries, Inc., to develop separate designs for evaluation and selection as the "city bus of the future." The final selection was scheduled to take place in mid-1974.

The three proposals for the so-called Transbus would have several common features, in addition to length. One was lower profile, plus further automatic lowering at stops, to speed loading and unloading and to facilitate use of the buses by the elderly and handicapped. The noise level was to be reduced from present levels by about 75% internally and 50% externally, and the emission level had to meet government standards. The 45–50-passenger buses also had to meet rigid safety standards while improving visibility through the use of large European-style windows. They would be radio controlled for more efficient dispatching.

The U.S. Department of Transportation (DOT) dedicated the nation's first personal rapid transit (PRT) system in Morgantown, W. Va., on Oct. 24, 1972. The first segment of a planned 2.2-mi automated transit system connecting two university campuses and the business district, it was officially opened and put into service for extensive tests. The small, pollution-free, rubber-tired cars, carrying 8 seated and up to 13 standing passengers, were automatically operated on two-way concrete guideways that could be heated to prevent icing. The cars were controlled by a computer, which caused them to respond to calls and directed them to passenger-selected station stops at speeds up to 30 mph. No operators were required, and the system was designed to move 1,100 passengers every 20 minutes. Total cost to DOT through the middle of 1973 was estimated at $43.7 million; additional funding would be required to provide the full service scheduled to begin in late 1974. DOT also made a grant of $11 million to help finance the start of a similar project in Denver, Colo.

Also well advanced were plans for construction of a $70 million–$80 million, 20-mi personal rapid transit system within Las Vegas, Nev., and extending to its airport. The privately financed system would be elevated and automated and would use small passenger cars moving at speeds between 25 and 35 mph. The expected one-way fare was $1.

—Frank A. Smith

..."it never kicks or bites,
never tires on long runs,
and never sweats in hot weather.
It does not require care in the stable,
and only eats while on the road."
Ransom E. Olds (1892)

Veterinary medicine

In the year or so prior to mid-1973 considerable attention was directed toward the environmental aspects of veterinary medicine by all segments of the profession. That many of the programs proposed were largely prospective in nature reaffirmed what was perhaps the most basic objective of veterinary medicine, the conservation of domestic animals as a natural resource. Several animal disease eradication programs were upgraded, and such people-related problems as animal waste disposal and control of the pet population were studied in depth. After an extensive survey the National Research Council issued a lengthy report that was expected to serve as a blueprint for development of the veterinary profession over the next decade.

Environmental veterinary medicine. In creating an executive committee on environmentology with past president E. W. Tucker as chairman, the American Veterinary Medical Association (AVMA) recognized the interrelationship of the environment with most activities engaged in by practitioners. The most immediate concerns specified by the committee were the use, disposal, and recycling of agricultural chemicals and animal wastes as they relate to food of animal origin for human consumption, and occupational health and safety as it relates to individual veterinarians in the clinic environment. Several veterinary schools had already added new dimensions to their public health disciplines by establishing staff positions in environmental health. In Colorado, where disposal of wastes from feed-lots had become a major problem, a veterinarian was named vice-chairman of that state's health planning agency.

The publicity being given to the street pollution caused by large numbers of dogs in metropolitan areas raised the problem of animal control to priority status in many cities. To combat overpopulation, largely by stray and unwanted animals, the AVMA urged development of a broad program throughout the country to encourage more responsible pet ownership. In addition to continued surgical sterilization of other than breeding animals, the program would require stricter enforcement of existing leash and licensing laws to include cats as well as dogs, a crash campaign of public information and education to help

Courtesy, National Cat Protection Society

Population explosion of cats and dogs in the United States is posing a serious problem. Only about one of every six kittens born in the U.S. finds a home, and city pounds are overburdened.

owners become more concerned, and cooperation with humane organizations and other groups to reduce the numbers of unneutered animals released for adoption. Some years earlier a hormonal contraceptive "pill" had been used on dogs, but proved to have undesirable side effects and was withdrawn from the market. The renewed emphasis on pet population control caused several manufacturers to reinvestigate the problem, and during the year one such product, said to be safe, was marketed in Britain. Veterinarians at Colorado State University studied the possibility of immunologic contraception—the so-called male pill—but this would require several more years of work.

Spay clinics. Veterinarians generally agreed that surgical sterilization by itself could not keep the burgeoning pet population in check. It was conservatively estimated that one promiscuously breeding female accounted for an average of 70 progeny during her lifetime; with breeding completely unrestricted the total would exceed 5,000. Nonetheless, the establishment of "spay clinics" by municipal and humane groups to provide for low-cost surgical sterilization of dogs and cats was looked upon as the most practical means to at least a temporary solution of the problem.

In 1971 the city of Los Angeles opened a Spay & Neuter Clinic that became the prototype for numerous others in the U.S. and elsewhere. In facilities equipped for this one purpose, veterinarians could neuter larger numbers of animals at substantially lower cost than in private hospitals, and without compromising the quality of medical and surgical attention. Other veterinarians, either as individuals or through their local associations, cooperated with humane groups and governmental agencies to provide similar services.

New horizons. After a two-year study of the directions the veterinary profession should take during the next decade, a committee of the National Academy of Sciences-National Research Council issued a 170-page report, "New Horizons for Veterinary Medicine." The main thrust of the report was to indicate that a work force of 42,000 veterinarians would be needed by 1980, some 4,000 more than the present educational system could provide. Of the approximately 25,000 active veterinarians in 1970, nearly 11,000 were engaged primarily or solely in small-animal practice. This number was expected to double by 1980 and would account for most of the new graduates, who increasingly preferred this area, leaving other specialties shorthanded. Livestock practice and food inspection occupied about 7,000 veterinarians in 1970 and were not expected to require increased numbers. Areas in which at least a doubling of personnel was expected to be necessary included equine practice, laboratory-animal medicine, zoo animal and wildlife practice, public health, industrial employment, and teaching and research.

Among its 40 specific recommendations the committee urged that food-animal medicine be made more attractive to new graduates through group practice and greater use of trained paramedical personnel, as had been common in small-animal medicine for some time. Upgrading of governmental food inspection, public health, and animal disease control services was recommended. Expansion of the existing veterinary schools or the building of new ones sufficient to provide for the anticipated need for veterinarians, together with more equitable provision for selecting students from states not having schools, was given high priority. Curriculum changes were proposed to permit greater flexibility, with some opportunity for specialization at the undergraduate level and for continuing postgraduate education.

Cuts in the federal budget proposed by U.S. Pres. Richard M. Nixon threatened considerable delay in implementing a substantial number of these recommendations, many of which depended upon some form of federal support. Nonetheless, the 19th veterinary school in the U.S., scheduled for completion at Louisiana State University in 1975, planned to start its first class in the fall of 1973, and several schools, notably those in California, Georgia, Illinois, and Missouri, had made curricular and administrative changes to facilitate such innovations as computer-based and autotutorial teaching, individualized study programs, and an improved efficiency that permitted as much as a 25% reduction in the time required to obtain the DVM degree, with greater emphasis on continuing education.

Relicensure. For several years the veterinary examining boards in Florida, Nebraska, and Tennessee had stipulated a certain level of attendance at professional meetings each year as a requirement for the relicensure of practitioners in their states. More recently, John R. McCoy, as president of the AVMA, urged adoption of a point-cumulative system of continuing education credits by all the U.S. states. In California voluntary establishment of such a system exempted veterinarians from a mandatory legislative requirement imposed on the other health professions.

In 1973 the U.S. Department of Health, Education, and Welfare proposed mandatory relicensure of all doctors, including veterinarians, to ensure their continuing competency. Veterinary schools in several states, notably in Michigan, Missouri, Georgia, and Alabama, had greatly expanded their continuing education programs, and a number of veterinary associations had established academies to further this aim. By 1973 there were also ten national academies of veterinary specialists that conferred diplomate status upon veterinarians who qualified for membership.

—J. F. Smithcors

Zoology

Zoologists during the year were involved in analyzing the materials brought back by the astronauts from the moon. Analyses of seven samples brought back by the crew of Apollo 14 showed that the amino acids glycine, aspartic acid, glutamic acid, and serine are found on the moon's surface. These are common amino acids found in the protein structure of all living things on earth. Stated simply, the theory of evolution proposes that simple amino acids fortuitously aggregated on the primitive earth millions of years ago from simpler substances and that over millions of years they eventually evolved into living things. What, therefore, does the discovery of amino acids on the moon's surface tell us about the beginning of the moon, the earth, and of life itself? Zoologists of the near and distant future will work with teams of other scientists to find out.

Cellular investigations. Simple questions relating to how cellular processes are performed, how energy transfers are made, what supplies the energy, what gives the directions, what kinds of regulatory processes are involved, and what orchestrates the whole show lead to the kinds of investigations many scientists perform in zoology. A great number of sophisticated technological maneuvers are required to provide the answers, and these activities occupied much of the practicing zoologist's time in 1973 and involved the skills of physicists, engineers, mathematicians, and other scientists as well.

Particular interest was shown during the year in learning more about every kind of cell membrane. Investigations covering a wide variety of natural membranes found molecules of protein partly inserted in a lipid bilayer and partly exposed to the aqueous medium outside. These findings tended to rule out the theory that membranes consisted of lipid bilayers with all the protein outside.

Scientists working with the readily available nerve cells in the brain of invertebrate animals found that most of the drugs they tested increased membrane potential (conductance of ions across the membrane) by increasing the permeability of potassium and decreasing that of chloride. They then found that there was good correlation between the ability to alter permeability and the effectiveness of the various drugs as pain killers. An increase in membrane potential, the authors concluded, probably explains why pain-killing drugs kill pain.

Evidence was also accumulating that the ratio of certain ions in cells of the brain may determine a "set point" to regulate hunger and body weight. Four neuropsychologists at Purdue University found that an excess of calcium ions in the cerebrum of a rat that had already eaten to satiation induced it to resume eating voraciously. The magnitude of the response depended on the concentration of ions injected.

One of the crucial cellular activities of the body is the conversion of foodstuffs to energy, a process involving the passage of electrons. Some of the key electron carriers are iron-containing proteins known as the cytochromes. A group of Japanese

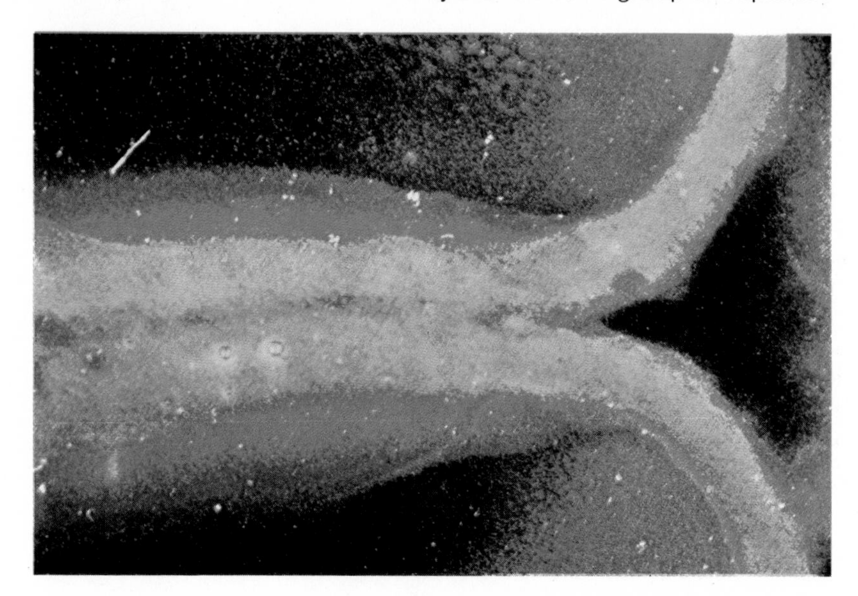

Electron micrograph of hydrated cell membrane has been converted from its standard black-and-white form to color by a new interferometry technique that employs special silver-free organometallic recording media. Layers of the membrane can be seen in color that are not distinguishable in shades of gray.

scientists reported that giving cytochrome c orally or by injection to old rats improved the ability of their livers to metabolize lipids. The authors do not suggest that cytochrome c might be used to correct lipid metabolism deficiencies known to occur in older people, but the implications are there.

In apparent contradiction of the long-held view that specialized tissues such as those in the central nervous system are capable of little or no regeneration in response to injury, Robert Y. Moore of the University of Chicago reported that cut axons in the adult brain and spinal cord were found to vigorously sprout new axons that would enter and supply nerves to structures not previously innervated by this group of axons. And, in areas where innervation had been removed, these particular axons appeared to sprout and re-innervate the vacated synaptic sites.

In a study of the effects of freezing on cells a group of zoologists took mouse embryos about 22 to 96 hours after conception and deep-froze them at hundreds of degrees below zero. Then the embryos, consisting of anywhere from 1 to 150 cells, were slowly thawed out and implanted in foster mothers. Approximately 45% survived to develop into perfectly normal mice.

Genetics. Since the middle 1950s one major effort in biology has been to document and add detail to what is called the central dogma of genetics: that DNA is the hereditary material, that its information is encoded in the sequences of its subunits that constitute the genes, and that this information is transcribed into RNA and then translated into protein. At the Oak Ridge (Tenn.) National Laboratory several investigators began trying to make electron micrographs of individual genes in action in 1967. By 1973 many of their pictures were strongly resembling what had been proposed by means of analyzing laboriously accumulated quantitative data.

Obtaining images at such a minute level of detail in biological specimens has often been considered an impossibly difficult task. Scientists thought that as the energy of the electron beam of the electron microscope rises, the damage it causes to biological materials would increase. But in recent months two scientists demonstrated an opposite effect. Based on preliminary work they believed that with an electron microscope they would be able to obtain at last a true molecular picture of the DNA and RNA chains, thus permitting a direct read-out of the genetic code. At high energy, they explained, delicate specimens of nucleic acids, enzymes, viruses, cell membranes, and other biological components could be imaged and their true molecular structure revealed, without the complications and limitations of staining.

Ecology. In recent years experimental laboratory studies have shown that the persistence of organochlorine insecticides (such as DDT) and polychlorinated biphenyls (PCBs) in the environment have a wide range of subacute effects on organisms at doses considerably below the acute ones; these include effects on enzyme induction, fine structure, metabolism, reproduction, and behavior. For many populations and species, especially the less abundant vertebrates, it is virtually impossible to carry out the crucial field experiment to determine whether the pesticide does or does not affect the population. This also spurs the search for specific pesticides that will affect only the target species and leave no harmful residues.

The use of viruses to control insect pests is attractive because of their undoubted economic

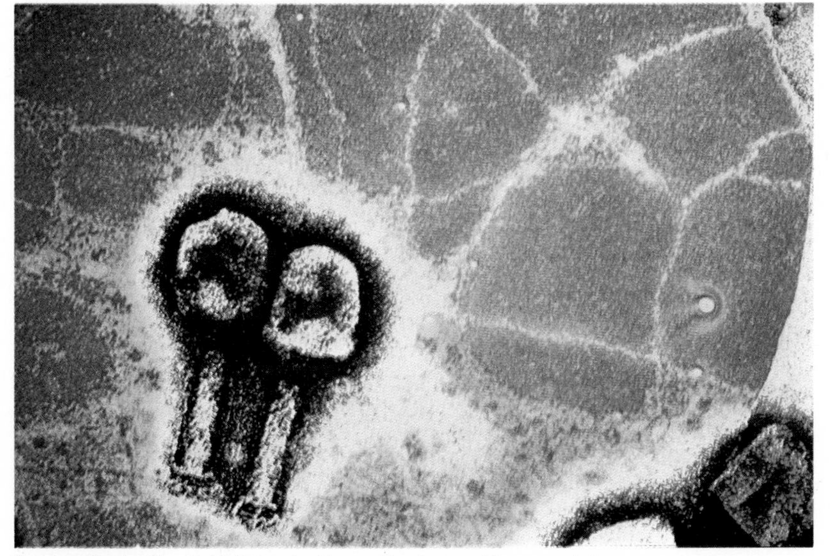

Courtesy Charles Hough, H. Fernandez-Moran, The Electron Microscope Laboratory, Division of Biological Sciences, The University of Chicago

Color electron micrograph of T4 bacteriophage and DNA fibers on ultra-thin carbon films was achieved by using the interferometric process described on the preceding page. Viruses such as T4 show great promise in controlling insect pests.

advantage and apparent very high degree of specificity. Until recently there were thought to be no records of transmission to animals outside the class Insecta. However, during the year naturally occurring antibodies to a virus that infects *Lepidoptera* (butterflies and moths) were found in several species of mammals, suggesting that those animals may have been exposed to the virus. Thus, while it is probably true that most insect viruses exhibit a high degree of host specificity and therefore have great potential in biological control programs, it is obvious that many researchers will be busy testing any potential viral insecticide for safety to humans, livestock, wildlife, and crops.

Insects of many species have evolved some visual or sonic signals for communication between potential mating partners. However, their most common means for such transfer of information is by odors and tastes. The chemicals involved are called sec pheromones. The study of this important aspect of insect life lagged until the realization that manipulation of insect behavior by means of sex pheromones offers an alternative to the use of conventional chemical insecticides or viruses. Numerous sex pheromones of *Lepidoptera* have now been identified and synthesized. In general, one or two sex pheromones per species were reported except that three were found in the silkworm moth *Bombyx mori.* If an insect contains more than one sex pheromone, it is important that they all be detected before undertaking structure determination and synthesis.

Researchers reported during the year that female codling moths *Laspeyresia pomonella* emit seven sex pheromones. At present, the role of all these is unknown. Other work revealed a similar number of sex pheromones in the zebra caterpillar moth *Ceramica picta.* Thus, this large number of sex pheromones could be common for *Lepidoptera* and might explain the weak attractancy of some of the synthesized sex pheromones.

The skin of the same varieties of apples contains one or more substances that are highly attractive to newly hatched codling moth larvae. There is now evidence that the attractant is an alpha farnesene, a compound present in the natural coating of several apple and pear varieties. When tested at a concentration equivalent to that in the apples, alpha farnesene caused the same response in these larvae. Meanwhile, another group identified the chemical structure of the repellent odor found in the secretion of the aphid *M. persicae* as trans-B-farnesene. This pheromone when mixed with an insecticide will cause those aphids avoiding a direct hit by a spray droplet to walk right over the plant and contact the insecticide.

General zoology. Approximately 1,500 entomologists from more than 60 countries met in Australia for the Fourteenth International Congress of Entomology. The program indicated interesting trends in contemporary entomology. Taxonomy, (the classification of species), which had dominated most earlier meetings, received little coverage even though all branches of entomology depended on the accurate determination of species. The shift in interest seemed to be toward economic problems and on pest management and integrated control. At the opening E. F. Knipling discussed the significance of his own pioneer work, involving the use of sterile males to control pests, in relation to biological control generally and showed how repeated releases of large numbers of "beneficial" insects could be used almost as living insecticides. Although there seemed to be fewer reports of

Mating cecropia moths are attracted to each other by chemicals called sex pheromones. Researchers hope to learn to manipulate pheromones in order to control the reproductive behavior of insect pests.

Kitty Kohout from Root Resources

J. Marlier from Bruce Coleman Inc.

Pollination of flowers by bees takes place successfully in a greenhouse made of polyethylene. In a glass greenhouse, however, the bees spend all their time trying to escape by crashing into the glass.

laboratory studies in insect physiology, an emerging discipline was insect behavior. For instance, one investigator studied aspects of the most basic feature of the sexual behavior of mosquitoes simply to find out whether or not they are monogamous—live in pairs having only one mate. He found that, depending on the conditions, some are and some are not.

In another study experimental snails trained by pairing touch with food chemicals soon acquired a classically conditioned feeding response to touch alone. After receiving electrical stimulation when they did not withdraw from touch rapidly, the snails learned to withdraw rather than to feed in response to touch alone. The learned responses persisted for up to two weeks without reinforcement before extinction, and could be demonstrated in the isolated nervous system. What happens in nerve cells during such a learning process is largely unknown because suitable experimental preparations have not been developed. This study was undertaken so that the primitive nervous systems of these animals could demonstrate the first case of associative learning that could be readily studied on the cellular level.

The remarkable jumping ability of the flea has always been a matter for wonder among naturalists. A detailed study of the exoskeleton and musculature of the oriental rat flea, *Xenopsylla cheopis,* aided by high-speed movies of jumping fleas, enabled a team of Israeli and British scientists to explain how a flea can jump so far so fast. The cause is a mass of resilin in the flea's third pair of legs (the pair used in jumping). Resilin is a highly elastic protein material that is compressed by muscle contractions and held by a series of "catches." When the muscles holding the catches in place relax, the energy stored in the compressed resilin is released, catapulting the insect into the air with an acceleration of about 140 units of gravity (G's). Resilin is also found in the wing-hinge ligament of flying insects. Though fleas have lost the ability to fly, they apparently retained several features associated with flight and adapted them to jumping.

Researchers found that bees will pollinate flowers in a greenhouse made of polyethylene. They will not pollinate flowers in a glass greenhouse but spend all their time trying to escape by crashing into the glass.

Several investigators calculated the amounts of energy three species of hummingbird expend in extracting nectar from three flower species and were able to judge the efficiency of each bird at each flower. If several flower species are equally available, a given hummingbird species selects those where it gets the most for its efforts.

The sense of smell is apparently not as well developed in birds as it is in other animals. But some biologists have suggested that albatrosses, shearwaters, and petrels can locate food by smell. To test this hypothesis, Thomas C. Grubb of the University of Wisconsin soaked identical sponges in cod liver oil and seawater and placed them on poles moored on the ocean. During the daytime he observed that one species of shearwater and two species of petrel approached the oil-soaked sponge. Other seabirds did not respond to either sponge. Nighttime testing showed similar results, at least for petrels. Grubb concluded that the

Norman Tomalin from Bruce Coleman Inc.

Hummingbird gathers nectar from a hibiscus. In a study of the energy expended by hummingbirds in extracting nectar from species of flowers, researchers found that the birds will choose those that yield the most for their efforts.

petrels and shearwaters studied can follow an attractive airborne odor to its source by smell alone, and that only these birds appear to have this ability.

A plague killed untold millions of frogs in 1972, and it was feared that millions more might die in 1973. No one knew the reason for this. Naturally, some environmentalists suggested the runoff of agricultural wastes into the frogs' watery habitat, but evidence was lacking. A more likely answer is infection caused by a common organism that the frogs are too weak to resist because the dry summer of 1971 produced a famine for them. Until someone finds out what it is that makes frogs die, science may be forced to turn to another animal to use in laboratory testing.

Whales have been hunted so ambitiously that many species are in danger of extinction and are on the endangered list published by the U.S. Department of the Interior. The United States banned all whaling, but nations such as the Soviet Union and Japan, which account for 80% of the total catch, would face loss of a major industry and opposed such bans. At its annual meeting in London the International Whaling Commission rejected a moratorium on whaling, but did set up quotas on total numbers of whales of each species and sex that could be caught each year. Thus, the quota for sperm whales, for example, was 6,000 males and 4,000 females for the North Pacific and 8,000 males and 5,000 females for the Southern Hemisphere. The commission also called for an International Decade of Cetacean Research, which was to begin in 1974 or 1975.

The opening address at the first international symposium on seals in 1973 pointed out that the world catch of fish of 69 million tons a year was causing the limit of fish resources to be approached. Thus, it might be argued that the deliberate reduction of seals, as important fish predators, would benefit fisheries. However, Victor Scheffer, a leading fur seal biologist, claimed that this was an oversimplification of a very complex situation.

Other reports at the symposium dealt with the existence of different "dialects" in the threat calls of male northern elephant seals from the colonies off the coast of California, and stressed the value of quantified comparative studies rather than the descriptive accounts that have predominated. A. V. Holden of Scotland referred to the importance of seals as indicators of pollution levels in the surrounding areas. High concentrations of certain environmental contaminants, particularly the organochlorine pesticides and PCBs, were found in seals. The amounts appear in some instances to be sufficient to have adverse effects on the animals.

Evolution. Marine arthropods called trilobites dominated the earth 500 million years ago, dying out in the late Paleozoic era but leaving a number of extremely well-preserved and abundant fossils. In 1973 Kenneth Towe of the Smithsonian Institution studied the lenses of a fossil trilobite's eyes. The lenses were found to be impregnated with calcite and had the optical properties of glass. These ancient lenses appeared to be better than the ones found in some living arthropods. Certainly no one knows what kinds of images trilobites saw, but it is possible to determine that they could see relatively sharp images over a depth ranging from a few millimeters to infinity.

Three groups of amphibians are living today: frogs, salamanders, and caecilians, the latter being legless, burrowing inhabitants of South America, Africa, and Asia. The fossil record of frogs and salamanders is relatively poor, and no authentic fossils had been recognized for the caecilians until 1973 when a single diagnostic vertebra from the Palaeocene of Brazil was described. The specimen most closely resembles vertebrae of the living genera *Geotrypetes* and *Dermophis.*

The coelacanth, *Latimeria chalumnae,* is considered a living fossil and until 1938 was thought to be extinct. Only 68 of these ocean fish have been caught since then, and very little is known about their physiology and biochemistry. A recent French-British-U.S. expedition to the Comoro Islands off the eastern coast of Africa captured two *Latimeria,* and several studies were reported that confirmed earlier notions about the placement of this fish on the phylogenetic tree. The respiratory properties of the whole blood and purified hemoglobin of the coelacanth were found to compare with those of other fishes, with *Latimeria* hemoglobin molecules being intermediate between the more primitive lampreys and modern bony fishes as expected. In another study the thyroid gland of *Latimeria* appears to be compact and encapsulated, thereby differing from the scattered pattern of modern bony fish, but the thyroid follicular structure of *Latimeria* is similar to other vertebrates.

The study of primates was expected to expand in a number of directions simultaneously. A detailed study of feeding behavior and tooth use was made in four living primates, *Tupaia, Galago, Saimiri,* and *Ateles,* using a combination of cinefluorography and occlusal analysis. These primates were chosen because they form a structural series parallel to a series of fossil primates, *Palenochtha, Peycodus,* and *Aegyptopithecus.* The study suggested that the observed changes in the structure of the jaw apparatus in both the living and fossil series occurred within a behavioral framework established early in primate evolution.

Two Harvard University zoologists determined that chimpanzees use the same amount of oxygen when running on two legs or four. This infers that when protohumans stood up and began to walk erect they may not have required more energy. (*See* Feature Article: THE NEW SCIENCE OF HUMAN EVOLUTION.)

The great extinction of large mammals in North America near the end of the Pleistocene puzzled paleontologists for decades. Thirty-one New World genera of large mammals disappeared in North America at the end of the last ice age (about 11,000 years ago), including ground sloths, giant beavers, glyptodonts, capybaras, saber-toothed cats, mammoths, mastodons, horses, peccaries, camelids, and certain antelopes, bears, and bovids. During recent months Paul S. Martin of the University of Arizona developed an explanatory model emphasizing ecological effects of a human invasion of North America via the Bering Strait land bridge perhaps 11,700 years ago. He postulated that human families moved southward between the Cordilleran and Laurentian ice sheets to find abundant prey inexperienced in avoidance and defense against this new enemy. Rapid extinction made preservation of the archaeologic record by normal geologic processes virtually impossible, but those archaeologic mammoth kill sites with dates cluster tightly around 11,200 years ago. By contrast, in Eurasia, which had long been inhabited by man, only four genera of large animals were lost in the last Pleistocene (mammoth, woolly rhinoceros, giant deer, and musk-ox).

—John G. Lepp

M. F. Soper from Bruce Coleman Inc.

Male northern elephant seal issues a threat call. Variations in such calls were noted by zoologists during the year, and further quantitative studies were urged.

A Pictorial Essay

Man's New Underwater Frontier

photographs by Flip Schulke

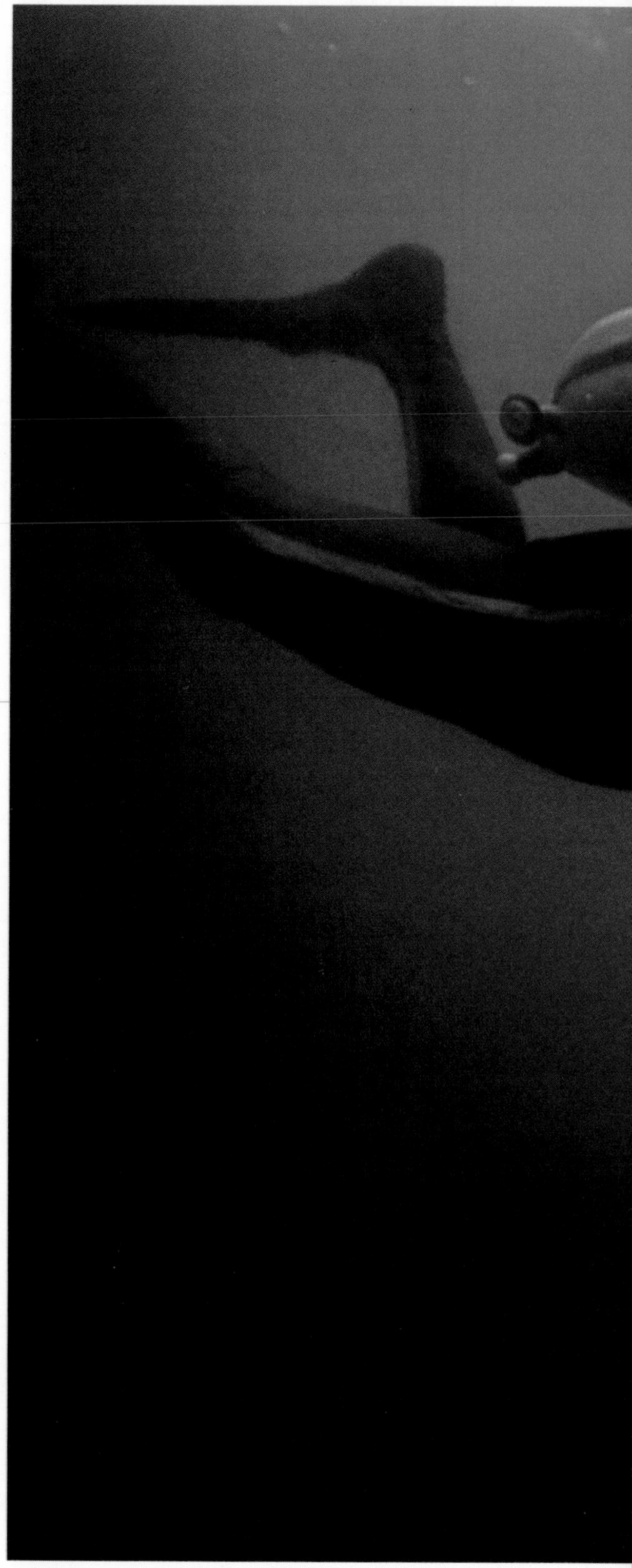

Man has long explored the seas and oceans of the world, using the life within them for everything from food to fuel to jewelry. But in recent years, as land-based natural resources have begun to dwindle and population pressure has increased, the sea has beckoned man as a reservoir of still-untapped riches, and he has started cultivating it with the tools of 20th-century technology. Fish are being "farmed" to increase their yield in a shorter time, and man-made reefs are being constructed on the sea floor in an effort to duplicate the rich marine life that surrounds natural coral reefs. Throughout the world engineers range the floor of the offshore continental shelves for oil, and they are at work to develop the technology that will allow them to explore at greater depths.

Scientists and engineers have developed many submersibles—underwater vehicles—that have aided greatly in the exploration of the deep waters and floors of the oceans. Newly designed navigation equipment has allowed divers to explore farther from a fixed point and yet find their way back with ease. Complete undersea environments have been built in which men and women have lived for as long as 60 days.

Flip Schulke has traveled to many parts of the world and joined the divers under water in order to photograph man at work beneath the sea. Using specially designed equipment, described in Year in Review: PHOTOGRAPHY, he has produced a visual record of pioneers on a new frontier.

Bernard Delmotte, a diver for the team of Jacques-Yves Cousteau, marine explorer, collects a rare species of lake frog in Lake Titicaca, Bolivia.

Scientist-aquanaut Morgan Wells (above) inspects a dome used to isolate a coral community in heavily silted water off the Florida Keys. Instruments within the dome try to determine the damage done to the metabolism rate of the coral by human pollutants. Sylvia Earle (right), one of the first women aquanauts, finds her way under water by transmitting sonic waves from the device she holds and bouncing them off a stable locator.

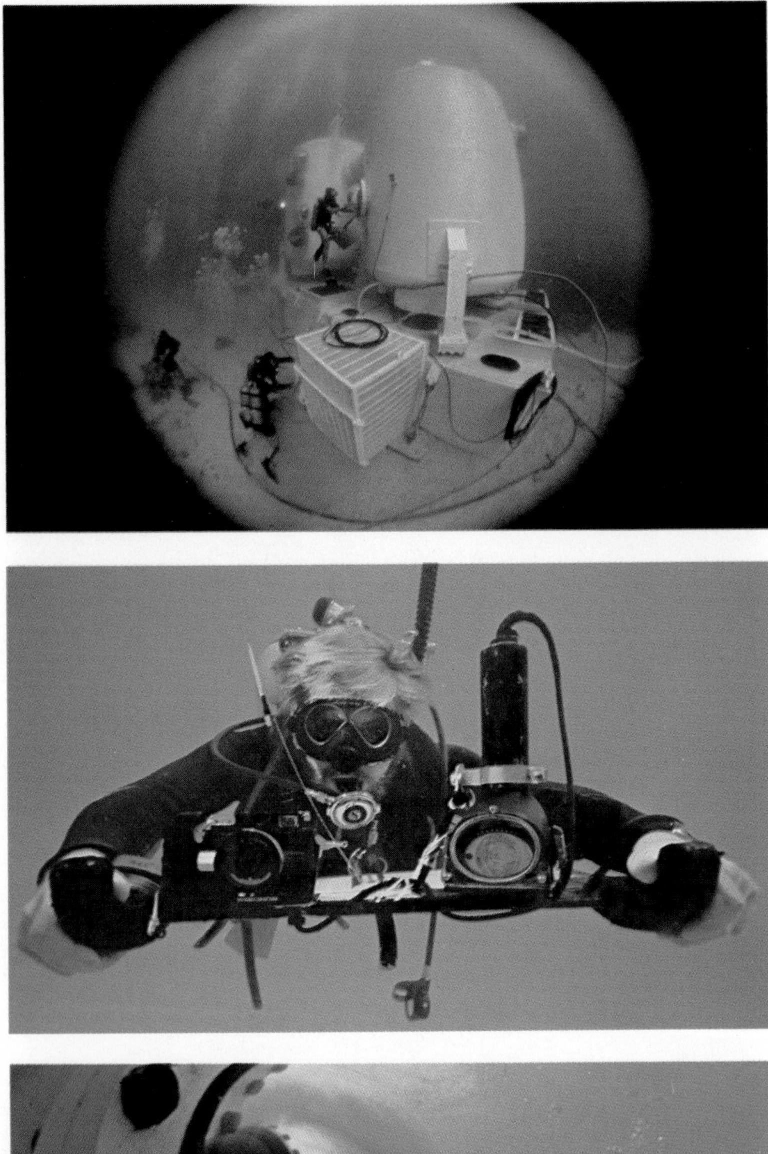

In Tektite I (top), a joint U.S.-General Electric Co. project, four aquanauts set a record of 60 days for consecutive time living on the ocean floor. The cylindrical-shaped compartments housed the men, who studied plant and animal life, sea-floor geology, and the chemistry of seawater. One of the aquanauts was Conrad Mahnken (bottom). The camera setup (center) was used to photograph fish in the Flare artificial reef project off the Florida Keys.

A diver holds the receiver of a newly designed underwater navigation system (above). The triangular shape of the receiver allows it to indicate with considerable precision the direction from which a stationary acoustic beacon beams signals. At the right, marine plants are collected for cultivation in a laboratory in order to study their method of reproduction.

Astronauts practice walking in space in an underwater test tank (left) at the Marshall Space Flight Center. The tank simulates the conditions the astronauts will encounter when leaving their spacecraft during orbital flight. A new type of underwater flare (below) provides a great amount of light for divers and weighs so little that it is easily portable. A battery-operated device that would provide the same illumination would be much more cumbersome.

The Perry "shelf diver" submersible (right) normally carries two men and was designed especially for exploring the continental shelf regions. The "Makakai" (below) is a two-man submersible with a transparent hull and an operating depth of 600 ft. It was the first submersible to use two sets of oppositely arranged cycloidal thrust units for its propulsion system.

A Plexiglas ball in the U.S. Navy submersible (left) allows occupants to descend to the ocean floor and see in all directions as they execute underwater activities. "Snoopy" (below, left) is an unmanned self-propelled television camera that sends its pictures back to a surface vessel. Two one-man "Sea Flea" submersibles of Jacques-Yves Cousteau (below) are used primarily for taking motion pictures.

The world's first permanent underwater wellhead chamber (opposite page) is installed on a "live" oil well 375 ft deep in the Gulf of Mexico. The 30-ft by 10-ft chamber houses workmen, who can complete the well in a dry, normal atmospheric environment, and eliminates the need for divers. The workmen descend to the wellhead chamber from their ship (above) by means of a service capsule (right).

Man's work in the sea covers an increasingly wide range of activities. Yellowtail, fish related to the tuna, are caught as infants in the coastal waters off Japan. They are then held in net cages (top) in protected areas of Japan's Inland Sea (center) and fed crushed mussels and inexpensive small fish. By means of this intensive fish farming, yellowtail caught in early spring will be of marketable size by November or December. Below, a diver lays an undersea telephone cable by plowing a furrow in the ocean floor, placing the cable in it, and covering the furrow. Old automobile tires (bottom, right) are built up into an artificial reef in an effort to form the rich fish communities that dwell near reefs.

Through the large circular windows of an underwater observation tower in Japan, people who are not divers are able to view a living coral reef. The tower contains stairs that descend from the water's surface to the floor of the reef.

Engines of the Future

by Donald N. Frey

As the price of gasoline soars and air pollution grows worse, the need to develop clean and economical engines for cars and trucks becomes increasingly urgent.

GAS

1.00

PER GALLON

The Wankel rotary engine is small, simple, and potentially inexpensive to produce. At present, it holds the most promise for use in small vehicles. The engine's cycle is shown (opposite) at the bottom, left to right. A fuel/air mixture is drawn through the intake port into the combustion chamber by the revolving rotor. No valves or valve-operating mechanism is required. The rotor continues revolving, reducing the space in the chamber containing the fuel/air mixture and thereby compressing it. When the mixture is fully compressed, the spark plug fires. This causes the gases to expand rapidly, driving the rotor around the cylinder and thereby providing power to the wheels. The rotor then expels the remaining gases through the exhaust port.

The reciprocating internal-combustion gasoline engine—the familiar combination of a carburetor, valves, cylinders, and pistons—triumphed over all rivals early in this century to become the dominant power plant for the automobile. Stanley Steamers and electric runabouts, once common to American streets, could not compete with the economy, efficiency, and convenience of the piston engine. In succeeding decades, the carbureted piston engine has had roads to itself everywhere there are cars, trucks, or buses.

The 1970s, however, brought a challenge. A Japanese firm introduced in a car called the Mazda a new type of engine. Although still a gasoline internal-combustion engine, it replaced the reciprocating pistons with an equilateral-triangular orbiting rotor that turns in a closed chamber. It is commonly called the Wankel engine after Felix Wankel, a German engineer specializing in the design of sealing devices, who first conceived it in the 1920s.

Challenges are coming not only from the Wankel. Electricity, steam, and turbines are being looked at with renewed interest and careful study. Why, after its many years of dominance, has the piston engine become apparently vulnerable to such competition?

Three basic criteria

Three basic considerations govern the U.S. automotive industry in its decision-making about engines. The first is cost. Manufacturers constantly search for the type of engine that is usable and least expensive to make. For decades the carbureted piston model best satisfied this criterion. The second consideration, of much more recent origin, concerns control of the air pollution caused by engine exhaust emissions. As ever more people throughout the world drive more and more cars, especially in large urban areas, pollution generated by automobiles becomes an increasingly serious and pressing problem.

The third fundamental consideration is the cost of fuel. Gasoline has risen in price in recent years, and the outlook is for more of the same. In fact, industry observers predict that the price per gallon to the consumer will be $1 by the year 2000. This has come about for several reasons. Exploration in the United States is not yielding as much new petroleum as in the past. Recent discoveries on the North Slope of Alaska and under the North Sea are sizable but probably not large enough to make a major difference in the price of gasoline. By far the greatest amount of proven oil reserves are in the Middle East, and the cost of petroleum from this region is rising for political reasons. The Organization of Petroleum Exporting Countries is demanding an increased share of the profits, forcing the oil companies that get their petroleum from those Middle Eastern nations to raise prices correspondingly.

The changing situation with regard to these three basic criteria has jeopardized the position of the carbureted piston engine. As for cost, the Wankel is smaller, simpler, and therefore cheaper to build. As for pollution control, an internal-combustion engine, because it is self-

***DONALD N. FREY** is chairman of the board of Bell & Howell Co. and a former vice-president of Ford Motor Co.*

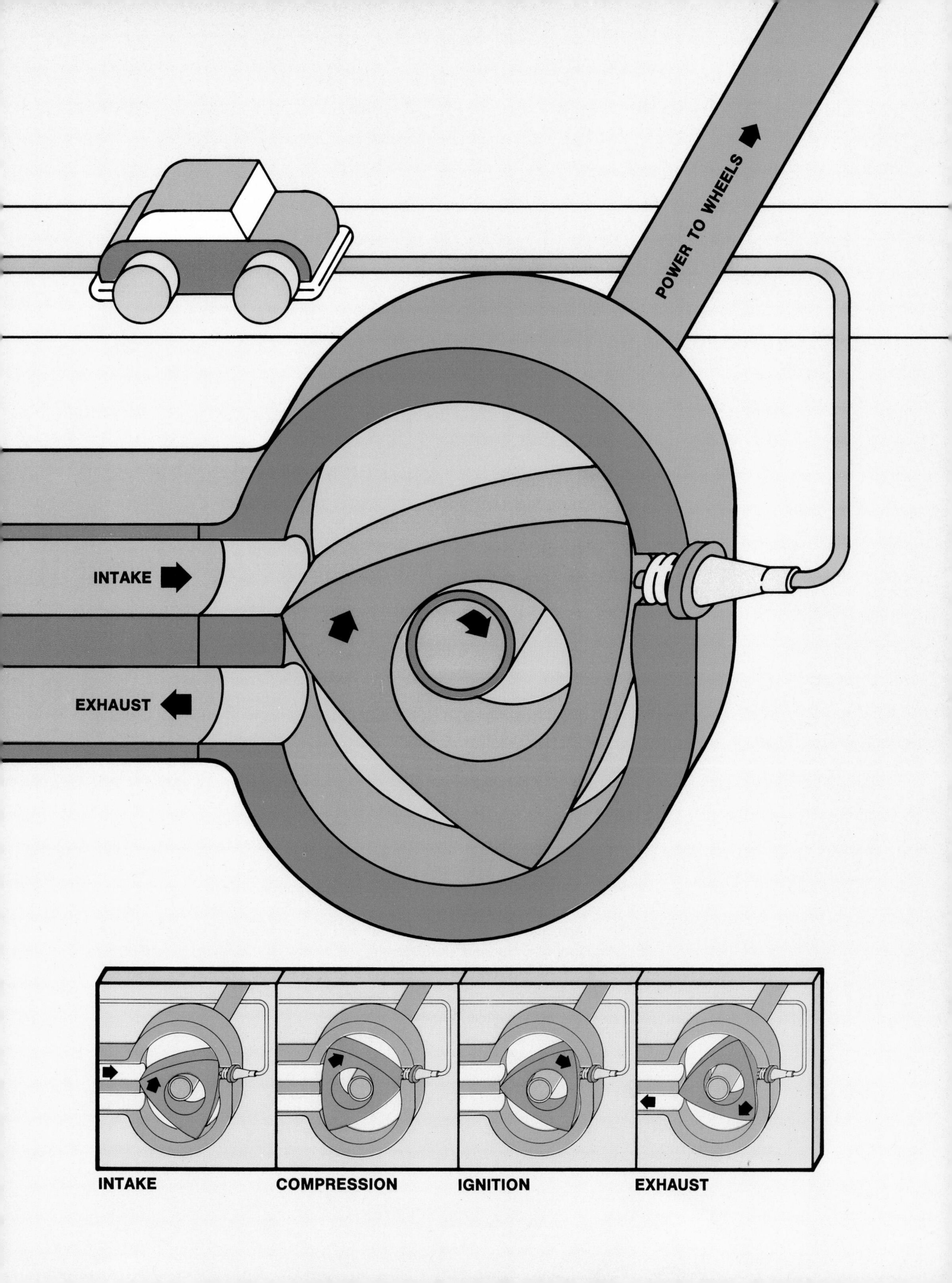
POWER TO WHEELS
INTAKE
EXHAUST
INTAKE
COMPRESSION
IGNITION
EXHAUST

Two types of battery cells supply power to an experimental electric car. The lead-acid power cells, mounted over the motor, accelerate the vehicle from 0 to 30 mph in less than 10 seconds. The zinc-air energy cells, generally placed at the opposite end of the vehicle from the motor, provide an extended driving range by emitting low voltage for a long period of time. Range remains a problem of the electric motor, however, along with expense and size. Town runabouts seem the best candidates for this power plant.

contained, is less accessible to the introduction of pollution control devices than are other possible types. The Wankel, though an internal-combustion engine and no less of a polluter than the piston model, has a "secondary" advantage in this area. Because it is both smaller and less expensive than its piston rival, the Wankel provides car makers with both the physical space and the economic margin to add pollution control equipment and still keep their automobiles from becoming larger and more expensive. The third consideration—cost of fuel—poses an obvious problem for any engine that burns gasoline. The introduction of pollution control equipment has further aggravated this situation by making the gasoline engine less efficient, in terms of miles per gallon, by a factor of 10–20%.

For these reasons the automotive industry has begun looking toward new sources of power and new designs for its engines. Much needs to be done and the situation is subject to change as new technology is introduced, but one certainty is that the piston engine will have to make room for rivals.

The Wankel rotary engine

Among these competitors the Wankel rotary engine was receiving the most attention in the early 1970s. Because of its low overall cost of production and the consequent advantage with regard to pollution control, the Wankel offers an increasingly attractive alternative to the piston engine.

At the same time, however, the Wankel has drawbacks that seem to preclude it from ever becoming the widely used power plant that the large V-8 engine is today. The Wankel does not adapt well to larger automobiles like the Cadillacs, Chryslers, and Lincolns of the 1970s. To power such automobiles, or trucks or buses, a multirotor Wankel system would have to be introduced. But then the Wankel would start to lose its greatest competitive advantage because it would no longer be cheaper to make. It would also become larger and no longer able to provide car makers with space and economic margin that would allow them to install antipollution equipment effectively.

Thus, it appears that the rotary engine will be limited to small vehicles. But this may not be as restricted an application as it would now seem. In the future the small-car portion of the market appears likely to expand considerably. This will happen for several reasons: increasing urban congestion, rising prices for fuel, and the increased initial and operating costs of the total automobile. Thus, at the expanding small end of the scale the Wankel seems likely to establish a strong position in the decades to come. The Wankel will also be made in very small sizes for outboard engines, garden tractors, and many other similar applications.

Other candidates

With the rotary engine apparently limited to small cars, are there any other alternatives to the piston engine that might better meet the

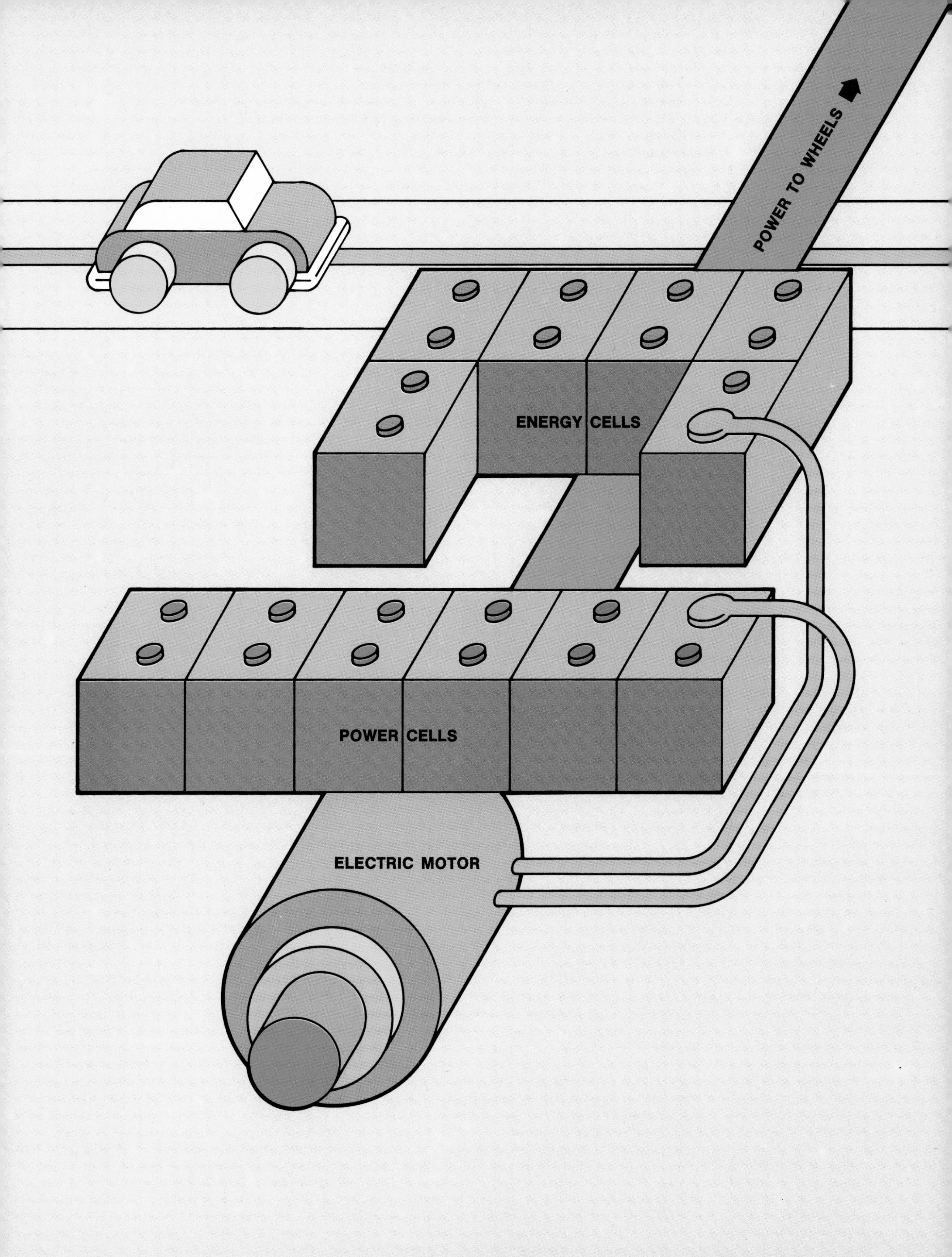
POWER TO WHEELS
ENERGY CELLS
POWER CELLS
ELECTRIC MOTOR

In a gas turbine engine, air is taken in, compressed, and forced into a combustion chamber, where it is mixed with fuel. Combustion causes the mixture to expand through nozzles aimed at two turbines. The first in line drives the compressor and engine accessories, while the second drives the output shaft, which furnishes power to the wheels. There is no mechanical connection between the turbines. This type of engine is comparatively clean and can use a wide variety of fuels, but the expense to mass-produce it and its inefficiency in use of fuel in hauling light loads through stop-and-go traffic seem likely to restrict its use to trucks and heavy luxury cars.

three criteria of low cost, low pollution, and fuel economy? Scientists and engineers are facing the problem and are experimenting with a number of different power plants. As of 1973 they had found no single alternative that could adequately meet all three criteria. Thus, the future will probably witness a mixture of engines, each particularly suitable for a specific type of vehicle.

Joining the Wankel in the competition for small cars is the battery-powered electric motor. The great advantages of an electric automobile are that it is nonpolluting, at least in a direct sense, and "fuel" could be as little as one-fourth as expensive as for a gasoline-powered car. The chief drawbacks include the expense, size, and weight of the batteries, limitations on range and speed because of the current state of development of the electric storage battery, and the pollution caused by the necessary increase in electric-power generating stations. It is possible that future engineering breakthroughs may reduce the cost of the batteries and allow greater range and speed, and it is also true that the pollution caused by a relatively small number of power stations can be much more effectively controlled than that from millions of automobiles. Nonetheless, the electric engine seems promising for use primarily on small limited-use vehicles, ideally town runabouts that would go no more than 40 miles per day at limited speeds. Given such vehicles, their engines could be small and simple enough so that the cost, including operating expense, would be reasonably competitive with the gasoline piston and rotary models.

A variation of the electric vehicle would use fuel cells instead of storage batteries to power the electric motor. To date the fuel cell is even less adaptable than the storage battery in terms of size, weight, and cost. Even more engineering development is needed and the likelihood of success is correspondingly limited.

At the large end of the motor vehicle spectrum—trucks, buses, and heavy automobiles—the gas turbine is a promising candidate to replace both the gasoline piston engine and the diesel engine. The latter, widely used in very heavy vehicles, is a heavy, noisy, but efficient piston engine using compression ignition of heavy fuel oil. In the gas turbine, air is compressed and heated with fuel in a combustion chamber and the gaseous products are then expanded through a bladed turbine wheel, which spins and creates power for the wheels. The gas turbine is vibrationless, has few moving parts, and boasts the considerable advantage of being able to use a wide variety of fuels, though the oft-mentioned peanut oil and perfume seem unlikely to be the final choices. Thus, the rising price of gasoline would have less effect on this engine. From the standpoint of pollution, it emits less carbon monoxide and unburned hydrocarbons than the piston and Wankel engines, but the emission of nitrogen oxides is too high and has become a target of current research.

The chief drawback of the gas turbine is that, in the parlance of engineers, it does not "scale down" (the reverse of the Wankel scale problem). It is difficult and expensive to mass-produce and also is

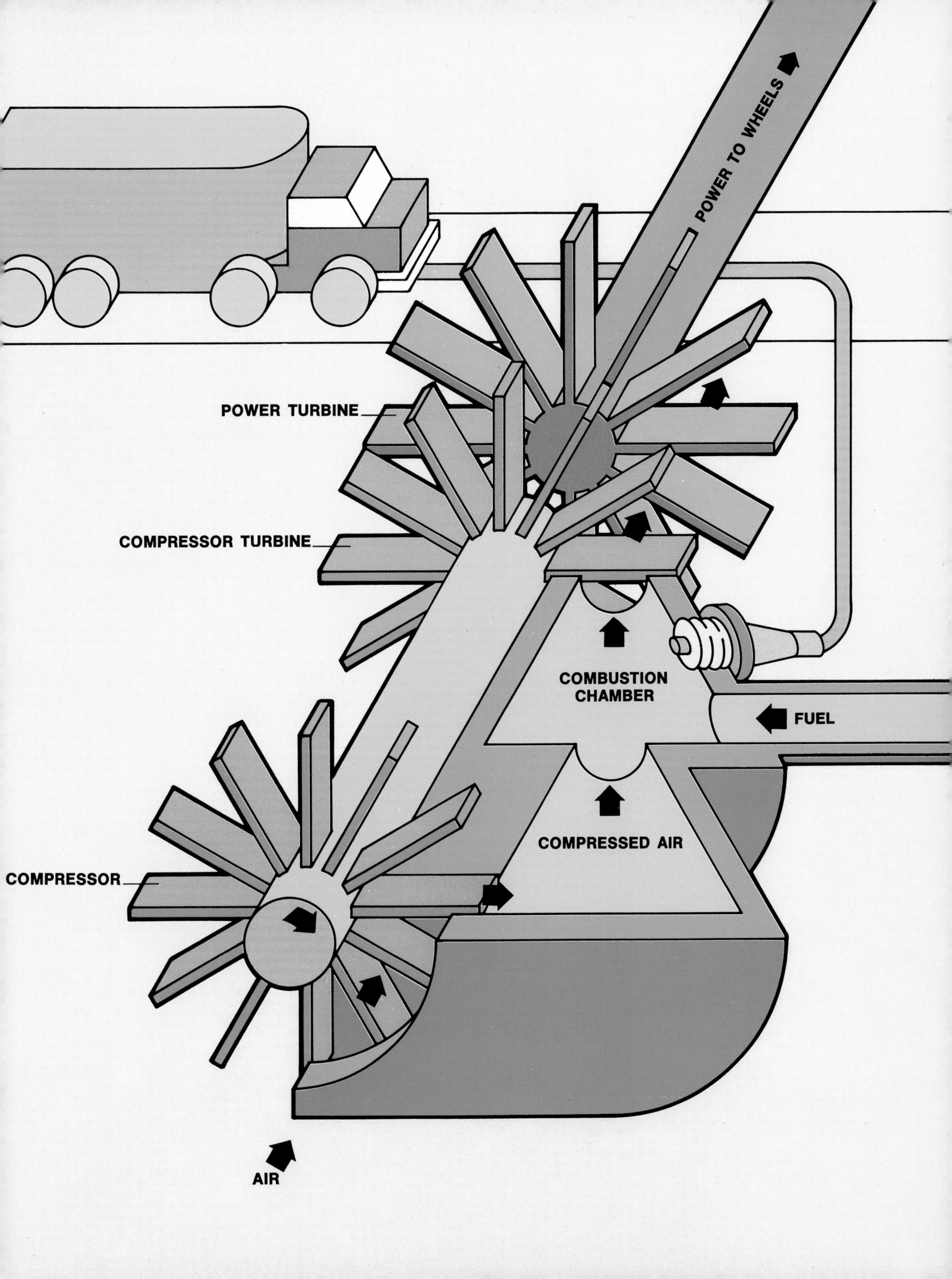
POWER TO WHEELS
POWER TURBINE
COMPRESSOR TURBINE
COMBUSTION CHAMBER
FUEL
COMPRESSED AIR
COMPRESSOR
AIR

"Steam" engines as presently conceived would not use water to make steam but would employ a fluid such as Freon that could easily be made gaseous and then be reconverted to a liquid. In the simplified drawing, opposite, the fluid is heated to a gaseous state in the boiler. The heated gas then expands at great pressure to one or more cylinders in which it pushes down a piston. The piston turns the crankshaft, furnishing power to the wheels. Meanwhile, the gas is expelled from the cylinder to a condenser coil, which must be large enough to dissipate the excess heat and reconvert the gas to liquid form for the cycle to begin again. The system is "closed" in that none of the fluid is lost. At present, such engines are too bulky and expensive to be practicable for use in cars and trucks.

not efficient in its usage of fuel when pulling a light load through stop-and-go traffic. Recent engineering advances, such as improved casting techniques that eliminate individual machining and the use of adjustable turbine nozzles, two-shaft designs, and exhaust heat recovery have reduced the manufacturing cost and improved the light-load efficiency to some extent. But the gas turbine as of 1973 promises to find its most practical and effective future applications in large vehicles.

Steam engines have received some attention as future alternatives. Scientists and engineers have recently experimented with Freon and other vapor closed-cycle systems to develop and improve on the steam engine. Since steam engines are of the external-combustion type, in which fuel is burned continuously rather than in short explosions, their exhaust products are low in pollutants and lower cost fuels can be used. But electric-powered motors are even cheaper to operate, and, furthermore, the steam engine continues to suffer from the failing that first caused it to yield to internal combusion early in the century—as an external-combustion engine, it is necessarily large, heavy, complex, expensive to build, and complex to operate. Though engineers continue their experiments to overcome these obstacles with such innovations as the closed-cycle model, they have not yet succeeded and the steam engine is not now a viable alternative for the future.

Natural gas is a more pollution-free piston engine fuel, but by the early 1970s it was rapidly becoming more expensive as world supplies dwindled. By the end of the century it promises to be extremely costly unless new deposits are found. But aside from this problem natural gas as an automotive fuel would need to be compressed considerably to a liquid form and carried in the vehicle under high pressure. Cars would require pressurized carburetors and would need to be refueled under pressurized conditions. It seems unlikely that, given any other feasible alternative, a car buyer would choose a vehicle that is more complex and costly and also confronts him with the potential safety hazard of high pressure.

The "great middle"

With the gas turbine showing promise for trucks, buses, and large automobiles and the Wankel and electric engines for small cars, what becomes of the great range of medium-sized vehicles of the type now being driven by the majority of Americans? Even assuming that this portion of the car population will gradually decline in favor of smaller models, a significant number will certainly remain. It is for these cars that the reciprocating piston engine seems likely to maintain its dominant position. Because the Wankel and electric engines cannot be scaled up and the turbine cannot be scaled down, and steam and natural gas are eliminated for other reasons, the conventional piston model will most probably hold onto the medium-sized market.

If the piston engine does continue in use, attempts must be made to deal with its drawbacks. Better pollution control and improved mile-

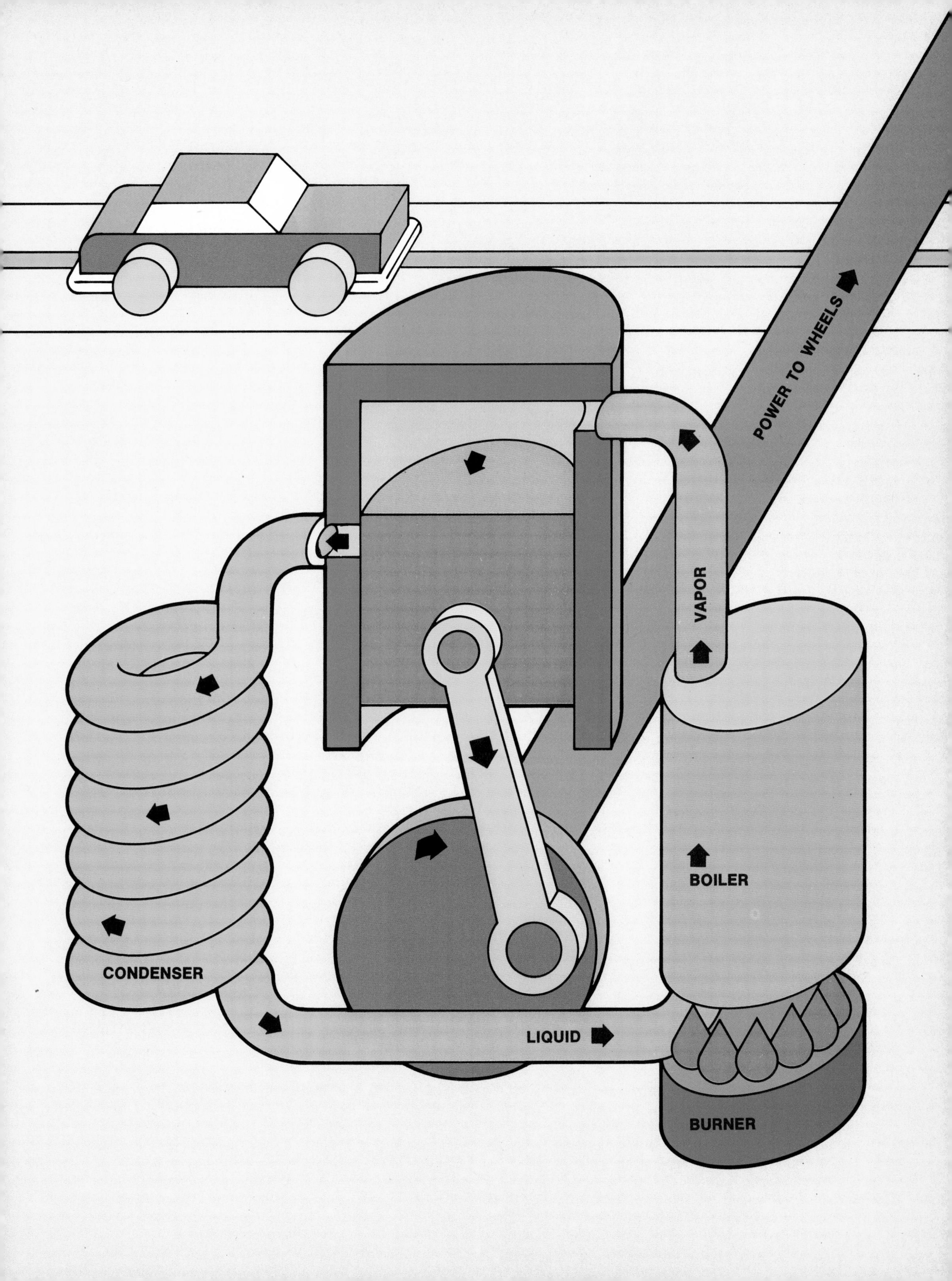
POWER TO WHEELS
VAPOR
BOILER
CONDENSER
LIQUID
BURNER

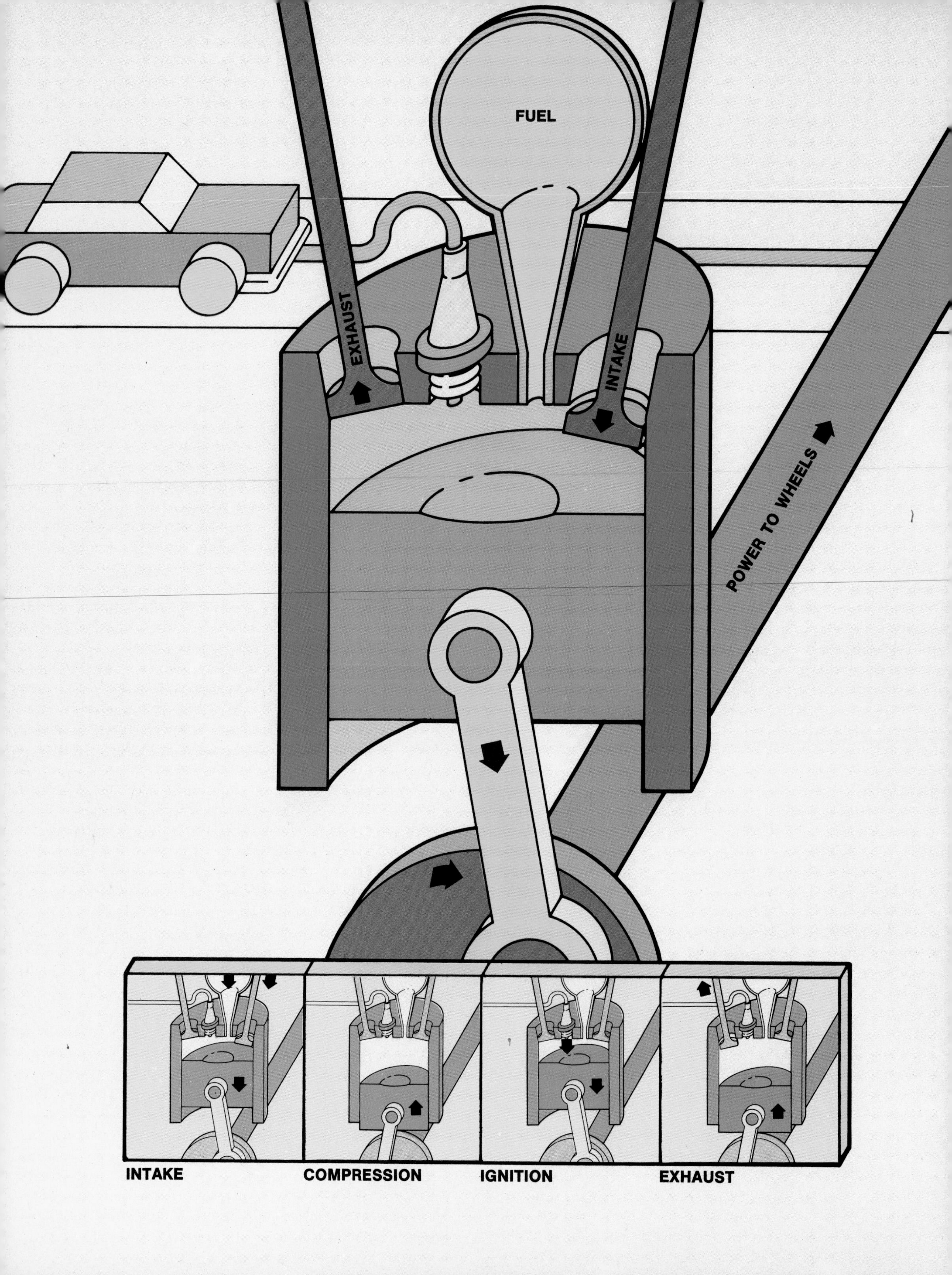
FUEL
EXHAUST
INTAKE
POWER TO WHEELS
INTAKE
COMPRESSION
IGNITION
EXHAUST

age must be achieved. To date, antipollution equipment has caused poorer mileage per gallon, and the two goals must be made compatible. In order to reduce pollution, engineers are experimenting with a variety of devices including catalytic converters, which are mufflerlike devices that contain a catalyst such as platinum to help oxidize the fumes escaping through the exhaust system. Automobiles alternatively may need a small reactor, a furnacelike device near the engine that burns fumes. To improve mileage, engineers are designing fuel-injection systems, replacing the carburetor, that will feed gasoline to the engine more accurately, and developing various shapes for the combustion chamber above the piston that improve fuel economy and reduce pollution. The diesel engine will also be considered for passenger cars, but success is unlikely due to noise, weight, and cost.

The effort to improve the piston engine must continue, but it does present a new problem for car manufacturers. As the engine is improved it will almost certainly become more sophisticated and complex and, therefore, more expensive to make. Thus, one of the great advantages of the piston model—its low cost of manufacture—stands in danger of being lost. Scientists and engineers face a considerable challenge as they try to maintain low manufacturing costs for engines while adding the necessary refinements for pollution control and fuel economy.

Although efforts are being made to induce people to use their cars less in dense metropolitan areas and change over to public mass transportation, the campaign has made little headway and does not seem likely to do so in the foreseeable future. People will continue to rely on automobiles for most of their transportation requirements. Consequently, the need to develop the cleanest and most economical cars possible is one of overriding importance for the last decades of the 20th century.

A gasoline piston engine designed to be cleaner and more economical than the current models is the stratified-charge, opposite. Air enters into the combustion chamber, which is a depression in the piston head, and swirls past the fuel injector. The tiny gasoline droplets introduced by the injector mix well with the air, and they swirl toward the spark plug. Combustion then begins, smoothly filling the entire chamber as the air, fuel, and burning mixture continue to swirl. Less gasoline per unit of air is required in this design, and the overabundance of air allows combustion to be nearly complete, thereby reducing exhaust emissions. This engine is now in the experimental phase.

Dan Morrill

The Rising Decibel Level

by Toba Cohen

Julius Caesar forbade chariot traffic after dark, never dreaming that a world beset by jet planes and motorcycles would wonder why such a minor noise bothered him.

The radio that brings pleasure to the owner, the horn that warns the pedestrian, the fire and police sirens that announce help are welcome sounds to those for whom they are meant and noise to those for whom they are not intended. One definition of noise, certainly, is sound without value, and it has been irritating man through the ages. In ancient Rome, Julius Caesar was so offended by the noise level that he enacted legislation forbidding chariot traffic after dark—possibly the first anti-noise ordinance on record.

Man has learned to make much more noise since Caesar's day and he has learned how to analyze it, but it is only recently that the noise problem has been recognized as a serious one warranting standardized control. Only now is noise emerging as a major consideration in industrialized countries.

Measuring noise

Noise is a normal by-product of man's activities, an offshoot of the process of energy production or expenditure called energy residual. Unlike the thermal pollution that threatens our air and water, noise represents only a very small amount of the total energy residual resulting from those activities. For example, the acoustic energy derived from a 10,000-hp propulsion unit of a jet engine is only on the order of one-hundredth of 1% of the total energy expended by the machine. The noise pollution problem arises because of the extreme sensitivity of the human ear.

The most common measurement for sound is the decibel (dB), which is approximately equal to the smallest change in the pressure level of the air that can be perceived by an acute human ear. (In technical terms, the decibel unit is defined as 20 times the logarithm to the base ten of the ratio of the noise pressure to a perception reference pressure. The normal scientific reference is 0.0002 dynes per square centimeter, which is approximately equivalent to 0.0000004 lb per sq ft.) However, the absolute level of pressure is not the only factor that determines how a human reacts to noise; the frequency of pressure fluctuations also plays an important part. Normal hearing is sensitive to a range of fre-

quencies between 20 and 20,000 Hz (Hz = cycles per second), with peak sensitivity around 3,000 Hz. In recent years many units have been developed for expressing the subjective intensity of sound, most of them based on the decibel concept but including such factors as duration of the noise and whether the noise contains distinctive tonal characteristics, such as whines and screeches, that cause great disturbance when produced at high frequencies.

One of the most common of these units, and one that correlates well with human response to a wide variety of noises, is the sound pressure level measured on the "A" scale, in which sounds of different frequencies are weighted in accordance with the sensitivity of the ear. On this scale, the rustling of leaves registers 0–10 dBA, a normal conversation 30–60, and heavy city traffic 60–90. A cocktail party can rise to 70–80, a power mower to 90–100, and construction noise to 110. A jet airliner overhead and a police siren both rate 110–120. A jet at takeoff can reach 140 and is rivaled by a rock band, which can register between 120 and 140.

Since the threshold of annoyance has been estimated to start at 50 dBA, man spends much of his time subjected to noise that is irritating at the least and a major health hazard at the extreme. Some generalized data suggest that exposures above 90 dBA produce physical symptoms (*e.g.*, dilation of the pupils of the eye, increased adrenalin production) and a lowering of efficiency. Ear discomfort occurs at about 120 dBA and pain between 130 and 140. Females are generally more sensitive to a high-frequency noise than males, and older people lose their ability to hear high frequencies, above about 10,000 Hz.

The decibel and the dBA scale measure, respectively, the physical intensity of noise and the subjective response of a single human, in terms of loudness, annoyance, noisiness, and interference with speech, when the subject is situated in a clinical or laboratory-type environment. Such measurements are an important aspect of noise control, since they enable the specialist to assess the hearing-damage potential of a given noise and the variability of response between individual subjects. They also afford a standard for measuring and assessing the degree of noise control necessary to protect the individual's physical and mental well-being. But it is the effect of noise on social well-being that provides the greatest challenge. When many human beings are involved, predicting their response to a particular noise or to their noise environment becomes a difficult task.

The usual technique for assessing the effect of noise on a large group of people is the social survey. A group or community is questioned in such a way that the attitudes of its members to a given noise environment can be interpreted statistically. These attitudes are then related to physical measurements of the noise exposure. (In acoustics, exposure refers to the measurement of noise levels plus the total number of noise exposure events; *e.g.*, in the case of an aircraft noise survey, the total number of takeoffs and landings in a given period.) The purpose of such surveys is to find out whether and how a given noise

***TOBA COHEN** is a medical writer at the American Medical Association.*

	noise level	response	hearing effects	conversational relationships
	150			
carrier deck jet operation	140	painfully loud	contribution to hearing impairment begins ↓	
	130	limit amplified speech		
jet takeoff (200 feet) discotheque auto horn (3 feet)	120	maximum vocal effort		
riveting machine jet takeoff (2,000 feet)	110			
garbage truck	100			shouting in ear
N.Y. subway station		very annoying		shouting at 2 feet
heavy truck (50 feet) pneumatic drill (50 feet)	90	hearing damage (8 hours)		very loud conversation, 2 feet
alarm clock	80	annoying		
freight train (50 feet)				loud conversation, 2 feet
freeway traffic (50 feet)	70	telephone use difficult		loud conversation, 4 feet
		intrusive		
air conditioning unit (20 feet)	60			normal conversation, 12 feet
light auto traffic (100 feet)	50	quiet		
living room				
bedroom	40			
library				
soft whisper (15 feet)	30	very quiet		
broadcasting studio	20			
	10	just audible		
	0	threshold of hearing		

environment interferes with normal living habits and to provide a basis for possible noise-control measures.

The problem of control

It has been estimated that between 6 million and 16 million Americans are going deaf as a result of exposure to noise in industry and in the military. For them, controls over noise will be too late. For those suffering from temporary hearing loss, called auditory fatigue and signaled by a ringing in the ears and a straining to hear voices immediately after exposure to loud noises, controls cannot come too soon. In the U.S. the cost of industrial noise alone, including accidents, compensation claims, inefficiency, and absenteeism, has been assessed at $4 billion a year. Nor can one escape possible hearing damage by choosing a safe occupation. In the average home, air conditioners, washing machines and clothes dryers, dishwashers, vacuum cleaners, mixers, and hair dryers all produce noise above the annoyance level. An exhaust fan or a blender can put out from 90 to 100 dBA.

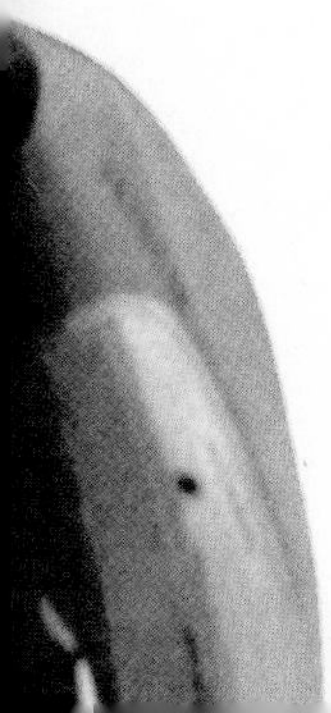

In recent years new laws have been introduced to control noise levels. Transportation, construction, and other noise-producing industries have been put under some restrictions, and mechanisms have been set up to process individual and community complaints. But here one encounters an anomaly. One would expect that since noise interferes with the normal living habits of a community, disturbing sleep, intruding on recreation, and straining family relationships, communities in high-noise areas would make their grievances known. Instead, there seems to be relatively little relationship between the measured noise exposure and the attitudes expressed by the community to this exposure.

The answer appears to lie in socioeconomic status. Wealthier and better-educated persons can and do articulate their grievances better than those who are lower on the socioeconomic scale, even though the grievances themselves may be less pressing. It may also be that population groups with high socioeconomic status have higher expectations and a greater investment in their environment, and are therefore more disposed to defend their life-style against intrusions. Furthermore, many segments of society may have become so used to living and working under conditions of constant stress that they are no longer aware of the intensity of the problem.

But whether they are aware of it or not, residents of communities where there is constant noise are affected in two ways: noise at high intensities or for excessively long periods can damage the hearing mechanism, while noise at lower intensities can produce mental stress. The latter may be characterized subjectively as annoyance, resulting, for example, from noise-induced sleep disturbance. The psychic damage produced by noise affects far more people than the physiological damage. Which is the more dangerous to society is, perhaps, an unanswerable question. From a scientific viewpoint, physiological damage is easier to observe and quantify. Usually it results from exposure to noise from industrial machinery, although repeated exposure to noise from household appliances is looming as a major cause. Noise-induced mental stress has many causes, some of which are readily apparent to any citizen of a modern, urbanized society.

The noisy skies

Of all noise-control problems, aircraft noise has received the most publicity and perhaps the greatest expenditure of money and effort. The term is used to describe the noise from overflying aircraft in communities around airports. There are other problems created by noise from aircraft, such as the effect on persons working close to operating aircraft engines and the noise levels inside passenger aircraft, but these are specialized technical matters affecting a comparatively small number of people.

The initial response to the aircraft noise problem was to deny that it was a problem. Communities near airports were told that it was their duty to accept this degradation of their environment in the name of progress or the national interest; for example, the U.S. Air Force coined the phrase "The Sound of Freedom." Since then, however, the air transportation industry has grown so rapidly that even the most devoted exponent of aviation can hardly expect such an approach to work. The world's commercial air traffic is doubling every ten years. It is estimated that about ten million people (1 in 20) in the United States are affected by aircraft noise; in the small and densely populated United Kingdom, the proportion rises to nearly 30%.

Compounding the problem, the growth in air traffic has paralleled the introduction of jet aircraft, with their extremely high noise levels. The original jet propulsion system was the turbojet engine, and today most large commercial airplanes are powered by turbofan units.

Considerable technological effort and money have been expended by manufacturers and government research agencies on both sides of the Atlantic in an effort to develop a quieter jet engine. In the United States, the Federal Aviation Administration has issued construction standards for new planes limiting takeoff and landing-approach noise, with a maximum of 108 dBA for the largest planes. There have been protracted international negotiations aimed at formulating a "noise certification" that new aircraft would have to meet before being allowed to operate commercially.

To some extent this effort has paid off. The engines fitted to the new wide-bodied jet transports have more than twice the thrust (over 40,000 lb) of the original turbojet engines used on the Boeing 707 and McDonnell Douglas DC-8, but they produce only about one-hundredth of the acoustic energy. Because of the logarithmic characteristics of the units used, this represents a reduction of only 10 to 20 dBA in noise energy, but even so it is a substantial improvement. At the same time, however, the number of aircraft movements is rising so rapidly—to about 500,000 a year at each of the world's four busiest airports: Chicago-O'Hare, New York-Kennedy, Los Angeles-International, and London-Heathrow—that the technological gains in noise reduction may well be eroded.

Even 108 dBA is far too high since, as we have seen, well-being is affected and efficiency drops at noise levels above 90 dBA. To remove aircraft noise as a major source of annoyance, an additional reduction of 20 dBA in individual aircraft noise levels would have to be achieved. Since this does not appear to be technically feasible in the near future, further alleviation seems possible only by siting airports away from population centers, restricting aircraft traffic and routes, soundproofing dwellings and work places, and, in the extreme situation, removing communities from areas of intolerable noise exposure. All these possibilities are under active consideration in the United States and elsewhere. Such programs as those proposed by the Environmental Protection Agency and other branches of the U.S. government and the aircraft noise standards enacted by the state of California give promise that relief from this source of disturbance will be a reality within the next decade.

The land area in the United States where the noise generated by aircraft and urban freeways makes residential living unacceptable increased from about 100 to 2,000 sq mi from 1955 to 1970.

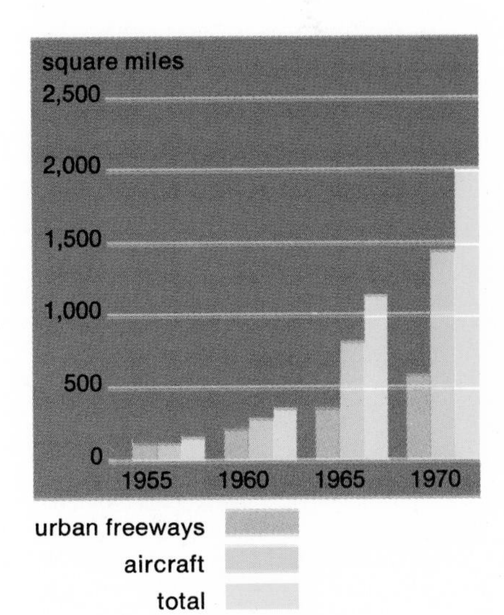

One type of aircraft noise pollution that has received extensive publicity is the sonic boom. This is caused by a series of pressure waves—similar to the waves produced by the bow of a ship as it cuts through the water—that are created by an aircraft moving through the atmosphere at a speed greater than the speed of sound. A supersonic transport cruising at an altitude of over 50,000 ft would produce pressure waves extending to the ground perhaps 20 to 30 mi on either side of the flight track. This means, for example, that a supersonic flight from Chicago to New York City would expose more than 30 million people to the boom, and this massive exposure would be repeated along all the major airways of the world.

Under these flight conditions, the amplitude of the pressure wave would be about two pounds per square foot (standard atmospheric pressure at sea level is 2,116 lb per sq ft) directly under the flight track and would decrease gradually on either side. Normally this overpressure is not enough to cause structural damage, but since the boom sounds like a distant explosion it can be extremely startling to someone who is not prepared for it. When a supersonic aircraft is maneuvering, decelerating, or accelerating, the amplitude can increase many times and there is danger of structural damage, such as broken windows and cracked plaster. A growing number of countries, including the United States, the United Kingdom, Sweden, and Switzerland, have taken steps to ban supersonic overflights, at least until more is known about the effect of the boom on man and his environment.

"The clamor of the crowded streets"

Another major source that exposes vast numbers of people to disturbing levels of noise is ground traffic. A recently published report estimated that in the United Kingdom between 19 and 46% of the country's 45 million urban population are likely to be exposed to high levels of traffic noise in their residential areas, and that by 1980 the percentages will be between 30 and 61. Extrapolating from these figures, it is not unreasonable to assume that up to 75 million people may be affected by traffic noise in North America.

Traffic noise is caused primarily by three categories of vehicles: heavy commercial vehicles, most probably diesel-powered; light commercial vehicles and private automobiles; and motor scooters and motorcycles. Vehicles in the first two groups cause most of the noise because they are the most numerous, although the distinctive sound of a motor scooter or motorcycle can be extremely annoying. A large motorcycle is normally about 10 dBA noisier than a family car at speeds over 50 mph.

The automobile designer concerned with noise control must take all aspects of the vehicle's design into consideration. The major noise producers on an automobile include the inlet and exhaust systems, engine cooling fan, transmission systems, brakes, and tires. Additional noise is produced when the engine is heavily loaded and when the vehicle is accelerated. The condition of the road surface must be considered, since a rough road produces more noise than a smooth one. Speed also plays an important part; here it is very difficult to generalize, but noise increases approximately six to ten dBA every time the vehicle's speed is doubled. A recent study has shown that there is a tremendous variation—over 30 dBA—in noise levels from different types of tires and from tires in different states of repair. Worn, smooth tires produce the least noise, while retreads and crossbar-patterned treads produce the most.

The traffic noise that most people experience arises on urban streets, where traffic moves at an irregular speed, starting and stopping at traffic lights, maneuvering around corners and up hills, and feeding from side streets. There, the major noise sources are engines and brakes, although tire squeal can also cause a problem. With the expansion of controlled-access highway networks, however, many populated areas are being exposed to noise from high-speed traffic. In this situation, the major sources of noise are tires and the roar from the engines and exhausts of large commercial vehicles, particularly heavy trucks. Nor have rural areas escaped. In recent years, once quiet forests and farmlands have been subjected to the sound of 108-dBA snowmobiles.

As in the case of aircraft noise, it does not appear that the traffic noise problem can be completely eliminated in the foreseeable future, given the current state of technological knowledge. Many helpful measures have been adopted, however, and more can and should be taken. Safety must not be compromised, of course, but it is possible to minimize noise from tires by specifying tread design and limiting speeds. Exhaust noise could be dramatically reduced by the application of existing technology.

It is harder to predict how much engine noise on highways can be controlled, but research indicates that radical engine-design changes could reduce such noise by 10 to 15 dBA within the next decade. Many states in the U.S. now have laws setting maximum permitted noise levels for commercial and passenger vehicles and motorcycles under normal driving conditions. The International Standards Organization proposed limits somewhat lower than those generally used, namely 86 dBA for commercial vehicles, 84 dBA for most motorcycles, and 80 dBA for passenger vehicles, and these should be adopted as goals for engine designers, to be achieved within two to four years.

The U.S. Federal Highway Administration has issued noise standards to be used on all new projects financed under the Federal-Aid Highway Act. Among the elements of highway design that reduce noise are fabricated or landscaped noise barriers, siting of buildings away from the road, depressed or underground routes, and regulation of traffic flows to prevent continual acceleration and deceleration. Naturally, economics plays a large part in this type of planning, however. Furthermore, in areas of high population density it is difficult to build a new system or to rearrange an existing one without exposing an additional sector of the population to increased noise.

Noise at work and play

The oldest acknowledged noise problem is that of industrial—or, more accurately, occupational—noise. Exposure to high noise levels of 90 dBA or more for extended periods can produce physiological damage to the ear, usually in the form of hearing loss. The first warning that the hearing mechanism is being affected is often tinnitus, or ringing in the ears, sometimes accompanied by a temporary loss of hearing in the upper audible-frequency range, around 5,000 Hz. If the exposure continues, the loss of hearing becomes permanent.

A recent study published by the U.S. Department of Commerce estimated that a third of the total labor force in the United States is exposed to noise levels that are potentially harmful, and the situation in most other industrial countries is probably similar. One method of dealing with the problem is through the use of ear plugs and ear defenders. In the U.S. the Walsh-Healey Public Contracts Act limits the noise level to which an unprotected worker can be exposed during an eight-hour day to a maximum of 90 dBA. Many other countries have enacted similar legislation.

Traditionally, hearing loss has been considered an occupational hazard, but recently a hearing danger has emerged in the form of a leisure-time activity. Electrically amplified music, which can register above 120 dBA in concert and on recordings, can cause temporary hearing loss in the same way as industrial noise. Some researchers believe that the "on-off" pattern of listening may allow sufficient time for the ear to recover, but conclusive evidence is not yet available.

Toward a quieter future

Vast amounts of money are being spent in an effort to alleviate the noise from airplanes, road vehicles, and trains. A reasonable estimate is that more than $100 million is being spent annually on aircraft noise control alone, about half of it by the U.S. government and the U.S. aerospace industry. Progress has been made, although there is still a long way to go before noise from this source ceases to be a major nuisance. In many other areas of industrial and domestic noise, the technology for noise reduction is available but the incentive is lacking.

It is not enough to produce data on noise pollution if legislators and administrators have not yet learned to interpret them for the benefit of all society. To this end, a pro-people noise policy must be enacted. Every piece of equipment must be designed with a quiet environment in mind. A whole new corps of noise-control experts must be trained and governments must support their research. Surely if space-age technology can build silent submarines it can provide silent trucks and soundproof buildings. We must know the facts about the cost of noise control and demand subsidies, if necessary, to cover that cost. If citizens, technicians, and designers work together to prod governments, perhaps we can return to the point where the most irritating sound on the city streets is no louder than a two-wheeled chariot.

AUDIOVISUAL MATERIALS FROM ENCYCLOPÆDIA BRITANNICA EDUCATIONAL CORPORATION:

Film: *Noise: Polluting the Environment.*

Cyclic AMP
by G. Alan Robison

Since its discovery in 1956 as the link between hormones and their influence on liver cells, the "miniature" nucleic acid, cyclic AMP, has been found to play many, always significant, roles in very different types of cells.

Living cells are capable of carrying out a bewildering variety of chemical reactions, all of which have to be carefully regulated in relation to each other if the cell is to survive. Some reactions occur in all living cells, whereas others occur in only one or a few types of highly specialized cells. A single cell is, therefore, a very complex system. Multicellular organisms are more complex still, by several orders of magnitude, because in them each type of cell has to be regulated in relation to all of the other types of cells in such a way that they all work together as a coordinated whole.

Two important systems have evolved in multicellular animals to ensure that this kind of coordination is maintained. These are the endocrine system and the nervous system, which together might be thought of as constituting a single system specializing in intercellular communication. Hormones and neurotransmitter agents are released from their cells of origin (endocrine glands in the case of hormones, neurons in the case of neurotransmitter agents) to influence the activities of other cells. What regulates the release of these substances and how do they act to influence other cells? Cyclic AMP was discovered in the course of attempts to answer one of these questions, and its role as a mediator of hormone action is still one of the most important roles of this nucleotide to have been recognized.

A molecular model of cyclic AMP, constructed for the Britannica Yearbook of Science and the Future *by N. C. Yang, professor of chemistry at the University of Chicago.*

"The role of RNA in transferring genetic information . . . is, by now, well known." Ribonucleic acid (RNA) consists of a subunit compound, or nucleoside, in which one half, the base moiety, is one of the four atomic side groups common to nucleic acids, while the other is the ribose moiety, a sugar. The ribonucleosides are linked together by phosphate groups. Each phosphate is attached to the third carbon position (called 3′) of the ribose of one nucleoside and to the fifth carbon (5′) of the next nucleoside. The repeated phosphate links form the backbone of the RNA structure. Letters used here and on the next page indicate atoms of carbon (C), hydrogen (H), oxygen (O), nitrogen (N), and phosphorus (P). The two bases here are adenine. Since nucleosides are named for their bases, these subunits are adenosines.

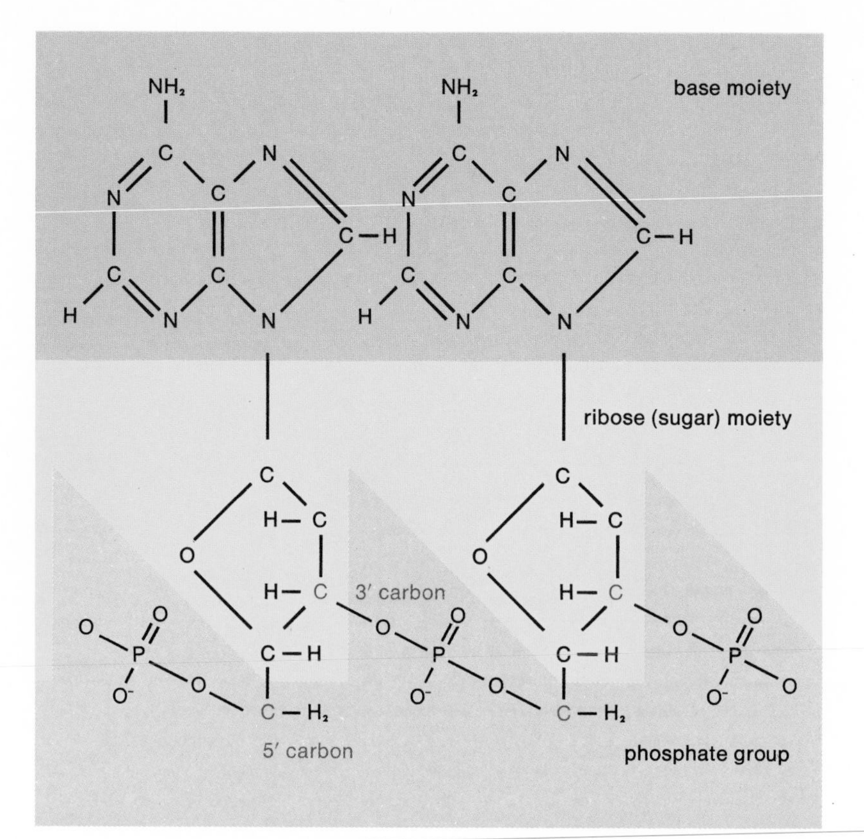

G. ALAN ROBISON, *co-author of* Cyclic AMP, *is professor of pharmacology and director of the program in pharmacology at the University of Texas Medical School.*

When ribonucleic acid (RNA) is broken down into its constituent structural units, or nucleotides, one of the products is adenosine 5′-monophosphate, otherwise known as 5′-AMP or 5′-adenylic acid. Adenosine is the subunit compound, or nucleoside, of a sugar molecule and adenine, one of the four bases or side groups found in nucleic acids. The 5′ designation means that the phosphate group is attached to the fifth carbon position of the ribose moiety, the sugar half of the adenosine compound. In an RNA chain, the phosphate will also be attached to the third carbon of the ribose moiety of the next nucleotide in the sequence. This repeating sequence of 3′ and 5′ linkages forms the backbone of RNA. It is also possible for the phosphate group to be attached to the third and fifth carbons of the *same* ribose moiety; this is adenosine 3′, 5′-monophosphate or 3′, 5′-adenylic acid. Since the phosphate connecting the 3′ and 5′ carbons forms something of a circular or cyclic structure, the compound is commonly referred to as cyclic adenosine monophosphate—cyclic AMP.

The role of RNA in transferring genetic information from the nucleus to the protein-synthesizing machinery in the cytoplasm is, by now, well known. Could a "miniature" nucleic acid consisting of only a single nucleotide also have an important role to play in the regulation of cell function? We are beginning to realize that cyclic AMP has a number of important roles to play, and the purpose of this article is to summarize what we presently know about them.

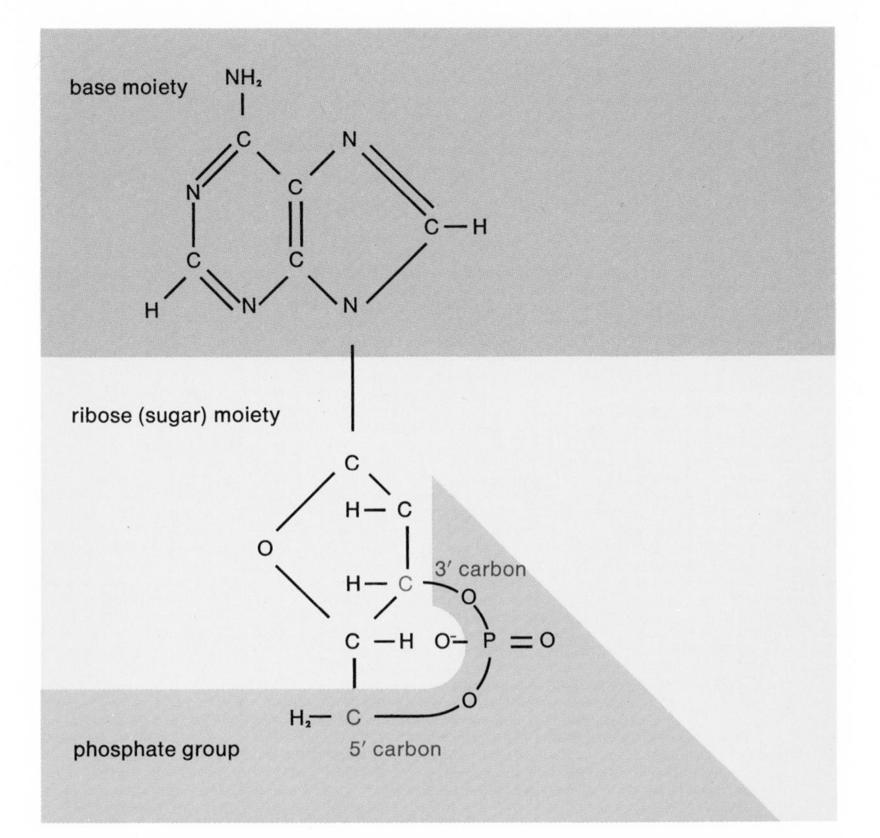

"Could a 'miniature' nucleic acid . . . also have an important role to play in the regulation of cell function?" Cyclic AMP is a single nucleotide, that is, one nucleoside (adenosine) with a phosphate group attached. The phosphate links to the 3′ and 5′ positions of the same ribose moiety in a circular, or cyclic fashion, a fact that gives the structure its common name. "Cyclic AMP" is actually an acronym for adenosine 3′, 5′-monophosphate.

Role in hepatic glycogenolysis

Cyclic AMP was discovered as the mediator of the hepatic glycogenolytic effect of glucagon and epinephrine by Earl W. Sutherland, Jr., and his colleagues at Western Reserve University, Cleveland, O., in 1956. Glucagon is a polypeptide produced by the alpha cells of the pancreas. Epinephrine, or adrenaline, is a smaller molecule, a derivative of the amino acid tyrosine, and is produced primarily in the medulla of the adrenal glands. Both hormones stimulate the liver to convert glycogen (animal starch) to glucose, which is a useful source of energy not only for bacteria but for many mammalian cells as well, and especially for brain cells. Claude Bernard, the great 19th-century French physiologist, had demonstrated that liver glycogen was the principal source of stored glucose in the animal body. In the late 1930s it was shown, by biochemists Carl and Gerty T. Cori, working at Washington University, St. Louis, Mo., and others, that an important step in the breakdown of glycogen involved a chemical disintegration in which a bond is split and phosphoric acid is added (phosphorolysis) to form glucose 1-phosphate, catalyzed by the enzyme phosphorylase. Sutherland showed that when glucagon or epinephrine was applied to liver slices, the result was a rapid increase in phosphorylase activity, sufficient to account for the conversion of glycogen to glucose 1-phosphate and subsequently to glucose itself. Luis Leloir and his colleagues in Argentina later showed that glycogen was synthesized from glucose 1-

phosphate by way of another intermediate, uridine diphosphate glucose (UDPG), under the catalytic influence of the enzyme glycogen synthetase.

Sutherland and his colleagues purified active phosphorylase from liver and then found another enzyme that would catalyze its inactivation. This latter enzyme was later shown to be a phosphatase, that is, an enzyme catalyzing the removal of a phosphate group, suggesting that the activation of phosphorylase might be associated with phosphorylation of the enzyme itself. It was then shown that phosphorylase activation in response to glucagon or epinephrine was indeed associated with the incorporation of phosphate into the enzyme molecule. It thus became clear that the activity of liver phosphorylase represented a balance between inactivation by a phosphatase and reactivation by a kinase, an enzyme that promotes incorporation of phosphate groups, and it appeared that glucagon and epinephrine were acting to somehow shift the balance in favor of the kinase. However, when the two hormones were added to the purified enzymes, or even to partially purified liver extracts, no effect was seen.

Then, in a classic series of experiments carried out in collaboration with Theodore W. Rall, Sutherland showed that if liver homogenates were fortified with ATP (adenosine triphosphate) and magnesium ions, the addition of glucagon or epinephrine did lead to the phosphorylation and, hence, activation of phosphorylase. This was the first physiologically important effect of a hormone to be demonstrated in a preparation not containing intact cells. When the particles containing fragments of the cell membrane were removed by centrifugation, Sutherland and Rall noted that the response to the hormones was lost. But if the particles were incubated separately with the hormones in the presence of ATP and magnesium ions, then a soluble factor was produced that did activate phosphorylase when added to liver extracts. This factor was soon identified as cyclic AMP. The enzyme in membrane fractions responsible for converting ATP to cyclic AMP was named adenyl cyclase. (It is now referred to as adenylyl cyclase or adenylate cyclase.) The presence of a phosphodiesterase, which inactivated cyclic AMP by converting it to 5′-AMP, was also established.

For these and other contributions to our understanding of cell regulatory mechanisms, Nobel Prizes were awarded to the Cori's in 1947, to Leloir in 1970, and to Sutherland in 1971. Our present understanding of the glycogenolytic action of glucagon and epinephrine, growing out of these and subsequent studies by other investigators, can be summarized briefly. The hormones first interact with specific receptors on the surface of the liver cell membrane. These receptors are specific in the sense that one type interacts only with glucagon, another only with epinephrine. The hormone-receptor interaction causes a conformational perturbation that leads, by a mechanism that is still not completely understood, to an increase in the activity of adenylyl cyclase on the inside of the cell membrane. As a result, the intracellular level of cyclic AMP begins to rise. Cyclic AMP then interacts with the regulatory

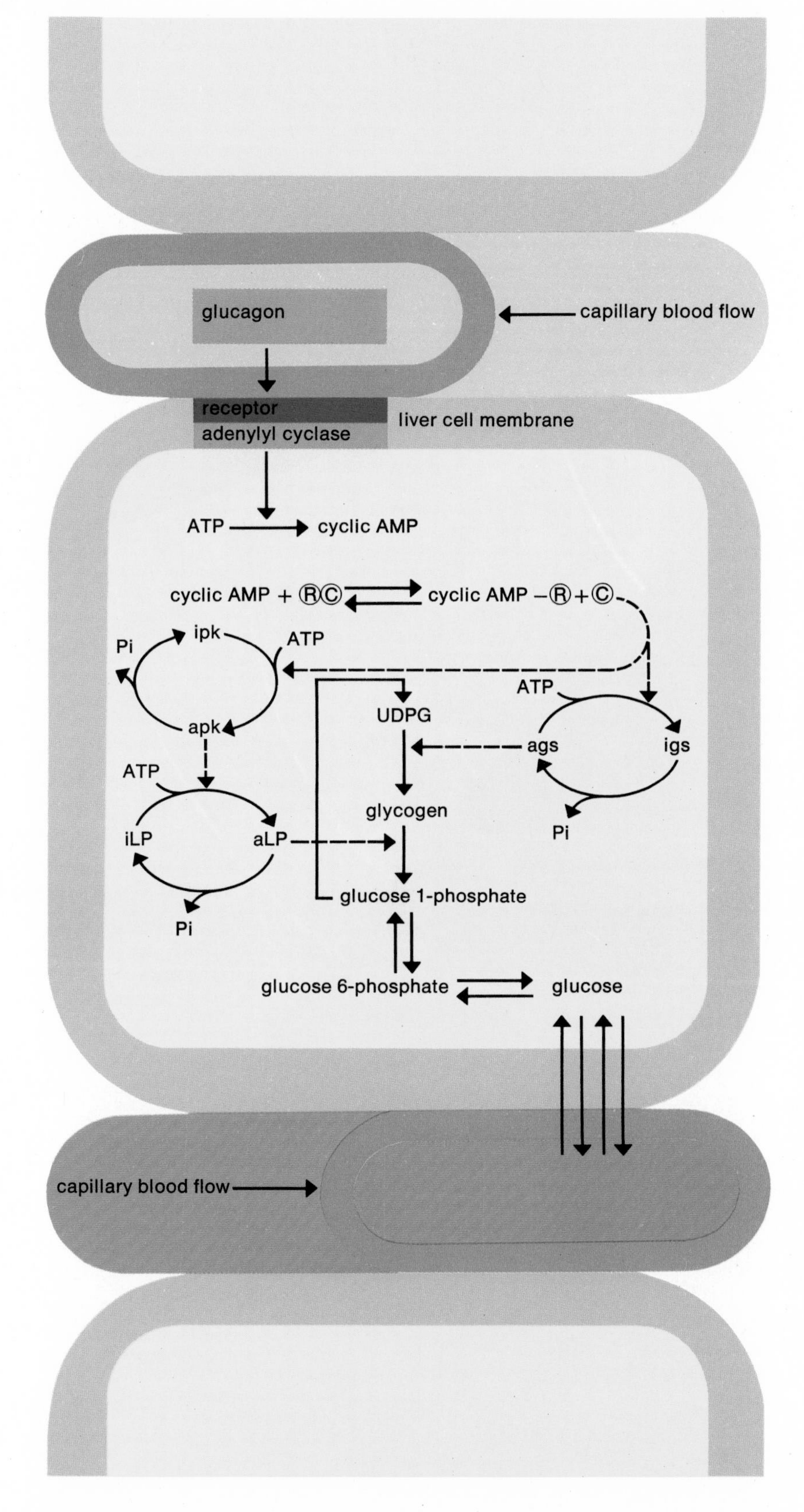

Cyclic AMP's "role as a mediator of hormone action is still one of the most important roles of this nucleotide to have been recognized." This action has been demonstrated in glycogenolysis, the conversion of stored starch (glycogen) into an energy source (glucose), which takes place in liver cells. The hormone glucagon interacts with a receptor site on the surface of the cell's membrane to cause the increased activity of adenylyl cyclase. This enzyme in the membrane converts adenosine triphosphate (ATP) into cyclic AMP, which then interacts with a regulatory subunit (R) of a protein kinase system, causing it to become dissociated from the catalytic subunit (C). This unit is free to catalyze the simultaneous phosphorylation of several proteins. Of these, glycogen synthetase (gs) becomes inactive (i) and ceases catalyzing glucogen production. Phosphorylase kinase (pk) becomes active (a) to cause phosphorylation of liver phosphorylase (LP), which breaks down glycogen into an intermediate form, glucose 1-phosphate. Two other intermediate forms are UDPG (uridine diphosphate glucose) and glucose 6-phosphate. The whole chain of reactions behaves much like a row of dominoes: almost immediately following the arrival of glucagon at the receptor site, glucose passes out through the liver cell membrane into the blood.

subunit of a protein kinase system. This subunit is inhibitory, such that in the absence of cyclic AMP it prevents the activity of the catalytic subunit. In the presence of cyclic AMP, however, the subunits become dissociated, and the catalytic subunit becomes free to catalyze the phosphorylation of several proteins. One such protein is phosphorylase kinase, which in turn catalyzes the phosphorylation of phosphorylase itself. Another substrate is glycogen synthetase. An interesting and important difference between these substrates is that when glycogen synthetase becomes phosphorylated, it becomes *less* active. Thus, the simultaneous phosphorylation of these proteins, leading to activation of phosphorylase and inactivation of glycogen synthetase, provides a powerful mechanism for promoting the breakdown of glycogen to glucose.

Role in other differentiated cells

Although it was initially thought by some investigators that the role of cyclic AMP might be limited to mediating the hepatic effects of glucagon and epinephrine, it was soon shown by Sutherland and his colleagues and others that this was not the case. Robert Haynes, then a member of Sutherland's department at Western Reserve, showed that adrenocorticotropic hormone (ACTH), but not glucagon or epinephrine, stimulated adenylyl cyclase in slices of adrenal cortex. Conversely, ACTH did not share the ability of glucagon and epinephrine to stimulate adenylyl cyclase in liver cells. Sutherland had shown that the application of high concentrations of cyclic AMP to liver slices led to the production of glucose, and Haynes showed that the result in adrenal cortical slices was to stimulate the production of steroid hormones. This was the same effect ordinarily produced by ACTH itself. Jack Orloff and Joseph Handler at the National Institutes of Health showed that vasopressin (the antidiuretic hormone) acted by way of cyclic AMP to increase the permeability of certain epithelial cell membranes. Subsequently, numerous other hormones were shown to affect their target cells by stimulating adenylyl cyclase. Some of these hormones or related agents and the tissues in which the effect has been demonstrated are:

Hormone	*Tissues*
Glucagon	Liver
Catecholamines	Many tissues
Adrenocorticotropic hormone	Adrenal cortex
Vasopressin	Kidney (medulla)
Hypothalamic releasing hormones	Anterior pituitary
Luteinizing hormone	Interstitial cells
Follicle-stimulating hormone	Seminiferous tubules
Melanocyte-stimulating hormone	Melanophores
Thyroid-stimulating hormone	Thyroid
Parathyroid hormone	Kidney (cortex) and bone
Secretin	Adipose
Histamine	Brain
Prostaglandins	Many tissues

The reason that different hormones affect different cells is that different cells are equipped with different receptors. The distribution of receptors is, in most cases, highly specific in the sense that most hormones affect only one or a few different types of cells, just as most cells respond to only one or a few hormones. Exceptions to these generalizations are known. Epinephrine and norepinephrine, for example, stimulate adenylyl cyclase in a great variety of cells and produce a corresponding variety of effects when injected into animals (or when released physiologically, as during stressful periods requiring the rapid mobilization of energy). Fat cells (adipose cells) are also exceptional in that they possess a variety of different types of receptors, and, therefore, respond to a variety of different hormones. The effect of cyclic AMP in these cells is to stimulate the conversion (by lipolysis) of stored triglycerides to free fatty acids, which can be used by other tissues as a source of energy. Many of the effects of cyclic AMP in multicellular animals, including glycogenolysis in the liver and lipolysis in adipose tissue, can be interpreted as ergotropic effects, that is, they enhance or promote the utilization of energy. Another example of this would be the increased force of contraction produced by epinephrine in the heart. It is interesting to note that certain other hormones, such as insulin, which act during periods of rest to promote the storage rather than the expenditure of energy, have now been shown to produce some of their effects in their target tissues by reducing the level of cyclic AMP. Thus, insulin opposes the actions of glucagon and epinephrine in the liver, and of epinephrine and other hormones in adipose tissue.

If all of these various hormones act by way of cyclic AMP, why do they not all produce the same effects? The answer to this question is that different cells are equipped with entirely different enzyme systems, in addition to the different receptors mentioned previously. Although each somatic cell of an organism contains the total genetic information for that organism, only parts of the total are expressed in differentiated cells—which parts differing from one type of cell to another. Liver cells, for example, do not contain the enzymes required for converting cholesterol to steroid hormones; therefore, they cannot possibly produce steroid hormones no matter how high the level of cyclic AMP may be raised, even though this is the characteristic effect of cyclic AMP in adrenal cortical cells. In general, in differentiated cells cyclic AMP alters the rates of basic processes that these cells are equipped to perform and does not endow them with properties that they previously did not have. Some of the known effects of cyclic AMP, in addition to those already mentioned, are:

Stimulation of gluconeogenesis
Stimulation of ketogenesis
Stimulation of amino acid uptake
Inhibition of lipogenesis
Increase of permeability
Stimulation of steroidogenesis
Maintenance of adrenal sufficiency
Stimulation of exocytosis
Inhibition of cell growth
Increase of force of contraction of cardiac muscle
Relaxation of smooth muscle
Inhibition of platelet aggregation
Inhibition of neuron firing

This list is very incomplete, but should serve to indicate the widespread influence that cyclic AMP can have as a regulator of differentiated cell function.

The mechanisms by which cyclic AMP acts to produce these various effects are not known in all cases. Evidence is accumulating, however, to suggest that many, and perhaps most, of these effects result from the phosphorylation of one or more proteins, as occurs in glycogenolysis. According to this hypothesis, cyclic AMP produces different effects in different cells because the cells contain different substrates for the protein kinase, which has been shown to exist in all animal cells studied. Among the substrates identified to date, besides phosphorylase kinase and glycogen synthetase, are adipose tissue lipase, a ribosomal protein in adrenal cortical cells, nuclear histones from many cells, and several membrane proteins.

Because hormones that stimulate adenylyl cyclase were seen as bringing information to the membranes of their target cells, with cyclic AMP carrying the information from the membrane to other intracellular systems, Sutherland and his colleagues suggested that the hormones might be thought of as first messengers, with cyclic AMP serving as a second messenger. The role of cyclic AMP as a second messenger in hormone action has now been established, and this is still the best understood role of this nucleotide in differentiated cells. Other roles may exist, however, and recent evidence has suggested an important role in sensory physiology. Light, for example, was found to inactivate adenylyl cyclase in preparations of rod outer segments (photoreceptor cells) of the retina of a frog's eye. Also bitter-tasting but not sweet-tasting compounds were found to inhibit phosphodiesterase in tongue papillary preparations.

It should be noted that not all hormones act by altering the level of cyclic AMP in cells. The steroid hormones, for example, seem to have an entirely different mechanism of action. Instead of interacting with receptors on the cell surface, they interact with cytoplasmic receptor proteins, which then become capable of entering the nucleus to influence gene transcription. The primary actions of thyroxine and growth hormone also do not seem to be closely related to cyclic AMP. Each of these hormones, however, produces effects that may secondarily have an important influence on the formation or action of cyclic AMP, and this may be the basis of many of the permissive effects of these hormones.

Role in unicellular organisms

After its discovery in the mammalian liver, cyclic AMP was shown by Sutherland and his colleagues to be widely distributed, not only in other mammalian cells and tissues but throughout the animal kingdom. The question of whether cyclic AMP exists and functions in the cells of higher plants, and, if so, under what conditions, has still not been satisfactorily answered. Its presence in bacteria, however, was established at an early date, and the demonstration that glucose could sup-

gene transcription

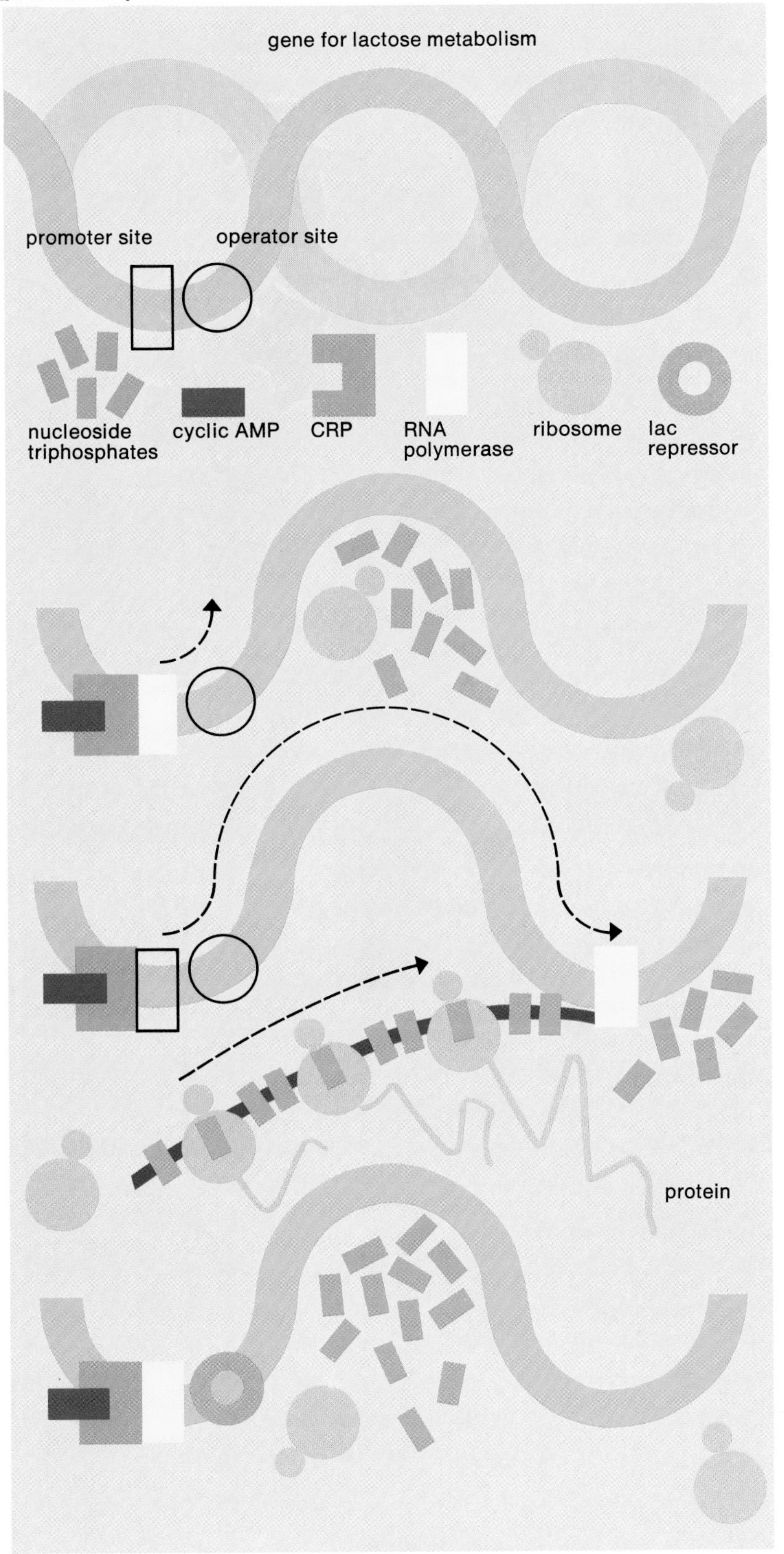

"Cyclic AMP only tells [a bacterial] system that it can make one or more enzymes—it does not specify which ones to make." Experiments showing cyclic AMP's role in transcribing genetic messages in bacteria can be made in cell-free mediums that are given all the necessary molecules. A DNA chain rich in genes for metabolizing the sugar lactose will begin transcribing its message when a unit of cyclic-AMP receptor protein (CRP) bound to cyclic AMP activates the promoter site on the DNA chain. RNA polymerase, an enzyme, will then bind with the promoter site and begin linking nucleoside triphosphates together in the sequence dictated by the DNA. Ribosomes then assemble the protein or enzyme called for in the message. If, however, another protein, the lac repressor, attaches to the operator site at the beginning of the gene chain, the polymerase will not bind and no genetic message can be sent, even though cyclic AMP is present.

press the formation of cyclic AMP in *Escherichia coli* led eventually to a better understanding of how these organisms function.

Although unicellular organisms are less complex than multicellular organisms, they are still complex; in fact, their regulation poses some special problems not encountered in higher organisms. Everything these organisms do has to be done by the same cell, and there is only a limited amount of space in which to do it. Consider the plight of a bacterium living in the human bowel. By bacterial standards, this might be regarded as a fairly luxurious environment, containing as it does a large number of carbohydrates and other potential sources of energy. Different enzymes are required for the utilization of each of these substrates, and the bacterium contains the genetic information for synthesizing many of them. But because the amount of intracellular space is so limited, it would be disadvantageous for all of these enzymes to be synthesized all at the same time. Which ones should be synthesized? The solution that has evolved in many species of bacteria is that the enzymes required for the utilization of glucose are synthesized at all times whereas those required for the utilization of other substrates are synthesized only when the substrate to be utilized is present, and only after all of the glucose has been used. The enzymes that are synthesized at all times are called constitutive enzymes; those made only in response to a particular substrate are called inducible enzymes. The mechanism by which glucose acts to prevent the synthesis of inducible enzymes was a puzzle for many years, but we now know it involves cyclic AMP in a very important way.

The ability of glucose to prevent the formation of cyclic AMP is important because cyclic AMP is needed for the synthesis of inducible enzymes. The way this works is that when cyclic AMP is made, it interacts with a special protein called cyclic AMP receptor protein (CRP), and this allows the CRP to interact with certain portions of DNA. As a result of this, another enzyme called RNA polymerase can "read" the DNA, leading to the synthesis of messenger RNA and hence of the appropriate enzyme. It should be noted that even when the cyclic AMP-CRP complex interacts with DNA, the genetic information needed for the synthesis of inducible enzymes will not be transcribed unless the substrate for the enzyme is also present. In other words, cyclic AMP only tells the system that it *can* make one or more new enzymes—it does not specify which ones to make. If lactose is present, then enzymes for the utilization of lactose will be made. If glycerol is present, then enzymes for using glycerol will be made.

Another effect that cyclic AMP has in bacteria is to stimulate the synthesis of flagella, which cause the cells to become motile. This is advantageous to the bacteria because it enables them to seek more congenial environments.

Still another microorganism in which the role of cyclic AMP has been studied is the cellular slime mold. These organisms are unicellular during part of their life cycle and multicellular during another. Since they replicate only during the unicellular stage and differentiate only during the multicellular stage, these cells have been of special interest to developmental biologists. The cells continue to grow and replicate during the unicellular stage so long as plenty of food is available, and during this phase the level of cyclic AMP is very low. It now appears that the reason for this is that the cells release large quantities of a phosphodiesterase, which rapidly destroys any cyclic AMP that might be produced. As it happens, the favorite food of these organisms seems to be bacteria. As soon as all of the bacteria have been consumed, then a remarkable series of events is set in motion. First, cyclic AMP begins to accumulate in the vicinity of one or a few cells. This causes neighboring cells to be attracted to the cyclic AMP-producing cells, and also causes the cells to become stickier, so that once attracted they tend to aggregate. Then, as each new cell joins the aggregate, it in turn begins to secrete cyclic AMP, with the result that new cells are attracted from an ever increasing distance. Eventually the aggregated mass turns into a migrating pseudoplasmodium. The distance moved by this slimy conglomeration of cells will depend on the relative humidity and other factors, and during this time cyclic AMP production falls. Later, after migration stops, the cells that initially began releasing cyclic AMP begin to release it again, causing neighboring cells to differentiate into stalk cells. The stalk cells push downward through the cell mass, causing the other cells to be lifted upward, leading ultimately to the formation of a fruiting body. Spores are released from the mature fruiting body, unicellular amoebas emerge, and the whole cycle starts anew.

Regulation of growth and development

Studies with microorganisms have suggested that cyclic AMP must have evolved as a regulatory agent at a very early stage in the evolutionary process. At least three features of the role of cyclic AMP in higher forms can be discerned even in bacteria. First, the level of cyclic AMP inside the cell depends on something in the extracellular environment. Second, the precise end result of the change in cyclic AMP may depend upon many other factors, such as which substrates are available, in the case of bacteria, or the type of cell involved, in the case of higher organisms. Third, cyclic AMP seems not to be required under all circumstances, but only under what might be construed as emergency conditions.

An unusual feature that is *not* commonly seen in higher forms emerges at the level of the slime mold. This is that cyclic AMP is released into the extracellular medium to influence other cells from the outside. This kind of role, involving intercellular communication, appears in higher forms to have been taken over by hormones and neurotransmitters, substances that do not exist in the cellular slime mold. But another aspect of slime mold physiology merits attention. This is the association between low levels of cyclic AMP and rapid replication, or between high levels and the promotion of differentiation. Might such an association exist at more advanced levels of evolution?

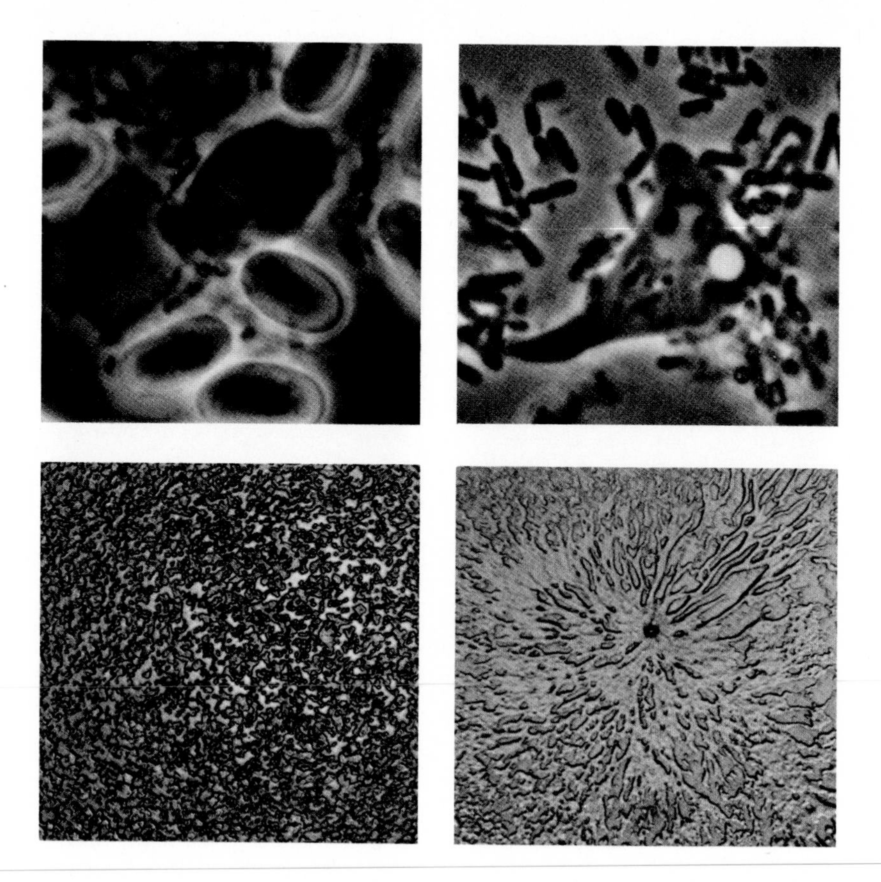

"An unusual feature that is not commonly seen in higher forms emerges at the level of the slime mold. . . . Cyclic AMP is released . . . to influence other cells." The life cycle of the cellular slime mold Dictyostelium discoideum *is revealed in micrographs here and on the next two pages. Undifferentiated cells feed, grow, and replicate (this page, top row) until their bacterial food is consumed. Then, cyclic AMP starts to accumulate near some cells, which begin to attract neighboring cells into an aggregate mass (bottom row and next page, top left). In time, the mass becomes a migrating pseudoplasmodium (next page, top right and bottom row). After migration, cyclic AMP is released again, starting a culmination phase (page 392) in which some cells differentiate into stalk cells that push down through the cell mass, lifting others upward into a fruiting body. Spores are released once the fruiting body has matured (page 392, bottom right), and the whole cycle begins again.*

Current studies with cultured mammalian cells suggest that such an association does indeed exist. Research in this area is so active that it is not possible just now to paint a truly coherent picture. What the accumulating evidence suggests, however, is that in most types of cells cyclic AMP acts to prevent replication and to promote or maintain the differentiated state. This, in turn, suggests that cyclic AMP may play a very important role during early growth and development, long before the receptors for most hormones are even present.

As so often happens when a new field of research opens up, questions seem to be coming faster than answers. An especially intriguing puzzle has been posed by a few cells that do not seem to fit this pattern at all. Far from being inhibited, the replication of certain lymphocytes (white blood cells) and hematopoietic stem cells (blood-producing cells), and apparently also of salivary gland cells, appears to be stimulated by cyclic AMP. Although puzzling, this situation also has a silver lining. Almost nothing is known about the factors determining whether a cell continues to replicate or whether it enters the pathway leading to differentiation. The knowledge that cyclic AMP influences this decision differently in different types of cells may provide an important clue to understanding the factors involved. Further research in this direction cannot help but be of value in our attempts to understand and conquer cancer (see *1973 Britannica Yearbook of Science and the Future* Feature Article: CANCER UNDER ATTACK).

(Top left, bottom right) Roman Vishniac; (Opposite page and others below) courtesy, David Drage, Department of Theoretical Biology, University of Chicago, supported by NSF and NIH

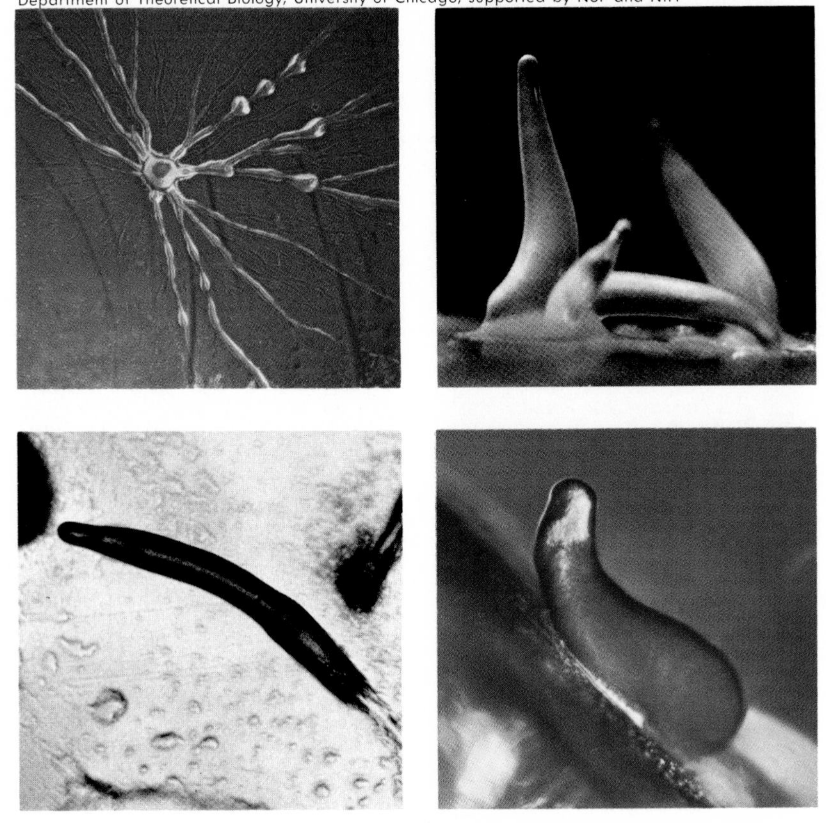

It is also possible that cyclic GMP (guanosine monophosphate) will prove to play an important role in regulating cell growth. The structure of cyclic GMP is similar to that of cyclic AMP, except that the base is guanine instead of adenine. It is the only 3′, 5′-mononucleotide other than cyclic AMP known to occur in nature, but, despite intensive investigation during the past few years, its biological role remains to be defined. Recent observations suggest that it may play an important role in the immune response. Antibody-producing cells begin to replicate rapidly in response to specific antigens, or foreign substances, and it has often been wondered why they ever stop. Research with cyclic GMP may help to answer this question.

Cyclic AMP and disease

Because cyclic AMP plays primarily a regulatory role, it might be expected, simply on theoretical grounds, to be involved in many human diseases. This is because a poorly regulated system may continue to function for long periods of time, however inadequately, whereas a system that lacks one or more essential components will not function at all. A poorly regulated human being will show up as a sick patient, but humans who cannot function at all are seen quite rarely.

Theory aside, it is now becoming clear that defects in the formation or action of cyclic AMP are responsible for a great deal of human illness. Down to a certain point, reductions in the level of cyclic AMP will

(Bottom right) Roman Vishniac; (others) courtesy, David Drage, Department of Theoretical Biology, University of Chicago, supported by NSF and NIH

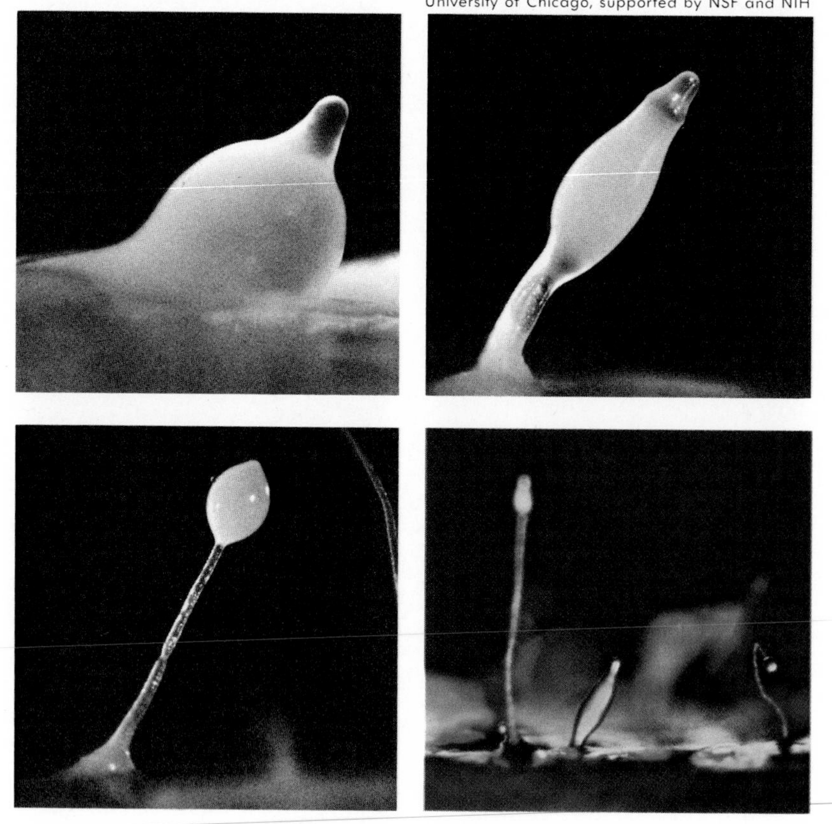

only alter cell function. For example, insufficient amounts of cyclic AMP in smooth muscle will lead to increased tone, such that structures composed of smooth muscle will be more resistant to pressure. When this happens in certain blood vessels, the result will be hypertension (high blood pressure), and when it happens in the bronchioles, the result will be asthma. Abnormally low levels of cyclic AMP in circulating blood platelets causes these cells to clump together more readily, leading to the formation of thrombi that can plug small vessels, causing strokes. A number of diseases are known in which hormone receptors are defective or absent. These conditions lead to abnormally low levels of cyclic AMP in the target cells involved, resulting in symptoms that are entirely predictable.

Reducing the level of cyclic AMP still further may in some cells lead to proliferative disorders. The skin disease psoriasis is a good example of this. Certain tumors may also develop as a result of low cyclic AMP, or at least this is now a distinct possibility. Whether all or only some of the properties of tumor cells are related to cyclic AMP remains to be established, and the possible role of cyclic AMP in cancer is currently a subject of intense research interest.

Just as abnormally low levels of cyclic AMP may be incompatible with health, abnormally high levels may also be dangerous. For example, many of the metabolic disorders seen in diabetic patients are the result of excessive cyclic AMP production in hepatic and adipose tissue. In adipose tissue this leads to excessive mobilization of free fatty acids, which are converted in the liver to ketone bodies. These latter compounds are responsible for the metabolic acidosis that was at one time the most common cause of death of diabetic patients.

More recently it has been discovered that cholera is caused by excessively high levels of cyclic AMP, in this case in the epithelial cells lining the intestines. Cholera has historically been one of mankind's greatest killers and it still remains as one of the most devastating diseases in large sections of the world. The characteristic signs and symptoms of cholera (weakness, apathy, cyanosis, inability to speak, undetectable pulse and heart sounds) are those of hypovolemic shock, resulting from the rapid loss of water and electrolytes from the gut. The causative organism is *Vibrio cholerae*, which produces an enterotoxin (a toxin, or poison, which acts upon cells of intestinal lining) that has an unusual effect on adenylyl cyclase. Unlike hormones, which affect adenylyl cyclase reversibly, such that cyclic AMP levels return to normal almost as soon as the hormone is removed, the bacterial enterotoxin exerts an insidious and essentially irreversible effect, as if it had destroyed some factor that normally acts to inhibit adenylyl cyclase. The result, in any event, is that cyclic AMP in intestinal cells reaches very high levels for long periods of time. This causes a striking increase in the rate at which water and electrolytes are pumped through the cells out into the gut.

Interestingly enough, in view of the enormous amount of human misery it has caused over the ages, cholera toxin now holds promise of becoming an extremely useful experimental tool. Although in patients it is ordinarily restricted to the gastrointestinal tract, recently it has been found to affect adenylyl cyclase in almost all cells, when applied to them experimentally. Continued research with this toxin may, therefore, teach us much about how adenylyl cyclase is ordinarily held within the cell membrane in a state of low activity, and how hormones act to temporarily increase this activity.

Could cholera toxin ever be used in the treatment of diseases caused by abnormally low levels of cyclic AMP? This is one of many questions about this miniature nucleic acid that can only be answered by further research.

See also *Encyclopædia Britannica* (1973): NUCLEIC ACIDS.

FOR ADDITIONAL READING

Bonner, J. T., "Hormones in Social Amoebae and Mammals," *Scientific American* (June 1969, pp. 78–91).

Hirschhorn, N., and Greenough III, W. B., "Cholera," *Scientific American* (August 1971, pp. 15–21).

Pastan, Ira, "Cyclic AMP," *Scientific American* (August 1972, pp. 97–105).

Robison, G. A., Butcher, R. W., and Sutherland, E. W., *Cyclic AMP* (Academic Press, 1971).

UHURU: The First Orbiting X-ray Laboratory

by Riccardo Giacconi

Bearing the Swahili word for "freedom," the *Uhuru* satellite has probed the sky for X-ray sources and in so doing has revolutionized man's understanding of these celestial phenomena.

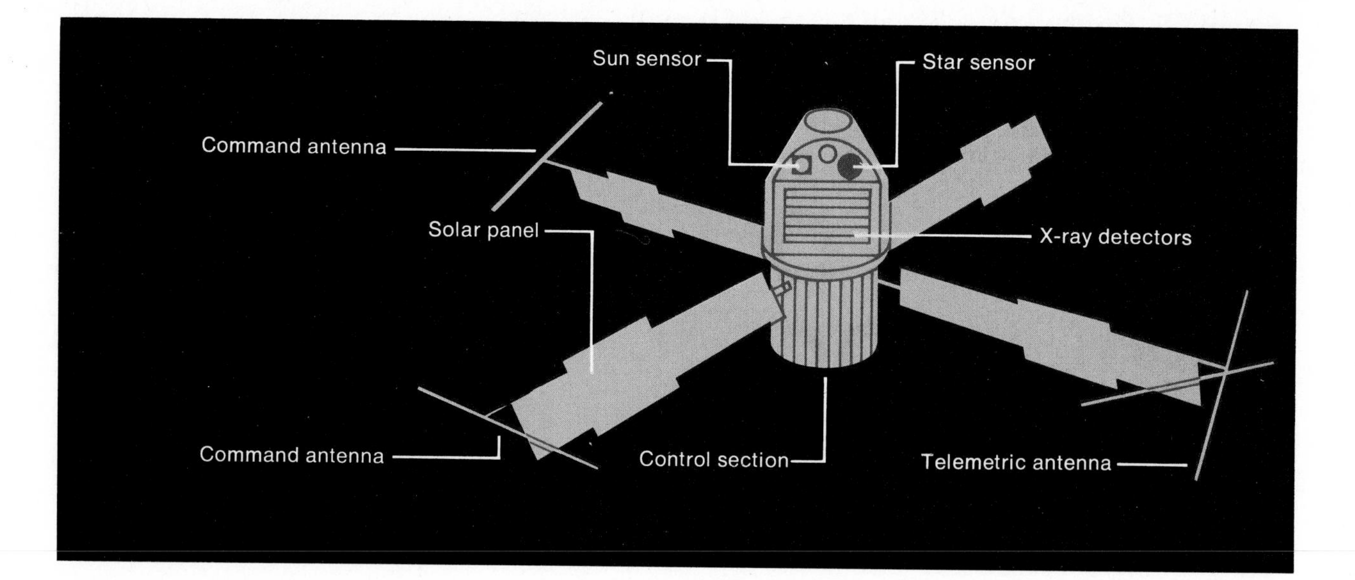

Uhuru, the first orbiting X-ray observatory, was launched off the coast of Kenya on Dec. 12, 1970. With the solar panels fully deployed it measures 13 ft from tip to tip. The 175-lb control section, about 2 ft long and 2 ft in diameter, contains all the systems necessary for attitude control, data and power storage, and communication. Uhuru *travels in a nearly circular orbit of about 335 mi in altitude and completes a full revolution every 96 minutes.*

RICCARDO GIACCONI, *a physicist and executive vice-president of American Science & Engineering, Inc., was one of the designers of the* Uhuru *satellite.*

On Dec. 12, 1972, a small group of scientists and technicians at the U.S. National Aeronautics and Space Administration (NASA) and American Science & Engineering, Inc., gathered to mark the second year of successful operation in orbit of the *Uhuru* satellite, the first orbiting X-ray observatory. This unpublicized, yet significant event was typical of the modest beginning and inexpensive execution of the Small Astronomy Satellite program of NASA. SAS-A, or *Uhuru,* as the operational satellite was called, was the first of three planned SAS launches. This small spacecraft, weighing about 300 lb, was carried into orbit by the simplest of NASA's arsenal of rockets, the Scout. Yet it was one of the most successful space experiments ever launched and has produced more exciting and unsuspected discoveries than any other astronomy space mission. The data from its X-ray detectors revolutionized man's understanding of the nature of cosmic X-ray sources and greatly expanded the scope of this new branch of observational astronomy.

X rays generated in space do not penetrate the atmosphere of the earth. Therefore, the beginning of X-ray astronomy had to await the development of space-borne instrumentation. The *Uhuru* story began in 1963, one year after the discovery of X-ray sources outside the solar system by means of a rocket-borne X-ray detector. This first discovery was accomplished during a rocket flight by a group of scientists from American Science & Engineering (Riccardo Giacconi, Herbert Gursky, and Frank Paolini) and MIT (Bruno Rossi). It was soon followed by a confirmation and improvement of these results in 1963 by a rocket experiment of a group from the Naval Research Laboratory under the direction of Herbert Friedman. It was already clear at that time that X-ray stars were among the most luminous objects in the heavens, but their distance from us was so great that even from the brightest of them, Sco X-1, only a few X-ray photons (parcels of radiant energy) would impinge on a 1-sq cm detector every second. Each individual

Courtesy, The Johns Hopkins University Applied Physics Laboratory

Uhuru is installed onto the tip of the Scout rocket that will launch it into orbit (top). The heat shield of the Scout without the nose tip is seen below. A U.S. vehicle, the Scout is 73.5 ft in length and weighs 39,600 lb at launch.

photon carries approximately 3,000 electron volts (eV) of energy (a photon of light has 2 or 3 eV of energy). The X-ray photons can be individually detected by means of a Geiger counter or proportional counter. These same counters, however, are also sensitive to cosmic ray particles, which traverse them with a frequency of about one per square centimeter per second. The signal from faint X-ray sources, therefore, becomes lost in this background "noise" unless a very large number of photons can be collected. This can be accomplished either by using detectors with an area much greater than the few tens of square centimeters used in the first rocket flights, or by collecting data

(Top) Engineers arrive at the Santa Rita control platform (right) off the coast of Kenya. The San Marco platform, controlled by Santa Rita and from which Uhuru *was launched, lies in the distance. (Bottom) The Scout rocket with* Uhuru *on its tip is readied for the launch by service workers on the adjacent gantry.*

Courtesy, American Science & Engineering, Inc.

for much longer times than the 300 seconds which a rocket payload spends above the atmosphere. There is clearly a limitation on how large a detector can be carried in a typical research rocket. (Detectors as large as one square foot were developed and flown in the late 1960s.) On the other hand, one can increase the time available for observation by a factor of 10,000 or more if the detectors can be placed in the earth's orbit and continue operating for several months.

Thus, in the fall of 1963 a group of scientists at American Science & Engineering proposed to NASA that an orbiting X-ray observatory be launched. The project was judged to be scientifically sound. The spacecraft needed was sufficiently similar to other requirements in the field of astronomy to suggest that it could fit in a Small Astronomy Satellite Program. From 1967 to 1970 the construction of *Uhuru* took place. Its designers included Riccardo Giacconi, Herbert Gursky, Harvey Tananbaum, and Edwin Kellogg.

Structure of the satellite

The satellite carries two proportional counters, which are the X-ray detectors. These detectors consist of argon gas at atmospheric pressure enclosed in a box with a thin window of beryllium on the side exposed to the X rays. An insulated anode wire crosses the gas and is maintained at about 1,000 V positive potential with respect to the container. An X-ray photon after traversing the thin window is absorbed in the gas, giving rise to a photoelectron. This photoelectron, which has almost the same energy as the initial photon, ionizes the gas in the detector. Many electrons and positive ions are thus formed. The electrons are accelerated toward the anode wire, creating more electrons along their path until they are all collected on the anode. For each X-ray photon an electrical pulse of several thousand electrons is registered. The number of electrons collected is proportional to the energy of the impinging photon. By counting how many pulses occur and the energy corresponding to each one, an observer can determine the flux of X rays impinging on the detector as well as their energy distribution.

The two detectors carried on *Uhuru* were each about 800 sq cm in area. In order to detect X-ray photons from a specific direction in space—in front of each detector—scientists placed mechanical collimators which confined the field of view of one detector to 0.5×5 degrees of arc and of the other to 5×5 degrees of arc. The satellite was designed to rotate slowly on its axis, allowing the detectors to sweep a band of the sky 5° wide each 12 minutes. The orientation of the spin axis of the satellite as well as the speed of rotation could be controlled by command from the ground. Approximately once a day the spin axis was centered in a new direction so that different portions of the sky could be explored in succession.

The data were stored on a magnetic tape and transmitted to the ground during each orbit. Early in the mission the tape recorder failed; however, information was then transmitted continuously and

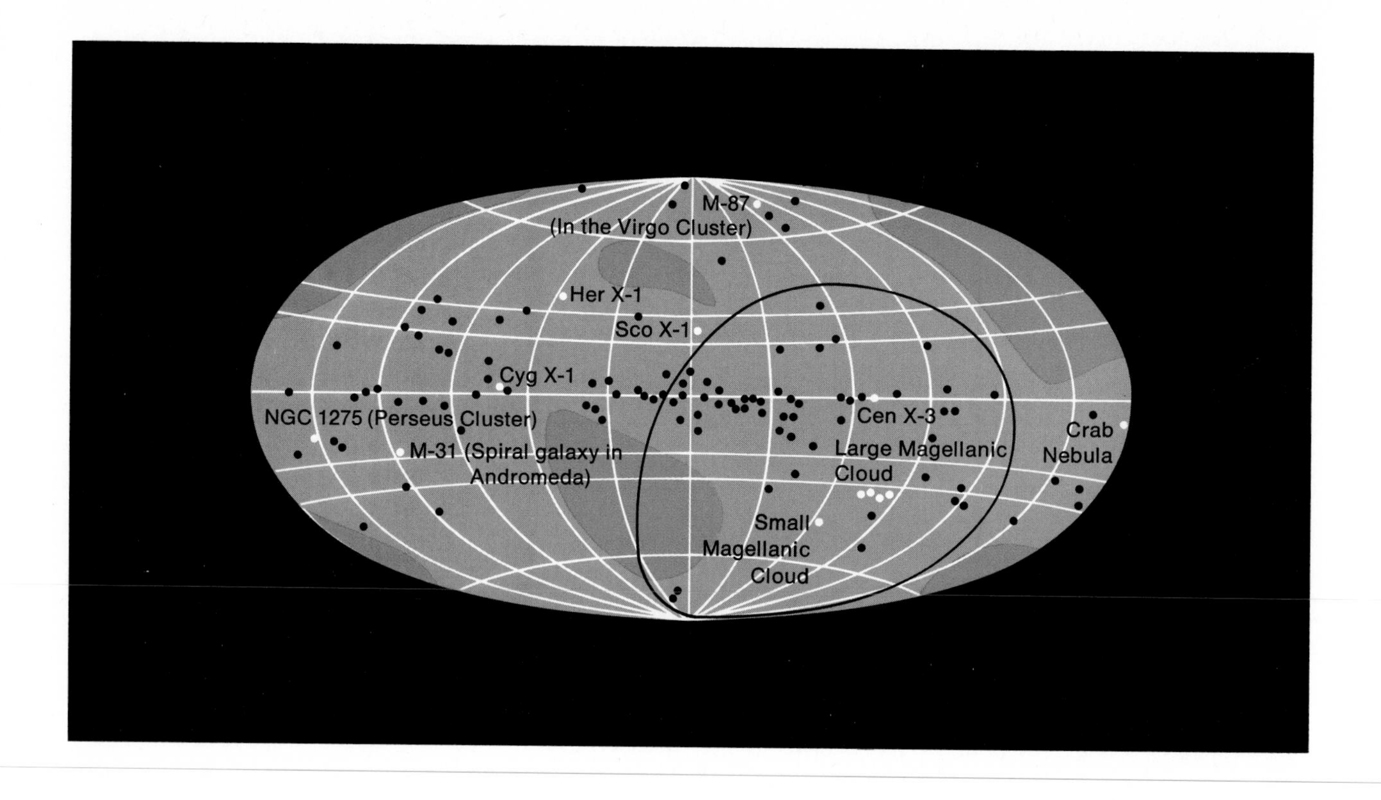

X-ray sources cluster near the horizontal axis of the celestial sphere, which represents the plane of our Milky Way galaxy. This indicates that more than half of the observed sources are in our galaxy.

received directly by several stations along the satellite path, permitting ground observers to recover 50% of all data. Star sensors, sun sensors, and magnetometers were provided so that the orientations of the detectors in the sky could accurately be determined at each instant of time. This permitted observers to locate the sources of X-ray emission with respect to visible or radio objects.

Several means of rejecting background counts caused by cosmic rays or particles trapped in the Van Allen belts of radiation were provided to diminish their effects on the experiment. In addition, it was found that one could avoid most of the trapped particle belts if one could launch the satellite in an equatorial orbit of 300 naut mi. In order to do this inexpensively it was necessary to launch the satellite from a station near the earth's equator. Cape Kennedy, Fla., at 28° latitude N, was not suited for this purpose. The Italian government on the other hand had built an equatorial rocket launch platform, the San Marco, which was positioned off the coast of Kenya and was suitable for the launch of Scout rockets. NASA, therefore, arranged to have the satellite placed in orbit from there. Since it happened that the launch occurred on the day on which the Kenyans celebrate their national independence, the operational satellite was named *Uhuru* (Swahili for "freedom") in honor of the Kenyan people and their hospitality.

The continuously successful operation of the spacecraft since its launching, although not unprecedented in space efforts, exceeded the hopes of its designers. While it was expected that with increased time available for observations, the knowledge of X-ray sources would be improved, no one expected the wealth of truly amazing discoveries that were revealed. These are described in the following sections.

Galactic X-ray sources

In the figure showing the distribution of cosmic X-ray sources detected by *Uhuru* in its first two months of operation, the galactic coordinates were chosen so that the horizontal axis represents the plane of our galaxy, the Milky Way. The projection of the celestial sphere in two dimensions is otherwise conventional. One immediately observes a great concentration of X-ray sources along the horizontal axis, which indicates that more than half of all X-ray sources observed (out of a total of 125) are in our own galaxy. Of these 125 X-ray sources, only about one-third were known prior to *Uhuru*.

The stars are not uniformly distributed along the axis, but are clustered in specific regions that correspond to the center of our galaxy in the Sagittarius-Scorpius region and to the spiral arms in Cygnus, Serpens, and Centaurus. The location of these sources in the galaxy, as well as other information concerning their intensity and position, and the energy distribution of the emitted radiation, allow us to derive some general conclusions: (1) While approximately 10% of all X-ray sources can be identified with supernova remnants, most of the remainder apparently show no correlation with previously known classes of stars or stellar systems. (2) X-ray sources are among the brightest stellar objects in the sky; their power output in X rays corresponds to about 10^{36} to 10^{38} ergs per second, or approximately 1,000 to 100,000 times the output of our own sun in all wavelengths. The fact that stellar X-ray sources could, in fact, be so powerful was definitely proven by the detection of individual sources in the Large and Small Magellanic Clouds, whose distance is precisely known. Since the clouds are very far from the earth, the mere detection of these sources implies that they are extremely powerful. (3) The total number of sources of this type, that is, with power output in excess of 10^{36} ergs per second, is estimated to be no more than about 100, of which about two-thirds have already been detected.

Although we know that our own sun emits X rays with an average power of about 10^{29} ergs per second and we therefore expect most stars to do so as well, the integrated contribution from all such stars is at most 1 or 2% of all X rays that are emitted by stars in our galaxy. Thus, most of the X-ray emission comes from sources other than common stars. Although *Uhuru*'s sensitivity has extended the range of observable X-ray fluxes, X-ray emission has been detected from no other common star except the sun.

X-ray stellar sources represent, therefore, extremely rare types of stars or extremely short-lived states in the stellar evolution of common stars. This view was strengthened by the study of individual stellar X-ray sources and, in particular, by the binary X-ray stars.

Clearly, the discovery that X-ray sources are associated with binary systems composed of a normal star and a collapsed object (either

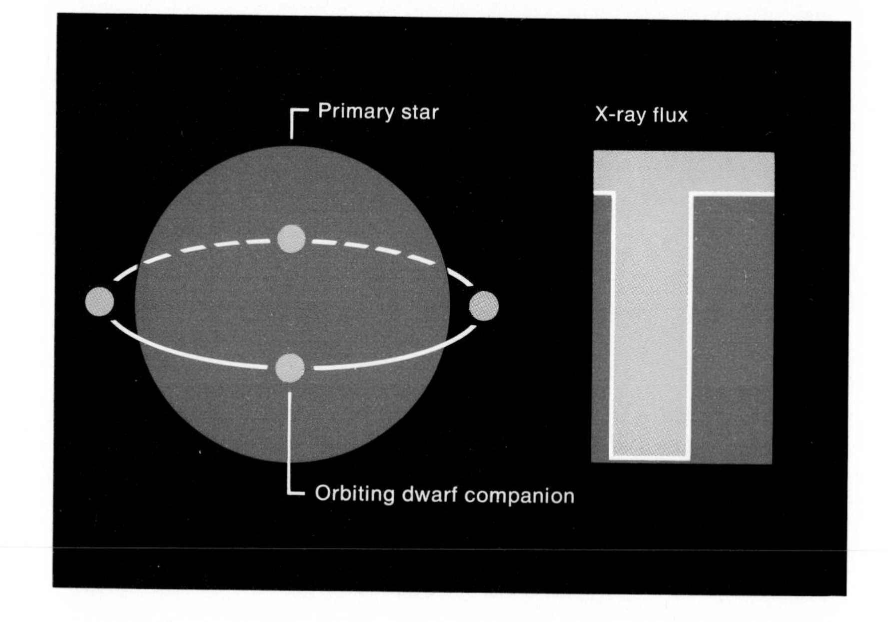

Certain X-ray sources were found by Uhuru *to be associated with binary systems. The source itself, a collapsed dwarf object, revolves around a large primary star (left). The eclipsing of the source by the primary object during each revolution accounts for the periodic variations in the intensities of the sources (right).*

a white dwarf, a neutron star, or a black hole) is one of the outstanding contributions made by *Uhuru*, and one of the great discoveries in astronomy of the last few years. Scientists have found that a number of X-ray sources exhibit periodic variations of their intensity, with periods ranging from 4.8 hours to 9 days. At regular intervals the X-ray flux decreases sharply, remains below detectable levels for a time, and then sharply returns to the previous value. Such light curves can only be interpreted as caused by eclipses.

In two of these objects, Cen X-3 and Her X-1, fast periodic pulsations occurring on the scale of seconds have also been observed. The Doppler shift of the frequency of these pulsations, caused by the orbital motions of the X-ray stars in the binary systems, gives direct compelling evidence of their binary nature. In the case of Her X-1 a visible light counterpart was discovered which also exhibits characteristic variations in visible light with the same period and phase as the X rays. In Cyg X-1 and 2U0900-40 (*Uhuru* Catalog designation), the fast flickering pulsations were not periodic, but their optical counterparts were found to be binaries from the nature of the variations in their emitted visible light spectra.

To illustrate the great import of these observations and their relevance to a host of astrophysical problems, no examples are better suited than Her X-1 and Cyg X-1. These X-ray sources seemed destined to join that select number of objects that have had great significance in the history of astronomy, such as the Crab Nebula, and CP 1919 (the first pulsar discovered).

Her X-1

Uhuru first observed Her X-1 to be a periodically pulsating X-ray source in November 1971. Astronomers were immediately alerted to the potential significance of this finding by the preceding discoveries of the

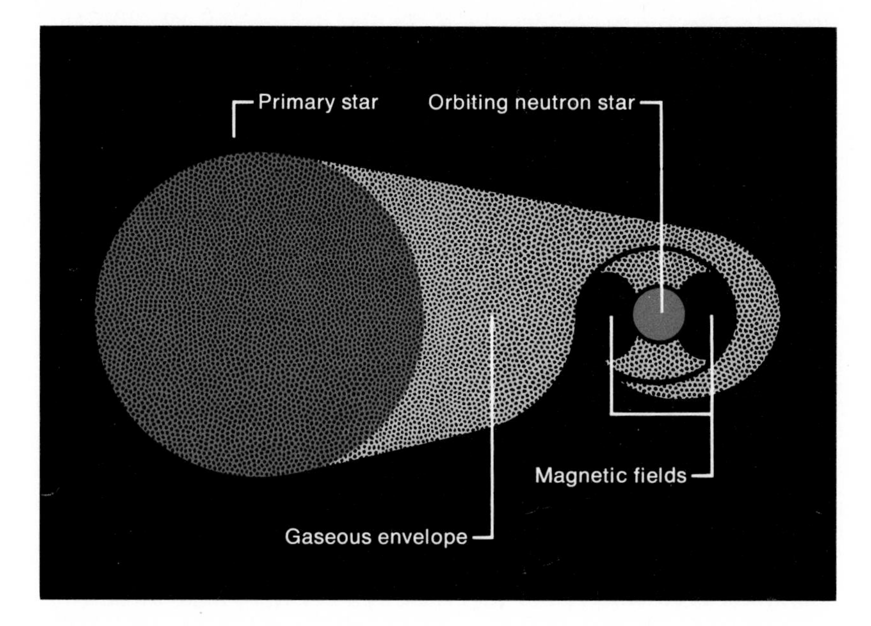

In the Her X-1 binary system gas in the envelope of the normal star escapes the gravitational attraction of that star, and falls on the neutron star. During this process the gas is heated to very high temperatures and rids itself of 10^{36} ergs per second by emitting X rays. Hot spots on the surface of the neutron star, perhaps corresponding to the poles of its very large magnetic field, may cause the periodic pulsations in the X-ray emission.

rapid flickering pulsations of Cyg X-1 and of the periodic pulsations in Cen X-3, both of which had been shown to be members of a binary system. The pulse shape of Her X-1 appeared to have a rapid rise and decay, the entire pulse lasting only about 20% of the cycle. A secondary peak (interpulse) could be observed. It was soon realized that the pulse shape could change within minutes.

Scientists carefully studied the stability of the frequency of these short pulsations and found that (just as in the case of Cen X-3) the frequency changed with a period of 1.7 days. Studying the behavior of the average X-ray intensity, they also found that the quantity of X-ray emission varied periodically every 1.7 days, with a full eclipse of the X rays lasting about one-fourth of a day. This immediately suggested that the X-ray source is part of a binary system. Accordingly, the changes in the frequency of the rapid pulsations are due to the Doppler shift that is caused by the orbital velocity of the X-ray star in its orbit. Thus, as the star comes toward the earth, the frequency of the pulsations appears to increase until it reaches a maximum and then, as the star starts moving away along its orbit, the frequency decreases until the X-ray star is hidden behind its companion.

The maximum difference in the time of arrival of the pulses from the point of the orbit closest to the earth to the point of the orbit farthest away was measured to be 26.38 ($\pm$0.06) seconds. If this number is multiplied by the velocity of light in a vacuum, one obtains directly the diameter of the orbit of Her X-1, approximately 7.5 million km. From knowledge of the radius of the orbit and the 1.7-day orbital period, scientists obtained the orbital velocity, 169.2 ($\pm$0.4) km per second. This direct measurement is finer than the best that can be achieved by rather laborious techniques in visible light. Also, the X-ray data yielded information about the nature and dimensions of the two stars composing the system. The radius of the normal star, or more precisely of its gaseous envelope, appeared to be of the order of three or four solar radii, while the radius of the star emitting X rays seemed to be much smaller than the radius of the earth.

Soon after the discovery of the binary nature of the system containing Her X-1, the optical star HZ Her was found to be also a periodic variable star showing a change of light intensity of about one magnitude with a period of 1.7 days. The X-ray period and the optical period, measured independently with a precision of about one part in a million, were shown to be identical so that there could be no doubt of the fact that HZ Her was the optical component of the system. Observations show that the visible light peaks in intensity when the X-ray source is in front. This effect is explained by X-ray heating of the hemisphere of the star facing the X-ray source, causing emission of more optical radiation. This results in a peculiar appearance of the star, which seems on one side to be at a temperature of 6,000 K and on the other at 10,000 K.

The most important feature of this binary system is that Her X-1 can only be plausibly understood if an object very similar to a pulsar (which is a magnetic rotating neutron star) exists within it. Gas in the envelope of the normal star escapes the gravitational attraction of that star and falls onto the neutron star. In this process the gas is heated to exceedingly high temperatures (greater than 10^8 degrees) and rids itself of 10^{36} ergs per second by emitting X rays. In this view the 1.24-second pulsations are caused by hot spots on the surface of the rotating neutron star. These may correspond to the poles of the star's very large magnetic field, where the material tends to be funnelled.

Marjorie Townsend, Uhuru *project manager, discusses the satellite's performance during preflight tests at the Goddard Space Flight Center, Greenbelt, Md.*

Courtesy, NASA

Finding what appears to be a neutron star in a binary system is important for two reasons. First, it apparently resolves the puzzle of never finding pulsars emitting radio waves in binary systems. The explanation would now be that when pulsars are in binaries they cease to be radio pulsars and become X-ray pulsars. The second is that having apparently been found in binaries, neutron stars for the first time can be directly measured for their mass and physical properties.

The current understanding of neutron stars is that they are stars that have exhausted their nuclear fuel and have collapsed under gravity to about a 10-km (6-mi) radius with densities exceeding 10^{13} g per cc. The further gravitational collapse of the star is stopped not by gas and radiation pressure arising from the burning of nuclear fuel but by the repulsive interactions between nucleons (protons or neutrons). It is clear that a study of the physics of neutron stars involves the study of matter in conditions outside the realm of our normal experience and beyond the range where physical laws, developed to describe matter at much lower densities and pressures, may be applied with certainty. Finding such objects in binary systems furnishes scientists with an ideal setting in which to study phenomena inaccessible to laboratory physics and to check particle-physics theories.

It should be pointed out that the sketch of the Her X-1 system given above is only a simplified model which, although believed to be correct, deliberately ignores a number of detailed experimental findings already known. For instance, the X-ray emission undergoes, in addition to the eclipsing cycle of 1.7 days, a longer cycle of 35 days in which the source appears to be switched off for 24 days and then on for 11. In visible light this effect is not seen; however, a similar on-off situation appears to occur at time intervals of years. Also additional sudden decreases of intensity were found in the X-ray emission that were interpreted as absorption by gas that is streaming in the system. Considering that most binaries in which mass is transferred demonstrate complex behavior, it is not surprising that such an unusual system as Her X-1 exhibits features that suggest even greater complexities and are not yet fully understood in detail.

Cyg X-1

Even more exciting in certain respects was the discovery by *Uhuru* of fast pulsations from Cyg X-1. The accurate X-ray position determination for this source led first to the discovery of a radio source, then to the identification of an optical counterpart, and finally to the present conclusion that Cyg X-1 is part of a binary system in which one member is the strangest of all objects, the often theorized but never previously observed black hole.

This conclusion stems from the strong evidence that the radio, optical, and X-ray sources are one and the same and that the X-ray source itself must be compact (about half the size of the moon) to allow for the fast time variations observed. From visible light observations one can determine that the system is binary with a period of 5.6 days and composed of a bright supergiant star and an unseen companion (the X-ray source). Using conventional techniques of analysis, scientists concluded that the unseen compact companion must have a mass greater than about ten solar masses, much greater than can be expected for white dwarfs or neutron stars. Such an object, therefore, would be a black hole, the object which is left behind when a star has undergone complete gravitational collapse.

The existence of black holes is predicted by Einstein's theory of general relativity. Because no light can escape from a black hole (hence the name), this object has proven among the most elusive to recognize. *Uhuru* appears to have provided the first definite indication that one may have been found. It must be pointed out that although the conclusion described above is by far the simplest interpretation of the data, it is not yet unanimously accepted. Cyg X-1, however, is not unique, and scientists hope that further study of similar objects may dispell any lingering doubt about this interpretation.

The significance of detecting and studying the properties of black holes can hardly be overemphasized. As John Wheeler and Remo Ruffini, scientists at Princeton University, have pointed out, "A black hole provides our laboratory model for the gravitational collapse of the universe itself." It is no surprise that the mere notion that such an object may indeed have been observed has created a flurry of excite-

The Perseus Cluster was discovered by Uhuru *to be a rich extragalactic X-ray source. X-ray emissions occur over almost the entire central region of the cluster, centered in the vicinity of the galaxy NGC 1275.*

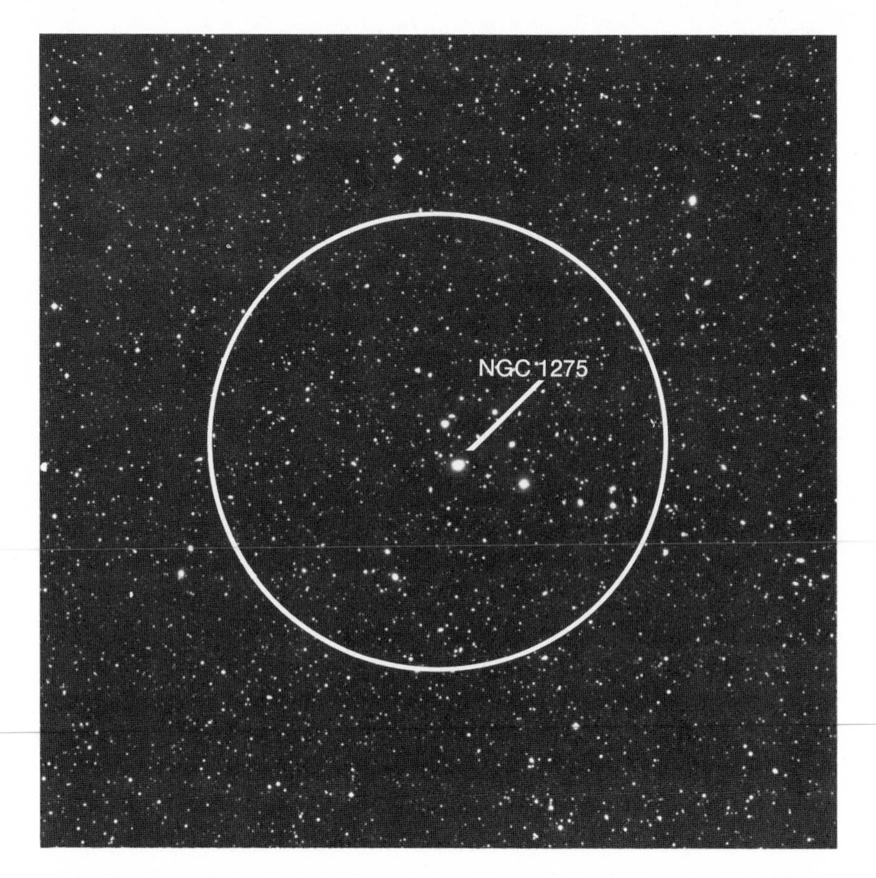

ment among astrophysicists and intensified a study of the properties of black holes. If the findings are confirmed, they will certainly be counted as among the most shining accomplishments of *Uhuru*.

Extragalactic X-ray sources

Looking beyond the confines of our galaxy with *Uhuru* proved an equally rewarding experience. From a few known X-ray sources of extragalactic origin that had been discovered by heroic efforts with rocket experiments, *Uhuru* helped increase the number to perhaps 60. Thus, scientists for the first time were in a position to attempt a classification of these sources. They found X rays to be emitted by all types of galaxies. In normal galaxies, such as our own, the Large and Small Magellanic Clouds, and the spiral galaxy in Andromeda, the X-ray emission can be explained in terms of the collected emission from individual X-ray sources. In active galaxies, such as radio galaxies, Seyferts, quasars, and N-type galaxies, the X-ray emission is much greater (a factor of 10^3 to 10^4) and appears to originate in activity taking place in small regions within the nucleus of the galaxy.

One of the most unexpected discoveries of *Uhuru* in extragalactic observations was the detection of intense and diffused regions of X-ray emission in the central region of rich clusters of galaxies. In Virgo and Perseus, extremely dense clusters of galaxies containing hundreds of objects are known to exist. In the central region of these clusters are the unusually active (in radio wave emissions) galaxies, M-87 and NGC 1275. X-ray emissions occur over a large region comprising almost the entire extent of the cluster centered on the active galaxies. In other parts of the clusters observers have found other large areas of X-ray emission but no evidence of individual unusually active galaxies.

The energy spectrum of the radiation emitted is what could be expected from a hot, very diluted gas at a temperature of 70 million degrees pervading the space between galaxies. It is not yet possible to distinguish between the different mechanisms that might be now, or have been in the past, at work to heat the gas. Nor can scientists relate directly the total mass of this gas, which appears to be about $^1/_{10}$ the total mass of the cluster, to the dynamic properties of clusters. Yet it is clear that scientists are for the first time "seeing" the intergalactic gas. Further study of these sources in the X-ray region may have profound consequences for theories concerning the origin and composition of the universe.

Future outlook

Both the galactic and extragalactic X-ray observations made from *Uhuru* provided tantalizing hints of further discoveries that might prove greatly significant. In galactic X-ray astronomy, it should be pointed out that binary X-ray sources and supernova remnants do not seem sufficient to explain the properties of all observed stars that emit X rays. For instance, the first discovered and brightest X-ray star, Sco X-1, does not easily fit into either category. On the basis of past experience, the data might be indicating that even more bizarre and interesting classes of objects are involved. In extragalactic X-ray astronomy well over half of all sources seem to belong to galaxies of a type yet unknown, in which the X-ray emission exceeds by a factor of at least ten all other forms of electromagnetic energy radiated. The study of these distant, powerful, and mysterious galaxies may in the future prove as rewarding as has the study, recently, of the weak radio sources.

Thus has *Uhuru*, a modest and technically rather unprepossessing X-ray observatory, made discoveries in which the most fundamental concepts of man's view of the universe are being tested.

Astronomy for the '70s

by Jesse L. Greenstein

Now is the time to spend three billion dollars and either find our place in the universe or (probably) face more awesome new mysteries.

The universe is even stranger than our fantasies. In the past decade it proved to be less understandable than expected because it is the seat of unexpected explosive violence. It may become more understandable because the range of modern technologies now available permits observation at all wavelengths, at all times, making astronomy a potentially experimental rather than a passive observational science. And the next decade may bring us close to detection of other intelligences on worlds circulating about distant stars.

The manned exploration of the moon has apparently ended with the successful Apollo 17. A new phase in astronomy in space must soon begin. The manned program brought back precious pictures, not only of the moon, but of the earth. For many, this view of our splendid blue globe, of our own small, isolated, delicate world, will remain the most humanly affecting result of the Apollo program. Rich as it was in insights on lunar geology, seismology, and evolution, the exploration of "space" it provides is fiercely limited, a quarter of a million miles, the distance traveled by light in 1.3 seconds. The sun is 500 light-seconds away, the nearest star 100 million light-seconds distant, the "edge" of the universe 400 million billion (4×10^{17}) light-seconds away, and ever receding.

Thus, from our light blue world, our vision soars through unimaginable spaces where our bodies can never travel. From the outer reaches of space, we must push back ignorance using the weak flow of photons, billions of years old. Poised on a delicate, closed, small world, we now, as we did thousands of years ago, try to find our place in a strange universe. And, in the process, our spirit must be enriched, our sense of human values deepened and strengthened.

I know I am but a man, a creature of the day. But when I contemplate the endless spirals of the stars, I am one with the Gods. (Ptolemaeus, *c.* A.D. 180)

Courtesy, Hale Observatories

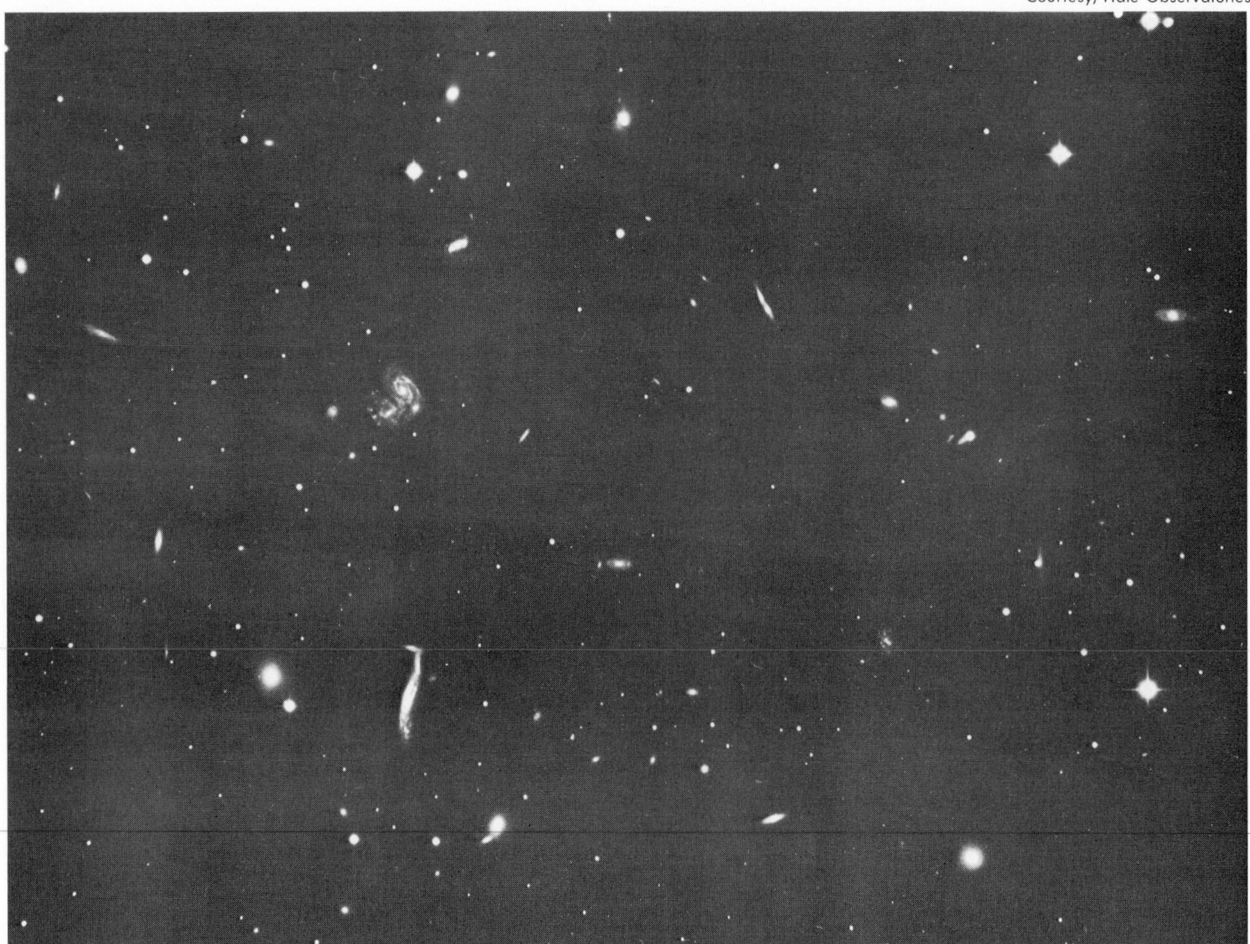

Cluster of galaxies in the constellation Hercules is revealed by the powerful 200-in. Hale telescope at Palomar Observatory. The large light sources having well-defined rays are stars in our galaxy.

Enough of poetry, now to the poetic prose of the future of astronomy. Astronomy has always been "every man's second science," the science that young people first love. It has been my own way of life, one that I value. Because of its emotional impact and, in part, communicability, it remains one of the few links between people and their scientists. Its specialized languages, mathematics, and technical jargon are only slightly less abstruse than those of the particle physicist, but, fortunately, many of its discoveries can be explained in normal language.

For the small, closely knit community of research astronomers, it is a full-time way of life, but one greatly dependent on public support through the federal government. For these reasons, I found myself in 1969 charged with the responsibility of assembling a group to review the entire scope and accomplishment of astronomy and astrophysics, to assess the importance and relevance of its subdisciplines, evaluate major projects, assess priorities, and write a report that might serve as a guide to federal expenditures over the next decade. I dragooned over 100 scientists into panels to form the Astronomy Survey Committee. They evaluated where their science stood, where it

***JESSE L. GREENSTEIN** is the Lee A. DuBridge professor of astrophysics at the California Institute of Technology and on the staff of the Hale Observatories.*

should go, and what new instruments it needed. Finally, the steering committee, of which I was chairman, evaluated the recommendations and assembled the best of them into a set of recommendations to the executive branch of the U.S. government (through the Office of Management and Budget), the National Science Foundation (NSF), and the National Aeronautics and Space Administration (NASA), which are charged with the support of basic research in astronomy. The committee recommended increased expenditures for basic research at a time of steady decrease in funding of science. Astronomy, a nearly useless basic science, spent approximately $250 million a year in the early 1970s. The committee suggested that during the remainder of the 1970s this rise to $335 million annually. How did they justify this?

What are the major discoveries, and what future lines of programs do they indicate? To quote from the group's report:

For thousands of years men have looked into the sky, long with wonder and fear but eventually with comprehension. The regular motions of the sun and stars, the wandering of the moon and the planets, provided early insights into cause and effect and the regularity of nature. With understanding came the hope of controlling nature and the beginning of science and of technology. . . . In this century, the rapid growth of science and technology has increased the depth of our insight and also our wonder; although we have found much, we still have too few explanations. . . . Navigators and explorers of the terrestrial globe found new continents inhabited by strange and different peoples. The explorers of the sky, however, have an almost unlimited sample of nature to study. They have found not merely interesting new details about individual stars or other objects but entirely new classes of objects undreamed of ten years ago. As each new technology was applied . . . new types of worlds were revealed. The previously well-organized universe, which for ancients was a planetary system centered on the earth, exploded into a bewildering universe of new types of objects, large and small, with exotic new names and marvelous new natures. Technology, theoretical insight, deeper understanding of the properties of matter, and the large computer, together with hard work, have made the last decade of astronomy one of the truly greatest periods in its history. . . .

Astronomy is a union of the science of the very small and the very large. Astronomers are interested in the properties of nuclei, atoms, molecules, solids, planets, interstellar matter, and stars. But stars are themselves units in larger aggregates—galaxies (Milky Ways) that agglomerate into clusters and extend throughout space as far as we can probe. There are more stars in our Milky Way (approximately 200 billion in number) than there are people who ever have lived on earth. . . . There are probably nearly as many galaxies in the observable universe as there are stars in our Milky Way. What strange new types of objects do they contain? Are there forces and energies at work that we do not yet know of? We are bathed from all directions by weak radio signals, apparently a remnant of the creation of the universe, degraded from an enormous burst of light at the beginning of time over ten billion years ago. What was it like then? Does time stretch backward forever, or was there a beginning? What, if anything, came before? Where do energy and matter come from? Is the total amount of energy and matter constant in time? The astronomer's daily life deals with such difficult questions. How many other planetary worlds are there, are they inhabitable, and are any inhabited? How long will the sun shine and the earth survive?

The exploration of the actual universe . . . is one of the nobler adventures of the human mind. Like a child at play, the astronomer busies himself with strange

Courtesy, Cerro Tololo Inter-American Observatory

Observatory at Cerro Tololo in the Andes Mountains of Chile nears completion. Its main instrument is to be a 158-in. optical telescope, a twin to one at Kitt Peak, Arizona. Because of its location in the Southern Hemisphere, the Cerro Tololo telescope will be able to provide astronomers with views of the Magellanic Clouds that can not be seen in the north.

toys—white dwarfs and red giants, pulsars and quasars—and in his theories moves the building blocks of atoms about to model something like the world he sees, with an imagination and a courage like a child's but with the resources of modern science and technology. It is fortunate that the pleasure he finds in . . . his discoveries are joys of science that are still communicable to the public. The optimism accompanying the exploration of our own West ended with the disappearance of the frontier; but exploring the external frontier of the heavens is endless, and its reward is knowledge, a more humbling wealth.

Fundamentally, the present confidence of astronomers stems from the revolutionary excitement and successes of the last decade of research. Science and technology combined to create major research disciplines for which ten years ago no scientific journal had an index listing. Quasars, pulsars, magnetized white dwarfs, complex interstellar molecules, and cool infrared stars were discovered or measured. From above the veil of our atmosphere the solar wind, the solar far ultraviolet and X rays, stellar or nebular X-ray sources, gamma-ray sources, and very cold infrared objects were all revealed. It was a decade of many Galileos. So many new unexpected types of astronomical objects were found that astronomy seemed likely to be at the beginning of an era of even greater discovery, and, possibly, of understanding. Many of the past discoveries had not been predicted; perhaps now that astronomers have a wider range of fact and theory, and have more open minds as to what is possible, they will be able to construct models that predict the nature of as yet undiscovered objects.

Policy planning

Attempts to plan broadly the scientific effort in astronomy are not new in the United States; in 1964 the National Academy of Sciences published the recommendations of a panel, headed by Albert E. Whitford, on the new facilities needed for ground-based optical astronomy. The Whitford study was a far-sighted one, recommending a large expansion of radio astronomy and a doubling of the collecting area of large optical telescopes. Its first recommendation, for new radio telescopes, went largely unheeded, and the rapid progress in radio astronomy occurred in spite of the lack of large new instruments in the U.S. The Australians and the British (who discovered the pulsars) mapped and counted the radio galaxies. Interstellar molecules were discovered in the U.S. because of its advanced high-frequency receivers. In addition, the U.S. led in very-long-baseline interferometers. The pressing need for very large optical telescopes, which do the most difficult work, was not met, even with the two 158-in. telescopes that were nearing completion at the Kitt Peak National Observatory near Tucson, Ariz., and at the Cerro Tololo Inter-American Observatory in Chile.

But planning can be more systematic, and governments responsive elsewhere. Under the Academy of Sciences of the U.S.S.R., Soviet astronomy became highly organized. The Soviets hoped to complete in 1973 the largest optical telescope in the world, a 236-in. reflector of novel altitude-azimuth, computer-driven design. In the U.K. a group of leading scientists, the Science Research Council, established needs and priorities for new large instruments on a firm budgetary basis; astronomy was selected as the physical science in which Great Britain could make large gains at relatively modest cost. Six European nations established a cooperative observatory in Chile, where a 142-in. reflector was under construction. The British and Australians were cooperating in building a 150-in. reflector in Australia.

Overplanning is naturally dangerous in any rapidly changing science. But the long lead times for telescope and spacecraft construction, and the high costs, require that scientists agree on priorities within their science and press forward with major instruments. It is possible to design large instruments that will be nearly all-purpose and carry a variety of auxiliary equipment as scientific needs change. Large size is necessary, fundamentally, because the rate of flow of information from the sky is so slow. There is no substitute for collecting area and high resolution; even the largest telescope in the U.S., the 200-in. Hale reflector in Palomar Observatory, receives only a few photons a second from a faint quasar. Thus, unlike many instruments used in physics, the large telescope, the workhorse of astronomy, cannot be

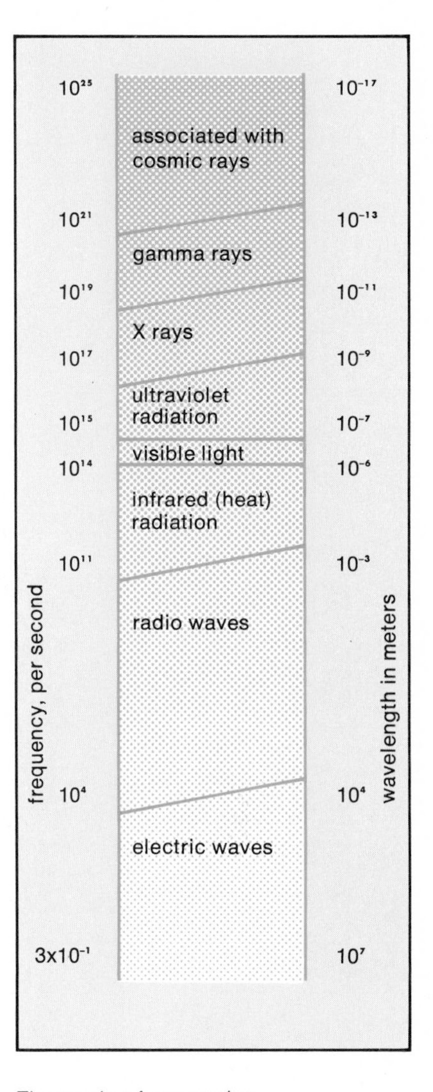

The varying frequencies and wavelengths of radiation in the electromagnetic spectrum are shown in the diagram above. As the slanting lines indicate, the boundaries between types of radiation are not hard and fast.

Courtesy, The National Radio Astronomy Observatory

Interferometer at the U.S. National Radio Astronomy Observatory at Green Bank, W.Va. (right) consists of three antennas, each 85 ft in diameter.

radio interferometer

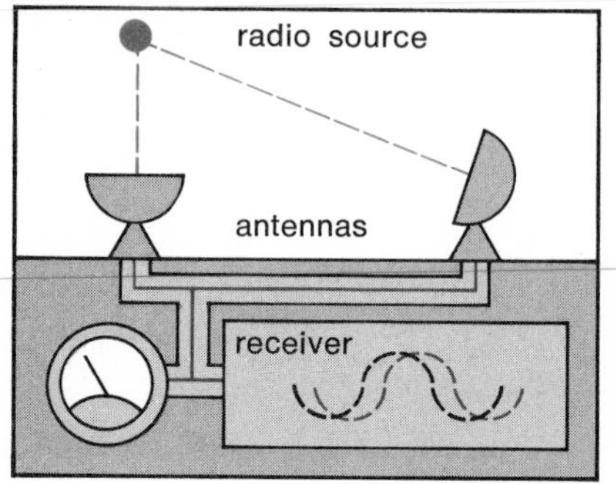

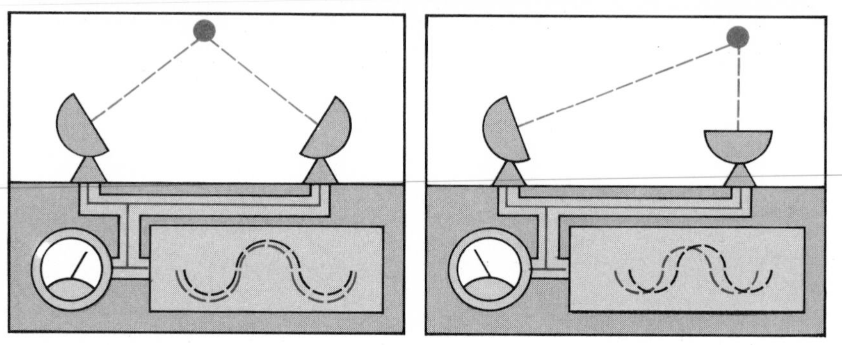

A two-element interferometer determines the location and angular diameter of a radio source by measuring the amount of interference between the voltages transmitted to the receiver by the antennas as the source passes through their beams. Because the sources are so far distant the angle shown in the diagram is much larger than actually occurs.

replaced, even by brains. In radio astronomy, angular resolution is the major need, to be attained either by very large size (a 400-ft diameter is a desirable goal) or by interferometry, the combination of signals received from two large antennas separated by miles or even by the diameter of the earth (the very-long-baseline interferometer).

In the U.S. astronomy was a science that the pioneers of education found exciting, and so the old private universities, along with land-grant and other state universities, established observatories. Private benefactors such as Charles Yerkes (University of Chicago), James Lick (University of California), the Carnegie Institution of Washington, D.C. (Mount Wilson), and the Rockefeller Foundation (Palomar) established the largest observatories in the country. The advent of radio astronomy and of research in space, however, came at times when private financial support of physical science was declining. These new fields also required capital investments beyond the reach of private benefactors. At the end of World War II, the benefits of scientific research shone far more clearly than in 1973, and the federal government established a pattern of support of basic and applied research through many agencies. The National Science Foundation was created to support basic research and took an active role in building U.S. astronomy. Reinforced by the shock of the Soviet satellite Sputnik in 1957, the National Aeronautics and Space Agency grew rapidly. While its major expenditures were for manned space flight, it was responsible for astronomy and astrophysics in space and gave valuable support to related ground-based research. The very large computers needed for orbital dynamics, flight control, and data recording not only helped space research but made possible large strides in theoretical astrophysics.

Courtesy, Commonwealth Scientific and Industrial Research Organization, Australia

Radio telescope at Parkes in Australia has a diameter of 210 ft. Used to detect radio waves ranging from 6 to 100 cm in length, it can be pointed to a source with an accuracy of 15 seconds of arc.

Size and finances of astronomy

In round numbers, there are about 3,000 members in the American Astronomical Society. Data for 1968 showed that, of a sample of 1,325, half were less than 32 years old, 60% were in universities, and 20% worked for the federal government. About 130 new astronomy Ph.D.s entered the field each year. But so many physicists transferred to astronomy that by 1972 half of the working astronomers in the U.S. had physics rather than astronomy Ph.D.s.

Academic institutions receive about $30 million a year from private and state funds, part for teaching and part for research in a wide range of astronomy. With an added federal budget of $40 million, the schools have been healthy and prosperous. Without it, most would collapse. Only a few privately endowed observatory and teaching departments could survive as research institutions if a rapid cutback of federal support of research should occur.

Another problem for astronomy is that a reduction of funding by the defense agencies, much of it for basic research, was not compensated for by an increase in money from the NSF. This resulted in about a 20% drop in astronomy dollar funding between fiscal 1968 and 1972. Combined with inflation, the real science "purchasing power" dropped by about 35% over those four years. The loss of potential research, the pressure on young scientists who face long delays in funding, the slowdown in needed facilities, and underemployment and the beginnings of unemployment of able young researchers combine to make the situation a sad one. If, indeed, astronomy stands at the threshold of what should be its greatest age, the situation becomes even more unfortunate.

A final issue concerns spending on space astronomy by NASA. Without allowing for manned space flight costs, NASA spends on astronomical satellites about $90 million per year; including other programs and especially the necessary "overhead" of launch sites and engineering and headquarter groups, its budget reaches $180 million to $235 million, five times the grants to universities from the federal government for ground-based research and eight times the NSF expenditures. In addition, manned experiments cost five to ten times as much as unmanned ones. The information obtained from the NASA flights has been rich and unexpected, and it would be a blow to astronomy if this source were cut off or curtailed. The cessation of astronomical

Evolution of a star varies according to its size. After forming from interstellar clouds of gas, all stars spend most of their existence in the "main sequence," during which they convert nuclei of hydrogen into those of helium in an energy-producing process called fusion. When the hydrogen in a star's core is depleted, the star begins to produce energy in its narrow outer shell. This causes it to expand and cool until it becomes a red giant. As all the hydrogen is used up, rapid gravitational contraction takes place. During this contraction stars must give off mass to maintain stability. Mass may be "blown off" through the release of gas, for smaller stars, or by means of explosions that become progressively more violent with more massive stars. The final state of any star is determined by its mass after shedding this excess matter. If the final mass is about that of our sun or less, the star will contract to become a small, dense white dwarf, about the size of our earth (top row). A larger star (center) may undergo a violent explosion called a supernova and end its days as a very small and dense pulsar. For an even larger star (bottom) the gravitational collapse might be so complete that it becomes a tiny, unimaginably dense object from which no signal of any kind can escape, called a "black hole."

science in space would have only minor effects on NASA expenditures, but it would be a disaster for the broad sweep of progress in the newer fields, such as X-ray and far infrared observation.

Unfortunately, the severe restrictions on budgets have resulted in the cancellation of the program of high-energy astronomical observatories, which was rated as the most important space program by the survey. Planetary science will continue, but astronomy and astrophysics have only one launch now actually scheduled before 1979.

Similarly, although the highest priority of all was given to a balanced program of radio astronomy, with emphasis on the very large array (VLA) and the expansion of radio astronomy observatories at the universities, in fact, in 1973, support of several university radio observatories was scheduled to be terminated by the NSF. It could only be hoped that these budgetary restrictions would not, in fact, cripple or badly unbalance the program needed for the astronomy of the 1970s.

The new discoveries

The findings of the past decade have been dramatic and often unexpected. One of the most significant is the nearly omnipresent "high-energy" universe; hot as the stars are, typical photons or particles there have energies corresponding to a few electron volts (5,000–50,000° K) at their surfaces and to a few thousand electron volts (ten million to at most a billion degrees) in their interiors. In interstellar space, typical temperatures are only a few degrees above absolute zero. However, charged particles, cosmic-ray protons, alpha particles, and heavy particles all have been found in interstellar space with energies from one million to over a billion billion electron volts (eV). Nature has an unknown, efficient, and almost ubiquitous way of creating high-energy particles and collecting magnetic fields over large volumes of space. It seems probable that cosmic rays enveloping the earth and solar system come from pulsars.

Another major development was the increasing application of the theory of general relativity. The importance of gravitation as an energy source was reestablished, both during the end stages of stellar evolution and in the collapse of supermassive stars. The core of a massive star may implosively collapse to densities of 10^{15} g per cc, where the gravitational redshift is 20%. (Gravitational redshift measures the energy lost to photons emitted by atoms located in an intense gravitational field.) Such a dense star is almost certainly the origin of the pulsar phenomenon. If its parent star was rotating, the condensed object can easily be left spinning hundreds of times per second. It has enormous rotational kinetic energy, derived from the gravitational potential energy during collapse. The pulsar's radiated energy arises from the gradual slowing of this spin, and is 100,000 times that of the sun. At such densities ordinary nuclei change into neutrons, the free electron-proton state being less stable than the combined, neutron state. At slightly higher densities the neutrons become stable. Thus, for the first time, gravitation and particle physics intertwine.

stellar evolution

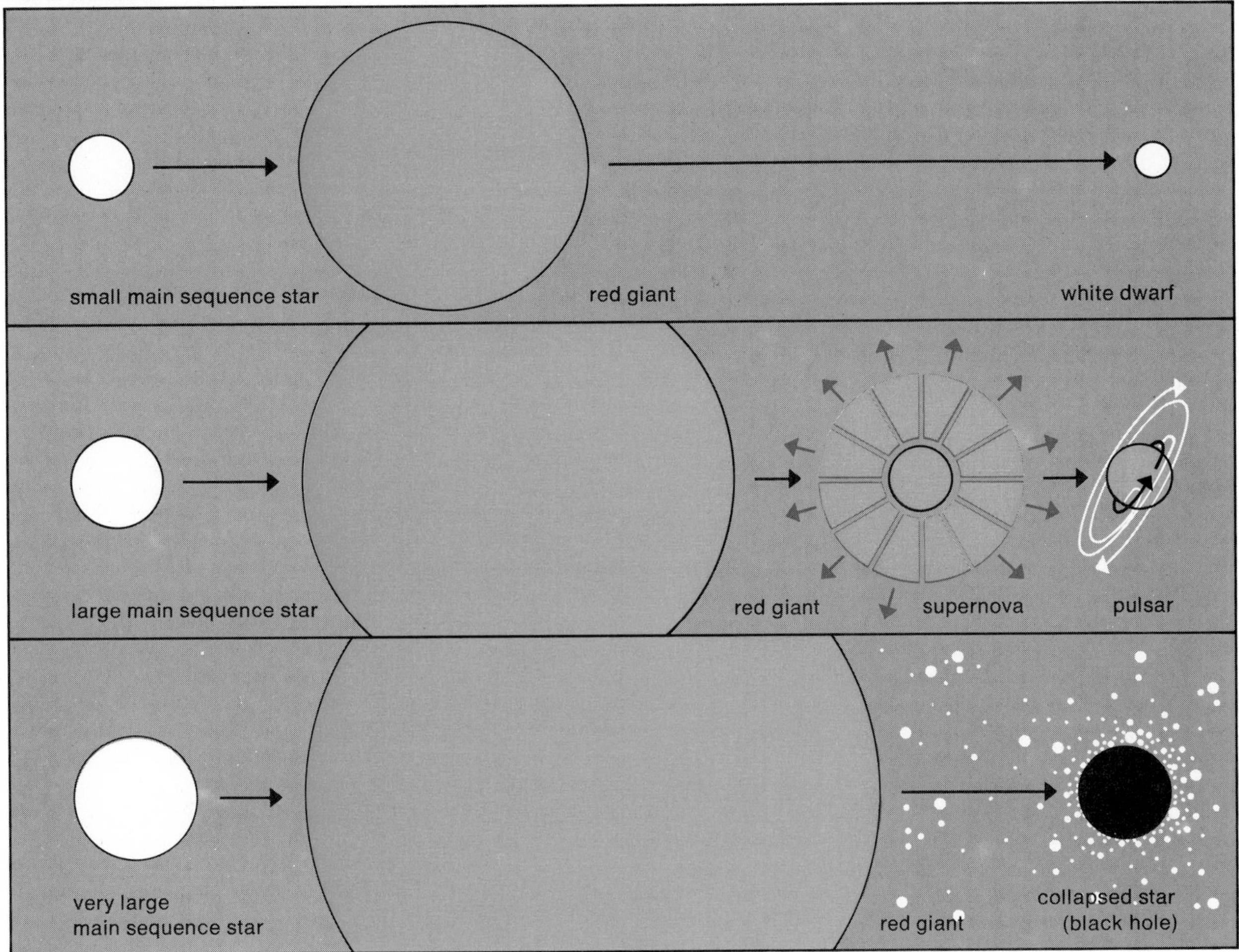

More speculatively, but quite probably, massive objects of a million to a billion solar masses may form in the nuclei of galaxies. Rotation complicates the evolution of such superstars, which is otherwise a catastrophic implosion into a black hole. The redshift of photons becomes infinite, so that photons are trapped in the closed geometry of what may be regarded as nearly a separate universe, cut off from ours except through its static gravitational field. During the process of collapse, especially if a rotating, nonuniform disk is formed, a significant portion of the potential energy of collapse may be emitted. Thus, gravitational collapse may be the ultimate source of the energy of the giant explosions found in the nuclei of galaxies.

The most direct evidence for the giant explosion in which our own expanding universe began is the "3° K blackbody" radiation. Predicted soon after the theory of general relativity explained the velocity-distance relation for galaxies, it arises from the early history of the explosion. When the universe was dense, all radiation was trapped by matter. About ten million years later it became sufficiently transparent for the photons, corresponding to ordinary light being absorbed and emitted by atoms, to leak out. These photons were red-shifted to

Courtesy, Hale Observatories

The galaxy NGC 5128 is viewed through the 200-in. Hale telescope at Palomar Observatory. Researchers found that the galaxy is also a source of radiation in the radio wavelengths.

long wavelengths by the universal expansion, and now appear in the radio-frequency range of the spectrum. They come from all directions, and are remarkably uniform in strength. They form a noise background measurable over nearly all frequencies, affecting the equilibrium of interstellar molecules everywhere and exciting radio-frequency transitions.

Radio astronomy is closely linked with another newer observing technique, the use of the near and far infrared regions of the electromagnetic spectrum. Infrared photons up to wavelengths of 20 microns (one micron = 0.001 mm) may be observed with favorably located (dry, high-altitude) telescopes, and more efficiently from very high mountains and from infrared-instrumented airplanes, balloons, and rockets. Almost all exploding galaxies and quasars had enormous infrared fluxes, peaking near 100 microns (rocket, balloon, airplane data) of 1 mm (radio and ground-based infrared techniques). Infrared surveys at the two-micron wavelength showed that thousands of objects were very bright in the near infrared. Some proved to be stars being born, wrapped in a blanket of solid dust grains. Others were dying red giant stars, in whose outer envelopes molecules formed and ultimately condensed into solids. Even the violent explosions of galaxy cores somehow produce large amounts of solid matter, which absorbs most of the light of the explosion and converts it into heat rays in the infrared.

A final topic, of quite human relevance, was the discovery in the radiowave spectrum of complex molecules. More than 30 molecular species were detected as well as isotopic forms with carbon (^{13}C) and oxygen (^{18}O). These discoveries have led to a revolution in our thinking about the probability that other planets, near other stars, have intelligent life forms.

The formation of the earth and the solar system was once believed to be an extraordinary event. Now most astronomers think that solar systems are commonplace, arising in a natural way during the formation of the stars themselves. How life begins on the planets is still obscure and is a subject best left to the biologist. How life becomes intelligent is perhaps the ultimate question. At present, astronomers believe that some molecules necessary for the development of life may have been present in the original dust cloud from which the solar system formed. The near omnipresence of dust (found by infrared) in young stars and the unexpectedly high concentration of complex molecules together suggest that if planets are formed they will contain an array of carbon-based molecules suitable as building blocks for prebiotic heavy molecules, like the amino acids.

The detection of other intelligent life forms by the radio-frequency signals they radiate, if they communicate with each other by radio, then becomes a tantalizing quest. Man could easily detect his own radiations (high-power radar, even high-frequency broadcasts) if they were used on a planet many light-years distant. The range for strong radio-emitters could be 1,000 light-years, within which there are a million stars. What would be the impact on human life, and self-image, of such a contact? The average star is more than 5,000,000,000 years older than the sun; if it has a still inhabitable planet near it, what type of creatures would the inhabitants be? Would we recognize them as intelligent? Would they so recognize us? If the large radio facilities recommended by astronomers are built, they may well permit man to listen to other worlds (but not to talk).

And it is not only in the new fields of space, infrared, and radio astronomy that technology has permitted enormous advances. The electronic age has changed the observing techniques and goals of those using conventional large telescopes. At the California Institute of Technology's 200-in. Hale reflector at Palomar Mountain, there is now an infrared photometer; a spectrophotometer that observes the spectrum of a star at 32 wavelengths simultaneously and then subtracts the sky background, recording data on computer tape; a television guider that permits guiding on stars a hundred times fainter than can be seen in the telescopes; image intensifier tubes on spectrographs of high and low dispersion; and computer control. The total cost of all these devices is about $1 million. The gain, for spectroscopy, is equivalent

Courtesy, Kitt Peak National Observatory

The 158-in. optical reflecting telescope at Kitt Peak National Observatory was unveiled in March. Second largest of its type in the U.S., the telescope will be used to see objects at the fringes of the universe, believed to be several billion light-years distant.

to that expected from a telescope of at least 600-in. aperture, which, if it could be built, would cost $600 million.

The recommended programs

If astronomy is to continue to progress, new facilities must be constructed, and new directions in research must be pursued. Federal expenditures in ground-based astronomy have averaged about $50 million and in space astronomy about $200 million annually. The entire program proposed by the Astronomy Survey Committee involves an additional $84.4 million a year over the next decade—about $51 million a year for new space projects and about $33.4 million a year for new ground-based programs. The committee defined four programs of highest priority for federal support. In order of importance, they are:

1. A very large radio array, designed to attain resolution equivalent to that of a single radio telescope 26 mi in diameter. This should be accompanied by increased support for existing smaller radio programs and advanced, new, small facilities at universities and research laboratories. Funding for the array began, but some university radio observations were scheduled to close.

2. An optical program that, by use of modern electronic auxiliaries, will vastly increase the efficiency of existing telescopes and also will create the new large telescopes necessary for research at the limits of the known universe.

3. A significant increase in support and development of the new field of infrared astronomy, including construction of a large ground-based infrared telescope, high-altitude balloon surveys, and design studies for a very large stratospheric telescope.

4. A program for X-ray and gamma-ray astronomy from a series of orbiting high-energy astronomical observatories (HEAOs), supported by construction of ground-based optical and infrared telescopes. (This is the HEAO program canceled by NASA; we can only hope it will be reinstated.)

The committee also identified several items of high scientific importance, which, although urgently needed, should not delay the funding of the other programs. These included: a very large millimeter wavelength antenna for the study of new complex molecules in space and of quasars in their early, explosive stages; upgrading existing solar observatories on the ground; continued support of the orbiting solar observatories; an expanded program of optical space astronomy; a large orbiting space telescope; a large steerable telescope for observing wavelengths of one centimeter and above with more resolution; and increased support for astrometry to provide additional needed information on the positions and apparent motions of stars.

What is the probable future? It is beyond the control of astronomers and probably beyond the control of the normal federal agencies. The fight against inflation, the Vietnam war, the problems of welfare, the cities, crime, the imbalance of payments have together created tight ceilings on federal expenditures, and, in fact, have caused almost a complete cessation of new ventures. Thus, while astronomy is now successfully tuned to all wavelengths, it is possibly out of tune with the one which limits funds—Congress and the executive branch. The outlook is hopeful only in that the brains, techniques, and plans exist to take advantage of whatever new can be started. The opportunities are unparalleled and exciting, and it would be sad indeed if they could not be fulfilled.

Engineering for Earthquakes

by George W. Housner and Paul C. Jennings

Designing buildings to be earthquake-resistant has become increasingly important as more and larger structures fill city skylines in earthquake-prone regions.

Destructive earthquakes are among the most frightening of natural disasters and have always roused deep fear in man. The psychologist William James was in San Francisco during the 1906 earthquake, and later wrote a graphic description of his feelings:

> I personified the earthquake as a permanent individual entity. . . . It stole in behind my back, and once inside the room had me all to itself, and could manifest itself convincingly. . . . It expressed intention, it was vicious, it was bent on destruction, it wanted to show its power. . . . I realize now, better than ever, how inevitable were men's earlier mythological versions of such catastrophes, and how artificial and against the grain of our spontaneous perceiving are the later habits into which science educates us. It was simply impossible for untutored man to take earthquakes into their minds as anything but supernatural warnings or retribution.

The most frightening aspects of an earthquake are its unseen and unexpected arrival and its sudden action that causes widespread death and destruction in a fraction of a minute. For at least the past 100 million years, earthquakes have been occurring, and for the past 10,000 years they have been destroying the buildings of man.

Each year more than 100 earthquakes of destructive intensity occur somewhere in the world. Usually, one or two of these are great disruptions that strongly shake large areas; for example, the Alaska earthquake of March 27, 1964, shook a region 600 mi long by 200 mi wide with destructive force. Most of the potentially destructive earthquakes that occur each year, however, are smaller shocks that affect relatively small areas with strong ground shaking. When these shocks occur in rural or remote areas, as they usually do, they cause little damage and receive little public attention. It is only when earthquakes occur in built-up areas, as they did in San Fernando, Calif., on Feb. 9, 1971, and in Managua, Nicaragua, on Dec. 23, 1972, that they do great damage and receive widespread publicity. These two earthquakes only affected areas about 20 mi square with intense ground shaking, but their locations were such that they produced great damage and loss of life.

The majority of earthquakes occur in two well-defined regions: the circum-Pacific belt, which has approximately 80% of the world's earthquakes, and the Himalaya-Mediterranean belt. Destructive earthquakes also occur in other parts of the world but much less frequently. Both Japan and the western United States, particularly California and Alaska, lie in the circum-Pacific belt and experience strong earthquakes relatively frequently. For a discussion of how earthquakes are generated, see *1972 Britannica Yearbook of Science and the Future* Feature Article: THE TREMBLING EARTH.

The effects of an earthquake depend both on the intensity of the shaking and on the resistance of the structures under stress. Most of the earthquakes that cause great destruction and loss of life occur in less developed countries in Central and South America and in the Middle East, where buildings are weaker. For example, the Nicaragua earthquake was about the same size as the San Fernando earthquake that struck the northern part of Los Angeles, but there were 100 times as many deaths in Managua as there were in Los Angeles (approximately 6,000 *v.* 62). This difference can be attributed to differences in engineering and construction in the two countries.

The overall safety of a city against earthquakes depends primarily on the minimum requirements specified in the building code. If the building code does not have adequate requirements for earthquake design, or if its requirements are not strictly enforced, a major earthquake hazard will exist. Therefore, the first condition for earthquake safety is to ensure that the building code has adequate requirements and that they are enforced. To do this requires a supply of highly trained engineers who are knowledgeable about earthquakes and their effects. It is also necessary to have an informed and enlightened

GEORGE W. HOUSNER *and* ***PAUL C. JENNINGS*** *are both professors in the division of engineering and applied science at California Institute of Technology.*

Courtesy, Albert C. Martin and Associates

The 52-story twin towers of Atlantic Richfield Plaza in Los Angeles were designed to withstand severe earthquakes. During construction (left) the horizontal steel beams of the towers were completely welded to the vertical steel columns to form joints that resist the lateral forces and distortion generated by an earthquake. Such forces are transferred around corners and resisted by the full ends of the frame. The completed buildings (right) have granite facades over their steel frames. Space was provided between each piece of granite, and windows were placed with gaps to the enclosing frames, so that the pieces will not jam against one another during an earthquake. A pliable material is caulked between the spaces to permit movement and yet keep the buildings waterproof.

local government, for the public must rely on it to foresee hazards and to take the necessary steps to protect against them.

Hazards from earthquakes

Strong earthquake ground shaking is a chaotic motion with an intensity that is greatest near the fault and that diminishes with distance away from it. (Faults are rupture surfaces in the rock of the earth's crust.) During the San Fernando earthquake, the ground shaking was of destructive intensity out to about 10 to 15 mi from the fault, and in the case of the 1964 Alaska earthquake it was strong enough to be destructive as far from the fault as 80 mi. In the region of very strong ground shaking during a large earthquake (magnitude 8 or greater), the ground may vibrate back and forth with maximum accelerations as great as 30 to 50% of the acceleration of gravity, or even higher in special circumstances. The strong shaking might last from 30 to 60 seconds, with perceptible motion lasting for 2 or 3 minutes. A person standing on the ground in the region of strong shaking will feel the ground moving back and forth in all directions simultaneously; that is, it will be shaking north-south, east-west, and up-down with approximately the same intensity. At larger distances from the source, the earthquake ground motion feels much smoother.

The response of a building shaken by the earthquake also modifies the perceived motions, and a person on the upper floor of a 50-story building near the center of an earthquake would feel a motion that is quite different from that of the ground. The earthquake would cause the building to sway back and forth with a motion much smoother than the rough jarring motion felt at ground level. The swaying might take five seconds to complete a cycle of motion, and during very strong ground shaking the top of the building might move back and forth as much as four feet.

The fault displacement can itself be hazardous during an earthquake. For example, during the great 1906 San Francisco earthquake fault rupture occurred over a distance of about 250 mi, and there was apparent on the surface of the ground a horizontal displacement across the two sides of the fault exceeding 15 ft. Any structure built across the fault would, of course, be pulled apart by such a relative displacement of its foundations. When the rock is covered by a deep layer of soil, the fault displacement in the rock can permanently deform the overlying ground without obvious rupture, and this deformation may break buried water pipes, sewer pipes, and gas mains, and may damage buildings by distorting their foundations.

When an earthquake occurs in hilly or mountainous regions, or other areas where landslides occur naturally, the shaking of the ground often will initiate landslides that do great damage. For example, during an earthquake in 1970 a landslide originating on a nearby mountain buried the towns of Ranrahirca and Yungay, Peru, killing 20,000 people. In earthquake country it is especially important to build where there is no hazard from landslides.

Cracked walls and dislodged floor tiles are typical of the damage inflicted by earthquakes on structures not designed to resist them effectively.

(Top) Noblitz from Pictorial Parade; (bottom) Fred Ward from Black Star

Sometimes great damage has been done during earthquakes as a result of a sudden weakening of the soil so that buildings and bridges settle unevenly into the ground with consequent damage. During the strong shaking, sandy, water-filled ground turns into a sort of quicksand with corresponding loss of strength. This behavior of the soil is called "liquefaction." The shaking of the ground loosens the contact between the sand grains, and because the soil is full of water it takes time for the sand particles to reestablish stable contacts in a more closely packed state. During this time the soil is very weak and cannot support the weight of buildings. In June 1964 an earthquake shook the Japanese port city of Niigata, which was underlain by saturated, sandy soil that liquefied, and about 500 buildings settled from 6 in. to 5 ft. Many of the buildings were severely damaged, and even those that settled without damage required costly repairs to make them usable again. Soils that may lose their strength during earthquake ground shaking usually can be identified by engineers ahead of time and can be avoided; if it is necessary to build on them, the structures can be put on foundations that extend to deeper and firmer ground.

Soil structures can also fail during earthquakes. Highways and railroads are often constructed on top of soil embankments; most large dams are now constructed of earth; and levees are built along the banks of rivers to keep the adjacent countryside from being flooded. Such earth structures have often failed in the past, but modern soil structures that have been engineered for earthquake resistance have survived very strong shaking.

Courtesy, Earthquake Engineering Research Center, University of California, Berkeley

Three-story steel frame structure is tested for earthquake resistance on a shaking table. Hydraulic jacks impart motion to the table so that it simulates an earthquake.

Accelerographs and shaking machines

Photographs of earthquake destruction often show buildings that are severely damaged, or that may even have collapsed completely. These buildings were deficient in an engineering sense; that is, their structural members did not have the requisite strength to resist the forces of the earthquake. To design structures that can resist earthquakes, it is necessary to know the nature of the ground shaking the structure might be subjected to, how structures will vibrate during earthquakes, and the resistance of steel, concrete, wood, and other construction materials to vibratory forces. The nature of destructive ground shaking is determined from actual records made of the motion of the ground during earthquakes. Special instruments, called accelerographs, have been developed for this purpose. By 1973 approximately 2,000 accelerographs were installed at selected points throughout the earthquake-prone areas of the world, such as Japan, India, Chile, Nicaragua, Mexico, United States, Canada, Italy, Yugoslavia, Turkey, and the Soviet Union.

The accelerograph is at rest until shaking of the ground is sensed, and this motion triggers the instrument into operation. In most modern accelerographs, the acceleration of the ground is transformed into the motion of a light beam, which makes a record on photographic film. Each instrument records three components of motion, typically north-south, east-west, and up-down, thus preserving a complete record of the ground accelerations during an earthquake. Accelerographs are also installed in buildings to record their swaying.

Since the occurrence of earthquakes is unpredictable, engineers had to develop other ways of generating motions in order to test buildings. In the U.S. and in Japan, special shaking tables were constructed. These consist of a movable platform, approximately 25 ft square, which is driven by electrically controlled hydraulic jacks; the jacks impart a motion to the platform like that recorded during destructive earthquakes. Large models of buildings, bridges, or dams are placed on the table and shaken as if they were experiencing an earthquake. In this way studies can be made of how structures and building materials can withstand vibratory stressing. For engineers studying the behavior of structures, such shaking tables are the best substitutes for an earthquake.

Another type of shaking machine is sometimes used to learn how real buildings vibrate. Bolted to the top floor of a building, it exerts forces that are in resonance with the natural vibration of the building. By this means, the structure is forced to vibrate and many of the important dynamic properties of the structure can be measured. One drawback of making such tests is that usually they must be carried out at night or on weekends so that the occupants of the buildings are not disturbed. The nine-story Engineering Building at the California Institute of Technology Jet Propulsion Laboratory was tested in this way, and later the motions of the building were recorded during the San Fernando earthquake. The good agreement of the two sets of measurements established the validity of such tests.

The design problem

When considering how to design buildings to resist earthquakes, several pertinent questions must be answered. First, what is the strongest shaking likely to occur in the region where the building is being constructed? Though very strong shaking has only a low probability of occurring, should it happen the building ought to be able to survive without collapsing. In the unlikely event that this maximum probable ground shaking does occur, it would be acceptable for the structure to suffer some damage so long as it did not collapse. For more moderate ground shaking, which does have a significant probability of happening during the lifetime of the structure, engineers must deal with the question of how much damage is economically acceptable. It would not be economically sound to spend $1,000 in order to prevent $500 of damage. For each structure there must be some balance point at which a small additional amount spent for increased earthquake resistance would be just equal to the expected cost of repairing the structure if the extra resistance were not provided.

The problem is complicated somewhat by the fact that the strength-producing parts of a building, such as the beams and columns of a high-rise office tower, involve only about one-third of the total cost of the structure. The other two-thirds provide exterior facing, heating and air-conditioning equipment, ceilings, partitions, floor coverings, electric lighting, elevators, and the like. It is, however, only the beams, columns, and other strength-giving members that provide safety against collapse. On the other hand, unless a structure is near collapse the costly damage to it is mainly to the architectural finish. For example, the first damage to be incurred is usually cracking of the plaster walls and ceilings. If the shaking is strong, there may be damage to the elevator equipment, to the stairs, and to the air-conditioning equipment. When designing for safety, the engineer must think about the performance of the structural frame of the building, but when designing to limit damage, he must consider the architectural features. The cost of repairing architectural finish can be high; extensive cracking of the interior walls and partitions in a high-rise building may cost several hundred thousand dollars to patch and repaint.

Unfortunately, it is easier and cheaper to construct hazardous buildings than safe ones. Thus, the basic question becomes, should society spend the money on eliminating earthquake hazards, or should the money be spent in other ways? Regrettably, it is only after a destructive earthquake that the average citizen is aware of earthquake hazards. During a time when there have been no earthquakes, there is a reluctance to take precautions.

Engineering for earthquake resistance

The cost of a high-rise building may be $20 million to $50 million, or more, and it is obviously desirable to protect this large investment against the possibility of severe damage or collapse during an earthquake. For a new high-rise in Los Angeles, for example, a study first is made of the possible earthquake motions that might occur at the proposed site of the structure. This is done by identifying nearby faults that are likely to generate earthquakes. Some of these, such as the San Andreas, are large, active faults capable of generating very large earthquakes. Others are smaller but might be more severe in their effects because they are closer to the building.

After potential sources of earthquake ground motion have been identified, the expected effects of each at the site are represented by accelerograph records. An engineer then uses these records to investigate how his proposed building will vibrate when subjected to the earthquake motions. He does this by employing a large digital computer that is programmed as to the size, shape, mass, and stiffness of the proposed building in order to derive its vibratory properties. The computer is then given the earthquake ground shaking at the base of the building, and from this calculates how the building will vibrate when subjected to the various ground motions; thus, in effect, the building experiences an earthquake before it is built. These computed

Courtesy, H. J. Degenkolb & Associates

Reinforced-concrete parking garage in Santa Rosa, Calif., contains two vertical shear walls on its street frontage. They were designed to help the building resist the lateral forces imposed on it by earthquakes.

vibrations allow the engineer to determine the stresses and strains in the columns and beams of the building frame, and then design the building so that none of the members are overstressed. Additional precautions are taken by the designer to make the structural frame of the building ductile, so that even should an unforeseen earthquake of greater intensity occur the structure can deform plastically without failing.

The performance of the high-rise buildings in Los Angeles during the San Fernando earthquake represented a notable success for earthquake engineering design. During the earthquake, the buildings vibrated back and forth with a much larger motion than that of the ground; for example, the 42-story Union Bank Building vibrated back and forth with an overall amplitude of motion of approximately two feet at the top of the building without overstressing the structure. As a result, the Los Angeles Building Department now requires all high-rise buildings to be designed in accordance with earthquake-resistance principles.

During the San Fernando earthquake, one of the hospital buildings at the Veterans Administration Hospital in Sylmar collapsed, killing 50 persons. This building had been constructed in 1928 before the building code required earthquake-resistant design. It was strong enough to support the weight of the floors and roof and the contents, but was weak in resisting vibrations produced by earthquakes. When the strong ground motion generated by the earthquake vibrated the old building, its columns were overstressed and they failed, and the entire structure collapsed to the ground. At the same location there were other similar hospital buildings that had been designed to resist earthquakes, and these buildings survived the strong shaking without serious damage and with no injury or loss of life. This again provided convincing evidence of the need for designing buildings to resist earthquakes.

Especially careful earthquake design is given to nuclear power plants. Because of the requirement for a high degree of safety, these installations are designed to withstand the maximum credible ground shaking without damage, even in the eastern and southern parts of the United States, where destructive earthquakes are very unlikely to occur. Not only are the buildings designed to resist earthquakes, but also the reactor itself and its fuel rods, as well as the boilers, piping, and control equipment.

When designing a nuclear power plant, engineers make it resistant to a so-called "design earthquake," which it must be able to withstand without damage. The design earthquake represents ground shaking stronger than the most severe shaking ever experienced in the region. The vibrations of the building and the equipment are computed for this shaking, and sufficient strength is provided to resist the stresses that are produced.

The recently completed $3 billion California Water Project, which brings water from the Feather River to the southern part of the state, also was especially designed to withstand earthquakes. Many of the dams and pumping plants of this project are located in the most highly earthquake-prone parts of the state, while the aqueduct over much of its length is near the San Andreas fault and, in fact, crosses it in three places. Since the fault is almost certain to move during the lifetime of the project, the aqueduct was designed with this in mind. Where the aqueduct crosses the fault, it does so on the surface of the ground, not underground in a tunnel. The aqueduct, which is like an elevated ditch, is formed of two earth embankments, with a concrete lining. The embankments are 20 ft wider than normal so that if the fault moves that much the water can still continue to flow. Provisions were also made so that it would be easy to make repairs in case the earthquake actually caused serious damage.

Courtesy, Brandow & Johnston Associates

Diagonal K-bracing in a steel-frame structure resists lateral forces caused by earthquakes and wind.

Reduction of earthquake hazard

It is not the earthquake that is hazardous, but rather it is the works of man that are hazardous. A nomad living in a tent in the desert need not fear earthquakes, for the worst that the ground shaking could do would be to knock down the tent, and this would not be dangerous to life or limb. It is only when man begins erecting buildings, bridges, and dams that the hazard appears. It has been demonstrated often that if these structures are not built properly the earthquake will cause them to collapse. It is, therefore, fair to say that the earthquake disaster is really a man-made disaster. Cities can be made safe against earthquakes, for structures can be designed that will not be a hazard during strong ground shaking.

Earthquake engineering as an organized activity did not begin in the U.S. until the late 1920s, and earthquake loading provisions were not placed in building codes until 1933. Therefore, all the cities in the seismic regions of the U.S. have many old, pre-1933 buildings that are very hazardous in the event of an earthquake. Although modern buildings are not perfect, they are much better than the old, weak structures. This was demonstrated by the San Fernando earthquake, in which only 3 people were killed by shaking in damaged new buildings whereas 59 were killed in old structures. The large death toll in the Nicaragua earthquake was the consequence of the collapse of many old, weak buildings in that city. This hazard from old buildings will be reduced in time, as the buildings deteriorate and are torn down and replaced. However, it is the responsibility of city and state governments to take measures to speed up this process. At present, every major city in earthquake-prone areas would suffer tremendously if subjected to a great earthquake. In time, all of the structures and facilities in the highly seismic regions of the U.S. and throughout the world should be designed to resist earthquakes.

FOR ADDITIONAL READING:

Freeman, John R., *Earthquake Damage and Earthquake Insurance* (McGraw-Hill, 1932).

Hodgson, John H., *Earthquakes and Earth Structure* (Prentice-Hall, 1964).

Jennings, Paul C. (ed.), *Engineering Features of the San Fernando Earthquake of February 9, 1971* (Earthquake Engineering Research Laboratory, California Institute of Technology, 1971).

Kawasumi, Hirosi (ed.), *General Report on the Niigata Earthquake of 1964* (Tokyo Electrical Engineering College Press, 1968).

Newmark, Nathan M., and Rosenblueth, Emilio, *Fundamentals of Earthquake Engineering* (Prentice-Hall, 1971).

Richter, Charles F., *Elementary Seismology* (W. H. Freeman, 1958).

Wiegel, Robert L. (ed.), *Earthquake Engineering* (Prentice-Hall, 1970).

AUDIOVISUAL MATERIALS FROM ENCYCLOPÆDIA BRITANNICA EDUCATIONAL CORPORATION:

Film: *Earthquakes: Lesson of a Disaster.*

Index

Index entries to feature and review articles in this and previous editions of the *Britannica Yearbook of Science and the Future* are set in boldface type, *e.g.*, **Astronomy.** Entries to other subjects are set in lightface type, *e.g.*, Radiation. Additional information on any of these subjects is identified with a subheading and indented under the entry heading. The numbers following headings and subheadings indicate the year (boldface) of the edition and the page number (lightface) on which the information appears.

Astronomy **74**-408, 183; **73**-184; **72**-187
  aurora polaris **74**-144
  Colonizing the Moon **72**-12
  honors **74**-249; **73**-251; **72**-260
  Orbiting Solar Observatory **73**-182
    solar corona il. **73**-184
  SAS-A Program **74**-396
  spectrograph **73**-175
  topography of Venus **73**-223

All entry headings, whether consisting of a single word or more, are treated for the purpose of alphabetization as single complete headings and are alphabetized letter by letter up to the punctuation. The abbreviation "il." indicates an illustration.

A

B

C

D

E

F

G

H

I

Q

R

S

T

U

V

W

X

Y

Z

Acknowledgments

6 Photographs by (top, left) Flip Schulke from Black Star; (top, right) Dan Morrill; (center, left) D. Durrance from Photo Researchers; (center, right) courtesy, Hale Observatories © California Institute of Technology and Carnegie Institution of Washington; (bottom, left) courtesy, Graphic Films; (bottom, right) John Kohout from Root Resources
18–29 Illustrations by John Craig
32 Photograph, The Naked Ape, courtesy, Playboy Productions
36–39 Illustrations by Dave Beckes
40–43 Illustrations by Jan Wills
50–51 Illustration, courtesy, Sobin Chemicals, Inc., Boston (photographed by Bill Arsenault)
82–83 Illustration by Peter Lloyd
128–143 Illustrations by Dave Beckes
144–145 Photograph by A. Lee Snyder
149 Illustration by Dave Beckes
158 Photographs by (from top to bottom) courtesy, Drs. Charles Hough and H. Fernandez-Moran, The Electron Microscope Laboratory, Division of Biological Sciences, University of Chicago; courtesy, Drs. Kenneth W. Adolph and Robert Haselkorn, University of Chicago; J. Marlier from Bruce Coleman Inc.; courtesy, *UNESCO Courier*
179, 182 Illustrations by Ben Kozak
180, 181, 192, 201, 205, 208, 210–216, 223, 224, 226, 228, 269, 278, 292, 295, 319 Illustrations by Dave Beckes
189, 219, 231, 245, 289, 335 Illustrations by Seper & Miller, Ltd.
356–367 Illustrations by John Craig
370–377 Photograph (airplane), courtesy, Boeing Co.
371, 373 Illustrations by Dave Beckes
378 Photographed by Bill Arsenault
380–387 Illustrations by Dave Beckes
394–395 Illustration by Ben Kozak
396, 400–403 Illustrations by Dave Beckes
408 Photograph, courtesy, Hale Observatories © California Institute of Technology and Carnegie Institution of Washington
413–417 Illustrations by Dave Beckes
422 Photograph by UPI Compix